Reinforced Polymer Composites

Reinforced Polymer Composites

Editor

Victor V. Tcherdyntsev

MDPI • Basel • Beijing • Wuhan • Barcelona • Belgrade • Manchester • Tokyo • Cluj • Tianjin

Editor
Victor V. Tcherdyntsev
National University of Science
and Technology "MISIS"
Russia

Editorial Office
MDPI
St. Alban-Anlage 66
4052 Basel, Switzerland

This is a reprint of articles from the Special Issue published online in the open access journal *Polymers* (ISSN 2073-4360) (available at: https://www.mdpi.com/journal/polymers/special_issues/Reinf_Polym_Compos).

For citation purposes, cite each article independently as indicated on the article page online and as indicated below:

LastName, A.A.; LastName, B.B.; LastName, C.C. Article Title. *Journal Name* **Year**, *Volume Number*, Page Range.

ISBN 978-3-0365-0968-6 (Hbk)
ISBN 978-3-0365-0969-3 (PDF)

Cover image courtesy of Victor Tcherdyntsev.

Contents

About the Editor

Victor V. Tcherdyntsev (Ph.D.) is a head of the Laboratory of Functional Polymer Materials National University of Science and Technology "MISIS", Moscow, Russia. His research activities are focused on the following themes: mechanical alloying and ball milling; synthesis of quasicrystalline phases and their application; glass and carbon fibers reinforced thermoplastics; polymer composites for radiation shielding; elaboration of wear resistance polymer composites; application of nanoparticles as reinforcers for polymer matrix; ultra high molecular weight polyethylene based materials for engineering and biomedical application; synthesis and characterization of thermal conductive polymer based composites. To date, he has authored and/or co-authored 108 papers indexed on Scopus, his H-index is of 23 and he has about 1500 total citations.

Preface to "Reinforced Polymer Composites"

This Special Issue focuses on the recent advances in reinforced polymer composites. Polymer materials are widely used in human life, medicine, industry, and so on. However, polymers have a lot of disadvantages, such as insufficient strength, stiffness, creep, and low usage temperature. That is why reinforcing fillers are widely used to improve polymer properties. The following factors should be taken into account to reach high mechanical properties when nanofillers are used: (a) uniform distribution of the filler in the polymer matrix of a bulk sample will result in a composite physical and chemical properties uniformity over its volume; (b) filler should not agglomerate inside the polymer bulk sample because it might act as stress concentrator; and (c) interaction between fillers and polymer matrix should result in a composite supramolecular structure improvement. Strong interfacial interaction between polymer matrix and filler surface can improve load transfer from matrix to reinforcing filler. This Special Issue covers all the fields related to the reinforced polymer composites, but special attention was given to the aspects related with effect of polymer-filler interfaces interaction on the composite properties and synthetic and natural fiber fillers for polymers.

Victor V. Tcherdyntsev

Editor

polymers MDPI

Editorial

Reinforced Polymer Composites

Victor V. Tcherdyntsev

Laboratory of Functional Polymer Materials, National University of Science and Technology "MISIS", Leninskii prosp, 4, 119049 Moscow, Russia; vvch@misis.ru; Tel.: +7-9104002369

Citation: Tcherdyntsev, V.V. Reinforced Polymer Composites. *Polymers* **2021**, *13*, 564. https://doi.org/10.3390/polym13040564

Received: 20 January 2021
Accepted: 1 February 2021
Published: 13 February 2021

Publisher's Note: MDPI stays neutral with regard to jurisdictional claims in published maps and institutional affiliations.

The development of modern technology requires the elaboration of new materials with improved operational and technological properties. At the same time, many of the currently known natural and artificial materials no longer meet the increasing requirements. The discovery of fundamentally new materials is an extremely rare phenomenon, which indicates that the overwhelming majority of "simple" materials have already been discovered, and there is no need to expect great achievements in this direction. Therefore, the main direction in the development of new materials is now the creation and improvement of composite materials.

The main advantages of composites over traditional types of materials are:

- a unique combination of properties (strength, deformation, elastic, electrical, frictional, thermophysical, and others), which cannot be achieved for "simple" materials;
- the ability to control the composites' properties over a wide range by simply changing their composition and production conditions.

In polymer composites reinforced with dispersive fillers, the matrix is the main element undergoing the mechanical loading, and the main purpose of introducing dispersed particles usually is to increase the elastic modulus of the material. Dispersed filling is also used to improve the thermophysical characteristics, electrical and magnetic properties, to reduce friction wear, and to reduce the flammability of the material. In the case of composites reinforced with continuous fillers, both the matrix and reinforcers play an important role in the composite mechanical behavior, and the characteristics of composites are associated strongly with the ability of the interface to transfer the mechanical loading from the matrix to reinforcers. Initially, the main goal of polymer filling was to reduce the cost of the material by using an inexpensive filler. Today, reinforced polymer composites are some of the most numerous and diverse types of materials that are promising for use in various fields of science and technology, where high requirements are imposed on the physical, mechanical, and tribological characteristics.

In industry, some polymers have become widespread as antifriction wear-resistant materials that can work in dry friction conditions and in aggressive environments. Antifriction polymers act as substitutes for such traditional materials as bronze, brass, steel, and antifriction cast iron. The combination of antifriction properties with high biocompatibility allows the use of polymers in the creation of implants in the musculoskeletal system. However, polymers materials in their original form have a number of disadvantages, which include low strength and hardness and low operating temperature, which significantly limits their use. To improve these properties, researchers are attempting to reinforce polymers with various fillers and create composite materials based on them. A special class among composite materials is nanocomposites, in which, due to the use of nanosized particles, a more uniform distribution of the reinforcing element in the matrix and strong interfacial interactions between the polymer and the filler are achieved. Nanoparticles are capable of influencing the crystallization mechanism of polymers, acting as a nucleating agent, on the surface of which the crystalline phase nucleates. Depending on the size, shape, and nucleation density of nanoparticles, the formation of various supramolecular structures in polymers is possible.

To obtain high-strength composites, it is necessary to provide a high transfer capacity of the load from the matrix to the filler. This ability is determined by the level of adhesion of the filler to the polymer matrix. The adhesion of the filler to the polymer can be achieved by:

- chemical interaction;
- mechanical adhesion of the filler and matrix;
- electrostatic and van der Waals forces.

The adhesion due to the mechanical adhesion of the filler and the matrix depends on the geometry of the filler surface and the properties of the polymer, which include the chemical structure of the polymer, the regularity of the molecular structure, conformational characteristics, and branching of polymer chains. Furthermore, mechanical adhesion is affected by the difference in the thermophysical properties of materials. For example, during the cooling of a composite material, the difference between the coefficients of the linear thermal expansion of the polymer and the filler can lead to the formation of residual thermal expansions. In the absence of chemical and mechanical interactions between the filler and the polymer matrix, the energy for pulling the filler out of the polymer is composed of electrostatic and van der Waals forces, which provide adhesive bonding.

The hardening mechanisms of reinforcing fillers are different and primarily depend on the type of filler. For example, when reinforced with continuous fibers, an increase in strength is achieved due to the redistribution of the load from the matrix to the fiber. The fiber takes on all the mechanical load, and accordingly, the strength of the composite is determined by the strength of the reinforcing fiber. The role of the matrix is to redistribute stresses between fibers and form a single whole composite.

This Special Issue consisting of 21 articles, including three review papers, written by research groups of experts in the field, considers recent research on reinforced polymer composites. First of all, it should be noted that all three review papers relate to the fiber reinforced polymer composites, which are a real hot topic in the field. Rajak et al. [1] give a general review of the present state of the fiber reinforced polymer composites. Classifications of composites along with the properties of their constituent elements are analyzed. Depending on the reinforcing fiber nature, such composites may be divided into synthetic and natural fiber reinforced ones. Synthetic fibers, such as carbon, glass, or basalt, provide more stiffness, while natural fibers, such as jute, flax, bamboo, kenaf, and other, are inexpensive and biodegradable, making them environmentally friendly. Due to the modern requirements of the environmental safety of the processes of both materials' manufacturing and the effective recycling or utilization of materials at the end of the life cycle, using natural materials as reinforcers for polymer composites is a second hot topic in the field.

As is noted in [1], to acquire the benefits of design flexibility and recycling possibilities, natural reinforcers can be hybridized with small amounts of synthetic fibers to make them more desirable for technical applications. One case of such hybrid composites, namely kenaf/glass reinforced ones, is considered in the review paper of Nadzri et al. [2]. It is summarized that reinforcing of kenaf fiber containing composites, which have great potential in automotive application, due to being lightweight, eco-friendly, and low-cost, with glass fiber, is a good way to enhance such composites because glass fiber has better mechanical and impact properties than kenaf fiber. It is noted that the optimum fiber content and fiber orientation in such composites is between 30% and 40% and 90°, respectively. Fiber surface modification also helps to improve the properties of kenaf/glass hybrid composites.

Mohamed et al. [3] review recent advances in the development of glass and carbon fiber reinforced polymer composite laminates used in structurally deficient flat slab floor systems to enhance the two-way shear capacity, flexural strength, stiffness, and ductility. It is summarized that vertically placed fiber reinforced polymer sheets/laminae/strips are more effective in resisting two-way shear force. It is noted that further research is needed to develop models and clear design guidelines for vertically placed fiber reinforced polymer retrofitting strips that considers the pattern, spacing, material, and concrete properties.

Additionally, the investigation of larger size specimens to predict the response of fiber reinforced polymer retrofitted slabs is required.

Looking through 18 regular research papers, published in this Special Issue, one can see that most of them relate to the above-mentioned hot topics. Among them, nine papers relate to polymer composites reinforced with synthetic fibers. Slonov and co-workers [4] study the effect of oligophenylene sulfone and polycarbonate additions on the rheological, mechanical, and thermal properties of polyetherimide reinforced with short carbon fibers. It is revealed that the addition of oligophenylene sulfone and polycarbonate significantly reduces the melt viscosity of polyetherimide based carbon filled composites, acting as a temporary plasticizer. The addition of oligophenylene sulfone results in a significant increase in the elastic modulus and strength of composites, whereas the introduction of polycarbonate leads to a decrease in toughness, while the elastic modulus and strength remain at the level of the initial composite. The introduction of both oligophenylene sulfone and polycarbonate leads to an increase in carbon fiber length in comparison with a composite without modifiers. For composites modified with oligophenylene sulfone, a higher adhesive interaction of the polymer matrix with the surface of the carbon fiber is observed, and the thermal stability and heat resistance of composites with oligophenylene sulfone melts are found to be significantly higher than for composites modified with polycarbonate.

Tian et al. [5] also use short carbon fiber as reinforcers for polymer composite; however, in this case, the modification is performed by the coating of the carbon fiber surface with a polydopamine layer. They prepare a high-performance rubber composite by mixing of dopamine-modified short carbon fiber with natural latex. It is observed that uniform and widely covered polydopamine coatings are formed on the carbon as a result of the modification, which significantly improves the interface adhesion between the carbon fiber and the rubber matrix. The modification of carbon fiber in the solution with the concentration of dopamine of 1.5 g/L for 6 h shows the best results among the treatment routes used. Natural fiber latex based composites reinforced with such carbon fibers show excellent thermal conductivity and dynamic mechanical properties, and their tensile strength is 10.6% higher than those of the composites containing unmodified fibers.

Zhu and coauthors [6] compare the wear resistance behavior of bearing bushes made of polyetheretherketone (PEEK), 30 wt % carbon fiber reinforced PEEK, 30 wt % glass fiber reinforced PEEK, and each 10 wt % of polytetrafluoroethylene, graphite, and carbon fiber filled PEEK. It is found that due to low thermal conductivity, unfilled PEEK and glass fiber reinforced PEEK present the much lowest articulating cycles to failure. The presence of graphite and polytetrafluoroethylene in the PEEK matrix not only reduces the shear force at the interface, but also minimizes the temperature increase in the bulk material. The wear mass loss of each 10 wt % of polytetrafluoroethylene, graphite, and carbon fiber filled PEEK is found to be 0.13×10^{-6} mm^3/Nm compared with 4.33×10^{-6} mm^3/Nm for PEEK filled with 20 wt % of carbon fiber. It is concluded that bushes made of PEEK composite formulated with polytetrafluoroethylene, graphite, and carbon fiber exhibit low friction, self-lubrication, and low temperature rise and therefore present superior bearing properties, including enhanced bearing life and reduced energy consumption in machinery.

In [7], the formation of polysulfone based composites reinforced with carbon fiber fabric via the solution impregnation method using N-methyl-2-pyrrolidone as the solvent is reported. To improve the adhesion between the polymer matrix and carbon fibers, thermal oxidation of carbon fibers is carried out. Such oxidation allows for a change in the carbon fibers' surface chemical composition, and a greater number of functional groups on the surface appear as a result of thermal oxidation, resulting in a strong bond between the polysulfone matrix and carbon fiber. The in-plane shear strength value of polysulfone reinforced with modified carbon fibers increases by more than 1.5 times compared with the composites containing untreated carbon fibers. Surface modification of carbon fibers results also in noticeable improvement in the flexural and tensile properties, as well as in the thermal stability of polysulfone based composites. Sherif et al. use the same solution impregnation technique to obtain polyethersulfone [8] and polysulfone [9] based composites reinforced with glass fiber fabrics. The preheating of glass fibers is used to remove the sizing coating of fabrics and to improve the adhesion between the matrix and reinforcer. It is shown that glass fiber preheating allows increasing the mechanical properties of composites by 20–40%. For both polymer matrices used, the best mechanical properties are achieved at a glass fiber-to-polymer weight ratio of 70/30.

Xu and co-workers [10] made a composite of the commercial proton-exchange Nafion membrane specially fabricated via electrospinning silica nanofibers with a three-dimensional network structure. The proton conductivity of the silica nanofiber–Nafion composite membrane at 110 °C is found to be almost doubled compared with that of a pristine Nafion membrane, while the mechanical stability of the composite Nafion membrane is enhanced by 44%. It is found that the silica nanofiber-Nafion composite membrane exhibits great high-temperature fuel cell performance with a 118 mW/cm^2 power density at low humidity, which is 38% higher than that of the pristine Nafion membrane.

Mao and co-workers [11] report investigations of the effect of the surface modification of polyethylene terephthalate fibers on their adhesion to a polypropylene matrix. Polyethylene terephthalate fiber is modified through solution dip-coating using a novel synthesized tetraethyl orthosilicate/silane coupling agent KH550/polypropylene maleic anhydride graft hybrid. As a result of such treatment, SiO_2 particles are designed to link with the maleated polypropylene molecular chain through silane coupling agent KH550, which forms an organic-inorganic film by the grain structure on the surface of polyethylene terephthalate fiber. As a result, the bending strength and modulus of the polypropylene reinforced with modified polyethylene terephthalate fiber increase by 21 and 34%, respectively, compared to the untreated fiber-filled composite.

Huang et al. [12] apply both computer-aided engineering simulation and experimental methods to investigate the glass fiber feature in a co-injection molding system. Fiber orientation distributions and their influence on the tensile properties for the single-shot and co-injection molding are discovered. Results show that based on the 60:40 skin/core ratio and the same materials, the tensile properties of the co-injection system, including the tensile stress and modulus, are a little weaker than those of the single-shot system. To discover and verify the influence of the fiber orientation features, the fiber orientation distributions of both the co-injection and single-shot systems are observed using micro-computerized tomography technology to scan the internal structures. It is found that the fiber orientation tensor in the flow direction of the co-injection system is about 89% of that of the single-shot system in the testing condition because the co-injection part has lower tensile properties.

Two papers in this Special Issue cover polymer composites reinforced with natural fibers. Yeh and Yang [13] test four types of waste bamboo fibers, differing in plant species, as fillers in polypropylene based composites. It is found that the composites reinforced with thorny bamboo fibers (*Bambusa stenostachya*) exhibit the highest moisture content and water absorption rate due the high hemicellulose content in these fibers, while the composites reinforced with Makino bamboo fibers (*Phyllostachys makinoi*) possess better tensile properties due to the high crystallinity and high lignin content of these fibers.

Roy et al. [14] investigate the effect of jute fiber surface treatment by alkali, stearic acid, or silane on the properties of natural rubber based composites. They show that combined alkali/silane treatment is the most efficient surface treatment method to develop strong interfacial adhesion between the natural rubber matrix and jute fibers. Composites reinforced with jute fibers modified by combined alkali/silane treatment show considerably higher torque difference, tensile modulus, hardness, and tensile strength as compared to either untreated or other surface treated jute fiber filled natural rubber.

The topic of polymer composites reinforced with hybrid fillers containing both synthetic and natural components in this Special Issue is covered by four papers. Shah and co-workers [15] investigate the impact behavior of epoxy based composites reinforced with woven glass fabric and bamboo powder and compare the obtained results with the data on the neat epoxy and epoxy reinforced with short bamboo fibers. Woven glass fibers are embedded at the outer most top and bottom layer of the bamboo powder filled epoxy composites, producing sandwich structured hybrid composites through lay-up and molding techniques. A significant improvement is observed with the inclusion of woven glass fibers in the composites. The non-hybrid composites break into pieces during the highest impact energy that is applied, while the hybrid composites experience only perforation, and the structure does not totally break. The non-hybrid composites can withstand an impact energy up to 10 J, while the hybrid composites show total failure at 35 J.

Calabrese et al. [16], using the vacuum-assisted resin infusion technique, prepare epoxy based composites reinforced with six layers of 2 × 2 twill weave woven flax fabric. Three layers of plain weave woven glass fabrics are used as external fiber reinforced skins, and the number of fully reinforced layers is 12. Samples with varying geometrical joint configurations are exposed to a salt-fog spray test up to 60 aging days, and the effect of the salt-fog environment on the mechanical behavior of the pinned hybrid glass-flax/epoxy composites is evaluated. A noticeable modification of the damage mode with increase of the aging exposition time is evidenced, with a reduction of the bearing phenomenon, thus favoring premature and catastrophic mechanisms such as shear out and net tension. As a consequence, the effective mechanical durability of the mechanical joint is limited.

Ibrahim et al. [17] investigate the isolation of lignin from empty oil palm fruit bunch fibers using formic acid at different concentrations. The isolated organosolv lignin is modified with graphene nanoplatelets and used as the filler reinforcement for photo-curable polyurethane in stereolithography 3D printing. Reinforcing of polyurethane with 0.6 wt % of graphene modified lignin provides tremendous enhancement of the hardness at 92.49 MPa, which means a 238% increment from the unmodified photo-curable polyurethane resin. Moreover, this composite shows higher tensile strength and resistance against the deformation of the material.

Hayeemasae and co-workers [18] utilize modified palm stearin as a mixed compatibilizer with maleated natural rubber for natural rubbed based composites reinforced with halloysite nanotubes. The main idea of the study is to use chip natural product, such as palm stearin, instead of silanes, which are considered to be expensive and require a high mixing temperature to obtain effective silanization. It is found that the overall properties of composites based on natural rubber and halloysite nanotubes are clearly improved by adding maleated natural rubber/modified palm stearin as a dual compatibilizer. Both maleated natural rubber and modified palm stearin have special functional groups that form hydrogen bonds with the hydroxyl groups on halloysite surfaces. Moreover, modified palm stearin can also promote the dispersion of halloysite nanotubes filler in the matrix due to its waxy character. It is observed that improvement in the filler-matrix interactions due to the above-mentioned dual compatibilizer addition results in the increase in the composites' mechanical properties, such as the tensile strength, modulus, and tear strength.

The remaining three papers in this Special Issue do not relate to polymer composites reinforced with fiber fillers. Chang et al. [19] investigate the effect of combining two approaches, namely the addition of nanoparticles and the crosslinking of two different polymers to create double-network hydrogels, on the mechanical properties of hydrogels. It is evaluated that the introduction of nanomaterials into the hydrogel network may allow the consideration of double-network hydrogels for new applications such as self-healing, shape memory, 3D printing, and dye removal. A mechanistic insight into the role of nanoparticles in the mechanical properties of double-network hydrogels is provided, including the elastic moduli and swellability. The existence of a "global" saturation point for double-network hydrogel nanocomposites is observed, beyond which it becomes less plausible to enhance the elastic modulus by simply increasing the concentration of the second network's hydrogel or nanoparticles.

The effect of Na_2CO_3 addition on the structure and properties of polyvinyl alcohol/carboxyl methyl cellulose sodium composite films is investigated in [20]. It is observed that the presence of Na_2CO_3 results in the hydrolysis of the vinyl acetate group of polyvinyl alcohol and in the decrease in the crystallinity of the composite film. At the same time, as the Na_2CO_3 content increases, more ester groups are hydrolyzed, and the amount of hydroxyl groups increases, which generates more intermolecular hydrogen bonds and, thus, increases the melting temperature of the composite. Additionally, the increase in the Na_2CO_3 content results in a significant increase in the water sorption properties of the composite films. The elaborated polyvinyl alcohol/carboxyl methyl cellulose sodium/Na_2CO_3 composites may have prospective applications as biodegradable superabsorbent resins and polymer electrolytes.

Finally, Lei et al. [21] study polymer blends containing nitrile rubber, brominated butyl rubber, and ethylene-vinyl acetate copolymer. It is shown that such blends exhibit excellent vulcanization plateaus and mechanical properties. Ethylene-vinyl acetate copolymer is observed to be the optimal polymer for improving the compatibility of the nitrile rubber/brominated butyl rubber blend. Hot air thermal aging tests show that the blends have good stability. Besides the studied blends, nitrile rubber/brominated butyl rubber/ethylene-vinyl acetate copolymer with a 50/50/30 blend ratio is found to be a comparatively ideal material for the purpose of high-damping isolation bearings. As the formation of the polymer blend can be applied to optimize the properties of polymer composites matrices, the obtained results may be of interest for the elaboration of advanced reinforced polymer composites.

Funding: This research received no external funding.

Data Availability Statement: No new data were created or analyzed in this study. Data sharing is not applicable to this article.

Acknowledgments: V.V.T. wishes to acknowledge the Ministry of Science and Higher Education of Russia for the funding in the framework of State Assignment Number 075-00268-20-02 dated 03/12/2020, and the state program of basic research "For the long-term development and ensuring the competitiveness of society and the state" (47 GP) on the basis of the universities, the plan for basic scientific research Number 718/20 dated 03/06/2020, Project Number 0718-2020-0036.

Conflicts of Interest: The author declares no conflict of interest.

References

1. Rajak, D.K.; Pagar, D.D.; Menezes, P.L.; Linul, E. Fiber-reinforced polymer composites: Manufacturing, properties, and applications. *Polymers* **2019**, *11*, 1667. [CrossRef]
2. Nadzri, S.N.Z.A.; Sultan, M.T.H.; Shah, A.U.M.; Safri, S.N.A.; Basri, A.A. A review on the kenaf/glass hybrid composites with limitations on mechanical and low velocity impact properties. *Polymers* **2020**, *12*, 1285. [CrossRef]
3. Mohamed, O.A.; Kewalramani, M.; Khattab, R. Fiber reinforced polymer laminates for strengthening of rc slabs against punching shear: A review. *Polymers* **2020**, *12*, 685. [CrossRef] [PubMed]
4. Slonov, A.; Musov, I.; Zhansitov, A.; Rzhevskaya, E.; Khakulova, D.; Khashirova, S. the effect of modification on the properties of polyetherimide and its carbon-filled composite. *Polymers* **2020**, *12*, 1056. [CrossRef]
5. Tian, X.L.; Han, S.; Zhuang, Q.X.; Bian, H.G.; Li, S.M.; Zhang, C.Q.; Wang, C.S.; Han, W.W. Surface modification of staple carbon fiber by dopamine to reinforce natural latex composite. *Polymers* **2020**, *12*, 988. [CrossRef]
6. Zhu, J.J.; Xie, F.; Dwyer-Joyce, R.S. PEEK composites as self-lubricating bush materials for articulating revolute pin joints. *Polymers* **2020**, *12*, 665. [CrossRef] [PubMed]
7. Chukov, D.I.; Nematulloev, S.G.; Tcherdyntsev, V.V.; Torokhov, V.G.; Stepashkin, A.A.; Zadorozhnyy, M.Y.; Zherebtsov, D.D.; Sherif, G. Structure and properties of polysulfone filled with modified twill weave carbon fabrics. *Polymers* **2020**, *12*, 50. [CrossRef] [PubMed]
8. Sherif, G.; Chukov, D.; Tcherdyntsev, V.; Torokhov, V. Effect of formation route on the mechanical properties of the polyethersulfone composites reinforced with glass fibers. *Polymers* **2019**, *11*, 1364. [CrossRef]
9. Sherif, G.; Chukov, D.I.; Tcherdyntsev, V.V.; Torokhov, V.G.; Zherebtsov, D.D. Effect of glass fibers thermal treatment on the mechanical and thermal behavior of polysulfone based composites. *Polymers* **2020**, *12*, 902. [CrossRef] [PubMed]
10. Xu, G.X.; Zou, J.; Guo, Z.; Li, J.; Ma, L.Y.; Li, Y.; Cai, W.W. Bi-functional composting the sulfonic acid based proton exchange membrane for high temperature fuel cell application. *Polymers* **2020**, *12*, 1000. [CrossRef]
11. Mao, Y.P.; Li, Q.Y.; Wu, C.F. Surface modification of pet fiber with hybrid coating and its effect on the properties of PP composites. *Polymers* **2019**, *11*, 1726. [CrossRef]
12. Huang, C.T.; Chen, X.W.; Fu, W.W. Investigation on the fiber orientation distributions and their influence on the mechanical property of the co-injection molding products. *Polymers* **2020**, *12*, 24. [CrossRef]

13. Yeh, C.H.; Yang, T.C. Utilization of waste bamboo fibers in thermoplastic composites: Influence of the chemical composition and thermal decomposition behavior. *Polymers* **2020**, *12*, 636. [CrossRef]
14. Roy, K.; Debnath, S.C.; Tzounis, L.; Pongwisuthiruchte, A.; Potiyaraj, P. Effect of various surface treatments on the performance of jute fibers filled natural rubber (NR) composites. *Polymers* **2020**, *12*, 369. [CrossRef]
15. Shah, A.U.M.; Sultan, M.T.H.; Safri, S.N.A. Experimental evaluation of low velocity impact properties and damage progression on bamboo/glass hybrid composites subjected to different impact energy levels. *Polymers* **2020**, *12*, 1288. [CrossRef]
16. Calabrese, L.; Fiore, V.; Bruzzaniti, P.; Scalici, T.; Valenza, A. Pinned hybrid glass-flax composite laminates aged in salt-fog environment: Mechanical durability. *Polymers* **2020**, *12*, 40. [CrossRef]
17. Ibrahim, F.; Mohan, D.; Sajab, M.S.; Bakarudin, S.B.; Kaco, H. Evaluation of the compatibility of organosolv lignin-graphene nanoplatelets with photo-curable polyurethane in stereolithography 3D printing. *Polymers* **2019**, *11*, 1544. [CrossRef]
18. Hayeemasae, N.; Sensem, Z.; Surya, I.; Sahakaro, K.; Ismail, H. Synergistic effect of maleated natural rubber and modified palm stearin as dual compatibilizers in composites based on natural rubber and halloysite nanotubes. *Polymers* **2020**, *12*, 766. [CrossRef]
19. Chang, A.; Babhadiashar, N.; Barrett-Catton, E.; Asuri, P. Role of nanoparticle-polymer interactions on the development of double-network hydrogel nanocomposites with high mechanical strength. *Polymers* **2020**, *12*, 470. [CrossRef]
20. Zhu, J.F.; Li, Q.Y.; Che, Y.C.; Liu, X.C.; Dong, C.C.; Chen, X.Y.; Wang, C. Effect of Na_2CO_3 on the microstructure and macroscopic properties and mechanism analysis of PVA/CMC composite film. *Polymers* **2020**, *12*, 453. [CrossRef]
21. Lei, T.; Zhang, Y.W.; Kuang, D.L.; Yang, Y.R. Preparation and properties of rubber blends for high-damping-isolation bearings. *Polymers* **2019**, *11*, 1374. [CrossRef]

polymers

MDPI

Review

Fiber-Reinforced Polymer Composites: Manufacturing, Properties, and Applications

Dipen Kumar Rajak [1,2,*], Durgesh D. Pagar [3], Pradeep L. Menezes [4] and Emanoil Linul [5,6,*]

1 Department of Mechanical Engineering, Sandip Institute of Technology & Research Centre, Nashik 422212, India
2 Department of Mining Machinery Engineering, Indian Institute of Technology (ISM), Dhanbad 826004, India
3 Department of Mechanical Engineering, K. K. Wagh Institute of Engineering Education & Research, Nashik 422003, India; durgeshpagar90@gmail.com
4 Department of Mechanical Engineering, University of Nevada, Reno, NV 89557, USA; pmenezes@unr.edu
5 Department of Mechanics and Strength of Materials, Politehnica University of Timisoara, 300 222 Timisoara, Romania
6 National Institute of Research for Electrochemistry and Condensed Matter, 300 569 Timisoara, Romania
* Correspondence: dipen.pukar@gmail.com (D.K.R.); emanoil.linul@upt.ro (E.L.); Tel.: +91-9470307646 (D.K.R.); +40-728-44-0886 (E.L.)

Received: 20 September 2019; Accepted: 8 October 2019; Published: 12 October 2019

Abstract: Composites have been found to be the most promising and discerning material available in this century. Presently, composites reinforced with fibers of synthetic or natural materials are gaining more importance as demands for lightweight materials with high strength for specific applications are growing in the market. Fiber-reinforced polymer composite offers not only high strength to weight ratio, but also reveals exceptional properties such as high durability; stiffness; damping property; flexural strength; and resistance to corrosion, wear, impact, and fire. These wide ranges of diverse features have led composite materials to find applications in mechanical, construction, aerospace, automobile, biomedical, marine, and many other manufacturing industries. Performance of composite materials predominantly depends on their constituent elements and manufacturing techniques, therefore, functional properties of various fibers available worldwide, their classifications, and the manufacturing techniques used to fabricate the composite materials need to be studied in order to figure out the optimized characteristic of the material for the desired application. An overview of a diverse range of fibers, their properties, functionality, classification, and various fiber composite manufacturing techniques is presented to discover the optimized fiber-reinforced composite material for significant applications. Their exceptional performance in the numerous fields of applications have made fiber-reinforced composite materials a promising alternative over solitary metals or alloys.

Keywords: fiber-reinforced polymer; composite materials; natural fibers; synthetic fibers

---

1. Introduction

Rapid growth in manufacturing industries has led to the need for the betterment of materials in terms of strength, stiffness, density, and lower cost with improved sustainability. Composite materials have emerged as one of the materials possessing such betterment in properties serving their potential in a variety of applications [1–4]. Composite materials are an amalgamation of two or more constituents, one of which is present in the matrix phase, and another one could be in particle or fiber form. The utilization of natural or synthetic fibers in the fabrication of composite materials has revealed significant applications in a variety of fields such as construction, mechanical, automobile, aerospace, biomedical, and marine [5–8].

Research studies from the past two decades have presented composites as an alternative over many conventional materials as there is a significant enhancement in the structural, mechanical, and tribological properties of fiber-reinforced composite (FRC) material [9–11]. Though composite materials succeeded in increasing the durability of the material, currently a strong concern regarding the accumulation of plastic waste in the environment has arisen [12]. This concern has compelled researchers around the world to develop environmentally friendly materials associated with cleaner manufacturing processes [13–15]. Several different composite recycling processes also have been developed to cope with the thousands of tons of composite waste generated in a year. Mechanical recycling includes pulverization, where decreased sized recyclates are being used as filler materials for sheet molding compounds. In thermal recycling, degradation of composite waste by pyrolysis is done or an enormous amount of heat energy is obtained by burning composite materials with a high calorific value. There also exist more efficient processes such as chemical recycling (solvolysis) and high-voltage fragmentation (HVF). The addition of natural fillers such as natural fibers, cellulose nanocrystals, and nanofibrillated cellulose in the polymers matrix to fabricate eco-friendly composites has improved material properties while minimizing the problem regarding residue accumulation [16–19].

Many researchers have reported advantages of cellulosic fibers, such as being abundantly available in nature, nontoxic, renewable, cost-effective, and also providing necessary bonding with the cement-based matrix for significant enhancements in properties such as ductility, toughness, flexural capacity, and impact resistance of a material [20–22]. In modern techniques, inclusion of fly ash, limestone powder, brick powder, and many other mineral additives are used to strengthen the composite structures. Fracture toughness has been enhanced with the addition of fly ash in a concrete composite for structural applications resulting in increased lifespan of the material [23,24]. Natural fibers are mainly classified as fibers that are plant-based, animal-based, and mineral-based. As the asbestos content in the mineral-based fibers is hazardous to human health, these are not well-explored fibers with respect to research into fiber-reinforced composite materials, while plant-based fibers provide promising characteristics such as lower cost, biodegradable nature, availability, and good physical and mechanical properties [25,26]. Plant fibers include leaf fibers (sisal and abaca), bast fibers (flax, jute, hemp, ramie, and kenaf), grass and reed fibers (rice husk), core fibers (hemp, jute, and kenaf), seed fibers (cotton, kapok, and coir), and all other types, which may include wood and roots. Polymer matrices are also divided into a natural matrix and a synthetic matrix, which is petrochemical-based and includes polyester, polypropylene (PP), polyethylene (PE), and epoxy [27].

The latest research contributes the development of hybrid composites with the combination of natural and synthetic fibers. The composite structures consisting of more than one type of fiber are defined as hybrid composites. There are methods to combine these fibers, which involve stacking layers of fibers, the intermingling of fibers, mixing two types of fibers in the same layer making interplay hybrid, selective placement of fiber where it is needed for better force, and placing each fiber according to specific orientation [28]. Among all these, stacking of fibers is the easiest procedure, and others introduce some complications in obtaining a positive hybridization effect. Many researchers got success by developing optimized composite materials for efficient use in particular applications by varying fiber content, its orientation, size, or manufacturing processes. It is necessary to understand the physical, mechanical, electrical, and thermal properties of FRCs for their effective application. FRCs are currently being employed in copious fields of applications due to their significant mechanical properties. These composite materials sometimes depart from their designed specifications as some defects, such as manufacturing defects, cause them to deviate from the expected enhancement in mechanical properties. These manufacturing defects involve misalignment, waviness, and sometimes breakage of fibers, fiber/matrix debonding, delamination, and formation of voids in the matrix of a composite material. An increase of 1% voids content in composites and leads to a decrease in tensile strength (10–20%), flexural strength (10%), and interlaminar shear strength (5–10%), respectively. It can be eradicated by manipulating the processing parameters of manufacturing processes [29].

Therefore, there is a need to understand and study different types of composite manufacturing techniques to implement optimized techniques that will avoid defects and give apposite self-sustaining, durable composite material that is efficient for the desired field of application. There are many conventional manufacturing techniques for fabrication of a composite material that have been in practice for the past few decades and some of the recently developed automated composite manufacturing techniques use robot assistance for processing, which leads to complete automation and an immense rise in productivity [30].

2. Classification

Composite materials are classified according to their content, i.e., base material and filler material. The base material, which binds or holds the filler material in structures, is termed as a matrix or a binder material, while filler material is present in the form of sheets, fragments, particles, fibers, or whiskers of natural or synthetic material. As represented in Figure 1, composites are classified into three main categories based on their structure [31].

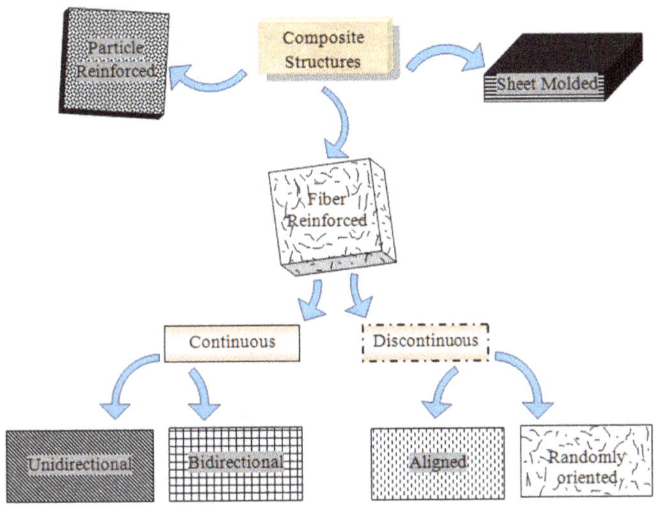

Figure 1. Classification of composites.

2.1. Fiber-Reinforced Composites

Composites consist of fibers in the matrix structure and can be classified according to fiber length. Composites with long fiber reinforcements are termed as continuous fiber reinforcement composites, while composites with short fiber reinforcements are termed as discontinuous fiber reinforcement composites. Hybrid fiber-reinforced composites are those where two or more types of fibers are reinforced in a single matrix structure [32]. Fibers can be placed unidirectionally or bidirectionally in the matrix structure of continuous fiber composites, and they take loads from the matrix to the fiber in a very easy and effective way. Discontinuous fibers must have sufficient length for effective load transfer and to restrain the growth of cracks from avoiding material failure in the case of brittle matrices. The arrangement and orientation of fibers define the properties and structural behavior of composite material [33,34]. Improvement in properties such as impact toughness and fatigue strength can be seen with the use of chemically treated natural fibers. Fibers of glass, carbon, basalt, and aramid in the dispersed phase were conventionally used in the matrix structure of a fiber-reinforced polymer (FRP) composite materials [35,36]. Significant properties of natural fiber polymer composites (NFPCs) have potential applications in the modern industry, as researchers currently are compelled towards the development of environmentally friendly materials due to stringent environmental laws.

There are numerous fibers available for composite materials and they are primarily categorized as natural or synthetic fibers. Further, recent studies have revealed unprecedented material properties when these two fibers are combined together, blending with a matrix material to form a hybrid composite. Some of the natural and synthetic fibers are shown in Figure 2.

Figure 2. Classification of fibers, reproduced from [37–53] under open access license.

2.1.1. Synthetic Fibers

Human-made fibers that are produced by chemical synthesis are called synthetic fibers and further classified as organic or inorganic based on their content [54]. Generally, the strength and stiffness of fiber materials are much higher than that of the matrix material, making them a load-bearing element in the composite structure [55–59].

Glass fibers (GFs) are most widely used among all the synthetic fibers as they offer excellent strength and durability, thermal stability, resistance to impact, chemical, friction, and wear properties. However, the machining of glass fiber-reinforced polymers (GFRPs) is relatively slow, challenging, and shows reduced tool life while working on conventional machining systems [60]. GFs also carry the disadvantage of disposal at the end of their service life [61].

However in some applications, more stiffness is required, so carbon fibers (CFs) are employed instead of GFs. Although some of the other types of synthetic fibers like aramid, basalt, polyacrylonitrile (PAN-F), polyethylene terephthalate (PET-F), or polypropylene fibers (PP-F) offer some advantages, they are rarely used in thermoplastic short-fiber-reinforced polymers (SFRP); they have been used for specific applications where their desired properties are applicable [62]. Carbon fiber-reinforced polymer (CFRP) composites have revealed numerous applications in aerospace, automobile, sports, and many other industries [63–65]. Young's modulus of solids and foams increased by 78% and 113%, respectively, when the weight percentage of carbon fibers increased from 10% to 30%. The improvement in the cellular structure resulted in the improvement of Young's modulus of the foams by 35% when carbon fiber/polypropylene (CF/PP) was used to make composite foams prepared by microcellular injection molding [66].

Graphene fibers are a new type of high-performance carbonaceous fibers that reflect high tensile strength with enhanced electrical conductivity when compared to carbon fibers. Several enhanced properties of graphene fibers show their potentiality in a variety of applications, such as lightweight conductive cables and wires, knittable supercapacitors, micromotors, solar cell textiles, actuators, etc. [67,68]. The molecular dynamics simulation of polymer composites with graphene reinforcements showed increases in Young's modulus, shear modulus, and hardness by 150%, 27.6%, and 35%, respectively. Furthermore, a reduction in the coefficient of friction and abrasion rate by 35% and 48% was achieved [69].

Basalt fiber (BF) possesses better physical and mechanical properties over fiberglass. In addition, BF is significantly cheaper than carbon fibers. The effect of temperature on basalt fiber-reinforced polymer (BFRP) composites has been investigated, where there was an increase in static strength and fatigue life at a certain maximum stress observed with a decrease in temperatures [70].

Thermal properties of Kevlar fiber-reinforced composites (KFRCs) are enhanced by hybridizing it with glass or carbon fibers, though there is less research on the hybridization of Kevlar fibers (KFs) with natural fibers. KFRCs show high impact strength with a high degree of tensile properties, but due to their anisotropic nature they possesses low compression strength compared to their glass and carbon fiber counterparts [71].

2.1.2. Natural Fibers

Natural fibers (NFs) are a very easy to obtain, extensively available material in nature. They reveal some outstanding material properties like biodegradability, low cost per unit volume, high strength, and specific stiffness. Composites made of NF reinforcements seem to carry some diverse properties over synthetic fibers, such as reduced weight, cost, toxicity, environmental pollution, and recyclability. These economic and environmental benefits of NF composites make them predominant over synthetic fiber-reinforced composites for modern applications [33]. Depending on the type, natural fibers have similar structures with different compositions. The inclusion of long and short natural fibers in thermoset matrices has manifested high-performance applications [72,73].

Sisal fiber (SF)-based composites are frequently being used for automobile interiors and upholstery in furniture due to their good tribological properties. When SFs were reinforced with polyester composites, the tensile strength increased with fiber volume and when reinforced with polyethylene (PE) composites, tensile strength of 12.5 MPa was observed in 6 mm long sisal fibers [74–76].

Hemp composite showed a 52% increase in specific flexural strength of a material when compared to GF-reinforced composite with a propylene matrix [77]. Composite material with 5% maleic anhydride-grafted polypropylene (MAPP) by weight mixed with polypropylene (PP) matrix that was reinforced with 15%, by weight, alkaline-treated hemp fibers manifested advancement in flexural and tensile strength by 37% and 68%, respectively [78].

Polylactic acid (PLA) thermoplastic composites with kenaf fiber reinforcement possess tensile and flexural strength of 223 MPa and 254 MPa, respectively [79]. Also, before laminating, removing absorbed water from the fibers results in the improvement of both flexural and tensile properties of kenaf fiber laminates [80,81]. Previously, polyester samples without any reinforcements showed flexural strength and flexural modulus of 42.24 MPa and 3.61 GPa respectively, while after reinforcement of 11.1% alkali-treated virgin kenaf fibers in unsaturated polyester matrix, composite material showed flexural strength and flexural modulus of 69.5 MPa and 7.11 GPa [82].

The sound and vibration behavior of flax fiber-reinforced polypropylene composites (FF/PPs) have been investigated using a sound transmission loss (STL) test. The results showed an increase in stiffness, damping ratio, and mass per unit area of the material due to increase in transmission loss, as the material possesses high sound absorption properties [83,84]. Use of short flax fiber (FF) laminates resulted in an enhancement in tensile properties of a material. Also, the material strength and shear modulus increased by 15% and 46%, respectively, with 45° fiber orientation [85].

The study on the free vibration characteristics of ramie fiber-reinforced polypropylene composites (RF/PPs) showed that higher fiber content in a polymer matrix leads to slippage between the fiber and the matrix, and this leads to an increase in the damping ratio during the flexural vibration. That means that an increase in fiber content results in enhancement in damping properties of RF/PP composite [86,87].

During the growth of a rice grain, a natural sheath forms around the grain, known as a rice husk (RH), which is treated as agricultural waste, but it is utilized as reinforcement in composite materials to investigate enhancement in material properties [88,89]. For the enhancement of the acoustic characteristics of the material, 5% of RH in polyurethane (PU) foam displayed optimum sound absorption performance [90].

Composite material consisting of 5% chicken feathers as reinforcement fibers with epoxy resin as matrix material showed optimum results following an impact test. Moreover, these chicken feathers used with 1% of carbon residuum (CR) fused with epoxy resin formed a hybrid composite, which displayed substantial enhancement in tensile, flexural, and impact strength of a material [91].

It has been seen that along with the length of a raw jute reed, tensile strength and bundle strength decrease from root to tip, with the root portion-based composite carrying 44% and 35% higher tensile and flexural strength, respectively, than that of the composites made from the tip portion of raw jute reed [92,93].

Randomly oriented coir fiber-reinforced polypropylene composites offers higher damping properties than synthetic fiber-reinforced composites. High resin content offers higher damping properties, therefore, lower fiber loading leads to more energy absorption. The maximum damping ratio of 0.4736 was obtained at 10% of fiber content in coir–PP composite, while further increasing fiber content to 30% showed improved natural frequency of material to 20.92 Hz [94,95].

Palm fibers (PFs) showed outstanding fiber-matrix interfacial interaction. Also, the addition of palm fibers in low-density polyethylene (LDPE) resulted in higher Young's modulus compared to homo-polymers [96].

Friction composites are fabricated using abaca fiber (AF) reinforcement, which offers excellent wear resistance property with a wear rate of 2.864×10^{-7} cm^3/Nm at 3% of fiber content. Also, the density decreased with increasing abaca fiber content [97].

The addition of luffa fibers (LFs) as a reinforcement constituent of composite material resulted in the advancement of the mechanical properties like tensile, compressive, flexural, impact strength, and water absorption characteristics of a material [98]. Adding a 9.6 wt % of LFs in epoxy matrix displayed a decrement in the density of the material by 3.12%, which further resulted in the reduction in material weight [99].

Energy absorption and load-carrying capacities of a tube material have been improved with the implementation of cotton fiber epoxy composite [100]. Manufacturing techniques and applications of some fibers with their matrix materials are depicted in Table 1.

Table 1. Matrix material used for some fibers with their applications and manufacturing techniques.

References	Materials Used		Application	Manufacturing Techniques
	Fiber Reinforcement	Matrix/Binder Material		
[64,65]	Carbon	PP, metals, ceramics, epoxy resin, Polyether ether ketone (PEEK)	Lightweight automotive products, fuel cells, satellite components, armor, sports.	Injection molding, filament winding, resin transfer molding (RTM)
[68]	Graphene	Polystyrene (PS), epoxy, Polyaniline (PANI)	Wind turbines, Gas tanks, aircraft/automotive parts.	CVD, pultrusion, hand/spray up method
[76]	Sisal	PP, PS, epoxy resin	Automobile body parts, roofing sheets	Hand lay-up, compression molding
[77]	Hemp	PE, PP, PU	Furniture, automotive.	RTM, compression molding
[80]	Kenaf	PLA, PP, epoxy resin	Tooling, bearings, automotive parts.	Compression molding, pultrusion
[83,84]	Flax	PP, polyester, epoxy	Structural, textile.	Compression molding RTM, spray/hand lay-up, vacuum infusion
[86,87]	Ramie	PP, Polyolefin, PLA	Bulletproof vests, socket prosthesis, civil.	Extrusion with injection molding
[89]	Rice Husk	PU, PE	Window/door frames, automotive structure.	Compression/injection molding
[92,93]	Jute	Polyester, PP	Ropes, roofing, door panels.	Hand lay-up, compression/ injection molding
[94,95]	Coir	PP, epoxy resin, PE	Automobile structural components, building boards, roofing sheets, insulation boards.	Extrusion, injection molding

2.1.3. Hybrid Fibers

Thermoplastic composites reinforced with natural fiber, in general, show poor strength performance when compared to thermoset composites. Therefore, to acquire benefits of design flexibility and recycling possibilities, these natural fiber composites are hybridized with small amounts of synthetic fibers to make them more desirable for technical applications. Hemp/glass fiber hybrid polypropylene composites exhibited flexural strength of 101 MPa and 5.5 GPa flexural modulus when filler content of 25% hemp and 15% glass was present in a composite structure by weight. An enhancement in impact strength and water absorption properties of the material was also perceived [101]. A scanning electron microscopy (SEM) study revealed excellent interfacial bonding between the fiber and the matrix of oil palm/kenaf fiber-reinforced epoxy hybrid composite that evince the improvement in the tensile and flexural properties of the material. Moreover, when compared to other composites, oil palm/kenaf fiber hybrid composite absorbs more energy during impact loading that makes the hybrid material a good competitor in the automotive sector [102]. A hybrid composite comprised of carbon and flax fibers reinforcement in the matrix of epoxy resin resulted in 17.98% reduction in the average weight of the material, and maximum interlaminar shear strength (ILSS) of 4.9 MPa and hardness of 77.66 HRC was observed [103]. Fiber hybridization is a promising strategy, where two or more types of fibers are combined in a matrix of composite material to mitigate the drawback of the type of fiber, keeping benefits of others. Synergetic effects of both the fibers aids to enhance properties of the composite material that neither of the constituents owned [104,105]. A hybrid composite made of epoxy resin as matrix material that had a reinforcement of 27% banana along with 9% jute fibers showed a tensile strength of 29.467 MPa. Another composite with the same matrix material that had reinforcement of 21.5% coconut sheath and 15.5% jute fibers showed a compressive strength of 33.87 MPa. An increasing amount of banana fiber reinforcements resulted in increased tensile strength of the composite material [106].

2.2. Particle-Reinforced Composite

Compared to FRC, particle-reinforced composite (PRC) is not that effective by means of material strength and fracture resistance property. However, ceramic, metal, or inorganic particles restrict the deformation and provide good material stiffness. In recent days, PRCs are also getting a bit of attention due to their isotropic properties and cost-effectiveness. Moreover, these composites are manufactured using similar techniques used for monolithic material [107,108]. PRCs are employed for civil applications such as roadways and concrete structures, where a high degree of wear resistance is expected. In concrete, cement acts as a binder material while aggregate of coarse rock or gravel as a filler material provides hardness and stiffness [109].

2.3. Sheet-Molded Composites

Sheet-molded composites (SMCs) are fabricated by bonding homogeneous layers of materials using a compression molding process to form nonhomogeneous composite laminates. The laminate is composed of layers and, in the case of FRP composites made of fiber sheets, buckling stability of the material improves with increasing the number of layers in the laminate [110]. SMC shows the application in large structural components like automotive body parts consisting of high strength to weight ratio [111–113]. Tensile properties of natural fibers can be defined by their chemical compositions. Tensile strength increases with an increase in cellulose content of the fibers, and decreases with increase in lignin content. Some of the properties of frequently used fibers are displayed in Table 2, and Table 3 depicted different properties offered by matrix material.

There are several factors, other than composite constituents and manufacturing processes, that influence the FRP composite performance.

Interphase: It is the region around the fiber in a matrix phase of a FRP composite structure. At the interphase stress, transfer from matrix to fiber takes place at loading conditions. Therefore, to evaluate the performance of composite, not only the properties of its constituent materials, but also understanding the behavior of interphase, is important [33].

Table 2. Some significant properties of frequently used fiber materials [114–117].

Fiber	Density (g/cm^3)	Elongation (%)	Tensile Strength (MPa)	Young's Modulus (GPa)
Aramid	1.4	3.3–3.7	3000–3150	63–67
E-glass	2.5	2.5–3	2000–3500	70
S-glass	2.5	2.8	4570	86
Cotton	1.5–1.6	3–10	287–597	5.5–12.6
Hemp	1.48	1.6	550–900	70
Jute	1.3–1.46	1.5–1.8	393–800	10–30
Flax	1.4–1.5	1.2–3.2	345–1500	27.6–80
Ramie	1.5	2–3.8	220–938	44–128
Sisal	1.33–1.5	2–14	400–700	9–38
Coir	1.2	15–30	175–220	4–6
Kenaf	0.6–1.5	1.6–4.3	223–1191	11–60
Bamboo	1.2–1.5	1.9–3.2	500–575	27–40
Oil palm	0.7–1.6	4–8	50–400	0.6–9
Betel nut	0.2–0.4	22–24	120–166	1.3–2.6
Sugarcane bagasse	1.1–1.6	6.3–7.9	170–350	5.1–6.2

Pretreatments: Physical or chemical treatments like preheating, alkalization, acetylation, and use of silane coupling agent on fibers to modify the fiber surface and its internal structure results in the improvement of adhesion at the interface and amalgamation of the matrix resin into the fibers [118].

Size effect: For FRP-confined cylindrical concrete columns, size effect depends on the mode of failure; there is no occurrence of size effect if failure is plasticity dominated. When failure is fracture-dominated, it occurs due to shear banding. While in large columns, cylinders of small size fail due to FRP rupture caused by plastic dilation in the concrete [119].

Confinement methods: FRP-confined high strength concrete (HSC), and ultra-high-strength concrete (UHSC) show highly ductile compressive behavior when sufficiently confined. On the other hand, if HSC or UHSC are inadequately confined, there is degradation of the axial compressive performance of the FRP tube-encased or FRP-wrapped specimen. FRP thickness and confinement method does not make much difference in the strain reduction factor ($k\varepsilon$), while for the concrete structures, $k\varepsilon$ decreases with an increase in concrete compressive strength [120].

Cross-section: Under concentric compression, the behavior of concrete-filled fiber-reinforced polymer tubes (CFFT) depends upon amount and type of tube material used, concrete strength, cross-sectional shape, specimen size, and manufacture method. When cross-sectional shape is taken into consideration, newly developed rectangular and square CFFT shows highly ductile behavior as a significant improvement with internal FRP reinforcement, when compared to conventional CFFTs [121]. Further studies have shown that specimen size does not influence the compressive behavior of CFFTs. Although a significant correlation has been observed between fiber elastic modulus and the strain reduction factor, fibers with a higher modulus of elasticity result in a decrease of the strain factor that further resembles concrete brittleness while manufacturing CFFTs [122].

Fiber volume: Maleic anhydride-grafted polypropylene (MA-g-PP) was used as a compatibilizer to improve adhesion between bamboo fiber and polypropylene matrix composite material. Composite with 5% MA-g-PP concentration and 50% fiber volume has increased impact strength by 37%, flexural strength by 81%, flexural modulus by 150%, tensile strength by 105%, and tensile modulus by 191%. When the fiber volume of chemically treated composite with MA-g-PP compatibilizer increased from 30% to 50%, it showed an increase in the heat deflection temperature (HDT) by 23 °C to 38 °C compared to virgin PP. Therefore, fiber volume of 50% fraction, 1–6 mm fiber length with 90–125 µm fiber diameters, coupled with MA-g-PP compatibilizer is the recommended optimized composition for bamboo fiber-reinforced polypropylene composites, which results in a maximum enhancement in the mechanical properties and a higher thermal stability is also achieved [123].

Table 3. Variety of available matrix materials.

References	Matrix Material	Properties	Applications
[2]	Polyethersulfone	Flame resistant	Automotive
[3]	Polyphenylene sulfide	Resistance to chemicals and high temperature	Electrical
[3,9]	Polysulfone	Low moisture absorption, high strength, low creep	Marine, food packaging
[6]	Polyethylene (PE)	Resistance to corrosion	Piping
[6,36,54,66,94,96,101]	Polypropylene (PP)	Resistance to chemicals	Packaging, automotive, construction
[6,13,79]	Polylactic acid (PLA)	Biodegradable, non-toxic	Food handling, bio-medical
[10,90]	Polyurethane (PU)	Wear resistance, low cost, sound and water-proof	Structural, acoustic
[16]	Poly(butylene adipate-co-terephthalate)-PBAT	Biodegradable, high stiffness	Coating, packaging
[19]	Cement	Durable	Structural
[28]	Poly(vinyl alcohol)	High tensile strength	Bio-medical
[33]	Natural rubber	Low density, low cost, biodegradable	Structural, automobile
[54,91,98,100,102]	Epoxy resin	High strength	Automotive, aerospace, marine
[82,92]	Polyester	Durable, resistance to water, chemicals	Structural

Fiber orientation: When CO_2 laser engraving was employed for material removal of GFRP, it was found that surface roughness and machined depth of the laser-engraved surface were hugely dependent upon the fiber direction [60]. T300 carbon fibers and 7901 epoxy resin as a matrix material were used to fabricate T300/7901 unidirectional (UD) fiber-reinforced composite to investigate mechanical properties in uniaxial tension/compression and torsional deformations. Micrographs of fiber matrix interface at different load levels were examined, which revealed that matrix plastic deformation has no significant effect on predicted ultimate load at failure. It also revealed a noticeable drop in the ultimate strength with the increase in fiber angle from 0° to 15°. Stress concentration factor (SCF) plays an important role while considering the failure prediction, without consideration of SCFs transverse strength will be overestimated [124]. Thermal buckling load for curvilinear fiber-reinforced composite laminates is more for antisymmetric laminates, while laminates with nonuniform temperature distribution exhibit high critical load carrying capacity [110].

3. Manufacturing Techniques

Manufacturing of FRP composite involves manufacturing of fiber preforms and then reinforcing these fibers with the matrix material by various techniques. Fiber preforms involve weaving, knitting, braiding, and stitching of fibers in long sheets or mat structure [125–127]. Preforms are used to achieve a high level of automation with the assistance of robotics, which offers control over the fiber angle and the fiber content on every zone of the part to be molded [128].

3.1. Conventional Manufacturing Processes

Prepregs are a combination of fibers and uncured resin, which are pre-impregnated with thermoplastic or a thermoset resin material that only needs the temperature to be activated. These prepregs are ready-to-use materials where the readily impregnated layers are cut and laid down into the open mold [128]. Dow Automotive Systems has developed VORAFUSE, a technique that combines epoxy resin with carbon fiber for prepreg applications to improve material handling and cycle time in the compression molding of composite structures. Working in collaboration with a variety of automotive companies, they have achieved significant weight reduction, which results in efficient manufacturing of CFRP composite structures [129].

Figure 3 shows the hand lay-up, which is the most common and widely used open mold composite manufacturing process. Initially, fiber preforms are placed in a mold where a thin layer of antiadhesive coat is applied for easy extraction. The resin material is poured or applied using a brush on a reinforcement material. The roller is used to force the resin into the fabrics to ensure an enhanced interaction between the successive layers of the reinforcement and the matrix materials [130–132].

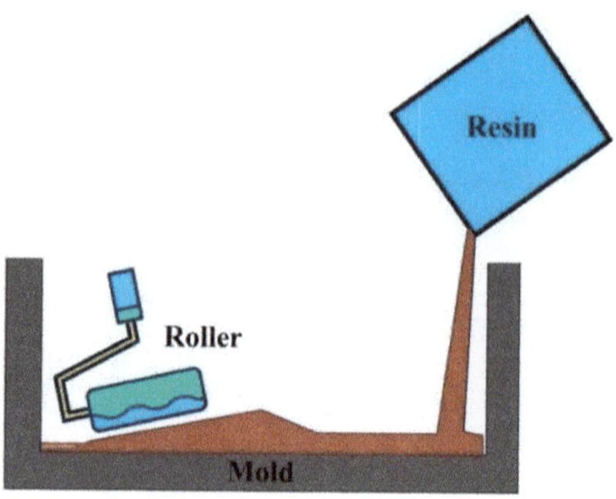

Figure 3. Hand layup process.

Spray-up technique is no different than hand lay-up. However, it uses a handgun that sprays resin and chopped fibers on a mold. Simultaneously, a roller is used to fuse these fibers into the matrix material. The process is illustrated in Figure 4. It is an open mold type of technique, where chopped fibers provide good conformability and quiet faster than hand lay-up [133,134].

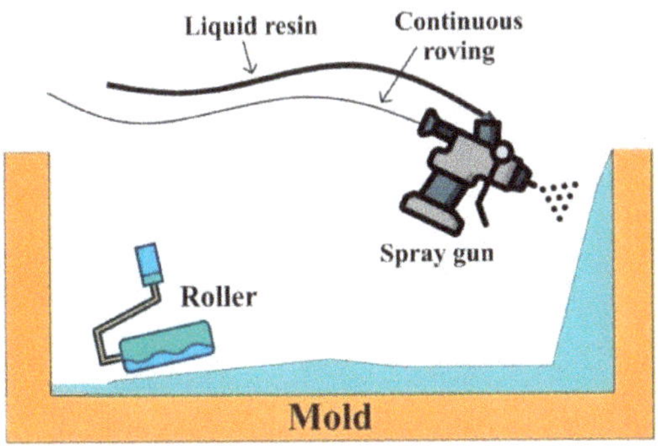

Figure 4. Spray-up process.

Vacuum bag molding uses a flexible film made of a material such as nylon polyethylene or polyvinyl alcohol (PVA) to enclose and seal the part from the outside air. Many times, the vacuum bag molding technique is performed with the assistance of the hand lay-up technique. Laminate is first made by using the hand lay-up technique, and then after it is placed between the vacuum bag and the mold to ensure fair infusion of fibers into the matrix material [135,136]. The air between the mold and the vacuum bag is then drawn out by a vacuum pump while atmospheric pressure compresses the part. The process can be well understood by Figure 5. Hierarchical composites were prepared with multiscale reinforcements of carbon fibers using a vacuum bagging process, which eliminated chances of detectable porosity and improper impregnation of dual reinforcements, with increases in flexural and interlaminar shear properties by 15% and 18%, respectively [137].

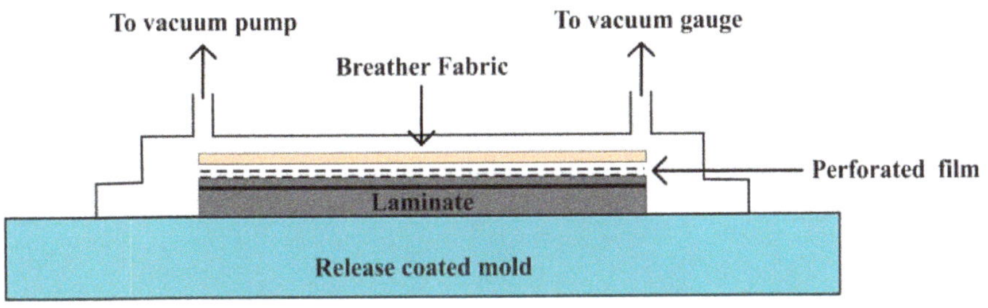

Figure 5. Vacuum bag molding process.

The preform fiber reinforcement mat or woven roving arranged at the bottom half of the mold and preheated resin is pumped under pressure through an injector [132]. The mechanism of the resin transfer molding (RTM) process can be understood with Figure 6. A variety of combinations of fiber material with its orientation, including 3D reinforcements, can be achieved by RTM [138,139]. It produces high-quality, high-strength composite structural parts with surface quality matching to the surface of the mold [140].

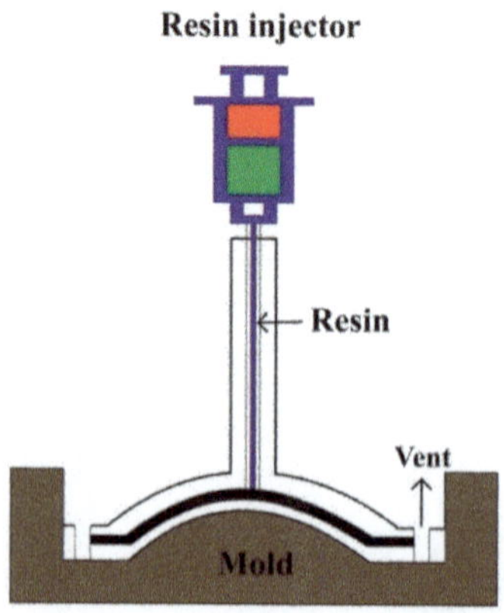

Figure 6. Resin transfer molding process.

Vacuum infusion or vacuum assisted resin transfer molding (VARTM) is a recent development, in which preform fibers are placed on a mold and a perforated tube is positioned between vacuum bag and resin container. Vacuum force causes the resin to be sucked through the perforated tubes over the fibers to consolidate the laminate structure, as shown in Figure 7. This process leaves no room for excess air in the composite structure, making it popular for manufacturing large objects like boat hulls and wind turbine blades [141,142]. For the improvement in the strength of textile composites, natural fibers are surface treated. Alkali treated flax fiber-reinforced epoxy acrylate resin composite fabricated using VARTM technique resulted in improvement of tensile strength by 19.7% [143].

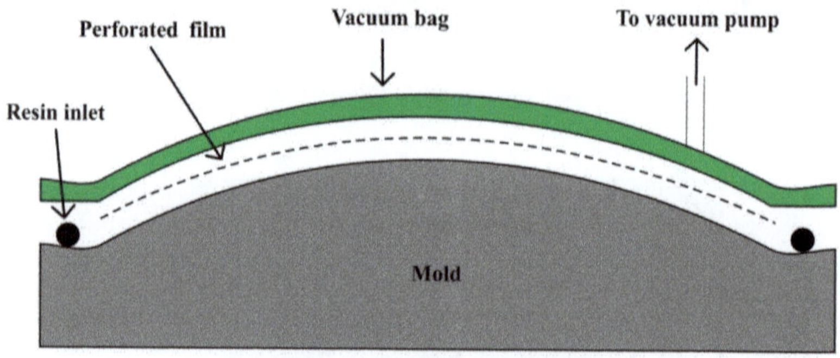

Figure 7. Vacuum infusion process.

It uses preheated molds mounted on a hydraulic or mechanical press. A prepared reinforcement package from prepreg is placed in between the two halves of the mold, which are then pressed against each other to get a desired shape of the mold. Figure 8 represents the stepwise processing of compression molding. It offers short cycle time, a high degree of productivity, and automation with dimensional stability, hence it finds diverse applications in the automobile industry [144–146]. Dispersion of 35% filler elements containing sisal fiber and zirconium dioxide (ZrO_2) particles in the matrix of unsaturated polyester (UP) was obtained by the compression molding technique, which displayed optimum mechanical properties when tested under SEM, X-ray diffraction, and Fourier transform infrared spectrometer (FTIR) [147]. Jute fiber-reinforced epoxy polymer matrix-based composite has been fabricated by using hand lay-up followed by the compression molding technique at s curing temperature ranging from 80 °C to 130 °C. Enhancement in the mechanical properties has been observed with the maximum tensile strength of 32.3 MPa, flexural strength of 41.8 MPa, and impact strength of 3.5 Joules [148].

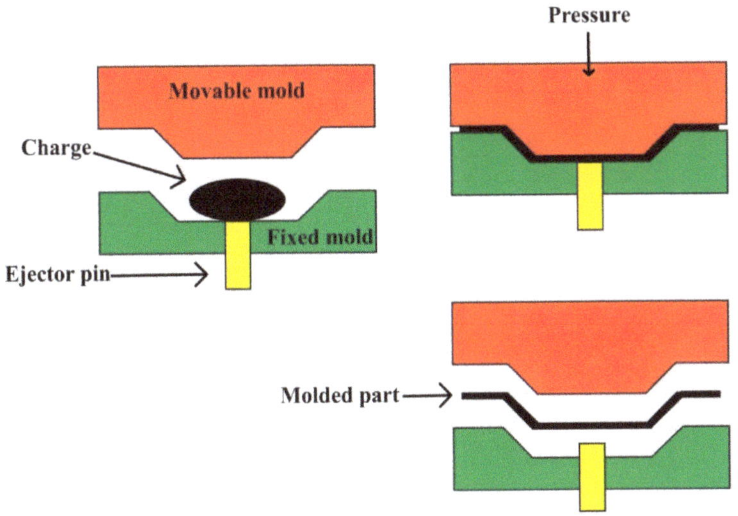

Figure 8. Compression molding process.

The pultrusion process can be explained (Figure 9) as strands of continuous fibers are pulled through a resin bath, which are further consolidated in a heated die. It is a continuous process, useful for fabrication of composites with a constant cross-section with a relatively longer length; it enables production with a high degree of automation and lower production cost [149–151].

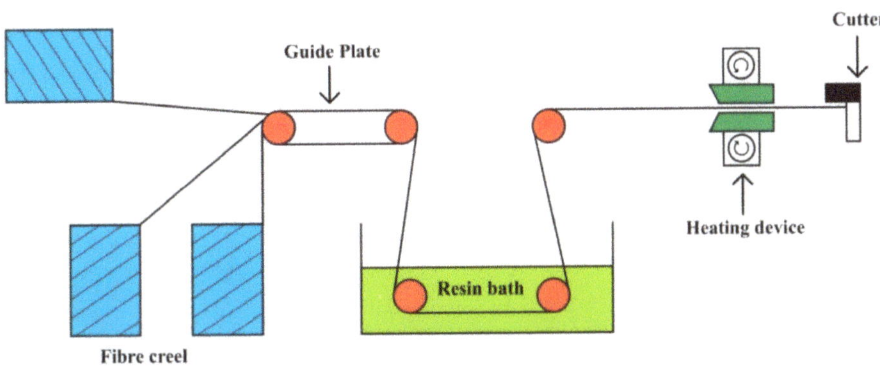

Figure 9. Pultrusion process.

Injection molding has the ability to fabricate composite parts with high precision and at very low cycle times. In a typical injection molding process, fiber composites in the form of pellets are fed through a hopper, and then they are conveyed by a screw with a heated barrel, as shown in Figure 10. Once the required amount of material is melted in a barrel, the screw injects the material through a nozzle into the mold. where it is cooled and acquires the desired shape [152]. Injection molding is found to be very effective for thermoplastic encapsulations of electronic products required in medical industries [153]. Improvement in fiber-matrix compatibility and uniformity in the dispersion of fibers in the matrix material is achieved during the surface treatments of biocomposites [154].

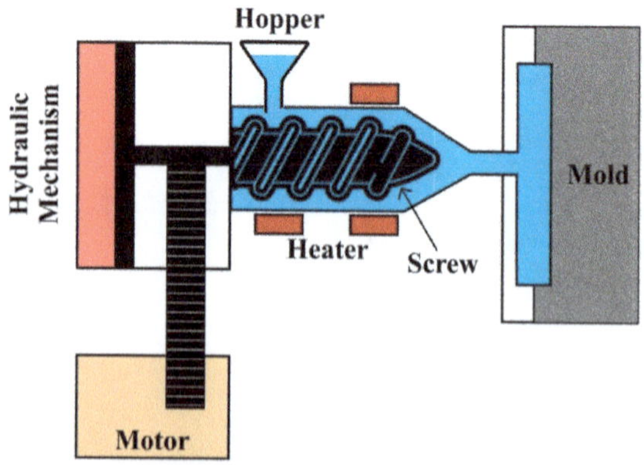

Figure 10. Injection molding process.

3.2. *Advance Manufacturing Processes*

The emerging nanotechnology has provoked researchers to seek out new nanoscale fiber manufacturing techniques for composite manufacturing. An electrostatic fiber fabrication technique called electrospinning uses electrical forces to generate continuous fibers of two nanometers to several micrometers. Polymer solution ejected through spinneret forms a continuous fiber, which is collected at the collector shown in Figure 11. It serves enhanced physical and mechanical properties, flexibility over process parameters, high surface area to volume ratio, and high porosity; therefore it finds potential in diverse fields of biomedical applications such as wound healing, tissue engineering scaffolds, drug delivery, as a membrane in biosensors, immobilization of enzymes, cosmetics, etc. [155,156].

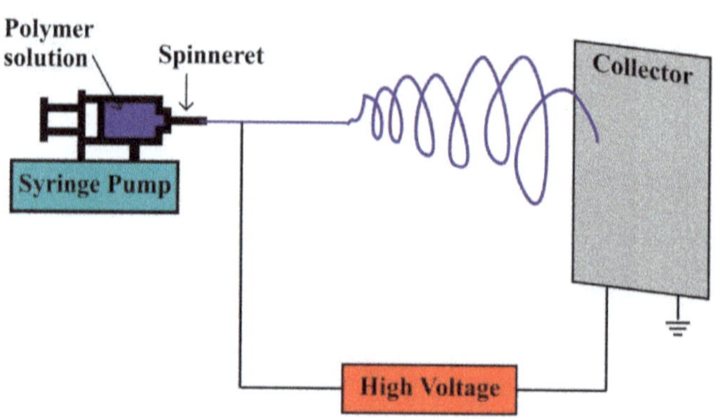

Figure 11. Electrospinning process.

Additive manufacturing (AM) offers a high level of geometrical complexity for the fabrication of fully customized objects as it takes advantage of computer-aided designing and also eliminates the requirement of molds, which saves cost and time of manufacturing process [157,158]. AM is one of the leading technologies in composite manufacturing as it provides wide range over the selection of fiber volume and fiber orientation. It has the ability to transverse design idea into the final product quickly without the wasting material and cycle time, which makes it ideal for prototyping and individualization [159–161].

Specially developed manufacturing techniques: The fabrication of carbon fiber-reinforced metal matrix composites (CF-MMC) involves powder metallurgy, diffusion bonding, melt stirring, squeeze casting, liquid infiltration, ion plating, and plasma spraying. Each one of them serves distinct benefits for manufacturing CF-MMC. Powder metallurgy and melt stirring being simplest and most economical; diffusion bonding uses specially designed tools, where carbon fiber preforms are prepared by infiltration in polymer binder and then stacked up with metal sheets. Slurry casting is carried out at the freezing temperature of the metal matrix material, eliminating the probability of interfacial reactions and degradations of the interface [162].

3.3. Automated Manufacturing Techniques

It is a continuous process that offers self-automation, which leads to reduced cost. Filament winding is useful to create axisymmetric, as well as some non-axisymmetric, composite parts, such as pipe bends [163]. Driven by several pulleys, continuous prepreg sheets, rovings, and monofilament are made to pass through a resin bath and collected over a rotating mandrel, as displayed in Figure 12. Then, after applying sufficient layers, mandrel, which has the desired shape of the product, is set for curing at the room temperature [164,165]. Recently developed robotic filament winding (RFW) technique is provided with an industrial robot equipped with a feed and deposition system. It yields advantages over process control, repeatability, and manufacturing time by replacing a human operator [166].

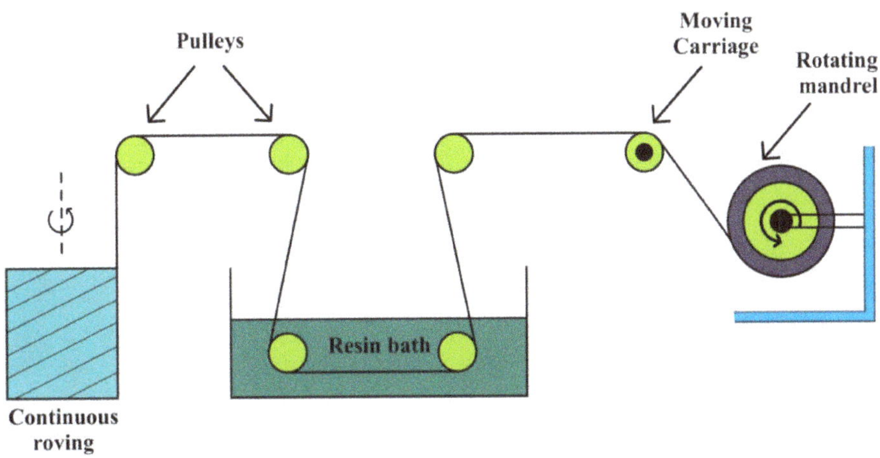

Figure 12. Filament winding.

Automated tape layup (ATL) and automated fiber placement (AFP) techniques are efficient for large, flat, or single curvature composite structures as it uses the assistance of a multiaxis articulating robot, where the material is deposited in accordance with a defined computer numerical control (CNC) path. The AFP process involves the individual prepreg lay-up of laminates onto a mandrel using a numerically controlled fiber placement machine, which are then further pulled off by holding spools [167]. Composite structures are fabricated quickly and accurately, but the expenses in employing required specialized equipment keep these technologies out of reach for small to medium scale manufacturers [168].

4. Applications

4.1. Civil

Fire resistant concrete: For many years, FRP composites have been widely used to strengthen the concrete structures and recent studies have introduced inorganic/cementitious materials to develop fiber-reinforced inorganic polymer (FRiP) composites. Phosphate cement-based FRiP is used to replace the epoxy in the FRP composite structure with improvement in fire resistance [169–173]. These inorganic cementitious materials consist of Portland cement, phosphate-based cement, alkali-activated cement, or magnesium oxy-chloride cement (MOC). FRiP retains about 47% of it strengthening efficiency when exposed to fire [174–176]. FRP sandwich material is a special form of laminated composite material, which offers high strength to weight ratio, thermal insulation, and service life benefits. Therefore, it has emerged as an excellent alternative to metallic skins for sandwich composites in structural engineering applications. Also, FRP sandwich systems provide more durable and cost-effective infrastructure in bridge beams, footbridges and bridge decks, multifunctional roofs, cladding and roofing systems for buildings, railway sleepers, and floating and protective structures [177].

Concrete beams: A significant improvement in flexural strength and load-carrying capacity is observed in FRP sheets bonded to the tension face of concrete beams, even when subjected to the harsh environment of wet and dry cycling [178]. To achieve higher means of strain levels, the anchorage of externally bonded FRP materials is applied prior to the premature debonding failure of reinforced concrete (RC) structures. Among the rest of the anchorage solutions, FRP anchors were found to be 46% more effective than vertically orientated U-jacket anchors, resulting in remarkably high anchorage efficiency. Simplicity, non-destructiveness, and ease of application are some other advantages for FRP to concrete applications [179]. The newly developed basalt microfibers are added longitudinally as reinforcement to the concrete structures to study its feasibility and flexural behavior; it exhibits improvement in curvature ductility with increased maximum moment capacity of the beams. Regardless of the type of concrete used, there is an enhancement in the flexural capacity of the beams with an increase in BFRP reinforcement ratio [180–182]. Figure 13a shows some RC beams. RC members can be strengthened by employing FRP anchors with varying fiber content and embedment angle to enhance the strain capacity of externally bonded FRP composites. As the anchor dowel angle increases relative to the direction of load, there is an increase in the strength of the joint with a decrease in ductility of joint [183].

(a)

(b)

Figure 13. Reinforced composite (RC) beams (**a**), concrete bridge (**b**), reproduced from [184,185] under open access license.

Bridge system: For applications such as constructing durable concrete structures and restoring aged structures like bridges and tunnels, sprayable ultra-high toughness cementitious composite (UHTCC) is implemented. The UHTCC improved the durability of concrete structures with higher compressive, tensile, and flexural strengths when compared to cast UHTCC. Also for RC–UHTCC beams with an increase in the thickness of UHTCC layer, there was an increase in the stiffness, effectively gaining control over the cracks occurring in the concrete layer of the beam specimens [186]. FRP composites have been proven as a viable structural material in bridge construction. Bridge systems use FRP or hybrid FRP–concrete as primary construction materials for the application of bridge components such as girders, bridge decks, and slab-on-girder bridge systems. When compared to RC decks, hybrid FRP concrete decks reveals higher durability with less stiffness deterioration under design truckloads [187]. A fixed concrete bridge on the Indian River, Florida has been displayed in Figure 13b. Unprecedented threats from terrorist activities or natural disasters impose danger to public civil infrastructure such as bridges, and therefore, the impact and blast resistance design of such structures has become a prominent requirement in the design process. FRP material has been employed to strengthen and improve the impact resistance properties of the structures, including RC beams, RC slabs, RC columns, and masonry walls. This also results in an increase in the load-carrying capacities, ductility, energy absorption, and tensile strength of the materials with an increase in strain rate [188].

Deck panels: Flexure and shear strength seemed to be higher in all FRP composites when compared with RC for the application of bridge deck panels [189]. Decks made of hybrid fiber-reinforced composite materials were found to effectively fit for their design requirements. Glass and jute fibers reinforced with vinyl ester as matrix were used to fabricate a hybrid composite by hand lay-up technique [190].

Earthquake-resistant columns: FRP composites find an important application as a confining material for concrete in the construction of concrete-filled FRP tubes as earthquake-resistant columns and in the seismic retrofit of existing RC columns [191].

Pile material: Composite pile materials are the best replacement for traditional piles such as concrete, steel, and timber, as composite piles serve longer service life, require less maintenance costs, and are environmentally friendly. Hollow FRP piles show high potential in load-bearing applications and also provide significant advantages in terms of cost efficiency and structural capabilities [192,193].

Concrete slabs: For both unreinforced and RC slabs, carbon epoxy and E-glass epoxy composite systems restored original capacity of the damaged slabs, as well as resulted in a remarkable increase of more than 540% in the strength of the repaired slabs. Moreover, with the use of FRP systems, unreinforced specimens revealed a 500% improvement, while steel-reinforced specimens showed a 200% upgrade, in the structural capacity for retrofitting applications [194].

Sensors: Due to severe damages and collapses in civil structures, the need for development and advancement in sensing technology and sensors has given rise to structural health monitoring (SHM) technology. This consists of sensors, data acquisition, and transmission systems that can be used to monitor structural behavior and performance of structures when subjected to natural disasters such as an earthquake. The SHM system can record real loads, responses, and predict environmental actions [195,196].

4.2. Mechanical

Mechanical gear pair: For the application of gear pair, polyoxymethylene (POM) with 28% glass fiber reinforcement revealed significant enhancement of about 50% in the load-carrying capacity, with lower specific wear rate when compared to unreinforced POM [197]. Gear pair made of carbon–epoxy prepreg laminate was comparable to steel for the evaluation of static transmission error (STE) and mesh stiffness curves. Results showed a significant reduction in STE peak-to-peak value, which further resulted in improved noise, vibration, and harshness (NVH) performance of the material [198–201].

Pressure vessel: In the automobile industry, there is remarkable growth in the demand for lightweight material to increase fuel efficiency with a reduction of emission. FRP composites are serving these demands, for example, for safe and efficient storage and transportation of gaseous fuels such as hydrogen, and natural gas pressure vessels are used [129]. Pressure vessels made of FRP composite materials, when compared to metallic vessels, provide high strength and rigidity, improved corrosion resistance, and improved fatigue strength, besides being light in weight [202,203]. A pressure vessel made of thermosetting resin and fiberglass reinforcement is displayed in Figure 14.

Figure 14. Pressure vessel made of thermosetting resin and fiberglass reinforcement, reproduced from [204] under open access license.

Hydraulic cylinder: For the transportation of soil material, a dump truck uses a hydraulic system consisting of an actuator made of a telescopic hydraulic cylinder. There is a 96% weight reduction when the steel cylinder is replaced with a carbon fiber-reinforced epoxy resin composite. When this telescopic cylinder made of composite was installed, there was a 50% reduction in the whole hydraulic system [205].

Headstock material: During a typical machining operation, interference between machine tool and workpiece produces high vibration in the cutting tool relative to the workpiece. Nearly half of the deflection in cutting tools comes from the headstock; therefore, headstock demands a high degree of damping property. A hybrid steel–composite headstock adhesively manufactured by glass fiber epoxy composite laminates served a 12% increase in stiffness and 212% increase in damping property for the application of a precision grinding machine [206].

Manipulator: A two-link flexible manipulator was developed using ionic polymer metal composite (IPMC), which manifests the potential of polymer-based composite materials for flexible joints and links in robotic assembly, as demonstrated in Figure 15. Sulfonated polyvinyl alcohol (SPVA), 1-ethyl-3-methylimidazolium tetrachloroaluminate (IL), and platinum (Pt) (SPVA/IL/Pt)-based IPMC manipulator links provide flexibility and compliant behavior during manipulating and handling of complex objects of different shapes and sizes [207].

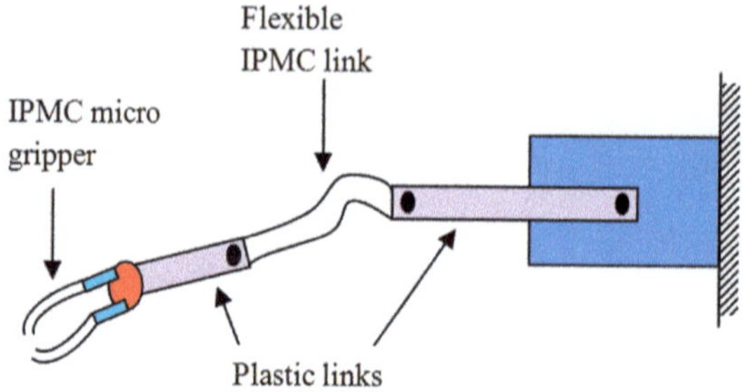

Figure 15. Flexible link manipulator.

Turbine blades: Turbine blades made of carbon fiber-reinforced silicon carbide (SiC) ceramic matrix composite (CMC) hold a bending strength of 350 MPa and fracture toughness of 4.49 $MPa\sqrt{m}$ when fiber content of 10–15% by volume is present [208].

4.3. Automobile

Braking system: In an automobile braking system, the temperature can reach up to thousands of degrees centigrade. A monolithic metal fails to perform well as they are not able to withstand these higher temperatures. Therefore, carbon fiber-reinforced silicon carbide (C-Si) finds applications in brake materials for heavy vehicles, high-speed trains, and emergency brakes in cranes [209]. Figure 16 shows a carbon–ceramic brake of a Chevrolet Corvette.

Figure 16. The braking system of corvette made of carbon–ceramic, which saved 4.9895 kg replacing iron, reproduced from [210] under open access license.

Trunk lid and body stiffener: In the transportation industry, CFRP fits as a reliable material for automobile body parts such as body stiffeners and engine hoods. As for this application, a higher strength to weight ratio is essential [211].

Bicycle: CFRP is replaced with hybridized carbon fiber with natural fibers, such as flax, to overcome the lower impact toughness and high cost of the material. A bicycle frame was fabricated using 70% flax fiber and 30% carbon fiber, which weighed just 2.1 kg and showed superior damping characteristics over aluminum, steel, and titanium [212].

Automobile body parts: Automobile body parts, such as engine hood, dashboards, and storage tanks, are manufactured by using reinforcements of natural fibers such as flax, hemp, jute, sisal, and ramie. For these composite structures VARTM manufacturing method was employed and liability is tested with structural testing and by using impact stress analysis. The result showed a reduction in the weight of the material with the enhancement in stability and strength. The improvement in safety features were measured under head impact criteria (HIC), and it was found that composite structures comprised of natural fiber reinforcements are reasonable for automobile body parts [213–217]. Figure 17 displays exterior body parts of a model Volkswagen x11 made of carbon fiber.

Figure 17. Volkswagen xl1 carbon fiber body parts, reproduced from [218] under open access license.

Door panels: Addition of bamboo fibers increases the cell wall thickness of polyurethane composite structures, leading to the improvement of the sound absorption coefficient in automobile door panels [219,220].

Engine hood: Improvement in tensile strength and wear resistance properties have been observed for the engine hood material of an excavator engine when epoxy resin composite with reinforcement of glass fiber has been used over aluminum sheet metal [221].

Interior structures: The composite structure comprises of biodegradable natural fibers which have found significant applications as sound and vibration absorption material in interior automobile components. Composite laminate with bamboo, cotton, and flax fibers with PLA fibers showed bending stiffness of 2.5 GPa, which is higher than all other composites [222,223]. Figure 18 shows the interior structure of a car.

Figure 18. Car interior, reproduced from [224] under open access license.

Engine frame: Steel engine subframe material, when replaced by carbon epoxy composite, displayed improvement in stiffness with a decrease in the maximum stress and weight from 16 kg to 5.5 kg [225].

T-joint: Epoxy resin composite with woven carbon fibers implemented for T-joint in the vehicle body revealed improvement in overall stiffness and strength behavior with a reduction in weight [226].

For the application of automobile bumper beam, glass/carbon mat thermoplastic (GCMT) composite has been designed and manufactured, which has shown improvement in impact performances with 33% weight reduction compared to the conventional glass mat thermoplastic (GMT) bumper beam [227].

4.4. Aerospace

Application of FRP composite materials in the field of aerospace industries can be seen with the implementation of highly durable, thermal-resistant, lightweight materials for the aircraft structure due to their outstanding mechanical, tribological, and electrical properties [69,74,228]. Natural fiber-reinforced thermoset and thermoplastic skins manifest the properties required for aircraft interior panels, such as resistance to heat and flame, serving easy recycling, and disposal of materials being cheaper and lightweight over conventional sandwich panels [229]. Though FRP composite shows a variety of applications in the aerospace industry due to their superior mechanical properties and lightweight structure, they face difficulty in recycling. To overcome this, natural fiber/biocomposite materials brought new prospects in the aerospace industry due to their biodegradability and lower cost [63].

Wireless signal transmission: Conductive fibers in the layer of fiber composite structure eliminate the requirement of separate wires for transceivers of communication devices. When voltage is applied to either layer of composite, it carries electric power to certain electric devices through the fibers [230].

The Hubble space telescope antenna: High stiffness with a lower coefficient of thermal expansion is achieved when P100 graphite fibers diffused in 6061 aluminum matrix composite material are employed to the high gain antenna of the Hubble space telescope [231].

Aircraft parts: Aircraft wing boxes made of ramie fiber composites revealed a 12–14% decrease in weight [232]. Hybrid kenaf/glass fiber-reinforced polymer composites showed enhanced mechanical properties with rain erosion resistance, suitable for aircraft application [233]. Carbon fiber-reinforced silicon carbide is applied for aircraft brakes to withstand temperatures up to 1200 °C [234].

Safety: The ablation method is carried out as one of the thermal protection methods for the spacecraft to ensure safety. An ablative composite material was used with zirconia fibers due to its significant mechanical properties and resistance to high-temperature ablation. It revealed that with 30% of zirconium fiber content composite material showed 19.33 MPa of bending strength; also at the higher temperature over 1400 °C, due to eutectic melting reaction, a ceramic protective layer forms which offers bending strength of 13.05 MPa [235].

4.5. Biomedical

Dentistry and orthopedic: Due to the strength characteristics and biocompatibility of fiber-reinforced composites, they are being used in the field of dentistry and orthopedics. Remarkable technological advances have been seen in the design of lower-limb sports prostheses [236]. For the reconstruction of craniofacial bone defects, new fiber-reinforced composite biomaterial replaces the material used for custom-made cranial implants [237]. A variety of aramid fibers display their biomedical applications in protein immobilization, for medical implants and devices, in modern orthopedic medicine, and as antimicrobial material. Typical polyamide (PA), i.e., nylon, is a synthetic polymer with high mechanical strength used in implants, and fibrous composites play a vital role in the manufacturing of dentures and suture materials. For antimicrobial applications, chitosan/m-aramid hybrids show enhancement in the surface area of assembled composites [238–241]. Biostable glass fibers reveal excellent load-bearing capacity in the implants, while antimicrobial properties are manifested by the dissolution of the bioactive glass particles that support bone-bonding [242].

Tissue engineering: Collagen–silk composite serves a promising application for reconstruction of lesioned tissues in tissue engineering. After fabricating the composite material by electrospinning, there is an increase in the ultimate tensile strength and elasticity of the material, with an increase in silk percentage [243]. Fibrous composite made of synthetic biodegradable polymers, polylactic co-glycolic acid (PLGA), gelatin, and elastin (PGE) scaffold can support dense cell growth and deliver tremendously high numbers of cells. This finds broad applicability in tissue engineering to meet design criteria necessary to generate scaffolds of natural and synthetic biomaterials [244,245]. A polyurethane cardiac patch loaded with nickel oxide (NiO) was fabricated using the electrospinning technique. When observed under SEM, PU/NiO nano-composite showed a reduction in the diameter of fibers and pores by 14% and 18%, respectively, compared to pure PU. Delayed blood clotting and a lower percentage of hemolytic revealed an improved antithrombogenic nature of PU/NiO nanocomposite, which plays a vital role in the repair of cardiac damage [246].

Wound healing: A fibrin–collagen filamentous polymer composite subjected to unconfined compression resulted in enhancement of elastic properties with increased node density and amalgamation between collagen and fibrin fibers. This led to the formation of a composite hydrogel, which further increased the modulus of shear storage at compressive strains. Fibrin has its active role in hemostasis and wound healing, while matrix gel based on collagen, gelatin, or elastin is utilized for scaffolds [247–249]. Biopolymers such as PLA, polyglycolic acid (PGA), PLGA, polycaprolactones (PCL), and polyesteramides (PEA) exhibit applications in biomedical fields to suture wounds, drug delivery, tissue engineering, fixing ligament/tendon/bone, dentistry, and surgical implants [250–257].

4.6. Marine

For marine applications, mechanical properties of materials get deteriorated in all types of metals, alloys, or composites due to seawater aging. Hybrid glass–carbon fiber-reinforced polymer composite $(GCG_2C)_s$ shows a high flexural strength of 462 MPa with the lowest water absorption tendency. Therefore retention of mechanical properties in hybrid $(GCG_2C)_s$ composite is more [258]. Moisture absorption properties exhibited by fiber composites are because of their structural or chemical composition, demonstrating various applications in the marine environment [259]. Due to the diffusion process, water molecules get absorbed in the material structure when it is exposed to the marine habitat. Diffusion in the material structure can be monitored by weight gain with respect to time. The number of water molecules that get absorbed is dependent upon the coefficient of diffusion of the material. Though the value of the coefficient of diffusion is lower in the composite materials, it is dependent upon various factors like the type of matrix material, the type of reinforcement material used, and the type of manufacturing process employed. Moisture absorption results in poor adhesion between the fiber and matrix in the composite structure, which ultimately deteriorates the properties of composite material [260–265].

Marine propeller: CFRP shows enhanced mechanical properties, such as high strength to weight ratio, resistance to corrosion, fatigue, and temperature changes with low cost of maintenance. These properties make CFRP a perfect fit for propeller material in marine applications [266].

Hull: Glass or carbon fiber skins with polymeric core sandwich composite panels have been used for the development of entire hulls and marine craft structures [267].

5. FRP Replacing Conventional Material

A variety of different fiber performances incorporated with composite materials, with the combination of distinct base materials and manufacturing techniques, offer an enhancement in properties of materials over pure metals, polymers, or alloys, which make FRP composites befitting for desired applications [268–270]. Composite materials with 5% MAPP by weight and 30% alkaline-treated hemp fibers by weight added to a PP matrix were found to be a replacement over pure PP, as an increment in flexural strength and tensile strength was found by 91% and 122%, respectively [78]. Flax/epoxy composite blades exhibit potential replacement characteristics, with respect to weight, structural safety, blade tip deflection, structural stability, and resonance, to replace glass/epoxy composite blades for small-scale horizontal axis wind turbine systems [271]. SEM morphology analysis revealed improvement in tensile and flexural strength due to good interface quality of RF/PP composite by 20.7% and 27.1%, respectively, when compared with pure PP [86]. A composite incorporated with

PP and bamboo fiber reinforcements that were extracted by using an eco-friendly technique called solvent extraction, provided excellent fiber flexibility. The PP composite made of 20% bamboo fibers revealed the highest modulus of rupture (MOR), resulting in a rise in its flexural strength, which is an 8.3% increase to that of neat PP [272]. Conventional GMT was substituted by GCMT composite for the application of automobile bumper beams, which saw a 33% weight reduction with improved impact performances [227].

Fibers as reinforcement in a matrix of a composite structure act as a load-carrying element. While the matrix material keeps fibers in their required position and orientation, it also facilitates stress transfer and protection from the environment. FRP materials have been found to be superior to metals for a variety of applications where higher strength to weight ratio is required [273–275]. In recent years, polymer composites have shown a great potentiality and superiority over a prevalent yet critical issue of friction and wear faced by conventional metals and alloys [276–278]. Besides the remarkable tribological characteristics, polymeric composites offer flexibility in multifunctioning by tuning their composition to provide a cost-effective way of developing new tribological materials [279]. For automobile and aerospace applications, CF-MMC is replacing existing unreinforced metals and alloys as it provides excellent mechanical, thermal, and electrical properties with enhanced wear and corrosion resistance to withstand harsh environments [97]. The most common types of FRP used as reinforcement in the concrete structures are CFRP, GFRP, and aramid fiber-reinforced polymer (AFRP). These FRPs shows good resistance to shear and flexural stresses [280–283]. For the concrete structures to withstand in a harsh environment, reinforcement materials need to be noncorrosive and nonmagnetic. FRP bars possess these properties, which makes them applicable for the RC structures over the conventional steel reinforcement [284–286]. Structural material aluminum 6061 is replaced with hybridized flax and carbon fiber composites, as they revealed improvement in vibration damping properties in a material. A 252% gain in tensile strength with 141% improvement in damping ratio has been observed. In addition, there was a 49% weight saving due to a reduction in material density [149]. Hybridized composite structures with jute and carbon fiber reinforcements offer economic and sustainable alternatives over CFRP, revealing outstanding damping properties [148]. Engine hood material made of an aluminum sheet metal of an excavator engine was replaced with black epoxy composite with aluminum tri-hydroxide reinforced with glass fibers [151].

6. Challenges

A major challenge in fabricating FRC material is the lack of fiber–matrix characterization cognition. For the application of FRPCs in variety of fields, understanding their constituent's significant material properties is necessary, with the basic constructs and the availability of manufacturing technology. For example, for the production of nanocomposites, one should acquire nanotechnology, including all the required tools and equipment. Also, the choice of manufacturing process eventually affects the final properties of material. Production volume influences the cost—the higher the volume of production, the less would be the cost of materials. Increasing production volume, in the case of the automobile industry, leads to greater risk of investing in raw materials while establishing manufacturing set-up according to the production rate and cycle time. Also the design complexity of the product augments the cycle time, slowing down the production rate.

Growing demand of high performance composites for aerospace and structural applications aggrandized the use of petroleum-based materials, leaving issue of composite waste disposal. However, nowadays, different researchers are developing various biocomposites using natural fibers and bio-based polymers, yet not all of these are completely biodegradable.

7. Conclusions

Composite materials are divulging numerous enhancements in distinct material properties since their invention in the last century. Copious amounts of research efforts have been made to discover optimized material to perform in a more effective way for desired applications. Over the past few decades, reinforcements of fibers or particles in the matrix structure of composite materials have revealed outstanding remarks, making them a popular choice for topmost applications.

Classifications of composite materials, along with the properties of their constituent elements, have been studied to understand the potentiality of different composite materials in various fields. Fiber-reinforced composite material was found to be one of the most promising and effective types of composites, as it claims dominance over the majority of applications from topmost fields.

There are numerous types of fibers available for fabrication of fiber-reinforced composites; those are categorized as natural and synthetic fibers. Synthetic fiber provide more stiffness, while natural fibers are cheap and biodegradable, making them environmentally friendly. Though both types of fibers have their efficacy in significant applications, latest research has revealed the exceptional performance of hybrid fiber-reinforced composite materials, as they gain the advantageous properties of both.

Composite materials are fabricated with a number of different techniques, among which every technique is applicable for certain material. Effectiveness of manufacturing technique is dependent on the combination of type and volume of matrix or fiber material used, as each material possesses different physical properties, such as melting point, stiffness, tensile strength, etc. Therefore, manufacturing techniques are defined as per the choice of material.

For distinct applications in a variety of fields, certain solitary materials might be replaced with composite materials, depending on the enhancement in its required property. Composite structures have shown improvement in strength and stiffness of material, while the reduction in weight is magnificent. Composites have also revealed some remarkable features such as resistance to impact, wear, corrosion, and chemicals, but these properties are dependent upon the composition of the material, type of fiber, and type of manufacturing technique employed to create it. In accordance with the properties required, composite materials find their applications in many desired fields.

More future research is intended to discover new composite structures with a combination of different variants and adopting new manufacturing techniques.

Author Contributions: Conceptualization, D.K.R.; methodology, D.D.P.; writing—original draft preparation, D.K.R., D.D.P., P.L.M. and E.L.; writing—review and editing, D.K.R., P.L.M. and E.L.

Funding: This research received no external funding.

Conflicts of Interest: The authors declare no conflict of interest.

References

1. Yashas Gowda, T.G.; Sanjay, M.R.; Subrahmanya Bhat, K.; Madhu, P.; Senthamaraikannan, P.; Yogesha, B. SSPolymer matrix-natural fiber composites: An overview. *Cogent. Eng.* **2018**, *5*, 1446667. [CrossRef]
2. Sherif, G.; Chukov, D.; Tcherdyntsev, V.; Torokhov, V. Effect of formation route on the mechanical properties of the polyethersulfone composites reinforced with glass fibers. *Polymers* **2019**, *11*, 1364. [CrossRef] [PubMed]
3. Chukov, D.; Nematulloev, S.; Zadorozhnyy, M.; Tcherdyntsev, V.; Stepashkin, A.; Zherebtsov, D. Structure, mechanical and thermal properties of polyphenylene sulfide and polysulfone impregnated carbon fiber composites. *Polymers* **2019**, *11*, 684. [CrossRef] [PubMed]
4. Linul, E.; Lell, D.; Movahedi, N.; Codrean, C.; Fiedler, T. Compressive properties of Zinc Syntactic Foams at elevated temperatures. *Compos. Part B Eng.* **2019**, *167*, 122–134. [CrossRef]
5. Clyne, T.W.; Hull, D. *An Introduction to Composite Materials*, 3rd ed.; Cambridge University Press: Cambridge, UK, 2019.
6. Zagho, M.M.; Hussein, E.A.; Elzatahry, A.A. Recent overviews in functional polymer composites for biomedical applications. *Polymers* **2018**, *10*, 739. [CrossRef] [PubMed]
7. Monteiro, S.N.; de Assis, F.S.; Ferreira, C.L.; Simonassi, N.T.; Weber, R.P.; Oliveira, M.S.; Colorado, H.A.; Pereira, A.C. Fique fabric: A promising reinforcement for polymer composites. *Polymers* **2018**, *10*, 246. [CrossRef] [PubMed]
8. Movahedi, N.; Linul, E. Quasi-static compressive behavior of the ex-situ aluminum-alloy foam-filled tubes under elevated temperature conditions. *Mater. Lett.* **2017**, *206*, 182–184. [CrossRef]

9. Chukov, D.; Nematulloev, S.; Torokhov, V.; Stepashkin, A.; Sherif, G.; Tcherdyntsev, V. Effect of carbon fiber surface modification on their interfacial interaction with polysulfone. *Results Phys.* **2019**, *15*, 102634. [CrossRef]
10. Linul, E.; Vălean, C.; Linul, P.A. Compressive behavior of aluminum microfibers reinforced semi-rigid polyurethane foams. *Polymers* **2018**, *10*, 1298. [CrossRef]
11. Yongxu, D.; Dong, L.; Libin, L.; Guangjie, G. Recent achievements of self-healing graphene/polymer composites. *Polymers* **2018**, *10*, 114.
12. Lebreton, L.C.M.; van der Zwet, J.; Damsteeg, J.W.; Slat, B.; Andrady, A.; Reisser, J. River plastic emissions to the world's oceans. *Nat. Commun.* **2017**, *8*, 15611. [CrossRef] [PubMed]
13. Scaffaro, R.; Maio, A.; Lopresti, F. Physical properties of green composites based on poly-lactic acid or Mater-Bi® filled with Posidonia Oceanica leaves. *Compos. Part. A Appl. S.* **2018**, *112*, 315–327. [CrossRef]
14. Scaffaro, R.; Maio, A. A green method to prepare nanosilica modified graphene oxide to inhibit nanoparticles re-aggregation during melt processing. *Chem. Eng. J.* **2017**, *308*, 1034–1047. [CrossRef]
15. Sun, M.; Sun, X.; Wang, Z.; Chang, M.; Li, H. The influence of shape memory alloy volume fraction on the impact behavior of polymer composites. *Polymers* **2018**, *10*, 1280. [CrossRef]
16. Ferreira, F.V.; Cividanes, L.S.; Gouveia, R.F.; Lona, L.M.F. An overview on properties and applications of poly (butylene adipate-co-terephthalate)-PBAT based composites. *Polym. Eng. Sci.* **2017**, *59*, E7–E15. [CrossRef]
17. Dufresne, A. Nanocellulose processing properties and potential applications. *Curr. For. Rep.* **2019**, *5*, 76–89. [CrossRef]
18. Habibi, Y.; Lucia, L.A.; Rojas, O.J. Cellulose nanocrystals: Chemistry, self-assembly, and applications. *Chem. Rev.* **2010**, *110*, 3479–3500. [CrossRef]
19. Ferreira, F.; Pinheiro, I.; de Souza, S.; Mei, L.; Lona, L. Polymer composites reinforced with natural fibers and nanocellulose in the automotive industry: A short review. *J. Compos. Sci.* **2019**, *3*, 51. [CrossRef]
20. Ardanuy, M.; Claramunt, J.; Toledo Filho, R.D. Cellulosic fiber reinforced cement-based composites: A review of recent research. *Constr. Build. Mater.* **2015**, *79*, 115–128. [CrossRef]
21. Ardanuy, M.; Claramunt, J.; García-Hortal, J.A.; Barra, M. Fiber-matrix interactions in cement mortar composites reinforced with cellulosic fibers. *Cellulose* **2011**, *18*, 281–289. [CrossRef]
22. Balea, A.; Fuente, E.; Blanco, A.; Negro, C. Nanocelluloses: Natural-based materials for fiber-reinforced cement composites. A critical review. *Polymers* **2019**, *11*, 518. [CrossRef] [PubMed]
23. Golewski, G.L. Determination of fracture toughness in concretes containing siliceous fly ash during mode III loading. *Struct. Eng. Mech.* **2017**, *62*, 1–9. [CrossRef]
24. Golewski, G.L. Effect of fly ash addition on the fracture toughness of plain concrete at third model of fracture. *J. Civ. Eng. Manag.* **2017**, *23*, 613–620. [CrossRef]
25. Pickering, K.L.; Efendy, M.G.A.; Le, T.M. A review of recent developments in natural fibre composites and their mechanical performance. *Compos. Part. A-Appl. S.* **2016**, *83*, 98–112. [CrossRef]
26. Alves Fidelis, M.E.; Pereira, T.V.C.; Gomes, O.F.M.; de Andrade Silva, F.; Toledo Filho, R.D. The effect of fiber morphology on the tensile strength of natural fibers. *J. Mater. Res. Technol.* **2013**, *2*, 149–157. [CrossRef]
27. Lotfi, A.; Li, H.; Dao, D.V.; Prusty, G. Natural fiber-reinforced composites: A review on material, manufacturing, and machinability. *J. Compos.* **2019**. [CrossRef]
28. Pegoretti, A.; Fabbri, E.; Migliaresi, C.; Pilati, F. Intraply and interply hybrid composites based on E-glass and poly (vinyl alcohol) woven fabrics: Tensile and impact properties. *Polym. Int.* **2004**, *53*, 1290–1297. [CrossRef]
29. Mehdikhani, M.; Gorbatikh, L.; Verpoest, I.; Lomov, S.V. Voids in fiber-reinforced polymer composites: A review on their formation, characteristics, and effects on mechanical performance. *J. Compos. Mater.* **2018**, *53*, 1579–1669. [CrossRef]
30. Dickson, A.N.; Ross, K.A.; Dowling, D.P. Additive manufacturing of woven carbon fibre polymer composites. *Compos. Struct.* **2018**, *206*, 637–643. [CrossRef]
31. Altenbach, H.; Altenbach, J.; Kissing, W. Classification of Composite Materials. In *Mechanics of composite Structural Elements*; Springer: Singapore, 2004; pp. 1–14. [CrossRef]
32. Panthapulakkal, S.; Raghunanan, L.; Sain, M.; Birat, K.C.; Tjong, J. Natural fiber and hybrid fiber thermoplastic composites. *Green Compos.* **2017**. [CrossRef]
33. Nair, A.B.; Joseph, R. Eco-friendly bio-composites using natural rubber (NR) matrices and natural fiber reinforcements. In *Chemistry, Manufacture and Applications of Natural Rubber*; Woodhead Publishing: Sawston, UK; Cambridge, UK, 2014. [CrossRef]

34. Agarwal, B.D.; Broutman, L.J.; Chandrashekhara, K. *Analysis and Performance of Fiber Composites*; John Wiley & Sons: Hoboken, NJ, USA, 2017.
35. Dixit, S.; Goel, R.; Dubey, A.; Ahivhare, P.R.; Bhalavi, T. Natural fibre reinforced polymer composite materials-A review. *Polym. Renew. Resour.* **2017**, *8*, 71–78. [CrossRef]
36. Arun Kumar, D.T.; Kaushik, V.P.; Raghavendra, R.P.S. Tensile and impact properties of jute/glass and jute/carbon fiber reinforced polypropylene. *J. Polym. Compos.* **2016**, *4*, 35–39.
37. Hempalaya. The Difference between Hemp and Linen Fibers. 2019. Available online: https://hempalaya.com/blogs/news/der-unterschied-zwischen-hanf-und-leinen-fasern (accessed on 8 May 2019).
38. Sunstrands. The Basics of Kenaf Fiber and Hurd. 2019. Available online: https://www.sunstrands.com/2019/the-uses-of-kenaf-fiber/ (accessed on 8 May 2019).
39. Handloom Policies & Research. Indira Gandhi Krishi Vishvavidyalaya (IGKV) Achieves a Breakthrough in Getting Linen Yarn Using the Flax Plant. 2017. Available online: https://www.unnatisilks.com/blog/indira-gandhi-krishi-vishvavidyalaya-igkv-achieves-a-breakthrough-in-getting-linen-yarn-using-the-flax-plant/ (accessed on 10 May 2019).
40. Textile School. Sampling from Cotton Bales. 2019. Available online: https://www.textileschool.com/469/sampling-from-cotton-bales/ (accessed on 12 May 2019).
41. Lidija Grozdanic. Students Use Rice Husks to Build Affordable Homes in the Philippines. 2017. Available online: https://inhabitat.com/students-use-rice-husks-to-built-affordable-homes-in-the-philippines/ (accessed on 13 May 2019).
42. Jute Geotextiles by Deyute, Geotextile Jute Fiber 732 gr/m^2 122 cm. 2019. Available online: https://www.deyute.com/product/geotextiles-natural-fibers/91 (accessed on 13 May 2019).
43. Backyard Poultry Contributor. Chicken Feather & Skin Development. 2019. Available online: https://backyardpoultry.iamcountryside.com/chickens-101/chicken-feather-skin-development/ (accessed on 13 May 2019).
44. Tanmay Halaye. Ramie Fiber Market Competitive Research and Precise Outlook 2019 to 2025. 2019. Available online: https://themarketresearchnews.com/2019/04/02/ramie-fiber-market-competitive-research-and-precise-outlook-2019-to-2025/ (accessed on 17 May 2019).
45. Textile Learner. Abaca Fiber (Manila Hemp) Uses/Application of Abaca Fiber. 2013. Available online: https://textilelearner.blogspot.com/2013/04/abaca-fiber-manila-hemp-usesapplication.html (accessed on 17 May 2019).
46. Fertilefibre Admin. How We Make Our Peat-Free Coir Composts. 2008. Available online: https://www.fertilefibre.com/blog/peat-free-environment/coir-composts/ (accessed on 21 May 2019).
47. Pond. Advantages of Using Natural Fibre Applications in Composites. 2017. Available online: https://pond.global/advantages-of-using-natural-fibre-applications-in-composites/ (accessed on 22 May 2019).
48. Black and Beautiful. The Luffa/Loofah Skincare Benefits. 2015. Available online: https://blackandbeautiful.fr/blog/en/2015/04/08/the-luffa-skincare-benefits/ (accessed on 24 May 2019).
49. BabaMu, Sisal Fibers. 2019. Available online: https://pixabay.com/photos/sisal-sisal-fibers-sisal-palm-4319997/ (accessed on 4 June 2019).
50. Textile Learner. Properties of Banana Fiber, Manufacturing Process of Banana Fiber, Application of Banana Fiber. 2014. Available online: https://textilelearner.blogspot.com/2014/01/properties-of-banana-fiber.html (accessed on 4 June 2019).
51. Beyond Materials, Basalt Fiber. Available online: https://beyondmaterials.com.au/2019/03/16/basalt-fiber/ (accessed on 11 June 2019).
52. Carbon Black, Carbon. Available online: http://carbon-website.000webhostapp.com (accessed on 15 June 2019).
53. Sanjay Impex. Fiberglass Scree. Available online: http://sanjayimpex.com/fiber-glass.html (accessed on 16 June 2019).
54. Sathishkumar, T.; Naveen, J.; Satheeshkumar, S. Hybrid fiber reinforced polymer composites—A review. *J. Reinf. Plast. Comp.* **2014**, *33*, 454–471. [CrossRef]
55. Rahman, R.; Zhafer Firdaus, S.P.S. Tensile properties of natural and synthetic fiber-reinforced polymer composites. In *Mechanical and Physical Testing of Biocomposites, Fibre-Reinforced Composites and Hybrid Composites*; Woodhead Publishing: Sawston, UK; Cambridge, UK, 2019; pp. 81–102. [CrossRef]

56. Jawaid, M.; Thariq, M.; Saba, N. *Mechanical and Physical Testing of Biocomposites, Fibre-Reinforced Composites and Hybrid Composites*; Woodhead Publishing: Sawston, UK; Cambridge, UK, 2019. [CrossRef]
57. Rajak, D.K.; Pagar, D.D.; Kumar, R.; Pruncu, C. Recent progress of reinforcement materials: A comprehensive overview of composite materials. *J. Mater. Res. Technol.* **2019**. [CrossRef]
58. Ghalia, M.A.; Abdelrasoul, A. Compressive and fracture toughness of natural and synthetic fiber-reinforced polymer. In *Mechanical and Physical Testing of Biocomposites, Fibre-Reinforced Composites and Hybrid Composites*; Woodhead Publishing: Sawston, UK; Cambridge, UK, 2019; pp. 123–140. [CrossRef]
59. Abdellaoui, H.; Raji, M.; Essabir, H.; Bouhfid, R.; Qaiss, A. Mechanical behavior of carbon/natural fiber-based hybrid composites. In *Mechanical and Physical Testing of Biocomposites, Fibre-Reinforced Composites and Hybrid Composites*; Woodhead Publishing: Sawston, UK; Cambridge, UK, 2019; pp. 103–122. [CrossRef]
60. Prakash, S. Experimental investigation of surface defects in low-power CO_2 laser engraving of glass fiber-reinforced polymer composite. *Polym. Compos.* **2019**. [CrossRef]
61. Chalmers, D.W. Experience in design and production of FRP marine structures. *Mar. Struct.* **1991**, *4*, 93–115. [CrossRef]
62. Unterweger, C.; Brüggemann, O.; Fürst, C. Synthetic fibers and thermoplastic short-fiber-reinforced polymers: Properties and characterization. *Polym. Compos.* **2013**, *35*, 227–236. [CrossRef]
63. Yi, X.S. Development of multifunctional composites for aerospace application. In *Multifunctionality of Polymer Composites*; William Andrew Publishing: Oxford, UK, 2015; pp. 367–418. [CrossRef]
64. Haim, A. Stability of composite stringer-stiffened panels. In *Stability and Vibrations of Thin Walled Composite Structures*; Woodhead Publishing: Sawston, UK; Cambridge, UK, 2017; pp. 461–507. [CrossRef]
65. Chung, D.D.L. Introduction to carbon composites. In *Carbon Composites: Composites with Carbon Fibers, Nanofibers, and Nanotubes*; Elsevier Science: Amsterdam, The Netherlands, 2017; pp. 88–160. [CrossRef]
66. Nobe, R.; Qiu, J.; Kudo, M.; Ito, K.; Kaneko, M. Effects of SCF content, injection speed, and CF content on the morphology and tensile properties of microcellular injection-molded CF/PP composites. *Polym. Eng. Sci.* **2019**, *59*, 1371–1380. [CrossRef]
67. Xu, Z.; Gao, C. Graphene fiber: A new trend in carbon fibers. *Mater. Today* **2015**, *18*, 480–492. [CrossRef]
68. Sreenivasulu, B.; Ramji, B.; Nagaral, M. A review on graphene reinforced polymer matrix composites. *Mater. Today: Proc.* **2018**, *5*, 2419–2428. [CrossRef]
69. Li, Y.; Wang, S.; Wang, Q. A molecular dynamics simulation study on enhancement of mechanical and tribological properties of polymer composites by introduction of graphene. *Carbon* **2017**, *111*, 538–545. [CrossRef]
70. Zhao, X.; Wang, X.; Wu, Z.; Keller, T.; Vassilopoulos, A.P. Temperature effect on fatigue behavior of basalt fiber-reinforced polymer composites. *Polym. Compos.* **2018**, *40*, 2273–2283. [CrossRef]
71. Singh, T.J.; Samanta, S. Characterization of Kevlar Fiber and its composites: A review. *Mater. Today: Proc.* **2015**, *2*, 1381–1387. [CrossRef]
72. Omrani, E.; Menezes, P.L.; Rohatgi, P.K. State of the art on tribological behavior of polymer matrix composites reinforced with natural fibers in the green materials world. *Eng. Sci. Technol. Int. J.* **2016**, *19*, 717–736. [CrossRef]
73. Ouarhim, W.; Zari, N.; Bouhfid, R.; Qaiss, A. Mechanical performance of natural fibers–based thermosetting composites. In *Mechanical and Physical Testing of Biocomposites, Fibre-Reinforced Composites and Hybrid Composites*; Woodhead Publishing: Sawston, UK; Cambridge, UK, 2019; pp. 43–60. [CrossRef]
74. Chand, N.; Fahim, M. Sisal reinforced polymer composites. *Tribol. Nat. Fiber Polym. Compos.* **2008**. [CrossRef]
75. Senthilkumar, K.; Saba, N.; Rajini, N.; Chandrasekar, M.; Jawaid, M.; Siengchin, S.; Alotman, O.Y. Mechanical properties evaluation of sisal fibre reinforced polymer composites: A review. *Constr. Build. Mater.* **2018**, *174*, 713–729. [CrossRef]
76. Saxena, M.; Pappu, A.; Haque, R.; Sharma, A. Sisal fiber based polymer composites and their applications. In *Cellulose Fibers: Bio-and Nano-Polymer Composites*; Springer: Berlin/Heidelberg, Germany, 2011; pp. 589–659. [CrossRef]
77. Shahzad, A. Hemp fiber and its composites—A review. *J. Compos. Mater.* **2011**, *46*, 973–986. [CrossRef]
78. Sullins, T.; Pillay, S.; Komus, A.; Ning, H. Hemp fiber reinforced polypropylene composites: The effects of material treatments. *Compos. Part. B-Eng.* **2017**, *114*, 15–22. [CrossRef]
79. Ochi, S. Mechanical properties of kenaf fibers and kenaf/PLA composites. *Mech. Mater.* **2008**, *40*, 446–452. [CrossRef]

80. Chin, C.W.; Yousif, B.F. Potential of kenaf fibres as reinforcement for tribological applications. *Wear* **2009**, *267*, 1550–1557. [CrossRef]
81. Abdi, B.; Azwan, S.; Abdullah, M.R.; Ayob, A. Flexural and tensile behaviour of kenaf fibre composite materials. *Mater. Res. Innov.* **2014**, *18*, S6–S184. [CrossRef]
82. Ben Mlik, Y.; Jaouadi, M.; Rezig, S.; Khoffi, F.; Slah, M.; Durand, B. Kenaf fibre-reinforced polyester composites: Flexural characterization and statistical analysis. *J. Text. Inst.* **2017**, *109*, 713–722. [CrossRef]
83. Huang, K.; Tran, L.Q.N.; Kureemun, U.; Teo, W.S.; Lee, H.P. Vibroacoustic behavior and noise control of flax fiber-reinforced polypropylene composites. *J. Nat. Fibers* **2019**, *16*, 729–743. [CrossRef]
84. Goutianos, S.; Peijs, T.; Nystrom, B.; Skrifvars, M. Development of flax fibre based textile reinforcements for composite applications. *Appl. Compos. Mater.* **2006**, *13*, 199–215. [CrossRef]
85. Habibi, M.; Laperrière, L.; Mahi Hassanabadi, H. Replacing stitching and weaving in natural fiber reinforcement manufacturing, part 2: Mechanical behavior of flax fiber composite laminates. *J. Nat. Fibers* **2018**. [CrossRef]
86. Chen, D.; Pi, C.; Chen, M.; He, L.; Xia, F.; Peng, S. Amplitude-dependent damping properties of ramie fiber-reinforced thermoplastic composites with varying fiber content. *Polym. Compos.* **2019**, *40*, 2681–2689. [CrossRef]
87. Du, Y.; Yan, N.; Kortschot, M.T. The use of ramie fibers as reinforcements in composites. *Biofiber Reinf. Compos. Mater.* **2015**. [CrossRef]
88. Majeed, K.; Arjmandi, R.; Al-Maadeed, M.A.; Hassan, A.; Ali, Z.; Khan, A.U.; Khurram, M.S.; Inuwa, I.M.; Khanam, P.N. Structural properties of rice husk and its polymer matrix composites. In *Lignocellulosic Fibre and Biomass-Based Composite Materials*; Woodhead Publishing: Sawston, UK; Cambridge, UK, 2017; pp. 473–490. [CrossRef]
89. Arjmandi, R.; Hassan, A.; Majeed, K.; Zakaria, Z. Rice husk filled polymer composites. *Int. J. Polym. Sci.* **2015**, *2015*, 1–32. [CrossRef]
90. Wang, Y.; Wu, H.; Zhang, C.; Ren, L.; Yu, H.; Galland, M.A.; Ichchou, M. Acoustic characteristics parameters of polyurethane/rice husk composites. *Polym. Compos.* **2018**, *40*, 2653–2661. [CrossRef]
91. Verma, A.; Negi, P.; Singh, V.K. Experimental analysis on carbon residuum transformed epoxy resin: Chicken feather fiber hybrid composite. *Polym. Compos.* **2018**, *40*, 2690–2699. [CrossRef]
92. Das, S.; Singha, A.K.; Chaudhuri, A.; Ganguly, P.K. Lengthwise jute fibre properties variation and its effect on jute–polyester composite. *J. Text. Inst.* **2019**, *110*, 1695–1702. [CrossRef]
93. Khan, J.A.; Khan, M.A. The use of jute fibers as reinforcements in composites. In *Biofiber Reinforcements in Composite Materials*; Woodhead Publishing: Sawston, UK; Cambridge, UK, 2015; pp. 3–34. [CrossRef]
94. Munde, Y.S.; Ingle, R.B.; Siva, I. Investigation to appraise the vibration and damping characteristics of coir fibre reinforced polypropylene composites. *Adv. Mater. Process. Technol.* **2018**, *4*, 639–650. [CrossRef]
95. Verma, D.; Shandilya, A.K.; Gupta, A. Coir fibre reinforcement and application in polymer composites: A Review. *J. Mater. Env. Sci.* **2013**, *4*, 263–276.
96. Chollakup, R.; Smitthipong, W.; Kongtud, W.; Tantatherdtam, R. Polyethylene green composites reinforced with cellulose fibers (coir and palm fibers): Effect of fiber surface treatment and fiber content. *J. Adhes. Sci. Technol.* **2013**, *27*, 1290–1300. [CrossRef]
97. Liu, Y.; Ma, Y.; Yu, J.; Zhuang, J.; Wu, S.; Tong, J. Development and characterization of alkali treated abaca fiber reinforced friction composites. *Compos. Interface* **2019**, *26*, 67–82. [CrossRef]
98. Panneerdhass, R.; Gnanavelbabu, A.; Rajkumar, K. Mechanical properties of luffa fiber and ground nut reinforced epoxy polymer hybrid composites. *Proced. Eng.* **2014**, *97*, 2042–2051. [CrossRef]
99. Bisen, H.B.; Hirwani, C.K.; Satankar, R.K.; Panda, S.K.; Mehar, K.; Patel, B. Numerical study of frequency and deflection responses of natural fiber (Luffa) reinforced polymer composite and experimental validation. *J. Nat. Fibers* **2018**. [CrossRef]
100. Laban, O.; Mahdi, E. Energy absorption capability of cotton fiber/epoxy composite square and rectangular tubes. *J. Nat. Fibers* **2016**, *13*, 726–736.
101. Panthapulakkal, S.; Sain, M. Injection-molded short hemp fiber/glass fiber-reinforced polypropylene hybrid composites-Mechanical, water absorption and thermal properties. *J. Appl. Polym. Sci.* **2006**, *103*, 2432–2441. [CrossRef]
102. Hanan, F.; Jawaid, M.; Md Tahir, P. Mechanical performance of oil palm/kenaf fiber-reinforced epoxy-based bilayer hybrid composites. *J. Nat. Fibers* **2018**. [CrossRef]

103. Ramesh, M.; Bhoopathi, R.; Deepa, C.; Sasikala, G. Experimental investigation on morphological, physical and shear properties of hybrid composite laminates reinforced with flax and carbon fibers. *J. Chin. Adv. Mater. Soc.* **2018**, *6*, 640–654. [CrossRef]
104. Swolfs, Y.; Gorbatikh, L.; Verpoest, I. Fibre hybridisation in polymer composites: A review. *Compos. Part. A-Appl. S.* **2014**, *67*, 181–200. [CrossRef]
105. Swolfs, Y.; Verpoest, I.; Gorbatikh, L. Recent advances in fibre-hybrid composites: Materials selection, opportunities and applications. *Int. Mater. Rev.* **2018**, *64*, 181–215. [CrossRef]
106. Abhemanyu, P.C.; Prassanth, E.; Kumar, T.N.; Vidhyasagar, R.; Marimuthu, K.P.; Pramod, R. Characterization of natural fiber reinforced polymer composites. In *AIP Conference Proceedings*; AIP Publishing: Melville, NY, USA, 2019; Volume 2080, p. 020005.
107. Chawla, N.; Shen, Y.L. Mechanical behavior of particle reinforced metal matrix composites. *Adv. Eng. Mater.* **2001**, *3*, 357–370. [CrossRef]
108. Mallick, P. *Fiber-Reinforced Composites: Materials, Manufacturing, and Design*, 3rd ed.; CRC Press: Boca Raton, FA, USA, 2007.
109. Tanzi, M.C.; Farè, S. *Foundations of Biomaterials Engineering*; Matthew, D., Ed.; Elsevier: Amsterdam, The Netherlands; Academic Press: London, UK, 2019.
110. Manickam, G.; Bharath, A.; Das, A.N.; Chandra, A.; Barua, P. Thermoelastic stability behavior of curvilinear fiber-reinforced composite laminates with different boundary conditions. *Polym. Compos.* **2018**, *40*, 2876–2890. [CrossRef]
111. Fang, K. *Encapsulation Technologies for Electronic Applications*, 2nd ed.; Elsevier: Amsterdam, The Netherlands, 2019; pp. 123–181.
112. Aditya Narayana, D.; Ganapathia, M.; Pradyumna, B. Investigation of thermo-elastic buckling of variable stiffness laminated composite shells using finite element approach based on higher-order theory. *Compos. Struct.* **2018**, *211*, 24–40. [CrossRef]
113. Kelly, J.; Mohammadi, M. Uniaxial tensile behavior of sheet molded composite car hoods with different fibre contents under quasi-static strain rates. *Mech. Res. Commun.* **2018**, *87*, 42–52. [CrossRef]
114. Balakrishnan, P.; John, M.J.; Pothen, L.; Sreekala, M.S.; Thomas, S. Natural fibre and polymer matrix composites and their applications in aerospace engineering. In *Advanced Composite Materials for Aerospace Engineering*; Woodhead Publishing: Sawston, UK; Cambridge, UK, 2016; pp. 365–383. [CrossRef]
115. Verma, D.; Senal, I. Natural fiber-reinforced polymer composites. *BiomassBiopolym.-Based Mater. Bioenergy* **2007**, *44*, 129. [CrossRef]
116. Kumar, R.; Ul Haq, M.I.; Raina, A.; Anand, A. Industrial applications of natural fibre-reinforced polymer composites–challenges and opportunities. *Int. J. Sustain. Eng.* **2019**, *12*, 212–220. [CrossRef]
117. Menezes, P.L.; Rohatgi, P.K.; Lovell, M.R. Studies on the tribological behavior of natural fiber reinforced polymer composite. In *Green Tribology*; Springer: Berlin/Heidelberg, Germany, 2012; pp. 329–345. [CrossRef]
118. Venkatachalam, N.; Navaneethakrishnan, P.; Rajsekar, R.; Shankar, S. Effect of pretreatment methods on properties of natural fiber composites: A review. *Polym. Polym. Compos.* **2016**, *24*, 555–566. [CrossRef]
119. Jamatia, R.; Deb, A. Size effect in FRP-confined concrete under axial compression. *J. Compos. Constr.* **2017**, *21*, 04017045. [CrossRef]
120. Vincent, T.; Ozbakkaloglu, T. Influence of fiber orientation and specimen end condition on axial compressive behavior of FRP-confined concrete. *Constr. Build. Mater.* **2013**, *47*, 814–826. [CrossRef]
121. Ozbakkaloglu, T. Compressive behavior of concrete-filled FRP tube columns: Assessment of critical column parameters. *Eng. Struct.* **2013**, *51*, 188–199. [CrossRef]
122. Ozbakkaloglu, T.; Vincent, T. Axial compressive behavior of circular high-strength concrete-filled frp tubes. *J. Compos. Constr.* **2014**, *18*, 04013037. [CrossRef]
123. Chattopadhyay, S.K.; Khandal, R.K.; Uppaluri, R.; Ghoshal, A.K. Bamboo fiber reinforced polypropylene composites and their mechanical, thermal, and morphological properties. *J. Appl. Polym. Sci.* **2010**, *119*, 1619–1626. [CrossRef]
124. Zhao, Y.Q.; Zhou, Y.; Huang, Z.M.; Batra, R.C. Experimental and micromechanical investigation of T300/7901 unidirectional composite strength. *Polym. Compos.* **2018**, *40*, 2639–2652. [CrossRef]
125. Chung, D.D.L. Polymer-matrix composites: Structure and processing. In *Carbon Composites: Composites with Carbon Fibers, Nanofibers, and Nanotubes*; Elsevier Science: Amsterdam, The Netherlands, 2017; pp. 161–217. [CrossRef]

126. Boisse, P. (Ed.) *Advances in Composites Manufacturing and Process Design*; Woodhead Publishing: Sawston, UK; Cambridge, UK, 2015.
127. Balasubramanian, K.; Sultan, M.T.H.; Rajeswari, N. Manufacturing techniques of composites for aerospace applications. In *Sustainable Composites for Aerospace Applications*; Woodhead Publishing: Sawston, UK; Cambridge, UK, 2018; pp. 55–67. [CrossRef]
128. Gascons, M.; Blanco, N.; Matthys, K. Evolution of manufacturing processes for fiber-reinforced thermoset tanks, vessels, and silos: A review. *Iie Trans.* **2012**, *44*, 476–489. [CrossRef]
129. Holmes, M. High volume composites for the automotive challenge. *Reinf. Plast.* **2017**, *61*, 294–298. [CrossRef]
130. Gunge, A.; Koppad, P.G.; Nagamadhu, M.; Kivade, S.; Murthy, K.V.S. Study on mechanical properties of alkali treated plain woven banana fabric reinforced biodegradable composites. *Compos. Commun.* **2019**, *13*, 47–51. [CrossRef]
131. Elkington, M.; Bloom, D.; Ward, C.; Chatzimichali, A.; Potter, K. Hand layup: Understanding the manual process. *Adv. Manuf. Polym. Compos. Sci.* **2015**, *1*, 138–151.
132. Jamir, M.R.M.; Majid, M.S.A.; Khasri, A. Natural lightweight hybrid composites for aircraft structural applications. In *Sustainable Composites for Aerospace Applications*; Woodhead Publishing: Sawston, UK; Cambridge, UK, 2018; pp. 155–170. [CrossRef]
133. Perna, A.S.; Viscusi, A.; Astarita, A.; Boccarusso, L.; Carrino, L.; Durante, M.; Sansone, R. Manufacturing of a metal matrix composite coating on a polymer matrix composite through cold gas dynamic spray technique. *J. Mater. Eng. Perform.* **2019**, *28*, 3211–3219. [CrossRef]
134. Marques, A.T. Fibrous materials reinforced composites production techniques. In *Fibrous and Composite Materials for Civil Engineering Applications*; Woodhead Publishing: Sawston, UK; Cambridge, UK, 2011; pp. 191–215. [CrossRef]
135. Ervina, J.; Ghaleb, Z.A.; Hamdan, S.; Mariatti, M. Colloidal Stability of Water-based Carbon Nanotube Suspensions in Electrophoretic Deposition Process: Effect of Applied Voltage and Deposition Time. *Compos. Part. A Appl. Sci.* **2018**, *117*, 1–10. [CrossRef]
136. Carruthers, J. Vacuum Bagging Process Overview, Coventive Composites. 2018. Available online: https://coventivecomposites.com/explainers/what-is-vacuum-bagging/ (accessed on 13 July 2019).
137. Awan, F.S.; Fakhar, M.A.; Khan, L.A.; Zaheer, U.; Khan, A.F.; Subhani, T. Interfacial mechanical properties of carbon nanotube-deposited carbon fiber epoxy matrix hierarchical composites. *Compos. Interface* **2018**, *25*, 681–699. [CrossRef]
138. Meola, C.; Boccardi, S.; Carlomagno, G. Composite Materials in the aeronautical industry. In *Infrared Thermography in the Evaluation of Aerospace Composite Materials: Infrared Thermography to Composites*; Elsevier Science: Amsterdam, The Netherlands, 2017; pp. 1–24. [CrossRef]
139. Ahmad, N.; Bilal, I.; Khattak, S. Polyester usage in manufacturing of electrical and mechanical products and assemblies. *Polyest.Prod. Charact. Innov. Appl.* **2018**. [CrossRef]
140. Davis, D.C.; Mensah, T.O. Fabrication and fatigue of fiber-reinforced polymer nanocomposites—a tool for quality control. In *Nanotechnology Commercialization: Manufacturing Processes and Products*; Mensah, T.O., Wang, B., Bothun, G., Winter, J., Davis, V., Eds.; John Wiley & Sons: Hoboken, NJ, USA, 2017. [CrossRef]
141. Yalcinkaya, M.A.; Guloglu, G.E.; Pishvar, M.; Amirkhosravi, M.; Sozer, M.; Altan, M.C. Pressurized Infusion (PI): A new and improved liquid composite molding process. *J. Manuf. Sci. Eng.* **2018**, *141*, 011007. [CrossRef]
142. Plummer, C.J.G.; Bourban, P.E.; Månson, J.A. Polymer matrix composites: Matrices and processing. *Ref. Modul. Mater. Sci. Mater. Eng.* **2016**. [CrossRef]
143. Ishikawa, H.; Takagi, H.; Nakagaito, A.N.; Yasuzawa, M.; Genta, H.; Saito, H. Effect of surface treatments on the mechanical properties of natural fiber textile composites made by VaRTM method. *Compos. Interface* **2014**, *21*, 329–336. [CrossRef]
144. Mitschang, P.; Hildebrandt, K. Polymer and composite moulding technologies for automotive applications. In *Advanced Materials in Automotive Engineering*; Woodhead Publishing: Sawston, UK; Cambridge, UK, 2012; pp. 210–229. [CrossRef]
145. Park, C.H.; Lee, W.I. Compression molding in polymer matrix composites. In *Manufacturing Techniques for Polymer Matrix Composites (PMCs)*; Woodhead Publishing: Sawston, UK; Cambridge, UK, 2012; pp. 47–94. [CrossRef]

146. Matveenko, V.P.; Kosheleva, N.A.; Shardakov, I.N.; Voronkov, A.A. Temperature and strain registration by fibre-optic strain sensor in the polymer composite materials manufacturing. *Int. J. Smart Nano Mater.* **2018**, *9*, 99–110. [CrossRef]
147. Biswas, B.; Hazra, B.; Sarkar, A.; Bandyopadhyay, N.R.; Mitra, B.C.; Sinha, A. Influence of ZrO^2 incorporation on sisal fiber reinforced unsaturated polyester composites. *Polym. Compos.* **2018**, *40*, 2790–2801. [CrossRef]
148. Singh, J.I.P.; Singh, S.; Dhawan, V. Effect of curing temperature on mechanical properties of natural fiber reinforced polymer composites. *J. Nat. Fibers* **2017**, *15*, 687–696. [CrossRef]
149. Ramôa Correia, J. Pultrusion of advanced fibre-reinforced polymer (FRP) composites. In *Advanced Fibre-Reinforced Polymer (FRP) Composites for Structural Applications*; Woodhead Publishing: Sawston, UK; Cambridge, UK, 2013; pp. 207–251. [CrossRef]
150. Verma, D.; Joshi, G.; Dabral, R.; Lakhera, A. Processing and evaluation of mechanical properties of epoxy-filled E-glass fiber–fly ash hybrid composites. In *Mechanical and Physical Testing of Biocomposites, Fibre-Reinforced Composites and Hybrid Composites*; Woodhead Publishing: Sawston, UK; Cambridge, UK, 2019; pp. 293–306. [CrossRef]
151. Joshi, S.C. The pultrusion process for polymer matrix composites. In *Manufacturing techniques for polymer matrix composites (PMCs)*; Woodhead Publishing: Sawston, UK; Cambridge, UK, 2012; pp. 381–413. [CrossRef]
152. Leong, Y.W.; Thitithanasarn, S.; Yamada, K.; Hamada, H. Compression and injection molding techniques for natural fiber composites. In *Natural Fibre Composites*; Woodhead Publishing: Sawston, UK; Cambridge, UK, 2014; pp. 216–232. [CrossRef]
153. Werner, V.M.K.; Krumpholz, R.; Rehekampff, C.; Scherzer, T.; Eblenkamp, M. Thermoplastic encapsulations of a sensor platform by high-temperature injection molding up to 360 °C. *Polym. Eng. Sci.* **2019**. [CrossRef]
154. González-López, M.E.; Pérez-Fonseca, A.A.; Manríquez-González, R.; Arellano, M.; Rodrigue, D.; Robledo-Ortíz, J.R. Effect of surface treatment on the physical and mechanical properties of injection molded poly(lactic acid)-coir fiber biocomposites. *Polym. Compos.* **2018**, *40*, 2132–2141. [CrossRef]
155. Bhardwaj, N.; Kundu, S.C. Electrospinning: A fascinating fiber fabrication technique. *Biotechnol. Adv.* **2010**, *28*, 325–347. [CrossRef]
156. Wang, G.; Yu, D.; Kelkar, A.D.; Zhang, L. Electrospun nanofiber: Emerging reinforcing filler in polymer matrix composite materials. *Prog. Polym. Sci.* **2017**, *75*, 73–107. [CrossRef]
157. Gonzalez-Henriquez, C.M.; Sarabia-Vallejos, M.A.; Rodriguez Hernandez, J. Polymers for additive manufacturing and 4D-printing: Materials, methodologies, and biomedical applications. *Prog. Polym. Sci.* **2019**, *94*, 57–116. [CrossRef]
158. Chua, C.K.; Leong, K.F. 3D Printing and Additive Manufacturing: Principles and Applications. In *Fifth Edition of Rapid Prototyping*, 5th ed.; World Scientific Publishing Co., Inc.: Singapore, 2017.
159. Goh, G.D.; Yap, Y.L.; Agarwala, S.; Yeong, W.Y. Recent progress in additive manufacturing of fiber reinforced polymer composite. *Adv. Mater. Technol.* **2018**, *4*, 1800271. [CrossRef]
160. Hu, C.; Sun, Z.; Xiao, Y.; Qin, Q. Recent patents in additive manufacturing of continuous fiber reinforced composites. *Recent Pat. Mech. Eng.* **2019**, *12*, 25–36. [CrossRef]
161. Parandoush, P.; Tucker, L.; Zhou, C.; Lin, D. Laser assisted additive manufacturing of continuous fiber reinforced thermoplastic composites. *Mater. Des.* **2017**, *131*, 186–195. [CrossRef]
162. Shirvanimoghaddam, K.; Hamim, S.U.; Akbari, M.K.; Fakhrhoseini, S.M.; Khayyam, H.; Pakseresht, A.H.; Ghasali, E.; Zabet, M.; Munir, K.S.; Jia, S.; et al. Carbon fiber reinforced metal matrix composites: Fabrication processes and properties. *Compos. Part. A-Appl. Sci.* **2017**, *92*, 70–96. [CrossRef]
163. Mantell, S.C.; Springer, G.S. Filament winding process models. *Compos. Struct.* **1994**, *27*, 141–147. [CrossRef]
164. Minsch, N.; Herrmann, F.H.; Gereke, T.; Nocke, A.; Cherif, C. Analysis of filament winding processes and potential equipment technologies. *Procedure Cirp* **2017**, *66*, 125–130. [CrossRef]
165. Hopmann, C.; Wruck, L.; Schneider, D.; Fischer, K. Automated winding of preforms directly from roving. *Lightweight Des. Worldw.* **2019**, *12*, 58–63. [CrossRef]
166. Sorrentino, L.; Anamateros, E.; Bellini, C.; Carrino, L.; Corcione, G.; Leone, A.; Paris, G. Robotic filament winding: An innovative technology to manufacture complex shape structural parts. *Compos. Struct.* **2019**, *220*, 699–707. [CrossRef]
167. Mouritz, A.P. Manufacturing of fibre–polymer composite materials. In *Introduction to Aerospace Materials*; Woodhead Publishing: Sawston, UK; Cambridge, UK, 2012; Volume 10, pp. 303–337. [CrossRef]

168. Frketic, J.; Dickens, T.; Ramakrishnan, S. Automated manufacturing and processing of fiber-reinforced polymer (FRP) composites: An additive review of contemporary and modern techniques for advanced materials manufacturing. *Addit. Manuf.* **2017**, *14*, 69–86. [CrossRef]
169. Toutanji, H.; Deng, Y. Comparison between Organic and Inorganic Matrices for RC Beams Strengthened with Carbon Fiber Sheets. *J. Compos. Constr.* **2015**, *11*, 507–513. [CrossRef]
170. Menna, C.; Asprone, D.; Ferone, C.; Colangelo, F.; Balsamo, A.; Prota, A.; Cioffi, R.; Manfredi, G. Use of geopolymers for composite external reinforcement of RC members. *Compos. Part. B-Eng.* **2013**, *45*, 1667–1676. [CrossRef]
171. Trapko, T. The effect of high temperature on the performance of CFRP and FRCM confined concrete elements. *Compos. Part. B-Eng.* **2013**, *54*, 138–145. [CrossRef]
172. Wang, K.; Young, B.; Smith, S.T. Mechanical properties of pultruded carbon fibre-reinforced polymer (CFRP) plates at elevated temperatures. *Eng. Struct.* **2011**, *33*, 2154–2161. [CrossRef]
173. Ding, Z.; Dai, J.G.; Munir, S. Study on an improved phosphate cement binder for the development of fiber-reinforced inorganic polymer composites. *Polymers* **2014**, *6*, 2819–2831. [CrossRef]
174. Fang, Y.; Cui, P.; Ding, Z.; Zhu, J.X. Properties of a magnesium phosphate cement based fire-retardant coating containing glass fiber or glass fiber powder. *Constr. Build. Mater.* **2018**, *162*, 553–560. [CrossRef]
175. Dai, J.G.; Munir, S.; Ding, Z. Comparative study of different cement-based inorganic pastes towards the development of FRIP strengthening technology. *J. Compos. Constr.* **2014**, *18*, A4013011. [CrossRef]
176. Ding, Z.; Xu, M.R.; Dai, J.G.; Dong, B.Q.; Zhang, M.J.; Hong, S.X.; Xing, F. Strengthening concrete using phosphate cement-based fiber-reinforced inorganic composites for improved fire resistance. *Constr. Build. Mater.* **2019**, *212*, 755–764. [CrossRef]
177. Manalo, A.; Aravinthan, T.; Fam, A.; Benmokrane, B. State-of-the-Art Review on FRP Sandwich Systems for Lightweight Civil Infrastructure. *J. Compos. Constr.* **2017**, *21*, 04016068. [CrossRef]
178. Toutanji, H.A.; Gómez, W. Durability characteristics of concrete beams externally bonded with FRP composite sheets. *Cem. Concr. Comp.* **1997**, *19*, 351–358. [CrossRef]
179. Kalfat, R.; Al-Mahaidi, R.; Smith, S.T. Anchorage devices used to improve the performance of reinforced concrete beams retrofitted with frp composites: State-of-the-Art Review. *J. Compos. Constr.* **2013**, *17*, 14–33. [CrossRef]
180. Elgabbas, F.; Ahmed, E.A.; Benmokrane, B. Flexural behavior of concrete beams reinforced with ribbed basalt-FRP bars under static loads. *J. Compos. Constr.* **2016**, *21*, 195–230. [CrossRef]
181. El Refai, A.; Abed, F. Concrete contribution to shear strength of beams reinforced with basalt fiber-reinforced bars. *J. Compos. Constr.* **2015**, *20*, 150–179. [CrossRef]
182. Abed, F.; Alhafiz, A.R. Effect of basalt fibers on the flexural behavior of concrete beams reinforced with BFRP bars. *Compos. Struct.* **2019**, *215*, 23–34. [CrossRef]
183. Zhang, H.W.; Smith, S.T. Influence of FRP anchor fan configuration and dowel angle on anchoring FRP plates. *Compos. Part. B-Eng.* **2012**, *43*, 3516–3527. [CrossRef]
184. Available online: https://theconstructor.org/concrete/prestressed-concrete-principles-advantages/28/ (accessed on 12 October 2019).
185. Scullybob. 2018. Available online: https://en.wikipedia.org/wiki/17th_Street_Bridge_(Vero_Beach,_Florida) (accessed on 20 June 2019).
186. Huang, B.T.; Li, Q.H.; Xu, S.L.; Zhou, B. Strengthening of reinforced concrete structure using sprayable fiber-reinforced cementitious composites with high ductility. *Compos. Struct.* **2019**, *220*, 940–952. [CrossRef]
187. Cheng, L.; Karbhari, V.M. New bridge systems using FRP composites and concrete: A state-of-the-art review. *Progr. Struct. Eng. Mater.* **2006**, *8*, 143–154. [CrossRef]
188. Pham, T.M.; Hao, H. Review of concrete structures strengthened with FRP against impact loading. *Structures* **2016**, *7*, 59–70. [CrossRef]
189. Alagusundaramoorthy, P.; Harik, I.E.; Choo, C.C. Structural behavior of FRP composite bridge deck panels. *J. Bridge. Eng.* **2006**, *11*, 384–393. [CrossRef]
190. Gopinath, R.; Poopathi, R.; Saravanakumar, S.S. Characterization and structural performance of hybrid fiber-reinforced composite deck panels. *Adv. Compos. Hybrid. Mater.* **2019**, *2*, 115–124. [CrossRef]
191. Ozbakkaloglu, T.; Lim, J.C.; Vincent, T. FRP-confined concrete in circular sections: Review and assessment of stress-strain models. *Eng. Struct.* **2013**, *49*, 1068–1088. [CrossRef]

192. Guades, E.; Aravinthan, T.; Islam, M.; Manalo, A. A review on the driving performance of FRP composite piles. *Compos. Struct.* **2012**, *94*, 1932–1942. [CrossRef]
193. Sen, R.; Mullins, G. Application of FRP composites for underwater piles repair. *Compos. Part. B-Eng.* **2007**, *38*, 751–758. [CrossRef]
194. Mosallam, A.S.; Mosalam, K.M. Strengthening of two-way concrete slabs with FRP composite laminates. *Constr Build. Mater.* **2003**, *17*, 43–54. [CrossRef]
195. Ou, J.; Li, H. Structural health monitoring in mainland China: Review and future trends. *Struct. Health Monit. Int. J.* **2010**, *9*, 219–231.
196. Li, H.N.; Li, D.S.; Song, G.B. Recent applications of fiber optic sensors to health monitoring in civil engineering. *Eng. Struct.* **2004**, *26*, 1647–1657. [CrossRef]
197. Mao, K.; Greenwood, D.; Ramakrishnan, R.; Goodship, V.; Shrouti, C.; Chetwynd, D.; Langlois, P. The wear resistance improvement of fibre reinforced polymer composite gears. *Wear* **2019**, *426*, 1033–1039. [CrossRef]
198. Catera, P.G.; Mundo, D.; Treviso, A.; Gagliardi, F.; Visrolia, A. On the design and simulation of hybrid metal-composite gears. *Springer Appl. Compos. Mater.* **2019**, *26*, 817–833. [CrossRef]
199. Bae, J.H.; Jung, K.C.; Yoo, S.H.; Chang, S.H.; Kim, M.; Lim, T. Design and fabrication of a metal composite hybrid wheel with a friction damping layer for enhancement of ride comfort. *Compos. Struct.* **2015**, *133*, 576–584. [CrossRef]
200. Shweiki, S.; Palermo, A.; Mundo, D. A study on the dynamic behaviour of lightweight gears. *Shock Vib.* **2017**, *2017*, 7982170. [CrossRef]
201. Rigaud, E.; Cornuault, P.H.; Bazin, B.; Grandais-Menant, E. Numerical and experimental analysis of the vibroacoustic behavior of an electric window-lift gear motor. *Arch. Appl. Mech.* **2018**, *88*, 1395–1410. [CrossRef]
202. Schäkel, M.; Janssen, H.; Brecher, C. Increased reliability for the manufacturing of composite pressure vessels. *Lightweight Des. Worldw.* **2019**, *12*, 10–17. [CrossRef]
203. Wilson, A. Vehicle weight is the key driver for automotive composites. *Reinf. Plast.* **2017**, *61*, 100–102. [CrossRef]
204. Future Pipe Industries. Tanks. 2019. Available online: https://www.futurepipe.com/products/tanks (accessed on 11 July 2019).
205. Solazzi, L.; Buffoli, A. Telescopic hydraulic cylinder made of composite material. *Appl. Compos. Mater.* **2019**, *26*, 1189–1206. [CrossRef]
206. Chang, S.H.; Kim, P.J.; Lee, D.G.; Choi, J.K. Steel-composite hybrid headstock for high-precision grinding machines. *Compos. Struct.* **2001**, *53*, 1–8. [CrossRef]
207. Jain, R.K.; Khan, A.; Inamuddin, I.; Asiri, A.M. Design and development of non-perfluorinated ionic polymer metal composite-based flexible link manipulator for robotics assembly. *Polym. Compos.* **2018**, *40*, 2582–2593. [CrossRef]
208. Lu, Z.L.; Lu, F.; Cao, J.W.; Li, D.C. Manufacturing properties of turbine blades of carbon fiber-reinforced SiC Composite Based on Stereolithography. *Mater. Manuf. Process.* **2014**, *29*, 201–209. [CrossRef]
209. Patel, M.; Saurabh, K.; Prasad, V.B.; Subrahmanyam, J. High temperature C/C–SiC composite by liquid silicon infiltration: A literature review. *B. Mater. Sci.* **2012**, *35*, 63–73. [CrossRef]
210. Stenquist, P. Superbrakes for Civilians? The Cost Is the Obstacle 2010. Available online: https://www.nytimes.com/2010/08/01/automobiles/01BRAKES.html (accessed on 2 July 2019).
211. Forintos, N.; Czigány, T. Multifunctional application of carbon fiber reinforced polymer composites: Electrical properties of the reinforcing carbon fibers—A short review. *Compos. Part. B Eng.* **2019**, *162*, 331–343. [CrossRef]
212. Amiri, A.; Krosbakken, T.; Schoen, W.; Theisen, D.; Ulven, C.A. Design and manufacturing of a hybrid flax/carbon fiber composite bicycle frame. *Proc. Inst. Mech. Eng. Part. P J. Sports Eng. Technol.* **2017**, *232*, 28–38. [CrossRef]
213. Kong, C.; Lee, H.; Park, H. Design and manufacturing of automobile hood using natural composite structure. *Compos. Part. B-Eng.* **2016**, *91*, 18–26. [CrossRef]
214. Hassan, S.M.; Amir, N.; Rahmati, A.M. Pedestrian safety investigation of the new inner structure of the hood to mitigate the impact injury of the head. *Thin Wall. Struct.* **2014**, *77*, 77–85.
215. Koronis, G.; Silva, A.; Fontul, M. Green composites: A review of adequate materials for automotive applications. *Compos. Part. B-Eng.* **2013**, *44*, 120–127. [CrossRef]

216. Kong, C.; Park, H.; Lee, J. Study on structural design and analysis of flax natural fiber composite tank manufactured by vacuum assisted resin transfer molding. *Mater. Lett.* **2014**, *130*, 21–25. [CrossRef]
217. Alves, C.; Ferrão, P.M.C.; Silva, A.J.; Reis, L.G.; Freitas, M.; Rodrigues, L.B. Ecodesign of automotive components making use of natural jute fiber composites. *J. Clean. Prod.* **2011**, *18*, 313–327. [CrossRef]
218. Belauto. Available online: https://belauto.com.my/2014-volkswagen-xl1-carbon-fiber-body-parts/ (accessed on 3 July 2019).
219. Ashworth, S.; Rongong, J.; Wilson, P.; Meredith, J. Mechanical and damping properties of resin transfer moulded jute-carbon hybrid composites. *Compos. Part. B-Eng.* **2016**, *105*, 60–66. [CrossRef]
220. Flynn, J.; Amiri, A.; Ulven, C. Hybridized carbon and flax fiber composites for tailored performance. *Mater. Des.* **2016**, *102*, 21–29. [CrossRef]
221. Wagh, P.H.; Pagar, D.D. Investigation of mechanical and tribological behavior of composite material filled with black epoxy resin and aluminium tri-hydroxide using reinforcement of glass fiber. *AIP Conf. Proc.* **2018**, *2018*, 020025. [CrossRef]
222. Zhang, J.; Khatibi, A.A.; Castanet, E.; Baum, T.; Komeily-Nia, Z.; Vroman, P.; Wang, X. Effect of natural fibre reinforcement on the sound and vibration damping properties of bio-composites compression moulded by nonwoven mats. *Compos. Commun.* **2019**, *13*, 12–17. [CrossRef]
223. Farid, M.; Purniawan, A.; Rasyida, A.; Ramadhani, M.; Komariyah, S. Improvement of acoustical characteristics: Wideband bamboo based polymer composite. *Iop Conf. Ser. Mater. Sci. Eng.* **2017**, *223*, 012021. [CrossRef]
224. Kelly, J. What Is the Ideal Interior Material for Cars? 2017. Available online: https://www.carconversation.com.au/opinions/what-is-the-ideal-interior-material-for-cars (accessed on 10 July 2019).
225. Belingardi, G.; Koricho, E.G. Design of a composite engine support sub-frame to achieve lightweight vehicles. *Int. J. Automot. Compos.* **2014**, *1*, 90. [CrossRef]
226. Hou, W.; Xu, X.; Han, X.; Wang, H.; Tong, L. Multi-objective and multi-constraint design optimization for hat-shaped composite T-joints in automobiles. *Thin Wall Struct.* **2019**, *143*, 106232. [CrossRef]
227. Kim, D.-H.; Kim, H.-G.; Kim, H.-S. Design optimization and manufacture of hybrid glass/carbon fiber reinforced composite bumper beam for automobile vehicle. *Compos. Struct.* **2015**, *131*, 742–752. [CrossRef]
228. Barile, C.; Casavola, C. Mechanical characterization of carbon fiber-reinforced plastic specimens for aerospace applications. In *Mechanical and Physical Testing of Biocomposites, Fibre-Reinforced Composites and Hybrid Composites*; Woodhead Publishing: Sawston, UK; Cambridge, UK, 2019; pp. 387–407. [CrossRef]
229. Alonso-Martin, P.P.; Gonzalez-Garcia, A.; Lapena-Rey, N.; Fita-Bravo, S.; Martinez-Sanz, V.; Marti-Ferrer, F. Green Aircraft Interior Panels and Method of Fabrication. European Patent EP2463083A2, 13 June 2012.
230. Maryanka, Y.; Meidar, M.I.; Curless, R.A. Method of Signal Transmission Using Fiber Composite Sandwich Panel. US Patent 8903311, 2 December 2014.
231. Rawal, S.P. Metal-matrix composites for space applications. *JOM* **2001**, *53*, 14–17. [CrossRef]
232. Boegler, O.; Kling, U.; Empl, D.; Isikveren, A. Potential of sustainable materials in wing structural design. In Proceedings of the Deutscher Luft- und Raumfahrtkongress, München, Germany, 16–18 August 2014.
233. Arockiam, N.J.; Jawaid, M.; Saba, N. Sustainable bio composites for aircraft components. In *Sustainable Composites for Aerospace Applications*; Woodhead Publishing: Sawston, UK; Cambridge, UK, 2018; pp. 109–123. [CrossRef]
234. Fan, S.; Yang, C.; He, L.; Du, Y.; Krenkel, W.; Greil, P.; Travitzky, N. Progress of ceramic matrix composites brake materials for aircraft application. *Rev. Adv. Mater. Sci.* **2016**, *44*, 313–325.
235. Zou, Z.; Qin, Y.; Tian, Q.; Huang, Z.; Zhao, Z. The influence of zirconia fibre on ablative composite materials. *Plast. Rubber Compos.* **2019**, *48*, 185–190. [CrossRef]
236. Scholz, M.-S.; Blanchfield, J.P.; Bloom, L.D.; Coburn, B.H.; Elkington, M.; Fuller, J.D.; Bond, I.P. The use of composite materials in modern orthopaedic medicine and prosthetic devices: A review. *Compos. Sci. Technol.* **2011**, *71*, 1791–1803. [CrossRef]
237. Lazar, M.A.; Rotaru, H.; Bâldea, I.; Boşca, A.B.; Berce, C.P.; Prejmerean, C.; Câmpian, R.S. Evaluation of the biocompatibility of new fiber-reinforced composite materials for craniofacial bone reconstruction. *J. Craniofac Surg.* **2016**, *27*, 1694–1699. [CrossRef] [PubMed]
238. Kowsari, E.; Haddadi-Asl, V.; Ajdari, F.B.; Hemmat, J. Aramid fibers composites to innovative sustainable materials for biomedical applications. In *Materials for Biomedical Engineering*; Elsevier: Amsterdam, The Netherlands, 2019; pp. 173–204. [CrossRef]

239. Teo, A.J.T.; Mishra, A.; Park, I.; Kim, Y.-J.; Park, W.-T.; Yoon, Y.-J. Polymeric biomaterials for medical implants and devices. *Acs Biomater. Sci. Eng.* **2016**, *2*, 454–472. [CrossRef]
240. Kim, S.S.; Lee, J. Antimicrobial polyacrylonitrile/m-aramid hybrid composite. *Ind. Eng. Chem. Res.* **2013**, *52*, 10297–10304. [CrossRef]
241. Kim, S.S.; Lee, J. Miscibility and antimicrobial properties of m-aramid/chitosan hybrid composite. *Ind. Eng. Chem. Res.* **2013**, *52*, 12703–12709. [CrossRef]
242. Vallittu, P.K.; Närhi, T.O.; Hupa, L. Fiber glass-bioactive glass composite for bone replacing and bone anchoring implants. *Dent. Mater.* **2015**, *31*, 371–381. [CrossRef]
243. Zhu, B.; Li, W.; Lewis, R.V.; Segre, C.U.; Wang, R. E-Spun composite fibers of collagen and dragline silk protein: Fiber mechanics, biocompatibility, and application in stem cell differentiation. *Biomacromolecules* **2014**, *16*, 202–213. [CrossRef]
244. Mengyan, L.; Mondrinos, M.J.; Xuesi, C.; Lelkes, P.I. Electrospun blends of natural and synthetic polymers as scaffolds for tissue engineering. In Proceedings of the 2005 IEEE Engineering in Medicine and Biology 27th Annual Conference, Shanghai, China, 17–18 January 2006. [CrossRef]
245. Lelkes, P.I.; Mengyan, L.; Mondrinos, M.; Ko, F. U.S. Patent No. US8048446B2. Available online: https://patents.google.com/patent/US8048446B2/en (accessed on 12 October 2019).
246. Jaganathan, S.K.; Mani, M.P. Enriched mechanical, thermal, and blood compatibility of single stage electrospun polyurethane nickel oxide nanocomposite for cardiac tissue engineering. *Polym. Compos.* **2018**, *40*, 2381–2390. [CrossRef]
247. Kim, O.V.; Litvinov, R.I.; Chen, J.; Chen, D.Z.; Weisel, J.W.; Alber, M.S. Compression-induced structural and mechanical changes of fibrin-collagen composites. *Matrix Biol.* **2017**, *60*, 141–156. [CrossRef] [PubMed]
248. Bensaıd, W.; Triffitt, J.; Blanchat, C.; Oudina, K.; Sedel, L.; Petite, H. A biodegradable fibrin scaffold for mesenchymal stem cell transplantation. *Biomaterials* **2003**, *24*, 2497–2502. [CrossRef]
249. Shevchenko, R.V.; James, S.L.; James, S.E. A review of tissue-engineered skin bioconstructs available for skin reconstruction. *J. R. Soc. Interface* **2010**, *7*, 229–258. [CrossRef] [PubMed]
250. Manvi, P.K.; Beckers, M.; Mohr, B.; Seide, G.; Gries, T.; Bunge, C.-A. Polymer fiber-based biocomposites for medical sensing applications. In *Materials for Biomedical Engineering*; Elsevier: Amsterdam, The Netherlands, 2019; pp. 57–88. [CrossRef]
251. Rebelo, R.; Fernandes, M.; Fangueiro, R. Biopolymers in medical implants: A brief review. *Procedure Eng.* **2017**, *200*, 236–243. [CrossRef]
252. Azimi, B.; Nourpanah, P.; Rabiee, M.; Arbab, S. Poly (lactide-*co*-glycolide) Fiber: An Overview. *J. Eng. Fiber Fabr.* **2014**, *9*, 155892501400900. [CrossRef]
253. Pivsa-Art, W.; Chaiyasat, A.; Pivsa-Art, S.; Yamane, H.; Ohara, H. Preparation of polymer blends between Poly(Lactic Acid) and Poly(Butylene adipate-*co*-terephthalate) and biodegradable polymers as compatibilizers. *Energy Proced.* **2013**, *34*, 549–554. [CrossRef]
254. Shanks, R.; Kong, I. Thermoplastic Elastomers. In *Applied Sciences*, 1st ed.; RMIT University: Melbourne, Australia, 2012; pp. 95–116.
255. Panwiriyarat, W.; Tanrattanakul, V.; Pilard, J.F.; Pasetto, P.; Khaokong, C. Preparation and Properties of Bio-based Polyurethane Containing Polycaprolactone and Natural Rubber. *J. Polym. Env.* **2013**, *21*, 807–815. [CrossRef]
256. Nicolae, A.; Grumezescu, A.M. Polymer fibers in biomedical engineering. *Mater. Biomed. Eng.* **2019**. [CrossRef]
257. Nandi, S.K.; Mahato, A.; Kundu, B.; Mukherjee, P. Organic-inorganic micro/nanofiber composites for biomedical applications. *Mater. Biomed. Eng.* **2019**. [CrossRef]
258. Jesthi, D.K.; Nayak, R.K. Improvement of mechanical properties of hybrid composite through interply rearrangement of glass and carbon woven fabrics for marine applications. *Compos. Part. B-Eng.* **2019**, *168*, 467–475. [CrossRef]
259. Dhakal, H.N.; MacMullen, J.; Zhang, Z.Y. Moisture measurement and effects on properties of marine composites. In *Marine Applications of Advanced Fibre-Reinforced Composites*; Woodhead Publishing: Sawston, UK; Cambridge, UK, 2016; pp. 103–124. [CrossRef]
260. Kootsookos, A.; Mouritz, A.P. Seawater durability of glass- and carbon-polymer composites. *Compos. Sci. Technol.* **2004**, *64*, 1503–1511. [CrossRef]

261. Yu, Y.; Yang, X.; Wang, L.; Liu, H. Hygrothermal ageing on pultruded carbon fibre/vinyl ester resin composite for sucker rod application. *J. Reinf. Plast. Compos.* **2006**, *25*, 149–160. [CrossRef]
262. Gellert, E.P.; Turley, D.M. Seawater immersion ageing of glass-fibre reinforced polymer laminates for marine applications. *Compos.A Appl. Sci. Manuf.* **1999**, *30*, 1259–1265. [CrossRef]
263. Siriruk, A.; Jack Weitsman, Y.; Penumadu, D. Polymeric foams and sandwich composites: Material properties, environmental effects, and shear-lag modelling. *Compos. Sci. Technol.* **2009**, *69*, 814–820. [CrossRef]
264. Siriruk, A.; Penumadu, D.; Weitsman, Y. Effect of sea environment on interfacial delamination behaviour of polymeric sandwich structures. *Compos. Sci. Technol.* **2009**, *69*, 821–828. [CrossRef]
265. Akbar, S.; Zhang, T. Moisture diffusion in carbon/epoxy composite and the effect of cyclic hygrothermal fluctuations: Characterization by dynamic mechanical analysis (DMA) and interlaminar shear strength (ILSS). *J. Adhes.* **2008**, *84*, 585–600. [CrossRef]
266. Kumar, A.; Lal, K.G.; Anantha, S.V. Design and Analysis of a Carbon Composite Propeller for Podded Propulsion. In *Proceedings of the Fourth International Conference in Ocean. Engineering (ICOE2018), Lecture Notes in Civil. Engineering 22*; Springer: Berlin/Heidelberg, Germany, 2019. [CrossRef]
267. Verma, D.; Goh, K.L. Natural fiber-reinforced polymer composites. In *Biomass, Biopolymer-Based Materials, and Bioenergy: Construction, Biomedical, and Other Industrial Applications*; Woodhead Publishing: Sawston, UK; Cambridge, UK, 2019; pp. 51–73. [CrossRef]
268. Kovácik, J.; Jerz, J.; Mináriková, N.; Marsavina, L.; Linul, E. Scaling of compression strength in disordered solids: Metallic foams. *Frat. Ed Integrita Strutt.* **2016**, *36*, 55–62. [CrossRef]
269. Movahedi, N.; Linul, E.; Marsavina, L. The Temperature effect on the compressive behavior of closed-cell aluminum-alloy foams. *J. Mater. Eng. Perform.* **2018**, *27*, 99–108. [CrossRef]
270. Taherishargh, M.; Linul, E.; Broxtermann, S.; Fiedler, T. The mechanical properties of expanded perlite-aluminium syntactic foam at elevated temperatures. *J. Alloy. Compd.* **2018**, *737*, 590–596. [CrossRef]
271. Park, H. A study on structural design and analysis of small wind turbine blade with natural fibre (flax) composite. *Adv. Compos. Mater.* **2015**, *25*, 125–142. [CrossRef]
272. Tang, Q.; Wang, Y.; Ren, Y.; Zhang, W.; Guo, W. A novel strategy for the extraction and preparation of bamboo fiber-reinforced polypropylene composites. *Polym. Compos.* **2019**, *40*, 2178–2186. [CrossRef]
273. Al-Mahaidi, R.; Kalfat, R. Fiber-reinforced polymers and their use in structural rehabilitation. In *Rehabilitation of Concrete Structures with Fiber-Reinforced Polymer*; Elsevier Science: Amsterdam, The Netherlands, 2018; pp. 15–20. [CrossRef]
274. Linul, E.; Marsavina, L. Prediction of fracture toughness for open cell polyurethane foams by finite element micromechanical analysis. *Iran. Polym. J.* **2011**, *20*, 736–746.
275. Linul, E.; Marsavina, L. Assesment of sandwich beams with rigid polyurethane foam core using failure-mode maps. *P. Rom. Acad. A* **2015**, *16*, 522–530.
276. Rajak, D.K.; Mahajan, N.N.; Linul, E. Crashworthiness performance and microstructural characteristics of foam-filled thin-walled tubes under diverse strain rate. *J. Alloy. Compd.* **2019**, *775*, 675–689. [CrossRef]
277. Marsavina, L.; Constantinescu, D.M.; Linul, E.; Voiconi, T.; Apostol, D.A.; Sadowski, T. Evaluation of mixed mode fracture for PUR foams. *Procedure Mater. Sci.* **2014**, *3*, 1342–1352. [CrossRef]
278. Linul, E.; Serban, D.A.; Voiconi, T.; Marsavina, L.; Sadowski, T. Energy-absorption and efficiency diagrams of rigid PUR foams. *Key Eng. Mater.* **2014**, *601*, 246–249. [CrossRef]
279. Pei, X.Q.; Friedrich, K. Friction and wear of polymer composites. In *Reference Module in Materials Science and Materials Engineering*; Elsevier: Amsterdam, The Netherlands, 2016. [CrossRef]
280. Habeeb, M.N.; Ashour, A.F. Flexural behavior of continuous GFRP reinforced concrete beams. *J. Compos. Constr.* **2008**, *12*, 115–124. [CrossRef]
281. Abed, F.; El-Chabib, H.; AlHamaydeh, M. Shear characteristics of GFRP-reinforced concrete deep beams without web reinforcement. *J. Reinf. Plast. Compos.* **2012**, *31*, 1063–1073. [CrossRef]
282. Rafi, M.M.; Nadjai, A.; Ali, F. Experimental testing of concrete beams reinforced with carbon FRP. *J. Compos. Mater.* **2007**, *41*, 2657–2673. [CrossRef]
283. Rashid, M.A.; Mansur, M.A.; Paramasivam, P. Behavior of aramid fiber-reinforced polymer reinforced high strength concrete beams under bending. *J. Compos. Constr.* **2005**, *9*, 117–127. [CrossRef]
284. Altalmas, A.; El Refai, A.; Abed, F. Bond degradation of basalt fiber-reinforced polymer (BFRP) bars exposed to accelerated aging conditions. *Constr. Build. Mater.* **2015**, *81*, 162–171. [CrossRef]

285. Al-tamimi, A.; Abed, F.H.; Al-rahmani, A. Effects of harsh environmental exposures on the bond capacity between concrete and GFRP reinforcing bars. *Adv. Concr. Constr.* **2014**, *2*, 1–11. [CrossRef]
286. El Refai, A.; Abed, F.; Altalmas, A. Bond durability of basalt fiber-reinforced polymer bars embedded in concrete under direct pullout conditions. *J. Compos. Constr.* **2014**, *19*, 1–11. [CrossRef]

polymers

MDPI

Review

A Review on the Kenaf/Glass Hybrid Composites with Limitations on Mechanical and Low Velocity Impact Properties

Seri Nur Zumaimi Ahmad Nadzri [1], Mohamed Thariq Hameed Sultan [1,2,3,*], Ain Umaira Md Shah [1,2], Syafiqah Nur Azrie Safri [1,2] and Adi Azriff Basri [2]

1 Laboratory of Biocomposite Technology, Institute of Tropical Forestry and Forest Products (INTROP), UPM Serdang 43400, Selangor Darul Ehsan, Malaysia; snzumaimi@gmail.com or seri_nurzumaimi@yahoo.com (S.N.Z.A.N.); ainumaira91@gmail.com (A.U.M.S.); snasafri@gmail.com (S.N.A.S.)

2 Department of Aerospace Engineering, Faculty of Engineering, Universiti Putra Malaysia, UPM Serdang 43400, Selangor Darul Ehsan, Malaysia; adiazriff@upm.edu.my

3 Aerospace Malaysia Innovation Centre (944751-A), Prime Minister's Department, MIGHT Partnership Hub, Jalan Impact, Cyberjaya 63000, Selangor Darul Ehsan, Malaysia

* Correspondence: thariq@upm.edu.my

Received: 5 February 2020; Accepted: 24 March 2020; Published: 4 June 2020

Abstract: Environmental awareness and trends to develop sustainable resources have directed much research attention towards kenaf fibre as an alternative reinforcement in composite manufacturing. Numerous studies have been conducted on kenaf and its hybrid composites. Most studies were conducted on kenaf/glass hybrid composites compared to other kenaf/synthetic hybrid composites. Similar with other materials, mechanical properties were the fundamental knowledge identified by the researcher. Limited studies conducted on other properties have restricted the use of kenaf composites to non-structural applications. To extend the potential of kenaf composites to automotive exterior or other critical applications, studies on impact properties can be a valuable contribution in the material field. This review discusses the mechanical and low velocity impact properties of kenaf/glass hybrid composites reported previously. Percentage loading of fibre, the angle of orientation in woven fibres and the chemical treatment applied to the fibre before compounding are the three major parameters that affect the mechanical and impact properties of the composites. This review provides insights into the mechanical and impact properties of kenaf/glass hybrid composites for future research.

Keywords: kenaf fibre; glass fibre; hybrid composites; low velocity impact

1. Introduction

The usage of natural fibres such as kenaf in automotive applications has the potential advantage of reducing the weight of the vehicle which helps to minimise the fuel consumption, thus reducing the emission of harmful gases [1,2]. However, due to several drawbacks such as strength of natural fibres, the application of these materials was limited to interior parts of vehicles. In expanding the potential applications of natural fibre composites to the exterior parts, the use of synthetic fibre as a hybrid material had been the interest of researchers around the world [3,4]. Pertaining to the chances of sudden impact to the exterior parts of vehicles, low velocity impact analysis is another vital aspect that needs to be addressed besides the main mechanical properties [5,6]. Therefore, this review was conducted to give an overview of the two main properties of kenaf, glass and their hybrid composites, which were mechanical and low velocity impact properties. Kenaf and glass fibre composites were seen to have big potential in various applications based on numerous studies reported to date.

The synthetic fibre such as glass fibre was found to enhance the mechanical properties of the composite [7]. Embedding woven E-glass as the outermost layer of three plied composites, also containing kenaf and jute, was reported to increase the tensile strength by more than 50% [8]. In the previous research, the tensile strength of bamboo/glass reinforcing polyester hybrid composite was 80.56 MPa compared to the tensile strength of bamboo reinforced polyester composite and bamboo/jute reinforced polyester hybrid composites, recorded as 46.67 and 60.32 MPa, respectively [9]. As reported in a previous study, the hybridization of hemp and glass composite presented the increasing trend of flexural strength changing from 97.5 to 101 MPa as the glass content in the composite increased from 0% to 15% [10].

Reducing the percentage of glass fibre in glass/polypropylene (PP) composites from 60% to 30% by replacing the reduced percentage with 30% twill woven bamboo fibres decreased the Charpy impact strength by less than 50%. However, in comparison with non-hybrid bamboo/PP composites, the glass/bamboo hybrid composites showed higher a Charpy impact strength, of 1129 J/m, compared to 530.9 J/m for the non-hybrid [11]. A composite with hybrid long fibres of both sugar palm and glass fibre, with a random orientation and 30:70 fibre ratio, respectively, in an epoxy matrix, showed its ability to withstand an impact energy up to 15 J in drop weight impact testing [12]. Various natural fibres have been used to be hybridised with glass fibres in an attempt to reduce the dependency on synthetic materials. On the other hand, the inclusion of glass fibre broadens the potential use of natural fibres in different applications.

This paper presents a short review on the mechanical and low velocity impact properties of kenaf/glass hybrid composites, to provide a reference for filling the gap in the respective research area. The aim is to enhance the potential of kenaf/glass hybrid composites to be used in automotive applications, based on their strength and impact properties.

2. Kenaf Plant

Kenaf plant, scientifically registered as *Hibiscus cannabinus L.*, belongs to the family of Malvaceae. This herbaceous dicotyledonous plant grows during the warm season, similarly to okra and cotton plants, which belong to the same family of Malvaceae [13,14]. Table 1 shows the taxonomy of the kenaf plant.

Table 1. Kenaf plant's taxonomy.

Kingdom	Plantae
Class	Magnoliopsida
Order	Malvales
Family	Malvaceae
Genus	Hibiscus
Species	*Hibiscus cannabinus L.*

Kenaf plant was introduced in the 1970s in Malaysia as an industrial crop in Kelantan, Pahang and Terengganu [15]. Then, kenaf plantations have been identified as an alternative to tobacco plantations [16]. Kenaf plant became popular in the 1990s, when it turned out to be an alternative resource for composite fabrications, such as in particle board, medium-density fibre board (MDF), wood plastic composites (WPC) and many more applications [17]. Besides being one of the alternative raw materials for replacing wood fibre in the pulp and paper industries, kenaf plant is also traditionally used to make ropes, canvas, and sacks [18].

3. Extraction of Kenaf Fibres

In engineering applications, specifically in the composites field, kenaf fibres need to be extracted from the respective parts of the kenaf plant. It has been reported that approximately 40% of the kenaf plant stem can be transformed into fibres. This high percentage makes kenaf economically comparable to other plants, such as jute, hemp, and flax [13]. The short growing cycle of the kenaf plant, in the range of 150 to 180 days, and its low water requirement, make it more suitable to be used in composites and other applications [19,20]. Figure 1 shows the illustration of bast fibre, which is the outermost tissue layer of the kenaf stem, as well as the core and pith, which are located in the centre of the kenaf stem [21,22].

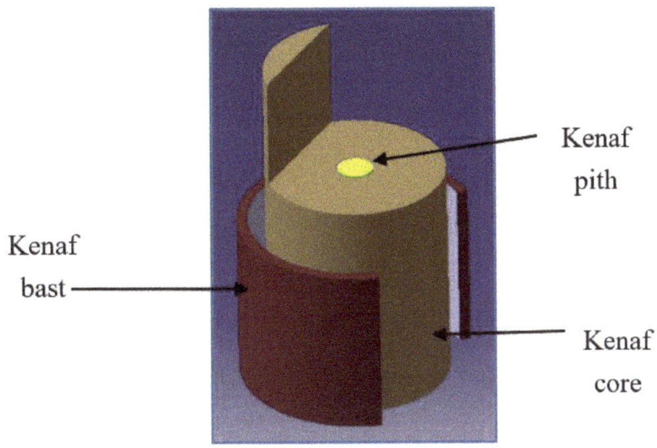

Figure 1. The illustration of kenaf stem.

The process of extracting plant fibres is also known as the retting or degumming process. There are a few retting methods applied to plant fibres reported to date, such as dew retting, water retting, enzymatic retting, chemical retting and mechanical retting [23]. Among them, the water, chemical, mechanical and enzymatic retting are the most commonly practiced processes in extracting kenaf fibres, as shown in Figure 2.

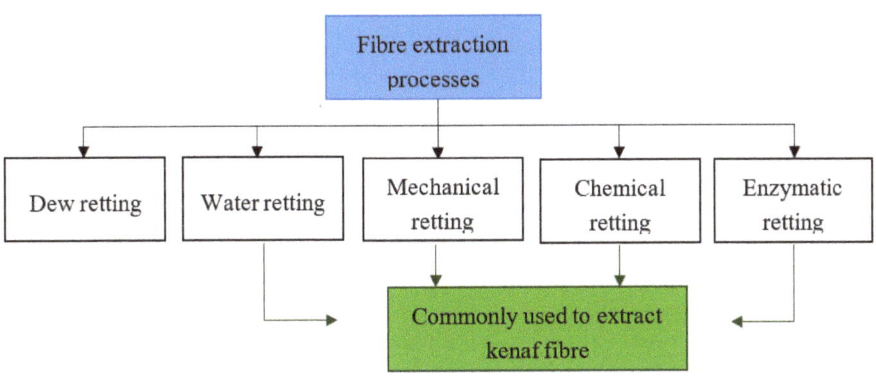

Figure 2. Different retting processes in extracting plant fibre.

Compared to mechanical and chemical retting, water retting takes a longer time to complete the process, which is about 22 days, including washing, drying and combing the fibre. This method begins by submerging the kenaf plant into water, such as in a pond, river or tank. Microbial actions will separate the kenaf fibres within the pith (Figure 1). This process, however, will lead to minor water pollution and produce a strong odour because of anaerobic bacterial fermentation [24,25]. The water retting process will be continued by washing the fibre with clean water to remove unwanted substances. The cleaned fibres will be oven dried at 60 °C for 24 h until the fibre achieves constant weight [26].

A traditional method, such as air drying, can be used, depending on the surrounding humidity [27]. The combing process is carried out to disentangle the fibre, and at the same time to further extract single fibres from fibre bundles [28].

Mechanical retting, also called decortication, is the process of isolating the bast and the pith (Figure 1) using a decorticator. This method is easier, cheaper, and faster, compared to the other retting methods. Kenaf stalk will be immersed into water for at least 5 days before separating the fibre through the decorticating process [29]. Commonly used decorticating processes involve hammer milling and roller milling. Hammer milling is the process of beating the kenaf stem until the bast and core are separated. The small fibre obtained after the hammering process will be passed through the meshes inside the hammer milling machine. The roller milling machine will roll and crush the kenaf stalk, producing minimal damage to the fibre. Using this machine, longer fibre is achieved. The process of extracting fibre continues by washing, drying, and combing, similarly to the procedures in the water retting process, as explained earlier [25,29]. Figure 3 illustrates the fibre decortication process applied to extract kenaf fibre.

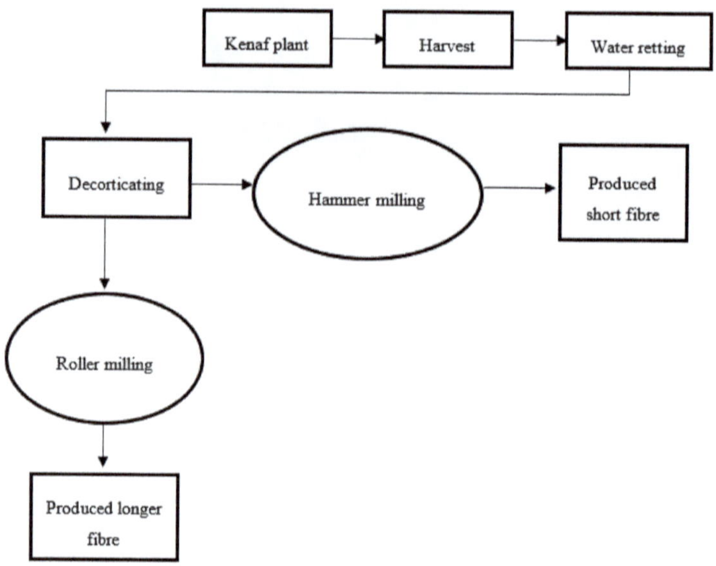

Figure 3. Fibre decortication process.

Physical retting, through steam explosion, is performed on the plant stalk before applying chemicals to soften the fibres. This process is eco-friendly and requires a shorter time to extract the fibres. Steam explosion is carried out at 0.5 MPa for 15 min, followed by alkali-oxygen treatment. These parameters have been suggested as the optimum pressure and time, respectively, to remove pectin, hemicellulose and lignin [24,30].

Chemical retting is the process where alkali hydrolysis takes place. The chemicals that are normally used are sodium hydroxide (NaOH), sodium benzoate ($C_7H_5NaO_2$), and hydrogen peroxide (H_2O_2). Smooth and long consistent fibre can be produced through this process in a short period of time. It has been suggested that these chemicals should be used in concentrations below 1%, as a more concentrated alkaline solution could lead to the degradation of the tensile strength [24].

Enzymatic retting is based on a similar concept to that of water retting, which uses bacteria that produce pectic enzyme to extract the fibre by dissolving pectin. As biotechnology has advanced, this enzyme can be now produced in the laboratory and is commercially available. Enzymatic retting can be controlled, is efficient and most importantly environmentally friendly [31]. A suitable temperature range between 40 and 50 °C has been established based on the most favorable working conditions for the enzyme [26,32]. However, this process involves higher costs, compared to water retting. The later washing, drying and combing processes are all similar to the procedure described for other types of retting.

4. Glass Fibre

Glass fibre is a commonly used synthetic fibre made from silicates, soda, clay, limestone, boric acid and various metallic oxides. All these constituents will undergo a heating process and will be refined in the narrow chamber [33]. Glass fibre is classified into several types, based on its main properties, and is denoted with letters C, D, E and S. C-glass fibre can be used in an acidic environment because it has the ability to resist corrosion. D-glass fibre has a low dielectric constant and it is commonly used in electrical applications. E-glass fibre is known as electrical glass, which is very good for electrical insulation. S-glass is a high strength glass fibre. It not only has high strength and stiffness, but also can be used in extreme conditions, such as in extreme temperatures or corrosive environments [34]. Table 2 shows the mechanical properties of different types of glass fibres.

Table 2. Mechanical properties of glass fibre types C, D, E and S.

	C-Glass	D-Glass	E-Glass	S-Glass	Ref.
Density (g/cm^3)	2.56	2.11	2.54	2.53	[34]
Tensile strength (MPa)	3300	2500	3400	4600	
Young's Modulus (MPa)	69	55	72	89	
Elongation (%)	4.8	4.5	4.7	5.2	

5. Mechanical Properties of Kenaf/Glass Hybrid Composites

Kenaf fibre has low density and good mechanical properties. Replacing part of synthetic fibre such as glass fibre with kenaf fibre in a composite will reduce the structure's weight and cost, and will be more environmentally friendly. The mechanical properties of kenaf and glass fibres are illustrated in Table 3.

Table 3. General mechanical properties of kenaf fibre and glass fibre.

	Kenaf	Glass
Density (g/m^3)	1.45	2.55
Tensile strength (MPa)	930	3400
Elastic modulus (GPa)	53	71
Elongation at break (%)	1.6	3.4
References	[18,35]	[36,37]

Kenaf/glass hybrid composites with 90° orientations was reported to achieve higher tensile strength compared to 0° orientations with values of 69.86 and 49.27 MPa, respectively. The load applied in the direction parallel to the fibre contributed to the higher strength of 90° orientations kenaf/glass hybrid composites compared to the load applied perpendicular to the orientation of fibre [38]. The contribution of synthetic fibre, which is known for its strength and stiffness, was seen in the increment of tensile and flexural strength of kenaf/glass hybrid composites to 65.29 and 115.71 MPa compared to non-hybrid kenaf composites with 49.48 and 77.63 MPa, respectively [39].

Table 4 summarises the findings reported from studies on the mechanical properties of kenaf/glass fibre hybrid composites.

Table 4. Reported research on mechanical properties of kenaf/glass fibre hybrid composites.

Materials	Mechanical Properties		Ref.
	Tensile	Flexural	
90° fibre orientation of kenaf + chopped strand glass fibre reinforced epoxy	69.86 MPa	162.566 MPa	[38]
kenaf fibre yarn + woven glass fibre reinforced epoxy	65.29 MPa	115.71 MPa	[39]
Woven kenaf fibre + glass fibre mat reinforced unsaturated polyester	85.49 MPa	124.07 MPa	[40]

Chemical treatment using alkaline solution is the most commonly used due to its low cost and high effectiveness [41]. Kenaf fibre treated with 9% NaOH solution for 12 h presented the highest flexural strength of 93.3 MPa compared to 3% NaOH and untreated kenaf fibre which were 63.2 and 25.1 MPa, respectively. The adequate surface roughness of kenaf fibre treated with 9% NaOH improved the fibre matrix bonding, while a 3% NaOH concentration was not sufficient to remove the impurities on the surface of fibre [42]. In a different study, maintaining the percentage of kenaf fibre loading at 25% in two types of polymer matrix, epoxy and polyester, results in higher tensile strength for the epoxy-based composites with a value of 93.59 MPa compared to polyester-based composites at 86.54 MPa [43]. The fact that epoxy resin has better properties (such as high strength and low shrinkage) compared to polyester resin which is poor in adhesive and high cure shrinkage, indirectly contributes to the overall strength of composites [44].

In a research project that investigated the mechanical properties of different amounts of kenaf (K) layers in glass (G) fibre composites, one layer of kenaf as the core between three layers of glass fibre each at the top and bottom of the sandwich structure (3G/K/3G) presented the highest tensile strength compared to 3G/2K/3G and 3G/3K/3G, which had two and three layers of kenaf as the core in the same sequences. The study suggested that the increment in the number of kenaf layers will reduce the tensile strength due to the low strength of kenaf fibre [45]. For different types of kenaf fibres, increasing the fibre loading from 10% to 40% increased the flexural strength of composites accordingly, whereas further increasing the fibre loading to 50% decreased the flexural strength due to the saturated mixture of fibre in the polymer matrix. The saturated mixture had decreased the ability of polymer to hold the fibre tightly, thus could not sustain the load applied to the whole structure [46]. The fibre size is also one of the important factors in determining the mechanical properties of composite. It was reported that 30% kenaf fibre in 20 mesh size had a higher tensile strength of 16 MPa compared to 40 mesh kenaf fibre at 13.6 MPa. A larger surface area of 20 mesh filler was suggested to enhance the surface contact and bonding between fibre and matrix [47].

6. Low Velocity Impact Properties of Kenaf Composites

Reaction such as rebounding, penetration and perforation will happen during a low velocity impact event [48]. The damage of the specimen can be analysed through the force against displacement graph. Closed loop graph curve indicated that the impactor rebounded when it hit the surface of the specimen [49]. The damages that happened during the low velocity impact event included the matrix cracking, delamination, fibre failure and penetration [50]. Matrix cracking is the initial stage of damage due to low velocity impact. The number of cracks will increase when a larger external load is applied leading to another failure which is delamination [51]. Delamination was found to be induced by matrix cracking when the high transverse shear stresses at the neighboring impacted matrix surface and later develops into a weak interfacial bond which develops into fibre fracture and fibre pullout [52,53]. The phenomena of shear cracking and bending cracking are often characterizations of pine tree damage. Pine tree damage usually exhibits on the stiff-thick composite and reserved pine tree damage exhibits on the flexible-thin composite [54,55]. Perforation is the damage when the impactor completely penetrated the specimen. The opened loop curve in force against displacement graph determined that the specimen experienced penetration by the impactor [49].

Researchers have reported that the highest energy absorbed by kenaf/epoxy composites was of 2.83 J at 15% fibre content. Meanwhile, a kenaf polyester composite absorbed the maximum energy of

4.53 J at 25% fibre content [56]. Kenaf fibre reinforced polyvinyl butyral (PVB) showed an increasing trend of impact strength when the fibre content was increased from 10% to 20% and 30%, while the highest impact strength result was achieved at 40% fibre content. However, the impact energy strength of the composite started decreasing at 50% and 60% fibre content [57]. The decreasing trend of the impact strength graph when the fibre content was at 50% and 60% also presented in the research on the hybridization of kenaf fibre reinforced recycled polypropylene/polyethelene (r-PP/PE) [58]. Fibre content is an important influencing factor applied not only to kenaf reinforced polymer composites, but also to kenaf hybrid composites. A study showed that the 50:50 fibre ratio in bamboo/kenaf hybrid composites led to the highest impact strength result, which was 45 J/m, compared to the only kenaf reinforced epoxy composite and to the 30:70 fibre ratio in bamboo-kenaf hybrid composites, which recorded 40.6 and 37.8 J/m, respectively, for impact strength [59].

In a study, three different fibre arrangements were used to fabricate kenaf/polyester composites, namely, perpendicular, anisotropic and isotropic. Pure polyester was used as a control. The impact strength of the obtained materials was tested on an Izod impact tester. Pure polyester showed the lowest result of impact strength, which is 1.42 kJ/m^2. Meanwhile, the highest impact strength result was recorded for the kenaf/polyester composite with the anisotropic fibre arrangement of 6.68 kJ/m^2 [60]. Fibre size also affects the impact energy absorbed by the composites [60]. A study focusing on kenaf fibre reinforced thermoplastic polyurethane (TPU) composites showed that the impact strength values of the materials prepared with the smallest size of kenaf fibre (<125 μm), medium size of kenaf fibre (125–300 μm) and the largest size of kenaf fibre (300–425 μm) were 2.1, 2.76 and 2.97 kJ/m^2, respectively [61].

Different types of matrix gave different impact strength results for woven kenaf fibre composites. In a previous study, the same type of fibre, namely woven kenaf fibre, and the same impact energy, but different fibre orientations, of 0°/90° and 45°/−45°, have been used to fabricate composites using three different types of matrix, specifically, epoxy, unsaturated polyester, and vinyl ester. Woven kenaf fibre reinforced unsaturated polyester showed the highest energy absorbed for both types of fibre orientation [57].

Kenaf fibre tends to absorb moisture because of its hydrophilic nature. Therefore, chemical treatment is applied to modify the fibre surface by removing the hydroxyl groups and impurities from the fibre [62]. A few chemical treatments can be performed to modify the kenaf fibre surface, such as alkaline treatment, silane treatment and a combination of alkaline and silane treatment [63]. The treatment using NaOH can be summarised in the chemical equation below [63]:

$$\text{Fibre-OH} + \text{NaOH} = \text{Fibre} + \text{Na}^+\text{O}^- + \text{H}_2\text{O} + \text{impurities} \quad (1)$$

The untreated fibre presented low impact strength due to the weak interfacial bond between fibre and matrix. The elimination of the hemicellulose, wax and –OH bonding during the fibre treatment enhanced the bonding between fibre and matrix. Treated kenaf fibre gave a higher impact strength to the composites, compared to the untreated one. This is because the NaOH solution that had been used in treating kenaf fibre improved the properties of kenaf-reinforced thermoplastic polyurethane composites [64]. Izod impact testing revealed that an untreated kenaf reinforced polyester composite showed the lowest value of impact strength, which was 2.61 kJ/m^2. Two different concentrations of NaOH solution, of 6% and 9%, had been used to treat kenaf fibre for 12 h. The impact strength of the composite comprising fibre treated with 6% NaOH concentration was 15.77 kJ/m^2, while that of the composite with fibre treated with 9% NaOH concentration was 6.92 kJ/m^2 [49]. Kenaf fibre treated with propionic anhydride gave better impact strength results, of 7.7 kJ/m^2, compared to kenaf fibre treated with succinic anhydride and untreated fibre, which recorded 7.3 and 5.4 kJ/m^2, respectively [62]. A previous study compared the impact strength of untreated and treated woven kenaf-banana hybrid composites. The authors concluded that the impact strength of the untreated woven kenaf-banana hybrid composite was 23 kJ/m^2, that of the alkali treated woven kenaf-banana material was 26 kJ/m^2 and that of the material treated with sodium lauryl sulfate (SLS) was 28 kJ/m^2 [65]. The findings of reported research studies about the effects of kenaf fibre treatment on the impact strength of the prepared composites are summarised in Table 5. Enzymatic treatment has been also used to modify the surface of fibre. In previous research, kenaf fibre treated with laccase enzyme showed a 26.31 J/m^2 impact strength, while untreated fibre showed a 22.51 J/m^2 impact strength [66].

Table 5. Previous studies on different kenaf fibre treatment.

Composites		Parameters	Impact Strength	Ref.
Kenaf + polyester	Chemical treatment	Untreated	2.61 kJ/m^2	[42]
		Treated with 6% NaOH	15.77 kJ/m^2	
		Treated with 9% NaOH	6.92 kJ/m^2	
Kenaf + polyester		Untreated	5.4 kJ/m^2	[62]
		Propionic anhydride	7.7 kJ/m^2	
		Succinic anhydride	7.3 kJ/m^2	
Kenaf/banana + unsaturated polyester		Untreated	23 kJ/m^2	[65]
		Treated with alkali	26 kJ/m^2	
		Treated with sodium lauryl sulfate	28 kJ/m^2	
Kenaf + recycled polypropylene	Enzymatic treatment	Untreated	22.51 J/m^2	[66]
		Treated with laccase enzyme	26.13 J/m^2	

Hybrid banana/kenaf/banana reinforced epoxy composites were evaluated by using the Charpy test and the maximum impact energy was found to be 12 J. This is the highest impact energy achieved in this study, among other hybrids, such as kenaf/kenaf/kenaf and neem/kenaf/neem composites, which reached values of 4 and 10 J, respectively [65]. The hybridization of kenaf fibre with synthetic fibre such as Kevlar enhanced the impact strength performance of the hybrid composite [67]. In a previous study, the highest impact strength of kenaf/Kevlar hybrid composite that been recorded was 50.1 kJ/m^2 which belonged to the Kevlar/kenaf/Kevlar/kenaf hybrid composite. This study also concluded that the number of fibre layers and fibre sequences became factors that affected the impact properties of the kenaf/Kevlar hybrid [68].

7. Low Velocity Impact of Glass Composites

Previous research reported that E-glass composites are more resistant to impact compared to C-glass, specifically, that smaller damage at a higher peak force was found on E-glass composites at the same impact energy level [69]. In terms of weaving pattern, woven E-glass fibre gives better mechanical properties, compared to chopped strand fibre mat. It was reported that woven E-glass fibre has an impact strength of 415 MPa, while chopped strand E-glass fibre presented 189 MPa [70].

Fibre content and fibre arrangement affect the impact properties of not only natural fibre composites, but also of glass fibre composites. A study showed that the maximum impact energy absorbed by composites made from one layer of plain-weave glass fibre and two layers of chopped strand glass mats, with 30% fibre content, was 112.105 kJ/m^2, while that of composites fabricated from two layers of plain-weave glass fibre and one layer of chopped strand glass mat, with 22% fibre content, was 77.141 kJ/m^2 [70]. Another study reported that a glass wool fibre reinforced epoxy material with a 40% fibre weight gave the highest impact strength value, which was 0.0862 J/m^2, compared to 30% and 50% glass wool fibre weight composites, which reached 0.0652 and 0.0421 J/m^2, respectively [71].

8. Low Velocity Impact of Kenaf/Glass Hybrid Composites

The impact strength properties of hybrid composites have been found to be affected by several factors, including volume fraction, fibre matrix adhesion, fibre orientation, fibre length, stress transfer and the thickness of the composites [72]. The reported research findings on kenaf fibre and glass fibre hybrid composites are tabulated in Table 6.

Table 6. Previous studies on kenaf/glass fibre hybrid composites.

Kenaf Fibre	Glass Fibre	Matrix	Fabrication Method	Ref.
Woven kenaf fiber	Woven E-glass fiber	Epoxy	Hand lay-up	[73]
Twisted long kenaf fiber	Glass fiber	Epoxy	Hand lay-up	[74]
Kenaf fiber (direction 0°, 90°)	Glass fiber direction (0°, 90°)	Epoxy	Hand lay-up	[38]
Twisted kenaf fiber	Woven glass fiber	Epoxy	Vacuum pump Compression moulding	[75]
Kenaf fibre	Woven fibre glass	Polyester	Hand lay-up Cold press	[76]
Kenaf mat (20 cm × 20 cm)	Glass fiber (20 cm × 20 cm)	Unsaturated polyester	Hand lay-up Compression	[77]
Kenaf fibre	Glass fibre	Unsaturated polyester	Cold pressure hand lay-up	[78]
Non- woven kenaf mat	E-glass fibre	Polypropylene	Compression moulding	[79]

The composite fabricated with a 90° kenaf/glass fibre orientation had a higher impact strength compared to the one with a 0° fibre orientation, namely, of 6.66 and 6 J, respectively. Due to its better impact strength, it was suggested that the hybrid composite can be used in structural applications [38]. Previous work reported that twisted neem/kenaf/neem composites, with a fibre orientation of 90°/45°/90°, embedded with glass fibre at the outermost top and bottom layers, recorded the maximum impact strength of 12.2 J, which was higher than the values achieved by neem/kenaf/neem composites with fibre orientations of 0°/0°/0° and 0°/90°/0° (11.23 and 11.64 J, respectively) [65].

At 9 J of impact energy, kenaf/glass hybrid composites with 3 mm thickness showed a major crack length of 52.92 mm, while kenaf/epoxy composites reached 100.61 mm for major crack length. The smaller damage on the kenaf/glass hybrid composites indicated that embedded woven glass hybridised with kenaf led to stronger impact resistance. Meanwhile, the glass/epoxy composite showed 16.02 mm radius of damage [79]. In another study, 75% glass fibre and 25% kenaf fibre composites showed almost similar results to those achieved for 100% glass fibre. The hybrid composites were selected to undergo low velocity testing. The hybrid composite with 10 layers of glass and kenaf fibre could absorb an impact energy of up to 40 J [49].

9. Conclusions

Kenaf fibre has great potential in automotive application as a reinforcement fibre, since it is light in weight, eco-friendly, low-cost and has good mechanical properties. Reinforcing kenaf fibre with glass fibre is one of the methods to enhance kenaf fibre composites because glass fibre has better mechanical and impact properties than kenaf fibre. There are several factors that affect the mechanical and impact properties of the kenaf/glass hybrid composites. Previous studies show that the optimum fibre content and fibre orientation were between 30% and 40% and 90°, where it can resist a higher impact strength and have better mechanical properties. Fibre surface modification also helps to improve the properties of kenaf/glass hybrid composites. The number of layers and the thickness of the composite also influences its mechanical and impact strength properties. From the review conducted, it can be suggested that kenaf/glass hybrid composites have higher tensile strength up to 85 MPa compared to kenaf composites. Other than that, kenaf/glass hybrid composite can also withstand low velocity impact energy up to 12 J. The mechanical and impact properties of kenaf/glass hybrid composites discussed in this manuscript show its potential for use in automotive applications. It was also found that limited studies were done on the mechanical and impact properties of the kenaf/glass hybrid composites, and further investigations were needed to maximise the potential of kenaf/glass hybrid composites.

Acknowledgments: The authors would like to thank Universiti Putra Malaysia for the financial support through the Fundamental Research Grant Scheme FGRS/1/2019/STG07/UPM/02/2 (5540320). The authors would like to thank the Department of Aerospace Engineering, Faculty of Engineering, Universiti Putra Malaysia and Laboratory of Biocomposite Technology, Institute of Tropical Forestry and Forest Product (INTROP), Universiti Putra Malaysia (HICOE) for the close collaboration in this research.

Conflicts of Interest: The authors declare no conflict of interest.

References

1. Zakikhani, P.; Zahari, R.; Sultan, M.T.; Majid, D.L. Thermal degradation of four bamboo species. *BioResources* **2016**, *11*, 414–425. [CrossRef]
2. Mohammed, L.; Ansari, M.N.; Pua, G.; Jawaid, M.; Islam, M.S. A review on natural fiber reinforced polymer composite and its applications. *Int. J. Polym. Sci.* **2015**, *2015*, 243947. [CrossRef]
3. Peças, P.; Carvalho, H.; Salman, H.; Leite, M. Natural fibre composites and their applications: A review. *J. Compos. Sci.* **2018**, *2*, 66. [CrossRef]
4. Khan, T.; Hameed Sultan, M.T.; Ariffin, A.H. The challenges of natural fiber in manufacturing, material selection, and technology application: A review. *J. Reinf. Plast. Compos.* **2018**, *37*, 770–779. [CrossRef]
5. Chee, S.S.; Jawaid, M.; Sultan, M.T.; Alothman, O.Y.; Abdullah, L.C. Thermomechanical and dynamic mechanical properties of bamboo/woven kenaf mat reinforced epoxy hybrid composites. *Compos. Part B Eng.* **2019**, *163*, 165–174. [CrossRef]
6. Venkateshwaran, N.; Elayaperumal, A.; Sathiya, G.K. Prediction of tensile properties of hybrid-natural fiber composites. *Compos. Part B Eng.* **2012**, *43*, 793–796. [CrossRef]
7. Sood, M.; Dwivedi, G. Effect of fiber treatment on flexural properties of natural fiber reinforced composites: A review. *Egypt. J. Pet.* **2018**, *27*, 775–783. [CrossRef]
8. Sanjay, M.R.; Yogesha, B. Studies on hybridization effect of jute/kenaf/E-glass woven fabric epoxy composites for potential applications: Effect of laminate stacking sequences. *J. Ind. Text.* **2018**, *47*, 1830–1848. [CrossRef]
9. Gangil, B.; Kumar, S. Comparative evaluation on mechanical properties of jute/bamboo-glass hybrid reinforced polyester composites. *Asian J. Sci. Technol.* **2017**, *8*, 5190–5194.
10. Panthapulakkal, S.; Sain, M. Injection-molded short hemp fiber/glass fiber-reinforced polypropylene hybrid composites—Mechanical, water absorption and thermal properties. *J. Appl. Polym. Sci.* **2007**, *103*, 2432–2441. [CrossRef]
11. Zuhudi, N.Z.; Lin, R.J.; Jayaraman, K. Flammability, thermal and dynamic mechanical properties of bamboo–glass hybrid composites. *J. Thermoplast. Compos. Mater.* **2016**, *29*, 1210–1228. [CrossRef]
12. Safri, S.N.; Sultan, M.T.; Jawaid, M.; Majid, M.A. Analysis of dynamic mechanical, low-velocity impact and compression after impact behaviour of benzoyl treated sugar palm/glass/epoxy composites. *Compos. Struct.* **2019**, *226*, 111308. [CrossRef]
13. Satya, P.; Karan, M.; Kar, C.S.; Mahapatra, A.K.; Mahapatra, B.S. Assessment of molecular diversity and evolutionary relationship of kenaf (*Hibiscus cannabinus* L.), roselle (*H. sabdariffa* L.) and their wild relatives. *Plant Syst. Evol.* **2013**, *299*, 619–629. [CrossRef]
14. Devadas, A.; Nirmal, U.; Hossen, J. Investigation into mechanical & tribological performance of kenaf fibre particle reinforced composite. *Cogent Eng.* **2018**, *5*, 1479210.
15. Mohd, H.; Arifin, A.; Nasima, J.; Hazandy, A.H.; Khalil, A. Journey of kenaf in Malaysia: A review. *Sci. Res. Essays* **2014**, *9*, 458–470. [CrossRef]
16. Ayadi, R.; Hanana, M.; Mzid, R.; Hamrouni, L.; Khouja, M.L.; Salhi Hanachi, A. Hibiscus cannabinus L.-Kenaf: A review paper. *J. Nat. Fibers* **2017**, *14*, 466–484.
17. Akil, H.; Omar, M.F.; Mazuki, A.A.; Safiee, S.Z.; Ishak, Z.M.; Bakar, A.A. Kenaf fiber reinforced composites: A review. *Mater. Des.* **2011**, *32*, 4107–4121. [CrossRef]
18. Kumar, R.; Hashmi, S.A.; Nimanpure, S.; Naik, A. Enhanced dynamic mechanical properties of kenaf epoxy composites. *Adv. Mater. Proceeding* **2017**, 749–757. [CrossRef]
19. Jeyanthi, S.; Rani, J.J. Improving mechanical properties by kenaf natural long fiber reinforced composite for automotive structures. *J. Appl. Sci. Eng.* **2012**, *15*, 275–280.
20. Saba, N.; Paridah, M.T.; Jawaid, M. Mechanical properties of kenaf fibre reinforced polymer composite: A review. *Constr. Build. Mater.* **2015**, *76*, 87–96. [CrossRef]
21. Adole, A.M.; Yatim, J.M.; Ramli, S.A.; Othman, A.; Mizal, N.A. Kenaf Fibre and Its Bio-Based Composites: A Conspectus. *Pertanika J. Sci. Technol.* **2019**, *27*, 297–329.
22. Juliana, A.H.; Paridah, M.T.; Rahim, S.; Nor Azowa, I.; Anwar, U.M. Effect of adhesion and properties of kenaf (*Hibiscus cannabinus* L.) stem in particleboard performance. *J. Adhes. Sci. Technol.* **2014**, *28*, 546–560. [CrossRef]

23. Kian, L.K.; Saba, N.; Jawaid, M.; Sultan, M.T. A review on processing techniques of bast fibers nanocellulose and its polylactic acid (PLA) nanocomposites. *Int. J. Biol. Macromol.* **2019**, *121*, 1314–1328. [CrossRef] [PubMed]
24. Paridah, M.T.; Basher, A.B.; SaifulAzry, S.; Ahmed, Z. Retting process of some bast plant fibres and its effect on fibre quality: A review. *BioResources* **2011**, *6*, 5260–5281.
25. Sisti, L.; Totaro, G.; Vannini, M.; Celli, A. Retting process as a pretreatment of natural fibers for the development of polymer composites. In *Lignocellulosic Composite Materials*; Springer International Publishing: Cham, Switzerland, 2018; pp. 97–135.
26. Bacci, L.; Di Lonardo, S.; Albanese, L.; Mastromei, G.; Perito, B. Effect of different extraction methods on fiber quality of nettle (*Urtica dioica* L.). *Text. Res. J.* **2011**, *81*, 827–837. [CrossRef]
27. Vetter, R.E.; Ribeiro, R.A.; Ribeiro, M.G.; Miranda, I.P. Studies on drying of imperial bamboo. *Eur. J. Wood Wood Prod.* **2015**, *73*, 411–414. [CrossRef]
28. Fan, M.; Weclawski, B. Long natural fibre composites. In *Advanced High Strength Natural Fibre Composites in Construction*; Woodhead Publishing: Cambriadge, UK, 2017; pp. 141–177.
29. Ayorinde, A.T.; Makanjuola, G.A.; Aluko, O.B.; Owolarafe, O.K.; Sanni, L.A. Performance evaluation of a kenaf decorticator. *Agric. Eng. Int. CIGR J.* **2019**, *21*, 192–202.
30. Gao, S.; Han, G.; Jiang, W.; Zhang, Y.; Zhang, X. Steam explosion and alkali-oxygen combined effect for degumming of kenaf fiber. *BioResources* **2015**, *10*, 5476–5488. [CrossRef]
31. Wong, L.Y.; Saad, W.Z.; Mohamad, R.; Tahir, P.M. Efficacy of Aspergillus fumigatus R6 Pectinase in Enzymatic Retting of Kenaf. *BioResources* **2016**, *11*, 10030–10041. [CrossRef]
32. Mohd Nurazzi, N.; Khalina, A.; Sapuan, S.M.; Dayang Laila, A.H.; Rahmah, M.; Hanafee, Z. A Review: Fibres, Polymer Matrices and Composites. *Pertanika J. Sci. Technol.* **2017**, *25*, 1085–1102.
33. Sharba, M.J.; Leman, Z.; Sultan, M.T.; Ishak, M.R.; Hanim, M.A. Partial replacement of glass fiber by woven kenaf in hybrid composites and its effect on monotonic and fatigue properties. *BioResources* **2016**, *11*, 2665–2683. [CrossRef]
34. Prince Engineering. 2018. Available online: https://www.build-on-prince.com/ (accessed on 12 July 2019).
35. Shah, A.U.; Sultan, M.T.; Jawaid, M.; Cardona, F.; Talib, A.R. A review on the tensile properties of bamboo fiber reinforced polymer composites. *BioResources* **2016**, *11*, 10654–10676.
36. Landesmann, A.; Seruti, C.A.; Batista, E.D. Mechanical properties of glass fiber reinforced polymers members for structural applications. *Mater. Res.* **2015**, *18*, 1372–1383. [CrossRef]
37. Gibhardt, D.; Doblies, A.; Meyer, L.; Fiedler, B. Effects of hygrothermal ageing on the interphase, fatigue, and mechanical properties of glass fibre reinforced epoxy. *Fibers* **2019**, *7*, 55. [CrossRef]
38. Ramesh, M.; Nijanthan, S. Mechanical property analysis of kenaf–glass fibre reinforced polymer composites using finite element analysis. *Bull. Mater. Sci.* **2016**, *39*, 147–157. [CrossRef]
39. Sapiai, N.; Jumahat, A.; Mahmud, J. Flexural and tensile properties of kenaf/glass fibres hybrid composites filled with carbon nanotubes. *J. Teknol.* **2015**, *76*, 115–120. [CrossRef]
40. Salleh, Z.; Berhan, M.N.; Hyie, K.M.; Isaac, D.H. Cold-pressed kenaf and fibreglass hybrid composites laminates: Effect of fibre types. *World Acad. Sci. Eng. Technol. Int. Sci. Index* **2012**, *71*, 11.
41. Amel, B.A.; Paridah, M.T.; Sudin, R.; Anwar, U.M.; Hussein, A.S. Effect of fiber extraction methods on some properties of kenaf bast fiber. *Ind. Crop. Prod.* **2013**, *46*, 117–123. [CrossRef]
42. Yuhazri, M.Y.; Phongsakorn, P.T.; Haeryip Sihombing, I.P.; Jeefferie, A.R.; Puvanasvaran, A.P.; Kamarul, A.M.; Kannan, R. Mechanical properties of kenaf/polyester composites. *Int. J. Eng. Technol.* **2011**, *11*, 127–131.
43. Bakar, N.H.; Hyie, K.M.; Jumahat, A.; Latip, E.N.; Kalam, A.N.; Salleh, Z. Influence of Different Matrices on The Tensile and Impact Properties of Treated Kenaf Composites. *Adv. Mater. Res.* **2016**, *1133*, 136–140. [CrossRef]
44. Bakar, N.H.; Hyie, K.M.; Mohamed, A.F.; Salleh, Z.; Kalam, A. Kenaf fibre composites using thermoset epoxy and polyester polymer resins: Energy absorbed versus tensile properties. *Mater. Res. Innov.* **2014**, *18*, 505–509. [CrossRef]
45. Salman, S.D.; Leman, Z.; Sultan, M.T.; Ishak, M.R.; Cardona, F. The effects of orientation on the mechanical and morphological properties of woven kenaf-reinforced poly vinyl butyral film. *BioResources* **2016**, *11*, 1176–1188. [CrossRef]

46. Salman, S.D.; Leman, Z.; Sultan, M.T.; Ishak, M.R.; Cardona, F. Influence of fiber content on mechanical and morphological properties of woven kenaf reinforced PVB film produced using a hot press technique. *Int. J. Polym. Sci.* **2016**. [CrossRef]
47. Mohd Radzuan, N.A.; Ismail, N.F.; Radzi, F.M.; Khairul, M.; Razak, Z.B.; Tharizi, I.B.; Sulong, A.B.; Che Haron, C.H.; Muhamad, N. Kenaf Composites for Automotive Components: Enhancement in Machinability and Moldability. *Polymers* **2019**, *11*, 1707. [CrossRef] [PubMed]
48. Pandian, A.; Sultan, M.T.H.; Marimuthu, U.; Shah, A.U.M. *Low Velocity Impact Studies on Fibre-Reinforced Polymer Composites and Their Hybrids–Review In Encyclopedia of Renewable and Sustainable Materials*; Elsevier: Oxford, UK, 2019; pp. 1–12.
49. Ismail, M.F.; Sultan, M.T.; Hamdan, A.; Shah, A.U.; Jawaid, M. Low velocity impact behaviour and post-impact characteristics of kenaf/glass hybrid composites with various weight ratios. *J. Mater. Res. Technol.* **2019**, *8*, 2662–2673. [CrossRef]
50. Razali, N.; Sultan, M.T.; Mustapha, F.; Yidris, N.; Ishak, M.R. Impact damage on composite structures—A review. *Int. J. Eng. Sci.* **2014**, *3*, 8–20.
51. Maimi, P.; Rodríguez, H.; Blanco, N.; Mayugo, J.A. Numerical modeling of matrix cracking and intralaminar failure in advanced composite materials. In *Numerical Modelling of Failure in Advanced Composite Materials*; Woodhead Publishing: Cambriadge, UK, 2015; pp. 175–192.
52. Boria, S.; Scattina, A.; Belingardi, G. Impact behavior of a fully thermoplastic composite. *Compos. Struct.* **2017**, *167*, 63–75. [CrossRef]
53. Tan, R.; Xu, J.; Sun, W.; Liu, Z.; Guan, Z.; Guo, X. Relationship Between Matrix Cracking and Delamination in CFRP Cross-Ply Laminates Subjected to Low Velocity Impact. *Materials* **2019**, *12*, 3990. [CrossRef]
54. Sarasini, F.; Tirillò, J.; D'Altilia, S.; Valente, T.; Santulli, C.; Touchard, F.; Chocinski-Arnault, L.; Mellier, D.; Lampani, L.; Gaudenzi, P. Damage tolerance of carbon/flax hybrid composites subjected to low velocity impact. *Compos. Part B Eng.* **2016**, *91*, 144–153. [CrossRef]
55. Sultan, M.T.; Worden, K.; Staszewski, W.J.; Pierce, S.G.; Dulieu-Barton, J.M.; Hodzic, A. Impact damage detection and quantification in CFRP laminates; a precursor to machine learning. In Proceedings of the 7th International Workshop on Structural Health Monitoring (IWSHM 2009), Stanford, CA, USA, 9–11 September 2009; pp. 1528–1537.
56. Wan Busu, W.N.; Anuar, H.; Ahmad, S.H.; Rasid, R.; Jamal, N.A. The mechanical and physical properties of thermoplastic natural rubber hybrid composites reinforced with Hibiscus cannabinus, L and short glass fiber. *Polym. Plast. Technol. Eng.* **2010**, *49*, 1315–1322. [CrossRef]
57. Salman, S.D.; Sharba, M.J.; Leman, Z.; Sultan, M.T.; Ishak, M.R.; Cardona, F. Physical, mechanical, and morphological properties of woven kenaf/polymer composites produced using a vacuum infusion technique. *Int. J. Polym. Sci.* **2015**, *2015*, 894565. [CrossRef] [PubMed]
58. Taufiq, M.J.; Mansor, M.R.; Mustafa, Z. Characterisation of wood plastic composite manufactured from kenaf fibre reinforced recycled-unused plastic blend. *Compos. Struct.* **2018**, *189*, 510–515. [CrossRef]
59. Ismail, A.S.; Jawaid, M.; Sultan, M.T.; Hassan, A. Physical and Mechanical Properties of Woven Kenaf/Bamboo Fiber Mat Reinforced Epoxy Hybrid Composites. *BioResources* **2019**, *14*, 1390–1404.
60. Yong, C.K.; Ching, Y.C.; Chuah, C.H.; Liou, N.S. Effect of fiber orientation on mechanical properties of kenaf-reinforced polymer composite. *BioResources* **2015**, *10*, 2597–2608. [CrossRef]
61. El-Shekeil, Y.A.; Salit, M.S.; Abdan, K.; Zainudin, E.S. Development of a new kenaf bast fiber-reinforced thermoplastic polyurethane composite. *BioResources* **2011**, *6*, 4662–4672.
62. Khalil, H.A.; Suraya, N.L.; Atiqah, N.; Jawaid, M.; Hassan, A. Mechanical and thermal properties of chemical treated kenaf fibres reinforced polyester composites. *J. Compos. Mater.* **2013**, *47*, 3343–3350. [CrossRef]
63. Hamidon, M.H.; Sultan, M.T.; Ariffin, A.H.; Shah, A.U. Effects of fibre treatment on mechanical properties of kenaf fibre reinforced composites: A review. *J. Mater. Res. Technol.* **2019**, *8*, 3327–3337. [CrossRef]
64. Noor Azammi, A.M.; Sapuan, S.M.; Ishak, M.R.; Sultan, M.T. Mechanical properties of kenaf fiber thermoplastic polyurethane-natural rubber composites. *Polimery* **2018**, *63*, 524–530. [CrossRef]
65. Alavudeen, A.; Rajini, N.; Karthikeyan, S.; Thiruchitrambalam, M.; Venkateshwaren, N. Mechanical properties of banana/kenaf fiber-reinforced hybrid polyester composites: Effect of woven fabric and random orientation. *Mater. Des.* **2015**, *66*, 246–257. [CrossRef]
66. Islam, M.R.; Beg, M.D.; Gupta, A. Characterization of laccase-treated kenaf fibre reinforced recycled polypropylene composites. *BioResources* **2013**, *8*, 3753–3770. [CrossRef]

67. Salman, S.D.; Leman, Z.; Sultan, M.T.; Ishak, M.R.; Cardona, F. Kenaf/synthetic and Kevlar®/cellulosic fiber-reinforced hybrid composites: A review. *BioResources* **2015**, *10*, 8580–8603. [CrossRef]
68. Yahaya, R.; Sapuan, S.M.; Jawaid, M.; Leman, Z.; Zainudin, E.S. Effect of layering sequence and chemical treatment on the mechanical properties of woven kenaf–aramid hybrid laminated composites. *Mater. Des.* **2015**, *67*, 173–179. [CrossRef]
69. Mostafa, N.H.; Ismarrubie, Z.N.; Sapuan, S.M.; Sultan, M.T. Effect of equi-biaxially fabric prestressing on the tensile performance of woven E-glass/polyester reinforced composites. *J. Reinf. Plast. Compos.* **2016**, *35*, 1093–1103. [CrossRef]
70. Heckadka, S.S.; Nayak, S.Y.; Narang, K.; Vardhan Pant, K. Chopped strand/plain weave E-glass as reinforcement in vacuum bagged epoxy composites. *J. Mater.* **2015**, *2015*, 957043. [CrossRef]
71. Thirukumaran, M.; Kannan, L.V.; Sankar, I. Study on mechanical properties of glass wool/epoxy reinforced composite. *Int. J. Comput. Aided Eng. Technol.* **2018**, *10*, 15–25. [CrossRef]
72. Salman, S.D.; Leman, Z.; Sultan, M.T.; Ishak, M.R.; Cardona, F. Effect of kenaf fibers on trauma penetration depth and ballistic impact resistance for laminated composites. *Text. Res. J.* **2017**, *87*, 2051–2065. [CrossRef]
73. Davoodi, M.M.; Sapuan, S.M.; Ahmad, D.; Aidy, A.; Khalina, A.; Jonoobi, M. Effect of polybutylene terephthalate (PBT) on impact property improvement of hybrid kenaf/glass epoxy composite. *Mater. Lett.* **2012**, *67*, 5–7. [CrossRef]
74. Mohammed, M.; Rozyanty, A.R.; Beta, B.O.; Adam, T.; Osman, A.F.; Salem, I.A.; Dahham, O.S.; Al-Samarrai, M.N.; Mohammed, A.M. Influence of weathering effect in natural environment on thermal properties hybrid kenaf blast/glass fibre and unsaturated polyester composite. *J. Phys. Conf. Ser.* **2017**, *908*, 012004. [CrossRef]
75. Ramnath, B.V.; Rajesh, S.; Elanchezhian, C.; Shankar, A.S.; Pandian, S.P.; Vickneshwaran, S.; Rajan, R.S. Investigation on mechanical behaviour of twisted natural fiber hybrid composite fabricated by vacuum assisted compression molding technique. *Fibers Polym.* **2016**, *17*, 80–87. [CrossRef]
76. Jamal, S.K.; Hassan, S.A.; Wong, K.J.; Yahya, M.Y. Flexural and interlaminar shear study of hybrid woven kenaf/recycled GFRP (rGFRP) composites subjected to bending load. *Int. J. Adv. Appl. Sci.* **2017**, *4*, 45–50. [CrossRef]
77. Elanchezhian, C.; Ramnath, B.V.; Kaosik, R.; Nellaiappan, T.K.; Santhosh Kumar, K.; Kavirajan, P.; Sughan, M.U. Evaluation of Mechanical Properties of Kenaf Based Hybrid Composite for Automotive Components Replacement. *ARPN J. Eng. Appl. Sci.* **2006**, *10*, 5518–5523.
78. Asumani, O.M.; Reid, R.G.; Paskaramoorthy, R. The effects of alkali–silane treatment on the tensile and flexural properties of short fibre non-woven kenaf reinforced polypropylene composites. *Compos. Part A: Appl. Sci. Manuf.* **2012**, *43*, 1431–1440. [CrossRef]
79. Majid, D.L.; Jamal, Q.M.; Manan, N.H. Low-velocity impact performance of glass fiber, kenaf fiber, and hybrid glass/kenaf fiber reinforced epoxy composite laminates. *BioResources* **2018**, *13*, 8839–8852. [CrossRef]

polymers

MDPI

Review

Fiber Reinforced Polymer Laminates for Strengthening of RC Slabs against Punching Shear: A Review

Osama Ahmed Mohamed *, Manish Kewalramani and Rania Khattab

College of Engineering, Abu Dhabi University, Abu Dhabi 59911, UAE; manish.kewalramani@adu.ac.ae (M.K.); r.nabil-adjunct@adu.ac.ae (R.K.)

* Correspondence: osama.mohamed@adu.ac.ae; Tel.: +971-50-1800-767

Received: 27 February 2020; Accepted: 16 March 2020; Published: 19 March 2020

Abstract: Reinforced concrete flat slabs or flat plates continue to be among the most popular floor systems due to speed of construction and inherent flexibility it offers in relation to locations of partitions. However, flat slab/plate floor systems that are deficient in two-way shear strength are susceptible to brittle failure at a slab–column junction that may propagate and lead to progressive collapse of a larger segment of the structural system. Deficiency in two-way shear strength may be due to design/construction errors, material under-strength, or overload. Fiber reinforced polymer (FRP) composite laminates in the form of sheets and/or strips are used in structurally deficient flat slab systems to enhance the two-way shear capacity, flexural strength, stiffness, and ductility. Glass FRP (GFRP) has been used successfully but carbon FRP (CFRP) sheets/strips/laminates are more commonly used as a practical alternative to other expensive and/or challenging methods such column enlargement. This article reviews the literature on the methodology and effectiveness of utilizing FRP sheets/strips and laminates at the column/slab intersection to enhance punching shear strength of flat slabs.

Keywords: flat slab; two-way shear; carbon fiber reinforced polymers; glass fiber reinforced polymers

1. Introduction

Reinforced concrete slabs that are not supported by beams, also known as flat slabs are the most popular floor system in the building construction industry due to architectural flexibility and speed of construction. Speed of constructing flat slab systems is owed primarily to savings in time that would have been needed to build formwork for supporting beams. They also offer designers the flexibility of placing heavy and light partitions anywhere on the floor slab without abiding by location of beams. However, flat slabs are susceptible to brittle two-way shear failure (i.e., punching shear failure) at the slab–column connection that is caused by the transfer of shear and unbalanced moments [1]. Although the mechanics of punching shear is not completely understood, many methods have been developed over the years to prevent this type of failure. Four basic stages in the punching failure of a slab–column connection are generally recognized. Firstly, flexural and shear cracks form in the tension zone of the slab near the face of the loaded area. Secondly, the slab tension steel close to the loaded area yields. Thirdly, flexural and shear cracks extend into what was the compression zone of the concrete. Finally, failure occurs before yielding of reinforcing steel extends beyond the vicinity of the loaded area. A possible reason for punching failure is the rupture of the reduced compression zone in the slab [2]. The failed surface forms a conical shape enclosed by an inclined critical crack pattern which meets a horizontal crack parallel to the reinforcing steel reaching the tension side [3]. When a slab–column connection is loaded, the initial response is linear-elastic, followed by cracking which reduces the connection stiffness. The deflected profile in the slab compression region can be considered as straight lines, while that in the tension region shows a slight discontinuity, especially when the shear crack intersects the reinforcement. Moe [4] tested 43 slabs and investigated the results of 140 footings in addition to 120 slabs that are reported in the literature and noted the common appearance of inclined cracks at 60% of the ultimate load. These inclined cracks started from bending cracks, then rapidly extended up to the proximity of the neutral axis and finally developed rather slowly but leaving only a very narrow depth of the compression zone unaffected. During the structural design phase, if the shear capacity is exceeded, shear reinforcement can be designed to enhance the capacity. Some studies indicated that using shear reinforcement in two-way slabs may double the punching shear strength [5].

Strengthening flat slabs at the column–slab juncture using externally applied material with high tensile strength did not start with the fiber reinforced polymer (FRP) sheets. Steel plates with steel anchor bolts were used successfully to increase the two-way shear strength of flat slabs. Ebead and Marzouk [6] strengthened slab–column junctions by installing on the tension side 6 mm thick American society for testing and materials (ASTM) A6 plates using 19 mm diameter ASTM A325 bolts. Steel plates increased the ultimate load of the flat slab by 45% when the applied load was central and by 122% when the specimens were subjected to both central loads and moments. The light weight, flexibility, and high tensile strength of FRP sheets and laminates make them viable alternative to steel plates. Externally bonded carbon fiber reinforced polymer (CFRP) sheets that are anchored to concrete at their ends were shown to increase punching shear strength, especially when applied skewed relative to the column orientation [7]. Flexural capacity of concrete flat slab can also be increased by bonding CFRP sheets, strips, or laminates to the tension side [8,9].

Steel reinforcing bars are generally used to carry flexural tensile stresses on tension side of reinforced concrete members. Studies by Rasha et al. [9] concluded that the flexural tensile reinforcement above the column in flat slab contributes to the punching capacity of the slab. Yield strength of flexural reinforcement is amongst the factors determining the necessary amount of tension steel. Nonetheless, Dilger et al. [10] noted that no conclusive evidence exists to indicate any effect of yield strength on punching shear capacity.

Experimental studies by Ebead and Marzouk [6] on centrally loaded 1.9 m × 1.9 m × 0.25 m flat slab specimens found that the ultimate load-carrying capacity in punching shear increased compared to reference unreinforced flat slab sample by 54% when the flexural reinforcement ratio over the column was 1%. Similarly, the ultimate load for punching shear increased by 36.5% compared to the unreinforced specimen when the flexural tensile capacity was 0.5%. When the area surrounding the column, where a central load is applied, was strengthened by steel plates using bolts, the stiffness of the strengthened flat slab specimens increased by 105% compared to the un-strengthened specimens. In addition, studies by Caldentey et al. [11] showed a reduction in punching shear strength when flexural reinforcement does not pass through the slab–column intersection, compared to similar flat slabs with flexural reinforcement passing through the intersection.

McHarg et al. [12] studied the effect of flexural top reinforcement distribution on punching shear capacity. The investigators noted a 14% increase in punching shear capacity when a portion of the top flexural steel is banded over the column compared to uniform distribution of reinforcing steel.

Perhaps one of the earliest studies that suggested punching shear strength in flat slabs is proportional to the cubic root of the compressive strength was presented in the 1956 in article by Elstner and Hognestad [13]. Inácio et al. [14] found that the punching shear capacity increases by 43% when high-strength concrete is used compared to normal-strength concrete, and failure is more brittle when compared to normal strength capacity.

Tests by Alexander and Simmonds [15] showed that increasing concrete cover thickness of top flexural reinforcement at interior slab–column connections increases two-way shear capacity by a small margin compared to smaller cover. However, larger cover makes behavior less stiff compared to a slab with smaller concrete covers which suffered larger bond deformation.

One of the early studies on the effect of column aspect ratio was conducted by Hawkins et al. [16]. The study concluded that the punching shear strength decreases markedly as the column aspect ratio increases beyond 2.0.

2. Punching Shear Strength in Selected Codes and Standards

Before examining the effect of FRP sheets and strips on two-way shear strength of concrete, it is useful to evaluate the inherent two-way strength of concrete as described in building codes and design standards. International codes and standards vary in their models for estimating punching shear capacity. In the following Table 1 the models of punching shear capacity of concrete in American Concrete Institute code ACI 318 [17] and Euro code 2 (EC 2) [18] are briefly compared and discussed.

Table 1. Comparison of two-way shear capacity in American Concrete Institute code ACI 318 and Eurocode 2 (EC 2).

Criteria	ACI 318-14	Eurocode 2 (EC2)
Two-way shear strength model	The least of, $0.33\lambda\sqrt{f_c'}$ (MPa) (a) $0.17\left(1+\frac{2}{\beta_0}\right)\lambda\sqrt{f_c'}$ (MPa) (b) $0.083\left(2+\frac{\alpha_s d}{b_0}\right)\lambda\sqrt{f_c'}$ (MPa) (c) b_0 is the critical perimeter for punching shear (mm). f_c' is the concrete compressive strength (MPA). Equation (a) represents the maximum punching shear capacity achievable without shear reinforcement. $\alpha_s = \begin{cases} 40 \text{ Interior columns} \\ 30 \text{ edge columns} \\ 20 \text{ corner columns} \end{cases}$	$0.18\xi\left(100\rho f_c'\right)^{1/3}$ ρ is the flexural reinforcement ratio, f_c' is the concrete compressive strength
Location of critical perimeter for punching shear	d/2 (d = effective depth, mm)	2d (d = effective depth, mm)
Effect of flexural reinforcement on punching shear capacity	Minimum flexural steel must be provided near tension face of the slab in two-perpendicular directions ($\rho_{min} = 0.18\%$, for grade 420 MPa reinforcement) (Section 8.6 [19] and commentary). Integrity reinforcement in column strip is required (8.7.4.2) to provide residual punching shear strength following punching shear failure. A minimum of two bottom bars shall pass in two directions through the region bounded by column vertical reinforcement and be anchored at exterior support.	Incorporated in punching shear strength formula. Clearly EC2 recognizes significant increase in punching shear strength in heavily reinforced slabs.
Effect of slab thickness on punching shear strength	It is incorporated in equation "c", indicated punching shear capacity increases with increase in the ratio of effective depth to critical perimeter $0.083\left(2+\frac{\alpha_s d}{b_0}\right)\lambda\sqrt{f_c'}$ (MPa) Implicitly addressed in the commentary by indicating that the minimum tension steel development length beyond column face on each side for slabs under predominantly uniform gravity load with clear span to depth ratio (l_n/h) at less than 15. For thicker slabs with (l_n/h), provide continuous tension steel to intercept punching shear cracks.	However, size effect parameter, limited by upper value of 2.0 is. $\xi = 1+\sqrt{\frac{200\text{ mm}}{d}}$
Effect of supporting member dimensions on punching shear strength	Incorporated in the parameter β, which represents the ratio of the long support side to the short side. Upper limit on punching shear strength is $0.33\lambda\sqrt{f_c'}$ (MPa) around corners of columns and decreases to $0.17\lambda\sqrt{f_c'}$ or less along long sides of the support between the two end sections	Not explicitly indicated in the capacity expression.

As indicated in Table 1, ACI 318 requires integrity longitudinal reinforcing steel to provide residual capacity in the event of loss or damage of vertical load-carrying member. A study by Weng et al. [20] showed that integrity reinforcement plays a significant role in post-punching behavior of flat slabs and recommends a minimum of 0.63% integrity reinforcement ratio to be provided. A typical concern in relation to punching shear failure in flat plates is that its brittle nature may lead to progressive collapse of large segment of the structural system. Mohamed et al. [21] discussed effect of two-way shear on triggering progressive collapse, along with recommended design/detailing measures to enhance the post-failure behavior of the structural system.

Additionally, Table 2 demonstrates the punching shear capacity of concrete slabs as described by American Concrete Institute code ACI 318 [17], the Canadian code CSA-S806-12 [22], the British standard BS-8110 [23], the European code Eurocode 2 [18] (including shear reinforcement), and Japanese standard JSCE [24].

Table 2. International code equations for design punching shear.

Code	Critical Section Location	Equations
ACI318-14 [17]	0.5 d	The least of, $0.33\lambda\sqrt{f_c}$ (MPa) (a) $0.17\left(1+\frac{2}{\beta_0}\right)\lambda\sqrt{f_c}$ (MPa) (b) $0.083\left(2+\frac{\alpha_s d}{b_0}\right)\lambda\sqrt{f_c}$ (MPa) (c)
CSA-S806-12 [22]	0.5 d	$V_c = min\begin{cases} 0.028\left(1+\frac{2}{\beta_c}\right)\lambda\varnothing_c\left(E_f\,\rho_f\,f'_c\right)^{\frac{1}{3}}\,b_{0.5}d \\ 0.147\left(0.19+\alpha_s\frac{d}{b_{0.5}}\right)\lambda\varnothing_c\left(E_f\,\rho_f\,f'_c\right)^{\frac{1}{3}}\,b_{0.5}d \\ 0.056\lambda\varnothing_c\left(E_f\,\rho_f\,f'_c\right)^{\frac{1}{3}}\,b_{0.5}d \end{cases}$ βc: is the ratio of the long side to short side of the column $\lambda = 1.00\ \varnothing_c = 0.65$ $b_{0.5}$: is the length of the critical shear perimeter $\alpha_s = \begin{cases} 4 \text{ Interior columns} \\ 3 \text{ edge columns} \\ 2 \text{ corner columns} \end{cases}$
BS-8110 [23]	1.5 d	$V_{cu} = 0.79\left(100\rho_f\frac{E_f}{E_s}\right)^{1/3}\left(\frac{400}{d}\right)^{1/4}\left(\frac{f_{ck}}{25}\right)^{1/3}b_{1.5}d$ $b_{1.5}$: is the length of the critical shear perimeter
Euro Code 2 [18]	2 d	$V_{Rd} = min\begin{cases} V_{Rdcs} = 0.75V_{Rdc} + A_{sw}\,\sigma_{wd}(1.5^{d}/_{s_r}) \leq 1.5 \\ V_{Rdout} = \frac{0.18}{\gamma_c}k\left(100\rho_f f_{ck}\right)^{\frac{1}{3}}u_{out}d \\ V_{Rdmax} = \frac{0.24}{\gamma_c}f_{ck}(1-f_{ck}/250)u_0 d \end{cases}$ $\sigma_{wd} = \begin{cases} (250+0.25d) \leq \frac{f_{yw}}{\gamma_s} & \text{(Steel Connectors)} \\ \frac{0.004E_{CFRP}}{\gamma_{CFRP}} \leq \frac{0.75\,\varepsilon_{fu}E_{CFRP}}{\gamma_{CFRP}} & \text{(CFRP)} \end{cases}$ $\gamma_s = 1.30$ [Value suggested by the authors in the absence of experimental calidation for each type of strengthening] $\gamma_{CFRP} = \begin{cases} 1.20 \text{ for applications with high degree of quality control} \\ 1.35 \text{ for applications with normal degree of quality control or under difficult on - site working conditions} \end{cases}$
JSCE [24]	0.5 d	$V_{pod} = \beta_d\beta_\rho\beta_r f_{pcd}b_w\frac{d}{\gamma_b}$ $f_{pcd} = \min\begin{cases} 0.2\sqrt{f_{cd}} \\ 1.2 \text{ MPa} \end{cases}$ $\beta_d = \min\begin{cases} 1.50 \\ \sqrt[4]{1000/d} \end{cases}$ $\beta_\rho = \min\begin{cases} 1.50 \\ \sqrt[3]{\frac{100\,\rho_f.\,E_f}{E_s}} \end{cases}$ $\beta_r = 1 + \frac{1}{(1+0.25\,u/d)}$ b_w: is the length of the critical shear perimeter $\gamma_b = 1.30$

3. Experimental Specimens, FRP Strengthening Patterns, and Methods

The following section includes discussion on details of experimental program, various FRP strengthening patterns and methods and some significant research findings. Binici and Bayrak [25] used 25-mm wide vertical CFRP strips for strengthening the flat slab specimens (2133 mm × 2133 mm × 152 mm) against punching shear. Strengthening of flat slab is achieved by punching 18-mm diameter vertical holes around the loading area. A practical downside to the method is the potential of cutting tension and/or integrity steel around the column which could contribute to failure. Two specimens had four holes in a line extending from each side of the loading area, one specimen had six holes on each side of the loading area, and one specimen had eight holes extending from each side of the loading area. The holes started at distance equals to $\frac{1}{4}$ of the effective depth (d/4) from the face of the loaded area and spaced at half of the effective depth (d/2) center-to-center. The authors clearly observed that that the use of diagonal CFRP strips between the holes containing the vertical strips as seen in Figure 1A1–A3 around the loading area is effective in preventing punching shear inside the shear-strengthened area compared to pattern B, with orthogonal strips. Extending CFRP shear-strengthening further along with diagonal strips to join vertical strips (A3) increases load-carrying capacity 1.51 times and the ductility 2.0 times the unstrengthening control specimens. In addition, extending the shear strengthening further from the loading area (A2 and A3), for the specimens with diagonal CFRP strips was effective in changing the failure mode from brittle punching to a more ductile failure mode.

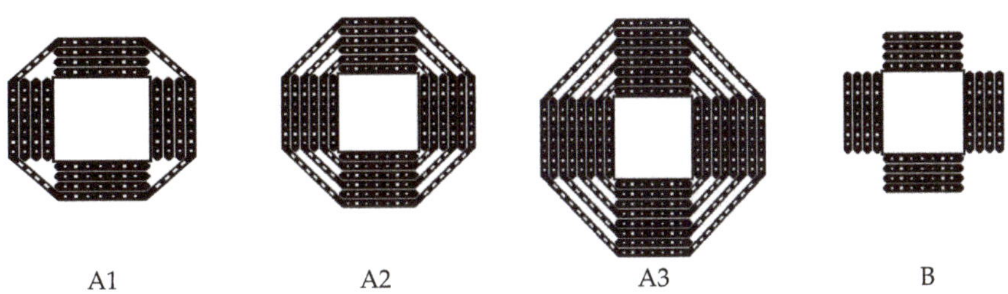

Figure 1. Carbon fiber reinforced polymer (CFRP) pattern A with diagonal strips provide higher two-way shear capacity compared to pattern B with orthogonal strips [25].

Applying CFRP vertically in the form of shear reinforcement is superior to bonding CFRP strips/laminate on the tension side in terms of enhancing punching shear strength. Sissakis and Sheikh [26] reported 80% increases in punching shear capacity by applying CFRP strips vertically through holes created at constant spacing perpendicular to the loading column in a manner similar to stitching of the slab.

Harajli and Soudki [27] demonstrated that providing reinforcing CFRP strips on the tension face of the slab increased the punching shear strength ranged from 17% to 45% compared to the control flat slab specimens that did not contain CFRP strips for strengthening purpose as seen in Figure 2. The mechanism by which this method of strengthening increases two-way shear strength is by restricting the growth of tension cracks or by increasing flexural strength at the connection. Therefore, strengthening the slab on the tension slide in the column zone may change flexural failure mode to flexure-shear failure or pure two-way shear failure. The authors tested 670 mm square specimens with tension steel reinforcement ranging from 1% to 2%. The loading column is 100 mm × 100 mm which was cast monolithic with the slab. Four CFRP strip widths (50, 100, 150, and 200 mm) were used. Using strips on the tension side only did not affect the punching shear cracking extent and remained at a distance 2.0 × h (h = overall slab depth) from the face of the loading column.

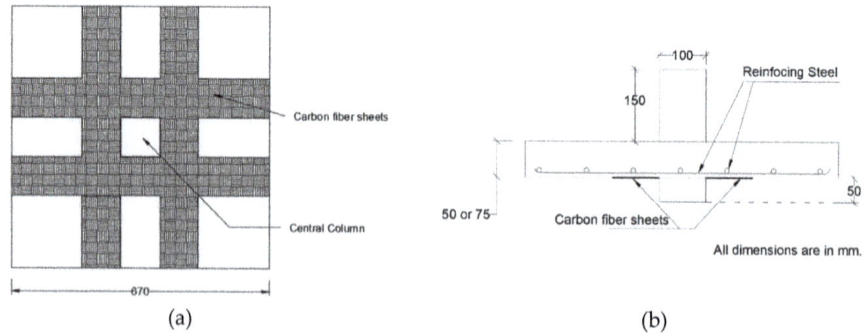

Figure 2. Test specimens for orthogonal CFRP strengthening strips, (**a**) CFRP strip pattern applied to tension side and (**b**) loading column and slab specimen dimensions [27].

El-Salakawy et al. [28] concluded that applying CFRP or glass FRP (GFRP) strips on the tension surface of flat slab around edge columns increases the flexural stiffness and delays opening of flexural cracks, and hence increases punching shear capacity. Depending on the area of FRP sheets and configuration, the increase of edge column punching shear strength ranges from 2% to 23% compared to unstrengthen specimens. The authors examined effectiveness of strengthening edge columns in flat slab systems. Specimens were 1540 mm × 1020 mm × 120 mm thick and the supporting edge column was 250 mm × 250 mm. The slabs were reinforced with an average of 0.75% tension steel in each direction and an average of 0.45% compression steel in each direction. L-shaped strips were used perpendicular to the free edge while straight strips were applied on the tension side parallel the main reinforcement in each direction. FRP strip width was 50 mm when CFRP was used for strengthening while 75 width strips were used when GFRP was used for strengthening as seen in Figure 3.

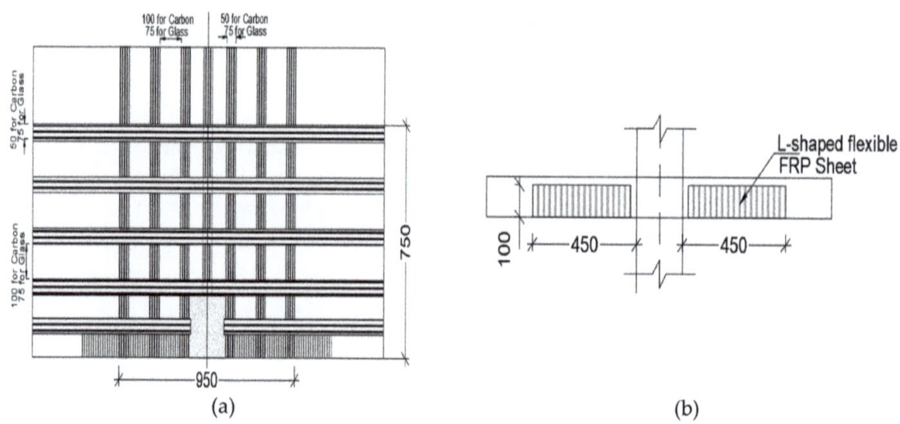

Figure 3. Orthogonal CFRP or glass fiber reinforced polymer (GFRP) strengthening strips for flat slab specimen supported by the edge column: (**a**) extend of strip application and (**b**) extension of strip into L-shape at the end of the slab [28].

A study by Chen and Li [29] confirms that GFRP strengthening sheet increases the punching shear capacity by acting as an external flexural reinforcement. The percentage increase in punching shear capacity is higher in lower grade concrete (16.9 MPa) than in concrete of normal strength (34.4 MPa). However, GFRP sheets change the failure mode from flexure-punching failure to brittle punching failure. This study included 1000 mm × 1000 mm × 150 mm specimens loaded centrally with 150 mm square column stub to simulate interior columns in flat slabs. The flexural tension steel reinforcement ratio was 0.59% in one direction and 1.31% in the perpendicular direction. The GFRP laminate in the form of fabric is applied as one layer (1.31 mm thick) or two layers (1.93 mm thick). Using one layer of GFRP laminate increased the ultimate load from 17% to 45% while using two layers increased the ultimate, depending on the concrete grade and reinforcement ratio. For identical conditions, using two layers of laminates always increased the ultimate load.

While the failure mode in punching shear created by the GFRP sheets is consistent with published literature reviewed in this paper, the investigators covered the area under the loading column with GFRP fabric, which is not typical for building structures, except for columns supporting the roof. For this study, the authors concluded that the two-way shear prediction formulas proposed by BS 8110 and JSCE produce consistent results with their experimental study while ACI 318 was found to be more conservative. The GFRP strengthening pattern used in the study is shown in Figure 4. Liberati et al. [30] attributed the conservatism in ACI318 prediction of punching shear capacity to the fact that ACI318 does not explicitly consider the positive contribution of flexural reinforcement.

Sharaf et al. [31] observed that strengthening flat slabs in the tension zone around column using externally bonded CFRP strips increases stiffness from 29% to 60% compared the reference unreinforced flat slab depending on amount of reinforcement. Similarly tension-side stiffening around the column increases punching shear strength from 6% to 16% depending on the area and pattern of the strengthening strips. CFRP strips, in general, delayed the initiation of flexural cracks and controlled their propagation. Tested slabs demonstrated failure in punching shear predominantly. Researchers also studied the effect of orthogonal application of CFRP strips (patterns O4 and O8), skewed application of strips (patterns S4 and S8) and combination of skewed and orthogonal strips (OS8) as shown in Figure 5.

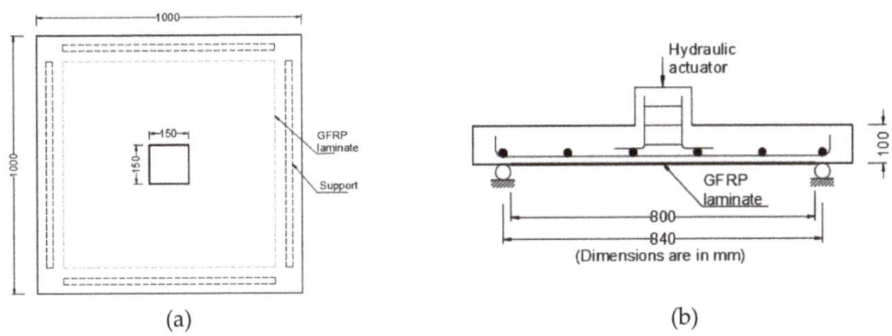

Figure 4. GFRP laminate applied to tension side of 1000 mm × 1000 mm specimen, (**a**) specimen and loading column dimension and (**b**) loading applied causing tension at the bottom with GFRP [29].

All strips had the same width of 100 mm, therefore the effect of CFRP strip area was studied by testing specimens reinforced with either four strips (O4 and S4) or eight strips (O8, S8, and OS8). Specimens with skewed pattern and higher CFRP strip area (S8 and OS8) carried the highest ultimate load compared to orthogonal strips.

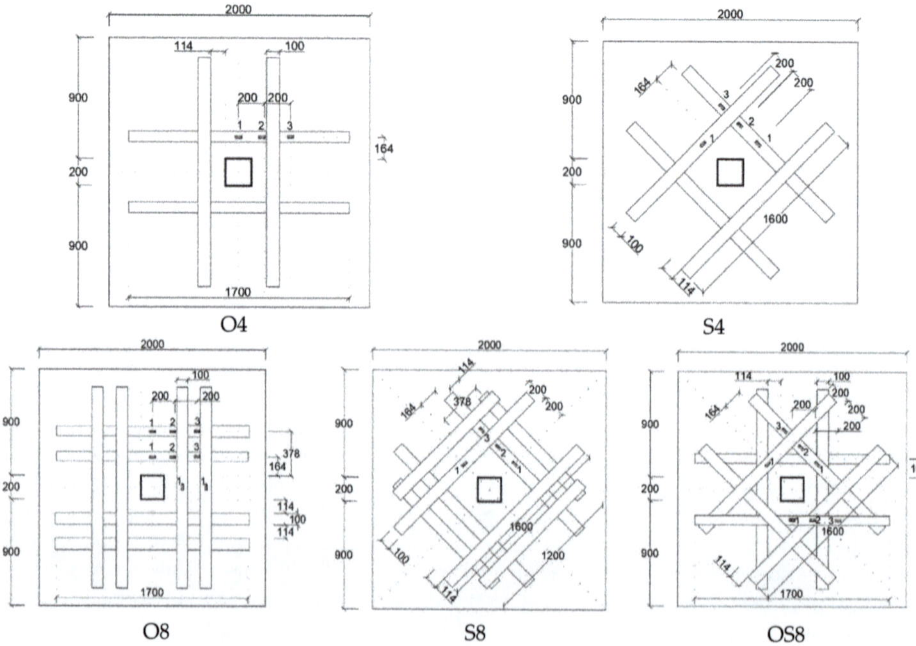

Figure 5. CFRP strips applied to tension side orthogonal to column orientation (O4 and O8), skewed with respect to column orientation (S4 and S8), or both orthogonal and skewed (OS8) [31].

Esfahani [32] compared effectiveness of CFRP strengthening applied on the tension side of flat slab when loading is monotonic versus cyclic. It was observed that CFRP strengthening of flat slabs on the tension side increases the punching shear strength under monotonic loading, but cyclic loading decreases the effectiveness of CFRP strips in resisting punching shear. The effectiveness of strips in enhancing punching shear strength under cyclic load was more pronounced with higher tension steel reinforcement ratio. He tested two sets of 1000 mm × 1000 mm × 100 mm slabs with tension steel reinforcement ratio of 0.84 and 1.59, respectively, as shown in Figure 6.

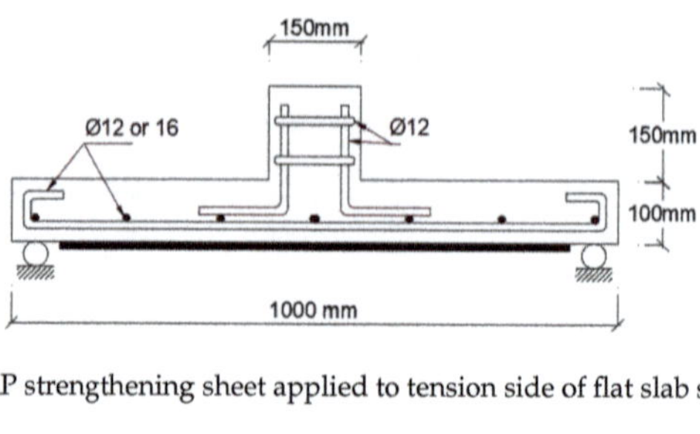

Figure 6. CFRP strengthening sheet applied to tension side of flat slab specimen [32].

Farghaly and Ueda [33] noted that applying 0.167 mm thick CFRP strips to the tension side of flat slab around the loading column increases punching shear strength by 20–40% compared to the unstrengthen flat slab depending on the area of the CFRP strips. Brittle punching shear failure was the dominant mode for the CFRP strengthened flat slabs. They studied 1600 mm × 1600 mm × 120 mm slab specimens with load applied through 100 mm × 100 mm. One set of slabs was strengthened with CFRP strips that are 50 mm wide (SF5) while a second set of slabs was strengthened with 100 mm wide (SF10) CFRP strips as shown in Figure 7. Steel reinforcement ratio was 1.4% designed to generate punching shear failure mode.

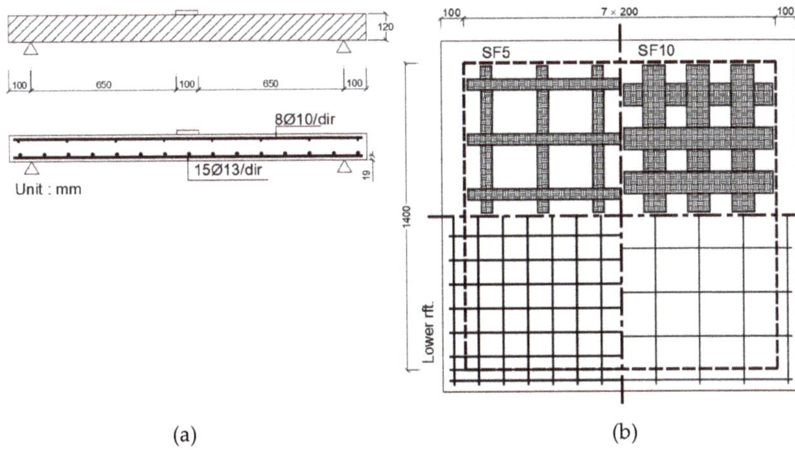

Figure 7. CFRP strips with different widths applied to the tension side of the slab, (**a**) two-way specimen cross section, (**b**) top and bottom reinforcement and two CFRP strip widths tested [33].

Erdogan [34] observed that vertical CFRP strips arranged in pairs perpendicular to each of the four sides of the column provide better resistance to punching shear compared to the arrangement of the vertical CFRP strips in a circular pattern around the column. In the case of perpendicular reinforcement arrangement, the failed surface was outside of the strengthened zone, while circular arrangement led to failure inside the strengthened zone. Furthermore, CFRP strengthening is more effective when the column is square than when the column is rectangular where a reduction in punching shear capacity was noticed. The investigator studied the effect of CFRP vertical strips (referred to as dowels) arranged around the interior column on punching shear strength as shown in Figure 8. The strips are applied vertically in pairs, perpendicular to each column dimension and through the depth of the slab. The pairs are spaced at d/2 (where d is the effective depth of tension steel) to ensure 45 degree cracks are intercepted. Vertical strips are intended to enhance resistance to punching shear by intercepting inclined punching shear cracks. The vertical CFRP strips were installed after concrete casting into vertical 14 mm diameter cylindrical openings. The openings were created by embedding 14 mm × 150 mm polyvinyl chloride (PVC) pipes prior to casting. Pipes were removed after casting.

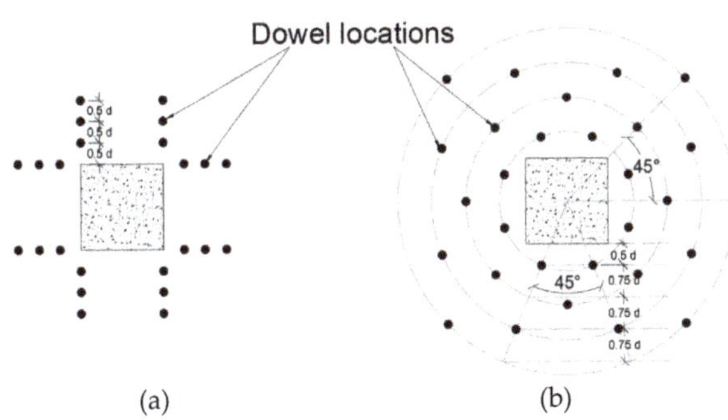

Figure 8. (**a**) Orthogonal "O" pattern of CFRP vertical shear reinforcement and (**b**) Circular "C" pattern of CFRP vertical reinforcement [34].

Erdogan [34] also studied the effect of column aspect ratio, as shown in Figure 9 on punching shear capacity. For columns having the same perimeter, changing the aspect ratio from 1 to 3 does not have significant effect on post-punching capacity. However, increasing the column aspect ratio decreases punching shear capacity. This behavior is recognized in ACI318 through the column aspect ratio β.

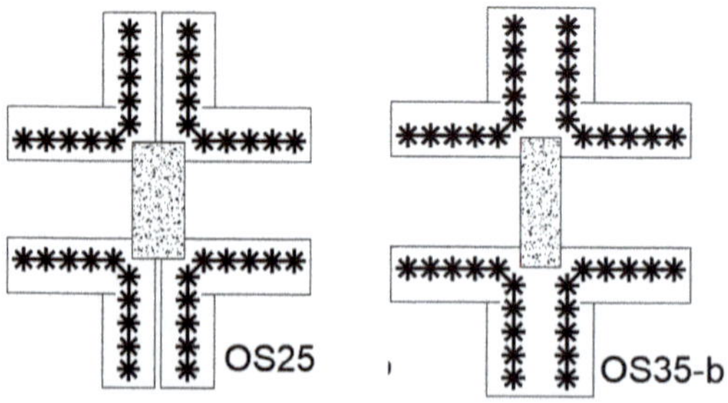

(a) (b)

Figure 9. CFRP vertical reinforcement orthogonal to column orientation five rows with (**a**) column aspect ratio of 2.0 (OS25) and (**b**) a column aspect ratio of 3.0 [34].

Urban and Tarka [35] observed an increase of 11% to 36% in punching shear capacity of flat slabs strengthened with CFRP strips around internal support columns. The percentage increase in punching shear capacity depends on the area of CFRP strips and whether or not anchor bolts were used to enhance bonding of the strips to the slab. The investigators studied 2300 mm × 2300 mm × 180 mm flat slab loaded with square 250 mm columns simulating internal flat slab–column intersection as shown in Figure 10a. The tensile steel reinforcement ratio for all specimens was 0.5%. Other than the control specimen, flat slabs were reinforced with 1.4 mm × 90 mm CFRP strips with varying areas (in terms of number of strips) applied around the loading column as shown in Figure 10. In addition to the adhesive, some samples were provided with additional 10 mm diameter anchoring bolts embedded 100 mm deep into the concrete slab in order to enhance bonding of the CFRP strips to concrete. Specimens strengthened using CFRP strips failed in a brittle explosive manner compared to the control specimen that was not strengthened. However, specimens in which CFRP strips were fixed to the slab using anchor bolts failed more gently than identical specimens without anchor bolts. The authors believe this "softer" failure is due to gradual pulling out of the anchor bolts from concrete as illustrated in Figure 10b.

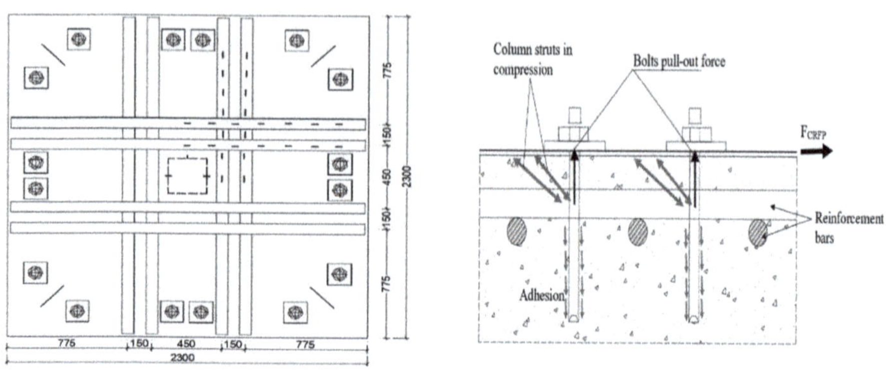

(a) (b)

Figure 10. (**a**) Specimens with 8 strips bonded to the tension side with M10 anchor bolts and (**b**) anchor bolt failure mechanisms [35].

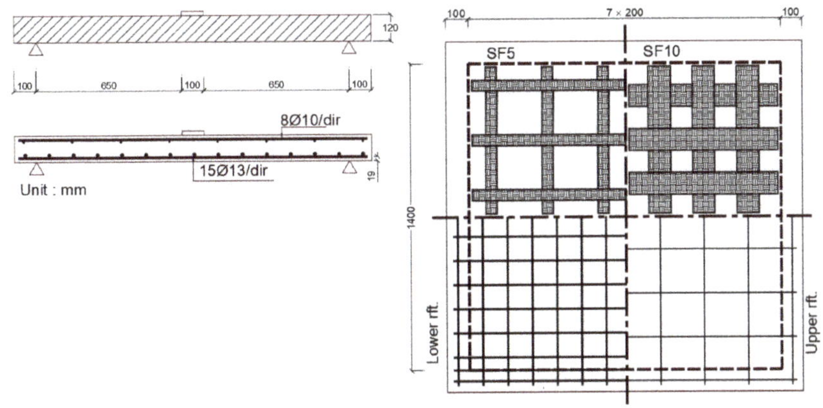

(a) (b)

Figure 7. CFRP strips with different widths applied to the tension side of the slab, (**a**) two-way specimen cross section, (**b**) top and bottom reinforcement and two CFRP strip widths tested [33].

Erdogan [34] observed that vertical CFRP strips arranged in pairs perpendicular to each of the four sides of the column provide better resistance to punching shear compared to the arrangement of the vertical CFRP strips in a circular pattern around the column. In the case of perpendicular reinforcement arrangement, the failed surface was outside of the strengthened zone, while circular arrangement led to failure inside the strengthened zone. Furthermore, CFRP strengthening is more effective when the column is square than when the column is rectangular where a reduction in punching shear capacity was noticed. The investigator studied the effect of CFRP vertical strips (referred to as dowels) arranged around the interior column on punching shear strength as shown in Figure 8. The strips are applied vertically in pairs, perpendicular to each column dimension and through the depth of the slab. The pairs are spaced at d/2 (where d is the effective depth of tension steel) to ensure 45 degree cracks are intercepted. Vertical strips are intended to enhance resistance to punching shear by intercepting inclined punching shear cracks. The vertical CFRP strips were installed after concrete casting into vertical 14 mm diameter cylindrical openings. The openings were created by embedding 14 mm × 150 mm polyvinyl chloride (PVC) pipes prior to casting. Pipes were removed after casting.

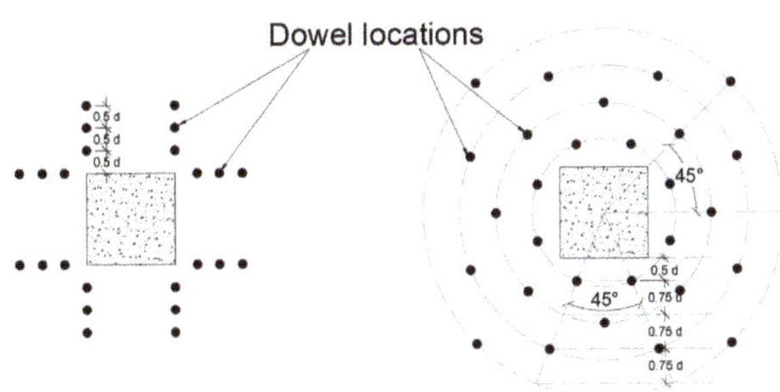

(a) (b)

Figure 8. (**a**) Orthogonal "O" pattern of CFRP vertical shear reinforcement and (**b**) Circular "C" pattern of CFRP vertical reinforcement [34].

Erdogan [34] also studied the effect of column aspect ratio, as shown in Figure 9 on punching shear capacity. For columns having the same perimeter, changing the aspect ratio from 1 to 3 does not have significant effect on post-punching capacity. However, increasing the column aspect ratio decreases punching shear capacity. This behavior is recognized in ACI318 through the column aspect ratio β.

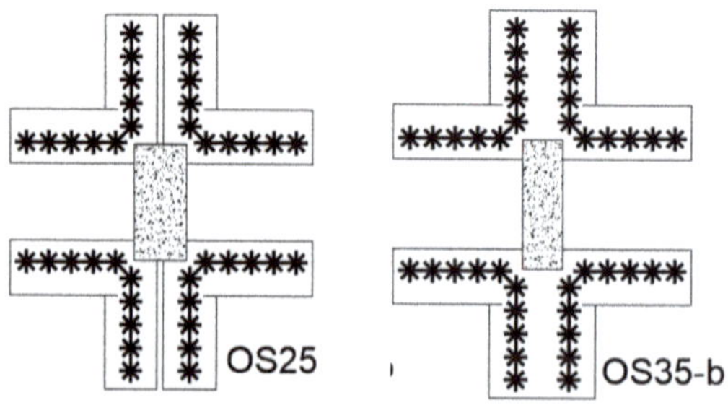

(a) (b)

Figure 9. CFRP vertical reinforcement orthogonal to column orientation five rows with (**a**) column aspect ratio of 2.0 (OS25) and (**b**) a column aspect ratio of 3.0 [34].

Urban and Tarka [35] observed an increase of 11% to 36% in punching shear capacity of flat slabs strengthened with CFRP strips around internal support columns. The percentage increase in punching shear capacity depends on the area of CFRP strips and whether or not anchor bolts were used to enhance bonding of the strips to the slab. The investigators studied 2300 mm × 2300 mm × 180 mm flat slab loaded with square 250 mm columns simulating internal flat slab–column intersection as shown in Figure 10a. The tensile steel reinforcement ratio for all specimens was 0.5%. Other than the control specimen, flat slabs were reinforced with 1.4 mm × 90 mm CFRP strips with varying areas (in terms of number of strips) applied around the loading column as shown in Figure 10. In addition to the adhesive, some samples were provided with additional 10 mm diameter anchoring bolts embedded 100 mm deep into the concrete slab in order to enhance bonding of the CFRP strips to concrete. Specimens strengthened using CFRP strips failed in a brittle explosive manner compared to the control specimen that was not strengthened. However, specimens in which CFRP strips were fixed to the slab using anchor bolts failed more gently than identical specimens without anchor bolts. The authors believe this "softer" failure is due to gradual pulling out of the anchor bolts from concrete as illustrated in Figure 10b.

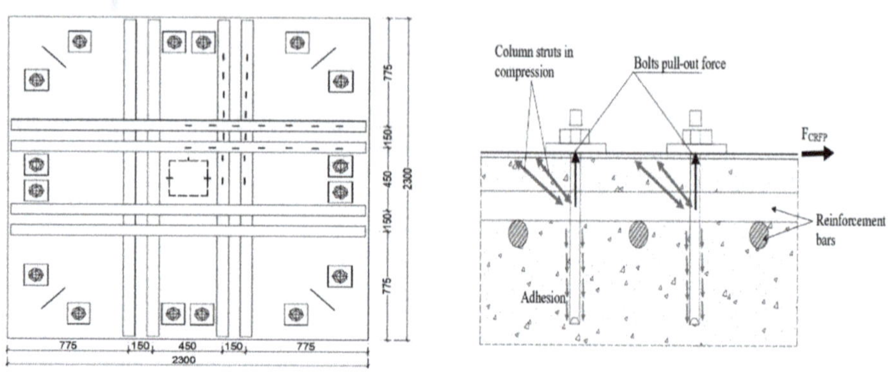

(a) (b)

Figure 10. (**a**) Specimens with 8 strips bonded to the tension side with M10 anchor bolts and (**b**) anchor bolt failure mechanisms [35].

When a smaller area of CFRP strips is applied to the tension side, the effect of using anchor bolts was negligible to crack width development under applied load. However, when a larger area of CFRP strips is used to strengthen the slab, anchor bolts decreased the crack width under the applied load compared to control specimen and identical specimen with CFRP strips without anchor bolts. CFRP strips increase slab stiffness, therefore, deflection of slab strengthened with CFRP strips was much lower than the control unstrengthen flat slab. In addition for the same CFRP area using anchor bolts may or may not decrease slab deflection, depending on the area of the CFRP strips. For a larger CFRP strip area, anchor bolts decreased deflection more significantly, especially for larger applied loads than control slab or CFRP slab without anchor bolts. Thus, anchor bolts have limited or no effect on deflection of slab strengthened with CFRP strips. However, as shown in Figure 11, CFRP strengthened specimens (WT-CF-8, WT-CF-K-8, and WT-CF-K-16) and were all much stiffer than the control (unstrengthend) specimen (s-2), as indicated by the smaller deflections for the same force.

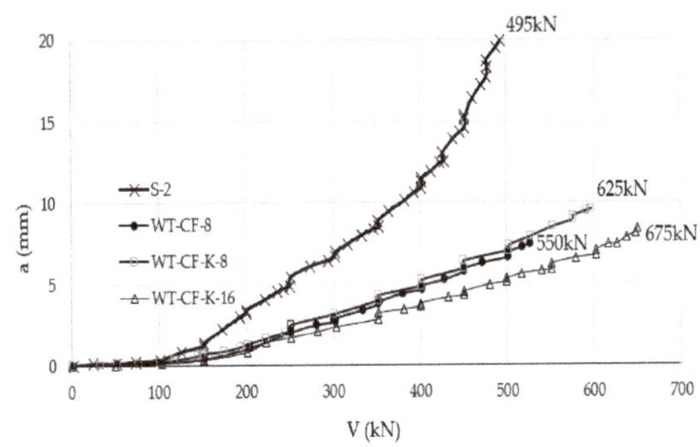

Figure 11. Load–deflection relationship for the control specimen (s), specimen with 8 CFRP strips specimen (WT-CF-8), specimen with 8-CFRP strips and anchor bolts (WT-CF-K-8), specimen with 8-CFRPs in double layers with anchor bolts [35].

The study by Halabi et al. [36] concluded that CFRP strengthening increases the ultimate failure load in one-way spanning flat slabs when the tension-side is strengthened with CFRP strip in the column zone. The investigators concluded that eccentric loading at flat slab–column connection strengthened with CFRP sheet decreases the ductility and ultimate load. Flat slabs designed for flexural failure in negative moment area above supporting column under concentric load, will transform into shear failure occurring at a lower applied load when the failure load is eccentric. The failure load decreases with eccentricity when the slab is strengthened with CFRP or remains un-strengthened.

The researchers studied a total of six 2000 mm × 1000 mm × 150 mm flat slab specimens shown in Figure 12a,b supported to span in one-way to simulate internal and external slab–beam connections. The loading was applied concentrically in some specimens and eccentrically in others through 250 mm square column stubs extending above and below the flat slab specimens as shown in Figure 13a,b. The tensile steel reinforcement ratio was 0.92% in longitudinal as well as transverse directions; on compression side of the flat slab specimens the steel reinforcement ratio was 0.5%. The reinforcement was chosen for flexural failure by steel yielding prior to concrete crushing.

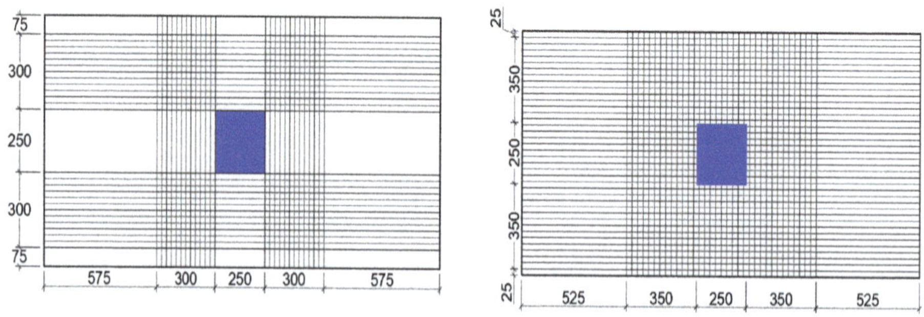

(a) (b)

Figure 12. (**a**) Configuration A with four 300 mm wide CFRP sheets and (**b**) configuration B with CFRP sheet covering entire width of slab in one direction [36].

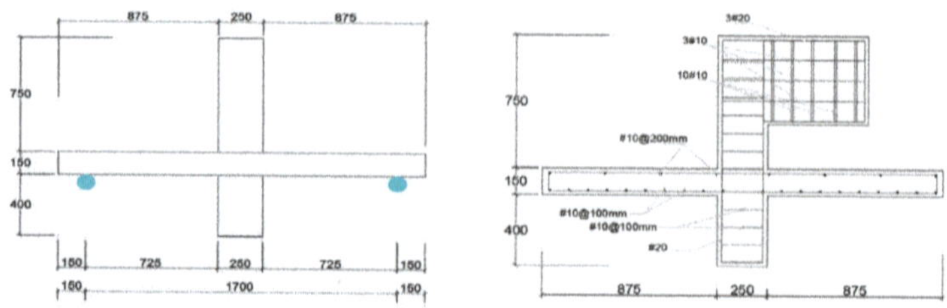

(a) (b)

Figure 13. (**a**) Test specimen for applying concentric gravity load and (**b**) test specimen for applying concentric and eccentric loading [36].

Abbas et al. [37] studied 600 mm × 600 mm × 90 mm flat slabs supported along two parallel sides to span in one-way. The steel reinforcement ratio on the tension face was 0.71%. The CFRP sheet were unidirectional applied to the slab on the tension side parallel to the direction of the span as shown in Figure 14. The loading was applied through a 40-mm diameter steel pipe.

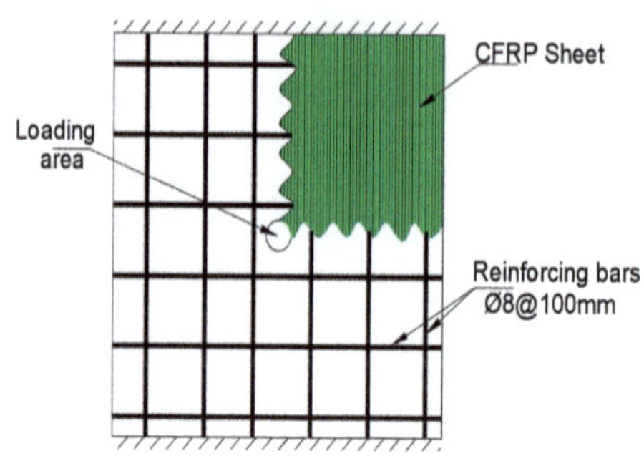

Figure 14. One-way supporting concrete specimen strengthened with CFRP sheet covering the entire specimen except the circular loading area [37].

It was observed that CFRP strengthening of one-way spanning flat slab increases the load-carrying capacity by 12.4% (for 39.9 MPa concrete) and 16.4% (for 63.2 MPa concrete) compared to un-strengthened control slab. However, in this study the investigators noted the presence of two peaks in the load-deflection curve. The first peak is associated with the punching failure of concrete as it loses shear capacity followed by a drop in the curve during which resistance is due to aggregate interlocked and reinforcing steel dowel action as shown in Figure 15.

After cracking the CFRP sheets are engaged in resisting the load leading to second peak in load-carrying capacity. In comparison to the control slab which would have experience significant drop in load-carrying capacity, the second peak of CFRP offers the slab much higher load-carrying capacity 189.6% (for 39.9 MPa concrete) and 275.5% (for 63.2 MPa concrete).

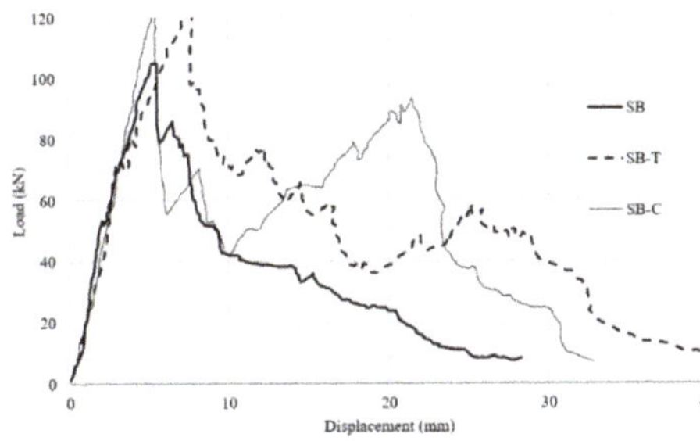

Figure 15. Load–displacement variation for punching of slabs for concrete grade B (SB: control slab; CFRP strengthened slab; and SB-T: TRM strengthened slab) [37].

Radik et al. [38] studied seven 1500 mm × 1500 mm × 152.4 mm flat slabs to evaluate the effect of strengthening on the tension side using 305 mm wide strips applied to the tension side of the slab in two perpendicular directions. Schematic of test apparatus and pattern of GFRP strips bonded to the tension face are shown in Figure 16.

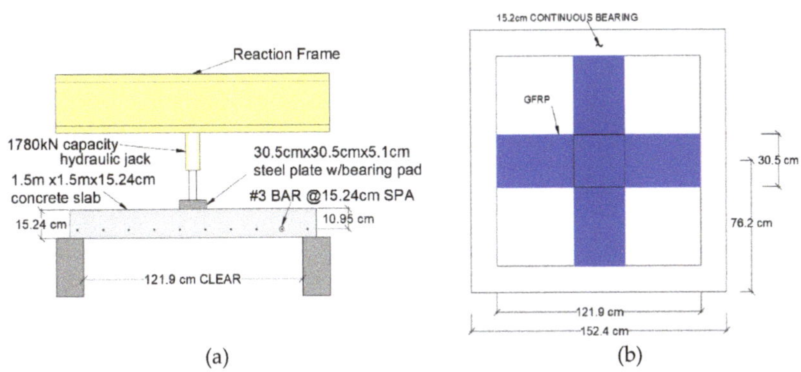

Figure 16. (**a**) Test apparatus with loading from top of the slab producing tension at the bottom, (**b**) 30.4 cm wide GFRP strips bonded to the tension face [38].

GFRP laminates bonded to the tension side increased both the ultimate load and ductility of the flat slab compared to the control specimen as shown in Figure 17.

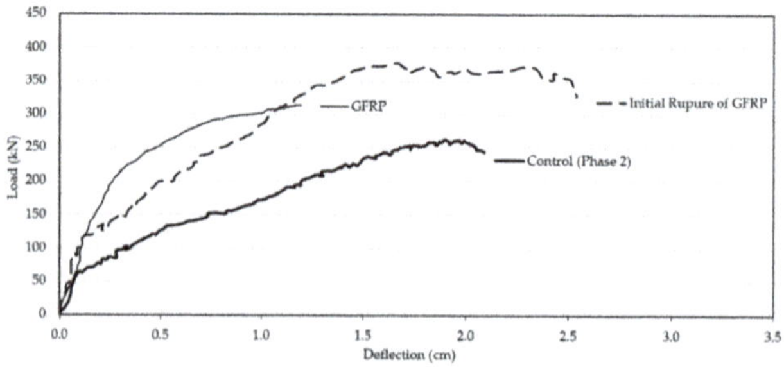

Figure 17. Load versus deflection plots for rehabilitated specimens [38].

Experimental study by Soudki et al. [39] concluded that applied CFRP strips on the tension side around internal columns of flat slab increases the punching shear strength by up to 29% compared to control un-strengthened slab. CFRP strengthened slabs are stiffer, hence, experience less deflections compared to the control slab. Applying strengthening strips near the column increased stiffness of the slab while applying the same strips offset from the column increased punching shear capacity. The most efficient configuration is skewed strips offset further from the column. This investigation was performed on 1220 mm × 1220 mm × 100 mm flat slab specimens made of 25.8 MPa concrete. The effect of CFRP was studied by reinforcing selected specimens using 100 mm wide × 1.2 mm thick strips applied on the tension side around the loading column in various configurations. The orientations and configurations of the CFRP strips studied includes orthogonal to column, skewed, offset from column, or adjacent to columns is shown in Figure 18.

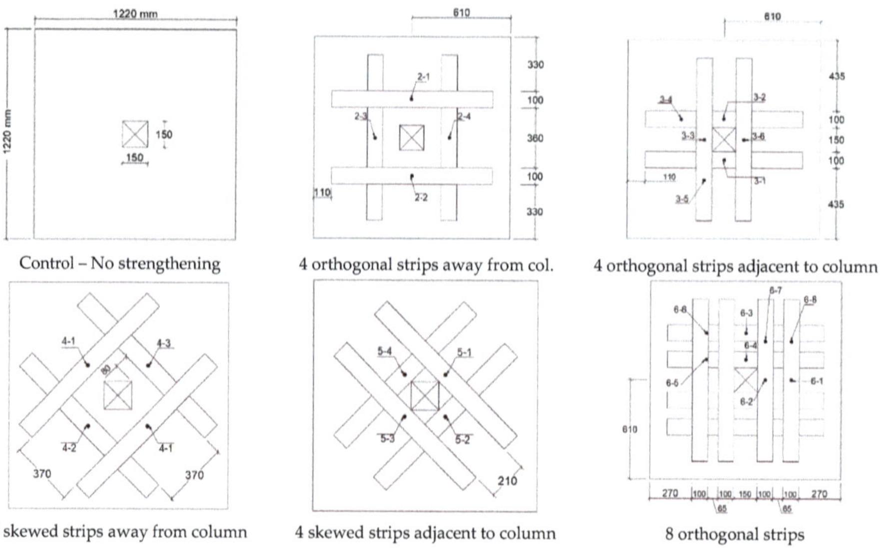

Figure 18. Effect of CFRP strip area, orientation with respect to column, and distance from face of column [39].

Erdogan et al. [40] examined a flat slab CFRP strengthening technique and concluded that it is capable of restoring the strength of flat slab specimen damaged through punching shear through interior column. It was also noted that rehabilitation of damaged flat slab using CFRP can restore strength to level higher than CFRP strengthened flat slab. They studied five 2130 mm × 2130 mm × 150 mm flat slabs loaded with 300 mm square plate representing internal column to flat slab connection. The goal of the study was to evaluate effectiveness of CFRP strengthening and repair of pre-loaded slabs that failed due to punching shear. The investigators indicated the dimensions chosen to represent 2/3 scale models of 8-m span flat slab that is 225-mm thick supported by 450-mm square columns. The flat slab specimens were reinforced with 19-mm bars at 140-mm spacing in each direction, without compression steel, which gives a tension reinforcement ratio of 1.86%. Specimen and column dimensions, loading points and eight CFRP rows around loading column are shown in Figure 19.

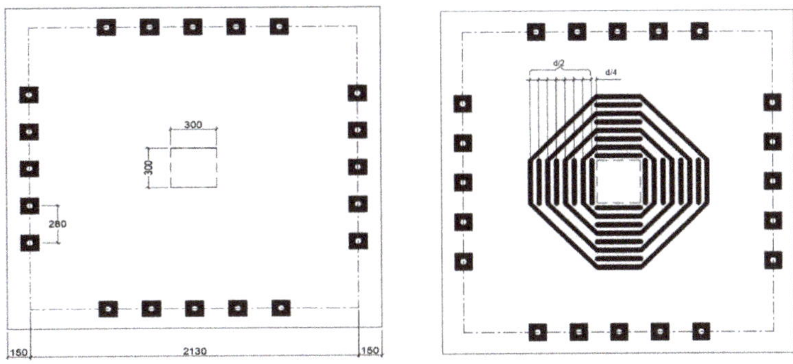

(a) (b)

Figure 19. (**a**) Specimen dimensions, column dimensions, and locations of perimeter loading points and (**b**) 8 CFRP rows around loading column [40].

Hussein and El-Salakawy [41] concluded that increasing the flexural tension GFRP reinforcement ratio of high-strength flat slab (80 MPa) from 1% to 1.5% (50% increase) increases the punching shear strength by 15%, while increasing the reinforcement ratio to 2% (100% increases) increases the punching shear strength by 27%. The longitudinal GFRP tension reinforcement was No. 16 (15.9 mm) having a tensile strength of 1685 MPa.

GFRP shear reinforcement curbed down widening of cracks and controlled its propagation. Control of cracking lead to enhancement of the stiffness and decrease of deflection in normal strength (40 MPa) slab. No. 13 (12.7 mm diameter) GFRP shear studs increased punching shear strength by 51% compared to the control slab without shear reinforcement. No. 10 (9.5 mm) corrugated GFRP shear reinforcement increased punching shear strength by 34% compared to the control slab.

High-strength (80 MPa) flat slabs with tension GFRP reinforcement from 0.5% to 1.5%, but without shear reinforcement, failed in punching shear in a brittle manner. Cracking begins at column corners then propagates radially from the column in all directions. When the load reached 45–50% of the ultimate load circumferential cracks appeared and connected radial cracks together in all directions. The normal-strength flat slab (40 MPa) failed in the same manner when there is no shear reinforcement. However, the normal-strength flat slab exhibited higher cracking at failure compared to the high-strength flat slab. It was noted that increasing the tension reinforcement ratio decreased the punching failure cone radius by making the inclination angel of the failure crack steeper.

The study was performed on six 2800 mm × 2800 mm × 200 mm flat slab specimens loaded by 300 mm square column. The specimens represent interior column–flat slab connection reinforced with flexural steel on the tension-side. One test series of three specimen was used to evaluate the effect of tension steel reinforcement ratio on punching shear capacity of high-strength concrete, and a second set of three specimens was used to examine the effect of GFRP shear reinforcement type on punching shear capacity of normal-strength concrete. The flat slab specimens (three in total) set made of high-strength concrete were reinforced with 1%, 1.5%, and 2% flexural steel. The normal-strength flat slab set tested for the effect of GFRP shear reinforcement consisted of one control specimen without shear reinforcement, a second specimen reinforced with GFRP shear studs, and a third specimen reinforcement with corrugated GFRP shear reinforcement as seen in Figure 20. The study by

Ferreira et al. [42] demonstrated that in the presence of studs, punching shear failure is initiated by loss of anchorage at the base of stud.

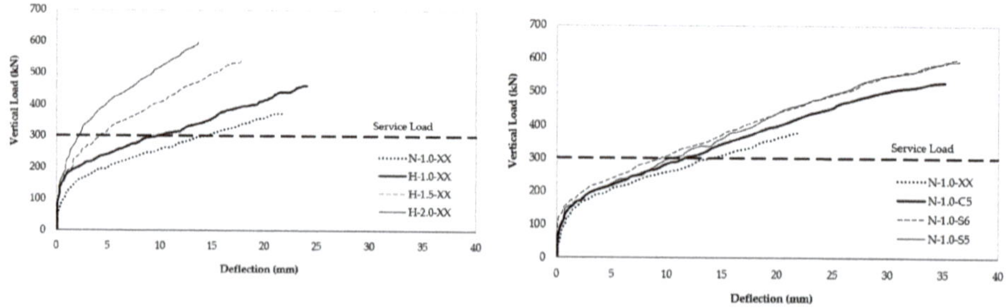

(a) (b)

Figure 20. (**a**) Load–deflection relationship for the high-strength (80 to 87 MPa) concrete, (**b**) load–deflection relationship for normal-strength (43 MPa) concrete [41].

Silva et al. [43] studied eight 1200 mm × 1200 mm × 100 mm flat slab specimens with loading through 100 mm square stub simulating internal column of flat slab. CFRP strengthening strips were 700 mm long × 100 mm wide × 1 mm thick. All strengthening strips, whether orthogonal (O) or skewed (S) were placed starting one effective depth (75 mm) from the face of the column as shown in Figure 21. End anchorage of CFRP strips, when applied, was done through 50 mm × 150 mm transverse steel plates attached to the slab through 10 mm diameter steel bolts. It was observed that the skewed arrangement of CFRP strips can increase punching shear strength flat slabs by nearly 46% compared to the control slabs. Using end anchorage at the end of the strips is particularly effective in enhancing the load-carrying capacity. The critical perimeter in control slabs is located 1.5- to 2.5-times the effective depth from the column face.

Failure in the control slab occurred by formation of radial cracks at distances from 1.5 to 2.5 × d (where d = effective depth) from the face of the column. The study reported that CFRP strengthened slab failed in flexure. All strengthened specimens without end anchorage failed by de-bonding and strips failed in rupture due to their high tensile strength.

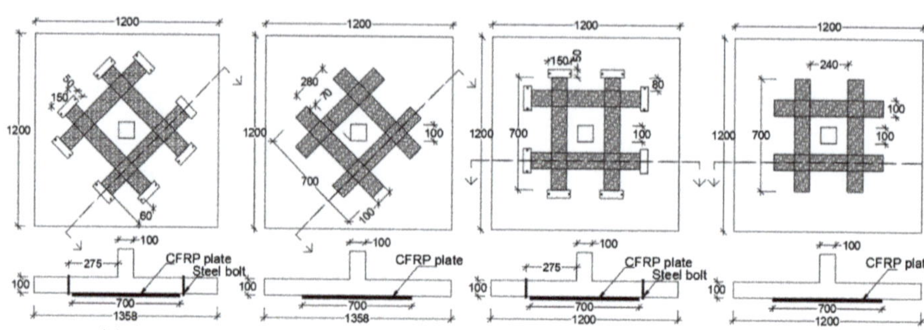

(a) (b) (c) (d)

Figure 21. (**a**) Skewed CFRP strips with end anchorage, (**b**) skewed CFRP strips without end anchorage, (**c**) orthogonal CFRP strips with end anchorage, and (**d**) orthogonal CFRP strips without end anchorage [43].

The study by Durucan and Anil [44] examined nine 2000 mm × 2000 mm × 120 mm flat slab specimens with concrete compressive strength of 20 MPa on average. The control flat slab did not have an opening while the remaining eight specimens contained openings at different locations with respect to the supporting columns, and strengthening with CFRP strips. The study included two (300 mm and 500 mm) square opening sized as shown in Figure 22.

It was demonstrated that CFRP strips were able to restore punching shear capacity of flat slabs affected by the presence of openings near the supporting columns, to nearly the same capacity of the control slabs without openings. Similarly, numerical studies by Mohamed et al. [45] observed that CFRP strengthening around openings located further from the supporting column enhances flexural capacity and overall stiffness of the slab. In fact, Lapi et al. [46] believes that the increase in punching shear strength through strengthening of the flat slab using FRP strips is the indirect effect of the increase in slab stiffness due to the strengthening strips.

A similar reduction in punching shear strength exists in voided flat slabs, where voids are incorporated in the slab to reduce self-weight [47], with applications in seismic areas. In such cases solutions for enhancing punching shear capacity may include use of a steel sheet in the critical area near the supporting column.

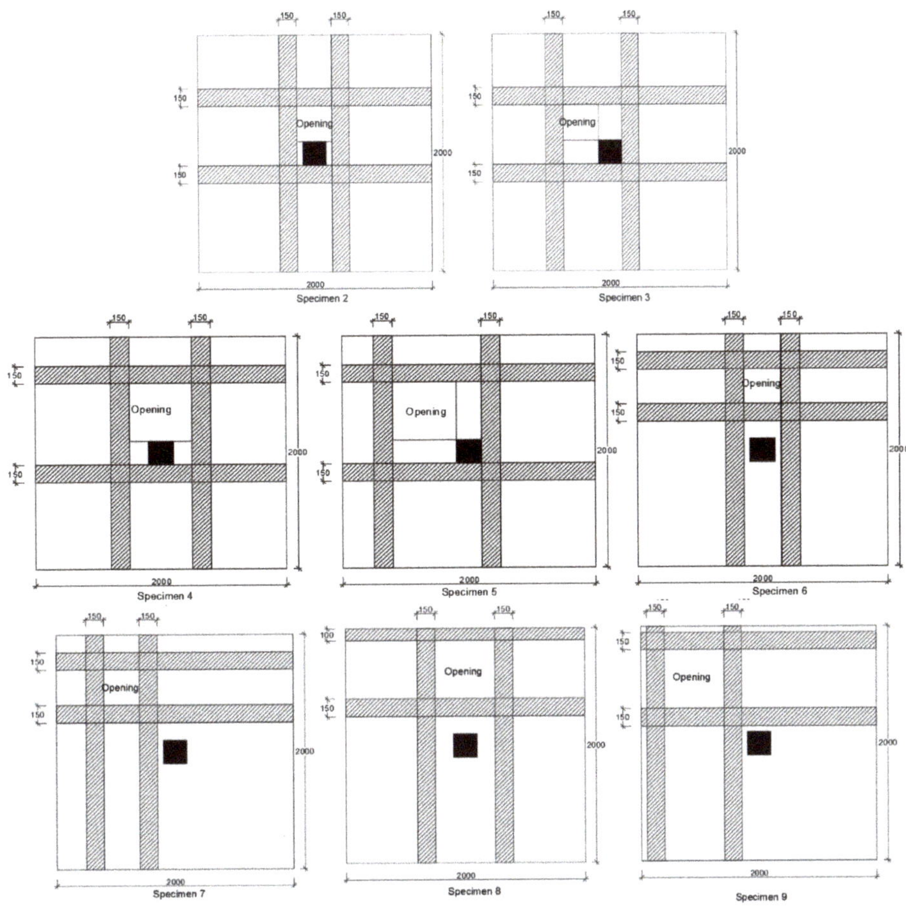

Figure 22. CFRP strips strengthening openings at various locations close-to or further-from the support [44].

Saleh et al. [48] carried out an experimental study on four 2300 mm × 2300 mm × 200 mm flat slabs loaded concentrically through 300 mm square plates. The control flat slab was prepared with 25 MPa concrete and was not reinforced in shear. Three specimens were reinforced in shear using L-shaped CFRP laminates installed through pre-drilled 25-mm holes as shown in Figure 23. The CFRP strips were fixed through the holes with part of it (to form L-shaped) to the top or bottom using epoxy resin. The L-shaped CFRP shear reinforcing strips were applied to the unloaded slab in three perimeters around the loading plate. Perimeter one is located at a distance d/2 (where d is the effective depth) from the loading plate. Perimeters 2 and 3 are located at distances 0.7d from the loading plate.

It was concluded that L-shaped CFRP strips are capable of increasing the ultimate load of flat slab specimens by 97–104% higher than the control flat slab without CFRP strips. Moreover, L-shaped strips increased the deflection at failure to 400% of the control specimen.

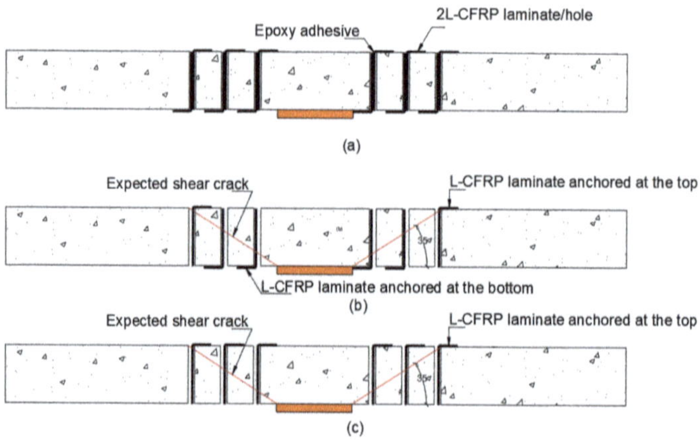

Figure 23. (**a**) Two L-shaped laminates per hole, (**b**) one L-shaped laminate per hole anchored at top or bottom alternately, and (**c**) one L-shaped laminate per hole anchored at the top [48].

4. Results of Experimental Studies and Major Findings

Many investigations have been conducted on strengthening the flat slab using both CFRP and GFRP. All have examined methods to delay or prevent punching shear failure. Table 3 shows summary of existing experimental work of FRP strengthening.

It is important to note that experimental studies are generally conducted on a scaled-down prototype specimens due to traditional laboratory limitations. Studies by Goh and Hrynyk [49] indicate that flat slab continuity and lateral restraint greatly affect the two-way shear capacity. Studies on multi-bay slabs showed improved two-way shear capacity compared to isolated column-zone specimens.

Table 3. Experimental results and numerical simulation of load-carrying capacity of reference reinforced concrete flat slab.

Reference/Size of Slab (mm)	Slab ID/Number of Samples	FRP Type	f_c' MPa	Ultimate Load (kN)	Conclusions
Binici and Bayrak [25] Specimen size (mm) 2133 × 2133 × 152	Control 1	No CFRP	28.3	494	1. Use of vertical CFRP strips inside pre-drilled holes, simulating stirrups is effective in enhancing punching shear capacity. Specimens with diagonal CFRP strips in addition to the vertical strips offered the highest enhancement in punching shear strength inside the shear reinforced zone. 2. The ultimate load carrying capacity and ductility of specimen A8, which was reinforced with vertical CFRP strips along with diagonal ones, were 1.51 and 2.0 times those of the control specimen Control (Control 1). 3. The enhancement in load-carrying capacity in specimen with highest CFRP strips and diagonal strips (A8) caused the yield length of steel reinforcement measured from the face of the loading plate to increase significantly compared to the control unstrengthen specimen (Control 1).
	Control 2			510	
	A4-1	4 CFRP layers		595	
	A4-2			668	
	A6	6 CFRP layers		721	
	A8	8 CFRP layers		744	
Harajli and Soudki [27] Specimen size (mm) 670 × 670 × 55 and 670 × 670 × 75	A1-SA1	Control CFRP sheet	31.9	49.2	1. Flat slab A1-SA1, A2-SA2, B1-SB1, and B2-SB2 are control specimens not strengthened with CFRP strips. Specimen names ending F5, F10, and F15 were reinforced with CFRP strips that are 50 mm, 100 mm, and 150 mm wide, respectively. 2. Increase in two-way shear strength ranged from 17% to 45%, depending on the area of CFRP sheets, slab thickness, and the reinforcement ratio of the slab. 3. In fours specimens, CFRP sheets experienced anchorage failure at the supports due to shearing of a thin concrete layer between the epoxy resin and the concrete surface. 4. Premature bond failure occurred when CFRP was applied in two layers, due to increase in horizontal shear between the CFRP layer and the concrete surface. 6. CFRP reinforcement improves shear strength of slab–column connections restricting the growth of the tensile cracks or increasing the flexural strength of the connections. 7. Use of CFRP sheets increases flexural strength and may modify the failure mode from pure flexural mode to combined flexural-shear mode or pure punching mode.
	A1-SA1F5		29.1	47.4	
	A1-SA1F10		34.3	65.4	
	A1-SA1F15		23.5	64.1	
	A2-SA2		35.5	60.5	
	A2-SA2F5		31.9	70.1	
	A2-SA2F10		35.5	77.7	
	A2-SA2F15		23.5	80.0	
	B1-SB1		35.5	78.8	
	B1-SB1F10		31.9	114.5	
	B1-SB1F15		33.0	104.0	
	B1-B1F10(2L)		34.3	107.5	
	B2-SB2		29.1	122.0	
	B2-SB2F10		29.1	142.3	
	B2-SB2F15		33.0	118.6	
	B2-B2F10(2L)		34.3	123.3	
El-Salakawy et al. [28] Specimen size (mm) 1540 × 1020 × 120	XXX	GFRP strips vertically around the column	33.0	125	1. Use of FRP sheets increased punching shear capacity, flexural stiffness, and delayed opening of flexural cracks. However, failure was mostly due to punching shear. 3. When shear bolts are used along with FRP strips, ductility at the slab–column connection increased and failure mode changed from punching to a flexure. Furthermore, combined FRP sheets and vertical steel bolts increased punching shear strength by 23% to 30%. 4. Punching shear capacities predicted by both ACI 318-02 (2002a) and CSA A23.3-94 (1994) were conservative.
	SF0		31.5	110	
	SX-GF		32.0	130	
	SX-CF		32.0	126	
	SX-GF-SB		40.2	170	
	SF-GF		32.0	135	
	SH-GF-SB		40.2	162	

Table 3. *Cont.*

Reference/Size of Slab (mm)	Slab ID/Number of Samples	FRP Type	f_c' MPa	Ultimate Load (kN)	Conclusions
Chen and Li [29] Specimen size (mm) 1000 × 1000 × 100	SR1-C1-F0	GFRP reinforced externally bonded		103.9	1. GFRP laminates increase the punching shear capacity of slab–column connections by functioning as external reinforcement. 2. GFRP laminates are more effective in enhancing the ultimate punching shear capacity in slabs with low concrete compressive strength and reinforcement ratio. 3. For lightly reinforced slabs, GFRP laminates applied to slab–column connection may change flexural punching failure into brittle punching shear failure. 4. Predictions of punching shear based on BS 8110 and the JSCE code were consistent with test results. However, ACI 318 is more conservative.
	SR1-C1-F1a			151.6	
	SR1-C1-F1b		16.9	144.4	
	SR1-C1-F2a			217.8	
	SR1-C1-F2b			186.4	
	SR1-C2-F0			123.8	
	SR1-C2-F1a			151.9	
	SR1-C2-F1b		34.4	208.0	
	SR1-C2-F2a			216.8	
	SR1-C2-F2b			220.7	
	SR2-C1-F0			146.1	
	SR2-C1-F1a			188.4	
	SR2-C1-F1b		16.9	190.8	
	SR2-C1-F2a			223.7	
	SR2-C1-F2b			224.7	
	SR2-C2-F0			225.7	
	SR2-C2-F1		34.4	263.9	
	SR2-C2-F2			289.4	
Sharaf et al. [31] Specimen size (mm) 1750 × 1750 × 120	Control	CFRP strips externally bonded	28	421	1. Strengthening of slabs with externally bonded CFRP strips delayed the initiation of cracks and controlled their propagation. Nonetheless, all strengthened specimens failed in punching shear mode. 3. The enhancement in punching shear load increased with increase in the area of CFRP reinforcement. 4. The presence of externally bonded CFRP strips reduced the strain in the internal steel reinforced bars. 5. ACI318, BS8110, and CSA codes offered very conservative estimation of punching shear compared to test results.
	4-O-CFRP		25	420	
	4-S-CFRP		28	451	
	8-O-CFRP		25	456	
	8-S-CFRP		25	462	
	8-O&S-CFRP		28	477	
Esfahani [32] Specimen size (mm) 1000 × 1000 × 100	R0.8-C25-F0	CFRP sheets externally bonded	23	138.0	1. Using CFRP sheets as flexural reinforcement can increase the punching shear strength of flat slabs, significantly. 2. Comparison between the results shows that cyclic loading decreases the enhancement of punching shear strength. 3. Equations proposed by Iranian Code ABA are similar to the equations presented by ACI Code and result in the same punching shear strength. Among the equations used for punching shear strength prediction, the equation proposed by BS 8110 Code predicted the punching shear strength most accurately with least scatter
	R0.8-C25-F10			191.0	
	R0.8-C25-F10-CL			172.0	
	R0.8-C25-F15			208.8	
	R0.8-C25-F15-CL			188.0	
	R1.6-C25-F0			210.0	
	R1.6-C25-F15			239.0	
	R1.6-C25-F15-CL			198.0	
	R1.6-C25-F30			245.0	
	R1.6-C25-F30-CL			210.5	

Table 3. *Cont.*

Reference/Size of Slab (mm)	Slab ID/Number of Samples	FRP Type	f_c' MPa	Ultimate Load (kN)	Conclusions
Farghaly et al. [33] Specimen size (mm) 1600 × 1600 × 120	SC	CFRP sheets externally bonded	44.7	NA	The study was a numerical simulation of CFRP strengthened slab using finite element software developed by the investigators. 1. Stiffness and punching shear capacity increases with area of CFRP strips. The increase in punching shear capacity ranged from 20% to 40% depending on the area of CFRP sheets. 2. The interface between concrete and CFRP strips was modeled with bond interfacial element that accounts for de-bonding failures. The element was used predict the slip profiles along the FRP-concrete interface. 3. Increasing the width of the CFRP sheet results in uniform stress transfer between the strengthening sheets and the concrete contact surface and decreases slip at the CFRP-concrete contact surface.
	SF5 SF10		33.5 39.6	215.3 260.6	
Erdogan [34] Specimen size (mm) 2000 × 2000 × 150	R1-A R1 R2 R3 OS13 OS14 OS15 OS25 OS25-b OS35-b CSWOP CSWP	CFRP	35 32 29 30 33 26 31 33 30 30 31 30	457 500 423 414 601 571 656 649 571 564 594 592	The study evaluated effects of CFRP strengthening of slab at supporting column on punching shear capacity and stiffness. Strips were applied vertically through the slab. CFRP strips were applied vertically at constant spacing from the four faces of the supporting column. 1. CFRP strengthening increasing Punching shear strength enhancement 31% to 53% depending on the strengthening scheme. 2. The arrangement and the spacing of the vertical CFRP dowels around the column stub were influential on the failure modes of the specimens 3. CFRP strengthening increasing the post-punching capacity of the specimens about 2.4 times the capacity of the un-strengthened test specimens. 4. ACI 318-08 provides safe estimations for the capacity of the strengthened test specimens in the database.
Urban and Tarka [35] Specimen size (mm) 2300 × 2300 × 180	S-2 WT-CF-8 WT-CF-K-8 WT-CF-K-16	CFRP strip externally bonded	38.8	495 550 625 675	The study is an experimental evaluation of the effect of CFRP sheets in enhancing punching shear resistance. 1. Applying bonded CFRP strips without end anchor bolts provides relatively small increase in two-way shear strength, not exceeding 10% compared to the un-strengthened specimen. 2. Adding anchor bolts at the end of the CFRP strips increases punching shear strength significantly compared to the same sample without anchor bolts and compared to un-strengthened specimen.
Radik et al. [38] Specimen size (mm) 1500 × 1500 × 152.4	Phase1-Control Phase1-2-0.5″ FRC Phase1-3-1.0″ FRC Phase1-4-GFRP Phase2-Control Phase2-3-.75″ FRC Phase2-4-GFRP	Polypropylene synthetic GFRP sheet	27.5	256 302 392 318 263 315 383	1. The goal of the study is to compare the effectiveness of applying fiber reinforced cement (FRC) layer on the tension side of flat slab in comparison to the effectiveness of GFRP sheet in increasing punching shear strength. The investigators concluded that FRC is more effective that GFRP in enhancing punching shear strength. 2. GFRP increased punching shear strength of flat slab by 24% compared to control flat slab without strengthening.

Table 3. *Cont.*

Reference/Size of Slab (mm)	Slab ID/Number of Samples	FRP Type	f_c' MPa	Ultimate Load (kN)	Conclusions
Soudki et al. [39] Specimen size (mm) 1220 × 1220 × 70	S	CFRP strip externally bonded	25.8	160.3	The investigators tested 6 flat slab specimens to determine the effectiveness of externally bonded CFRP strips in enhancing punching shear. 1. All flat slabs failed in punching shear with the CFRP strengthened specimens experiencing ultimate as high as 29% compared to the control flat slab. 2. Higher punching shear capacity was demonstrated by specimens strengthened near the supporting column while higher stiffness is obtained when the CFRP strips are placed near the supporting column. 3. The most effective CFRP strip configuration in enhancing punching shear strength is when the strips are placed further from the supporting column and skewed (with respect to column orientation. 5. Increasing the amount of CFRP strips did not significantly increase the capacity of the slabs.
	S-4-O-O			181.0	
	S-4-O-A			163.8	
	S-4-S-O			206.9	
	S-4-S-A			173.7	
	S-8-O-AO			192.9	
Erdogan et al. [40] Specimen size (mm) 2130 × 2130 × 150	LC	CFRP stirrups externally bonded	15.6	401	1. The application of external CFRP strips proved to be successful in enhancing the punching resistance of rehabilitated and strengthened specimens. 2. It was shown that CFRP-rehabilitated slab–column connections with partial pre-damage may have the potential to exhibit higher shear resistance and stiffness compared to their CFRP strengthened counterparts with no pre-damage. 3. ACI 318 (2011) could be safely used for the design of CFRP rehabilitation of damaged slab–column connections.
	LS			696	
	LF-R			465	
	NC			547	
	NS		32.3	750	
	ND-R			940	
	NF-R		32.3	683	
Hussein and El-Salakawy [41] Specimen size (mm) 2000 × 1000 × 150	CON-0	CFRP sheets externally bonded	37.3	152.2	1. The increase of the ultimate strength of CFRP strengthened slab–column connections decrease with the increase of the applied load eccentricity 2. Increasing the applied loading eccentricity markedly decreased the ultimate capacity, pre-yielding stiffness, and ductility of slab–column connections strengthened with external CFRP laminates 3. High reinforcement ratios of external CFRP laminates can result in punching shear failure of slab–column connections subjected to eccentric concentrated loads.
	STR-0			300.0	
	CON-25			135.0	
	STR-25		47.9	210.0	
	CON-35		53.4	130.0	
	STR-35			240.0	
Durucan and Anil [44] Specimen size (mm) 2000 × 2000 × 120	1 (control- no CFRP)	CFRP strips externally bonded	20.83	193.03	The investigators tested 9 flat slab specimens with openings at various locations in the vicinity of the supporting column. The control specimen was not strengthened while 8 specimens were strengthened with CFRP. 1. CFRP strengthening increases the punching shear capacity in comparison with the control (unstrengthen) specimen by 55% on average. 2. The adverse effect on punching shear capacity of openings near supporting columns is nearly eliminated by using CFRP strips around openings. 3. Highest ultimate load was demonstrated when opening was diagonal with respect to rectangular column, CFRP strips completely surrounded the opening, opening no adjacent to column, and opening not the largest of the test specimens. 4. CFRP strips increased initial elastic stiffness of flat slab specimens by nearly 486% on average, compared to unstrengthen specimens. CFRP strips also increased the maximum displacement at failure.
	2		20.56	161.18	
	3		19.96	186.08	
	4		21.23	157.71	
	5		19.78	173.31	
	6		20.12	197.42	
	7		21.45	219.36	
	8		20.03	190.86	
	9		21.09	201.84	

Table 3. *Cont.*

Reference/Size of Slab (mm)	Slab ID/Number of Samples	FRP Type	f_c' MPa	Ultimate Load (kN)	Conclusions
Abbas et al. [37] Specimen size (mm) 600 × 600 × 90	SA	CFRP sheets externally bonded	39.9	88.4	The investigators studied 12 flat slabs supported to span in one direction. Control slabs (SA and SB) were not strengthened. CFRP strengthening with done using single sheet with fibers parallel to the span direction. 1. The load-deflection response curve of CFRP Strengthened one-way specimens was characterized by two peaks compared to the control un-strengthened specimen. The investigators believe the second peak is due to the combined effect of dowel action of rebars, aggregate interlock, and unidirectional strengthening sheet. 2. The CFRP strengthened slabs showed nominal increase in the first peak load (9–18%) but there was significant increase in the second peak load.
	SA-C			99.4	
	SA-T			96.4	
	SB		63.2	106.0	
	SB-C			123.6	
	SB-T			125.2	
Saleh et al. [48] Specimen size (mm) 2300 × 2300 × 200	CS (Control slab)	No CFRP	25	569	1. Flat slabs strengthened with L-shaped CFRP strips failed at peak loads that are 97% to 104% higher than the control specimen. 2. CFRP L-shaped strips with configuration SS3 reached a deflection at failure load that is 400% higher than the control slab that was not strengthened. 3. All CFRP strengthened flat slab failed by punching shear at perimeter outside of the reinforced perimeter. Strengthening with L-shaped CFRP strip shifted the failure crack outside of the reinforced zone. 4. EC2 (2004) and ACI318 (2014) were both capable of predicting the capacity the strengthened specimens with reasonable accuracy.
	Configuration: SS1	Top & Bottom	28	1121	
	Configuration: SS2	Bottom/Top	27	1087	
	Configuration: SS3	Top	28	1163	
Hussein and El-Salakawy [41] Specimen size (mm) 2800 × 2800 × 200	H-1.0-XX	GFRP headed studs GFRP bars corrugated	80	461	The instigators tested 7 flat slab specimens including three high strength specimens (H-1.0-xx, H-1.5-xx, and H-2.0), and four normal strength specimens (N-1.0-S5, N-1.0-S6, N-1.0-C5, and N-1.0-xx). The high strength specimens were reinforced with variable GFRP bar ratios (1.0%, 1.5%, and 2.0%) and without shear reinforcement. 1. The high-strength concrete specimens (without shear reinforcement) failed in punching shear, without signs of flexural failure. Increase in reinforcement ratio from 1.0%, 1.5%, and 2.0% increased the punching shear capacity by 15% (H-1.5-xx stronger than H-1.0xx) and 27% (H-2.0-xx stronger than H-1.5-xx). 2. The three high strength specimens (80, 84, 87 and MPa) demonstrated better pre-cracking behavior and higher punching shear strength compared to normal-strength concrete specimens (43 MPa). 3. GFRP shear reinforcement controlled widening and propagation of shear cracks and enhanced the post-cracking stiffness which decreased the deflection. This was true for samples reinforced for shear resistance by headed studs (N-10.0-S6, N-1.0-C5) or stirrups (N-1.0-xx). 5. GFRP shear reinforcement not only increased the punching shear capacity, but also increased the failure load and corresponding deflection at failure.
	H-1.5-XX		84	541	
	H-2.0-XX		87	604	
	N-1.0-S5		43	595	
	N-1.0-S6		43	583	
	N-1.0-C5		43	527	
	N-1.0-XX		38	378	
Silva et al. [43] Specimen size (mm) 1200 × 1200 × 100	C-a	CFRP Stirrups Externally Bonded	28.0	130	The investigators studied 8 flat specimens including two control specimens (C-a and C-b) and 6 CFRP strengthened specimens (SE-a, S-a, OE-a, O-a, SE-b, S-b). 1. Skewed placement of CFRP strips with respect to column orientation (SE-a, S-a, SE-b, S-b) at the shear critical area is more effective in enhancing punching shear compared to orthogonal placement. In addition, usage of end anchorage strips (SE-a, SE-b) on skewed CFRP strips enhances strength even more. The highest improvement in punching shear strength was 46% compared to the control specimen without strengthening. 2. Failure patterns suggest that critical punching shear perimeter is located 1.5 to 2.0 times the effective depth from the face of the column, which is consistent with Euro Code 2. 3. Whether CFRP strips are orthogonal or skewed with respect to column orientation, the provision of end anchorage to CFRP strips increases the load carrying capacity. De-bonding of CFRP strips was the failure mode of specimens where end-anchorage of CFRP was not provided.
	SE-a			147.15	
	S-a			103.0	
	OE-a			137.34	
	O-a			137.34	
	C-b			98.64	
	SE-b			147.15	
	S-b			122.60	
	OE-b			127.53	
	O-b			122.63	

5. Conclusions

This article reviewed the literature published in the past two decades on the effect of strengthening flat slabs/plates using fiber reinforced polymer sheets/strips in the supporting column region to enhance two-way/punching shear capacity. While CFRP and GFRP are both used for strengthening, CFRP is the dominant material, therefore, covered extensively in the literature and in this study. Models for predicting punching shear capacity in selected international codes and standards were reviewed to develop an understanding of inherent capacity and factors influencing punching shear strength. Experimental studies by various investigators were presented and discussed.

1. Several building codes and standards such as the Canadian (CSA) and Euro code 2 (EC2) recognize the contributing of flexural reinforcement in enhancing the two-way shear capacity of flat slab systems. Punching shear capacity models in ACI 318 do not incorporate the influence of flexural reinforcing steel.

2. The two most commonly used techniques to strengthen the area near a supporting column using CFRP sheets/strips include: (1) bonding (gluing) CFRP sheets/strips on the tension side of the slab around the column, (2) installing the CFRP vertically in various ways that that mimic shear reinforcement. One-way to use vertical strips is to insert them vertically leaving part of strips extended outside of the slab to be bonded horizontally for anchorage purposes.

3. Bonding CFRP sheets/trips/laminates on the tension side around supporting columns of flat slabs increases the punching shear capacity of concrete as well as the ultimate load. The effectiveness and extent of capacity improvement depends on various factors such as orientation of strips with respect to support columns, end anchorage, area of strips/sheets, etc.

4. Increase in punching shear capacity by bonding CFRP sheets/strips to the tension side of test specimens at the support location is best when the strengthening sheets/strips are anchored at the ends. Steel anchor bolts installed vertically through the slab at the ends of the strips were used successfully to increase punching shear capacity as much as 46% compared control specimen without CFRP strengthening when the load is applied concentrically. Similar improvements in punching shear capacity was observed when end anchorage is done using CFRP strips applied at the end and perpendicular to the main strengthening strips. Improvement in punching shear capacity when CFRP is bonded to tension side is lower when the load is eccentric.

5. Installing CFRP strips vertically, resembling shear reinforcement perpendicular to the slab at selected spacing from column face is more effective in enhancing punching shear capacity compared to bonding CFRP strips/sheets to the tension side of the slab.

6. Need for Future Research

The literature review revealed that additional research is needed to address critical issues and fill gaps in the current state-of-knowledge in relation to FRP strengthened slabs. The following are amongst the pressing research needs:

1. Studies have shown that vertically placed FRP sheets/laminae/strips are more effective in resisting two-way shear force. It is necessary to investigate further methods of installation that induce minimal disturbance of the flat slab that is being retrofitted.

2. Further research is needed to develop models and clear design guidelines for vertically placed FRP retrofitting strips that takes into account pattern, spacing, FRP material properties, concrete properties, etc.

3. Most studies in the past focused on flat slab specimen sizes that are feasible within typical laboratories studies. However, the limited studies that considered larger mutli-span flat slab specimens indicated possible size effects on results. There is a need to investigate larger size specimens to predict the response of FRP retrofitted slabs.

Author Contributions: Conceptualization, O.A.M.; methodology, O.A.M. and R.K.; investigation, M.K.; resources, R.K; writing—original draft preparation, O.A.M. and M.K.; writing—review and editing, O.A.M. and M.K.; visualization, M.K.; supervision, R.K.; project administration, O.A.M. and R.K.; funding acquisition, O.A.M. All authors have read and agreed to the published version of the manuscript.

Funding: The authors gratefully acknowledge the financial support provided by the Office of Research and Sponsored Programs (ORSP) at Abu Dhabi University through grant No. 19300376.

Conflicts of Interest: The authors declare no conflict of interest.

References

1. Zaghlal, M.Y.A. Punch Resistance of Slab Column Connection under Lateral Loads. Master's Thesis, Zagazig University, Zagazig, Egypt, 2009.
2. Alexander, S.; Simmonds, S. Shear-moment transfer in slab–column connections. In *Structural Engineering Report*; University of Alberta: Edmonton, AB, Canada, 1989; p. 95.
3. Guandalini, S.; Burdet, O.L.; Muttoni, A. Punching tests of slabs with low reinforcement ratios. *ACI Struct. J.* **2009**, *106*, 87–95.
4. Moe, J. *Shearing Strength of Reinforced Concrete Slabs and Footings under Concentrated Loads*; Portland Cement Association: Skokie, IL, USA, 1961; Volume D47, p. 135.
5. Gomes, R.B.; Regan, P.E. Punching resistance of RC flat slabs with shear reinforcement. *ACI Struct. J.* **1999**, *125*, 684–692. [CrossRef]
6. Ebead, U.; Marzouk, H. Strengthening of two-way slabs subjected to moment and cyclic loading. *ACI Struct. J.* **2002**, *99*, 435–444.
7. Silva, M.A.L.; Madushanka, W.I.; Ariyasena, P.S.I.; Gamage, J.C.P.H. *Punching Shear Capacity Enhancement of Flat Slabs Using End Anchored Externally Bonded CFRP Strips*; Society of Structural Engineers: Colombo, Sri Lanka, 2018; Volume 28.
8. Malalanayake, M.L.V.P.; Gamage, J.C.P.H.; Silva, M.A.L. Experimental investigation on enhancing punching shear capacity of flat slabs using CFRP. In Proceedings of the 8th International Conference on Structural Engineering and Construction Management (ICSECM2017), Kandy, Sri Lanka, 7–9 December 2017.
9. Rasha, T.S.M.; Amr, B.; Hany, A. Effect of flexural and shear reinforcement on the punching behavior of reinforced concrete flat slabs. *Alex. Eng. J.* **2017**, *56*, 591–599.
10. Dilger, W.; Birkle, G.; Mitchell, D. Effect of flexural reinforcement on punching shear resistance. *Spec. Publ.* **2005**, *232*, 57–74.
11. Caldentey, A.P.; Lavaselli, P.P.; Peiretti, H.C.; Fernández, F.A. Influence of stirrup detailing on punching shear strength of flat slabs. *Eng. Struct.* **2013**, *49*, 855–865. [CrossRef]
12. McHarg, P.J.; Cook, W.D.; Mitchell, D.; Yoon, Y.S. Benefits of concentrated slab reinforcement and steel fibres on performance of slab–column connections. *ACI Struct. J.* **2000**, *97*, 225–234.
13. Elstner, R.C.; Hognestad, E. Shearing strength of reinforced concrete slabs. *Int. Assoc. Bridges Struct. Eng.* **1956**, *53*, 29–58.
14. Inácio, M.; Ramos, A.; Lúcio, V.; Faria, D. Punching of High-Strength Concrete Flat Slabs-Experimental Investigation. In Proceedings of the Fib Symposium Tel Aviv 2013, Tel Aviv, Israel, 22–24 April 2013; Volume 293, p. 500.
15. Alexander, S.D.; Simmonds, S.H. Tests of column-flat plate connections. *ACI Struct. J.* **1992**, *89*, 495–502.
16. Hawkins, N.M.; Fallsen, H.B.; Hinojosa, R.C. Influence of Column Rectangularity on the Behavior of Flat Plate Structures. *Spec. Publ.* **1971**, *30*, 127–146.
17. ACI Committee 318. *Building Code Requirements for Structural Concrete and Commentary: ACI 318-14*; American Concrete Institute: Detroit, MI, USA, 2011.
18. European Committee for Standardization. *EN 1992-1-1Eurocode 2: Design of Concrete Structures—Part 1-1: General Rules and Rules for Buildings*; The European Union per Regulation 305/2011; European Committee for Standardization: Brussels, Belgium, 2004.
19. ACI Committee 440. *Guide for the Design and Construction of Structural Concrete Reinforced with Fiber-Reinforced Polymer Bars: ACI 440.1R-15*; American Concrete Institute: Detroit, MI, USA, 2015.
20. Weng, Y.; Qian, K.; Fu, F.; Fang, Q. Numerical investigation on load redistribution capacity of flat slab substructures to resist progressive collapse. *J. Build. Eng.* **2020**, *29*, 101109. [CrossRef]

21. Mohamed, O.; Khattab, R.A.; Mishra, A.; Isam, F. Recommendations for reducing progressive collapse potential in flat slab structural systems. *IOP Conf. Ser. Mater. Sci. Eng.* **2019**, *471*, 052069. [CrossRef]
22. Canadian Standards Association. *Design and Construction of Building Structures with Fiber-Reinforced Polymer 2012*; CAN/CSA S806-12; Canadian Standards Association: Toronto, ON, Canada, 2012.
23. British Standards Institution. *Structural Use of Concrete—Code of Practice for Design and Construction*; British Standards Institution: London, UK, 1997.
24. Japan Society of Civil Engineering. *Recommendation for Design and Construction of Concrete Structures Using Continuous Fiber Reinforcing Materials*; Concrete Engineering Series 23; Japan Society of Civil Engineering: Tokyo, Japan, 1997.
25. Binici, B.; Bayrak, O. Punching shear strengthening of reinforced concrete flat plates using carbon fiber reinforced polymers. *J. Struct. Eng. ASCE* **2003**, *229*, 2273–2282. [CrossRef]
26. Sissakis, K.; Sheikh, S.A. Strengthening concrete slabs for punching shear with carbon fiber-reinforced polymer laminates. *ACI Struct. J.* **2007**, *104*, 49–59.
27. Harajli, M.H.; Soudki, K.A. Shear strengthening of interior slab–column connections using carbon fiber-reinforced polymer sheets. *J. Compos. Constr.* **2003**, *7*, 145–153. [CrossRef]
28. El-Salakawy, E.; Soudki, K.A.; Polak, M.A. Punching shear behavior of flat slabs strengthened with fiber reinforced polymer laminates. *J. Compos. Constr.* **2004**, *8*, 384–392. [CrossRef]
29. Chen, C.C.; Li, C.Y. Punching shear strength of reinforced concrete slabs strengthened with glass fiber-reinforced polymer laminates. *ACI Struct. J.* **2005**, *202*, 535–542.
30. Liberati, E.A.P.; Marques, M.G.; Leonel, E.D.; Almeida, L.C.; Trautwein, L.M. Failure analysis of punching in reinforced concrete flat slabs with openings adjacent to the column. *Eng. Struct.* **2019**, *182*, 331–343. [CrossRef]
31. Sharaf, M.H.; Soudki, K.A.; Dusen, M.V. CFRP strengthening for punching shear of interior slab–column connections. *J. Compos. Constr.* **2006**, *10*, 410–418. [CrossRef]
32. Esfahani, M.R. Effect of cyclic loading on punching shear strength of slabs strengthened with carbon fiber polymer sheets. *Int. J. Civ. Eng.* **2008**, *6*, 208–215.
33. Farghaly, A.S.; Ueda, T. Punching strength of two-way slabs strengthened externally with FRP sheets. *JCI Proc. Jpn. Concr. Inst.* **2009**, *31*, 493–498.
34. Erdogan, H. Improvement of Punching Strength of Flat Plates by Using Carbon Fiber Reinforced Polymer (CFRP) Dowels. Ph.D. Thesis, Middle East Technical University, Ankara, Turkey, 2010.
35. Urban, T.; Tarka, J. Strengthening of slab–column connections with CFRP strips. *Arch. Civ. Eng.* **2010**, *56*, 193–212. [CrossRef]
36. Halabi, Z.; Ghrib, F.; El-Ragaby, A.M.; Sennah, K. Behavior of RC slab–column connections strengthened with external cfrp sheets and subjected to eccentric loading. *J. Compos. Constr.* **2013**, *17*, 488–496. [CrossRef]
37. Abbas, H.; Abadel, A.A.; Almusallam, T.; Al Salloum, Y. Effect of CFRP and TRM strengthening of RC slabs on punching shear strength. *Lat. Am. J. Solids Struct.* **2015**, *12*, 1616–1640. [CrossRef]
38. Radik, M.J.; Erdogmus, E.; Schafer, T. Strengthening two-way reinforced concrete floor slabs using polypropylene fiber reinforcement. *J. Mater. Civ. Eng. ASCE* **2011**, *23*, 562–571. [CrossRef]
39. Soudki, K.; El-Sayed, A.K.; Vanzwol, T. Strengthening of concrete slab–column connections using CFRP strips. *J. King Saud Univ. Eng. Sci.* **2012**, *24*, 25–33. [CrossRef]
40. Erdogan, H.; Zohrevand, P.; Mirmiran, A. Effectiveness of externally applied CFRP stirrups for rehabilitation of slab–column connections. *J. Compos. Constr. ASCE* **2013**, *17*, 04013008. [CrossRef]
41. Hussein, A.H.; El-Salakawy, E.F. Punching shear behavior of glass fiber-reinforced polymer-reinforced concrete slab–column interior connections. *ACI Struct. J.* **2018**, *115*, 1075–1088. [CrossRef]
42. Ferreira, M.D.; Oliveira, M.H.; Sales, G.; Melo, S.A. Tests on the punching resistance of flat slabs with unbalanced moments. *Eng. Struct.* **2019**, *196*, 109311. [CrossRef]
43. Silva, M.A.L.; Gamagea, J.C.P.H.; Fawzia, S. Performance of slab–column connections of flat slabs strengthened with carbon fiber reinforced polymers. *Case Stud. Constr. Mater.* **2019**, *11*, e00275. [CrossRef]
44. Durucan, C.; Anil, O. Effect of opening size and location on punching shear behavior of interior slab–column connections strengthened with CFRP strips. *Eng. Struct.* **2015**, *105*, 22–36. [CrossRef]
45. Mohamed, O.A.; Khattab, R.; Okasha, R. Numerical analysis of RC slab with opening strengthened with CFRP laminates. *IOP Conf. Ser. Mater. Sci. Eng.* **2019**, *603*, 042038. [CrossRef]

46. Lapi, M.; Ramos, A.P.; Orlando, M. Flat slab strengthening techniques against punching-shear. *Eng. Struct.* **2019**, *180*, 160–180. [CrossRef]
47. Al-Gasham, T.S.; Mhalhal, J.M.; Jabir, H.A. Improving punching behavior of interior voided slab–column connections using steel sheets. *Eng. Struct.* **2019**, *199*, 109614. [CrossRef]
48. Saleh, H.; Kalfat, R.; Abdouka, K.; Al-Mahaidi, R. Punching shear strengthening of RC slabs using L-CFRP laminates. *Eng. Struct.* **2019**, *194*, 274–289. [CrossRef]
49. Goh, C.Y.M.; Hrynyk, T.D. Numerical investigation of the punching resistance of reinforced concrete flat plates. *J. Struct. Eng. ASCE* **2018**, *144*, 04018166. [CrossRef]

polymers

MDPI

Article

The Effect of Modification on the Properties of Polyetherimide and Its Carbon-Filled Composite

Azamat Slonov, Ismel Musov, Azamat Zhansitov *, Elena Rzhevskaya, Diana Khakulova and Svetlana Khashirova

Kabardino-Balkarian State University Named after H.M. Berbekov, st. Chernyshevsky, 173, 360004 Nalchik, Russia; azamatslonov@yandex.ru (A.S.); ismel@mail.ru (I.M.); elena.r-1382@mail.ru (E.R.); dianakhakulova@mail.ru (D.K.); new_kompozit@mail.ru (S.K.)

* Correspondence: azamat-z@mail.ru; Tel.: +7-928-718-9227

Received: 26 March 2020; Accepted: 1 May 2020; Published: 4 May 2020

Abstract: The effect of oligophenylene sulfone (OPSU) and polycarbonate (PC) on the rheological, mechanical and thermal properties of polyetherimide (PEI) and a carbon-filled composite based on it was studied. It is shown that the introduction of OPSU and PC leads to a decrease in the melt viscosity of PEI and a carbon-filled composite based on it with the preservation of their mechanical properties and heat resistance at a sufficiently high level. It was found that composites with OPSU have higher mechanical and thermal properties compared with composites with PC. Samples from modified carbon-filled PEI were printed by the fused deposit modeling (FDM) method. Three-dimensionally printed samples from carbon-filled PEI modified by OPSU showed significantly higher mechanical properties than composites with PC.

Keywords: polyetherimide; polycarbonate; polyphenylene sulfone; carbon fibers; modification; mechanical properties; 3D printing

1. Introduction

One of the classes of polymers used in various branches of modern technology with increased requirements for mechanical properties and thermal stability are high performance polymers. In recent years, these polymers have come to the fore as promising matrices for composite materials. High-performance polymers and composites based on them, in addition to increased strength and heat resistance, also have resistance to shock cyclic loads and cracking, high atmospheric and chemical resistance [1], which, in some cases, allows them to be used instead of metals [2].

Moreover, these materials are of significant interest to one of the fastest growing industries as additive manufacturing or 3D printing. Fused deposit modeling (FDM) is the most widely used 3D technology for polymer materials. Most of the work in the field of 3D printing by FDM is devoted to the study of standard and engineering plastics [3], such as acrylonitrile butadiene styrene (ABS) [4,5], poly-lactic acid (PLA) [6,7], nylon [8], polycarbonate (PC) [9,10], as well as their combinations [11–13], but high performance plastics are also actively explored in 3D printing [14–16].

The enhanced properties of high-performance polymers are ensured by their rigid chain structure and strong intermolecular interactions. On the other hand, these features greatly complicate their processing and obtaining composite materials based on them. A particularly difficult task increasing their manufacturability, which provides processing by high-speed injection molding and 3D printing methods for the manufacture of large-sized products of complex design with reduced wall thickness.

At present, the most promising amorphous high-performance thermoplastic used in 3D printing is polyetherimide (PEI). It is amorphous thermoplastic with high thermal stability and remarkable mechanical properties. It is used, among other applications, for advanced parts in electrical and electronics industry, in aircraft applications and in the automotive industry [17].

Basically, works on 3D printing of PEI are related to the study of the influence of printing parameters, such as a temperature, an angle of raster laying, an air gap between rasters, a width of the rasters, etc., on mechanical properties of samples from this material. For example, the effect of printing parameters on the mechanical properties of PEI was studied in [18]. It was shown that the raster orientation and nozzle temperature were closely related to the mechanical properties of printed samples. At the nozzle, the temperature is 370 °C, and 0° raster orientation reached mechanical properties close to the injection-molded parts. In [19], the influence of printing parameters on the mechanical properties of samples was also studied, where Stratasys PEI ULTEM 9085 was used. This material is a blend of PEI with PC. It is shown that the highest properties are achieved when using a negative gap between rasters. Moreover, in [20], the same material was used for study and it was found that the most optimal combination of properties is achieved with an air gap of 0 mm and a raster angle of 0°. In [21], the authors carried out the modification of PEI by PC and compared with the properties of the industrial brand for printing ULTEM 9085. It was revealed that the introduction of PC leads to an increase in the rheological properties of PEI. The viscosity of the obtained composite with a content of 10% PC is close to the viscosity of the industrial brand, while exceeding it in strength and elasticity modulus.

However, of particular interest for 3D printing are composites based on PEI with carbon fibers (CF), since they are characterized by high mechanical properties [22–24] and can be used for aircraft detail printing. In particular, [25] shows the results of printing PEI with a content of 10 % CF, compared to the ULTEM 9085. Both materials have been used to print aircraft engine components. It is shown that the printed samples of the carbon-filled composite are inferior in properties to the samples obtained on the basis of industrial material. The authors consider the reason for this to be the porosity and increased viscosity of the carbon-filled material.

It is known that the introduction of fibrous fillers significantly increases the polymer melt viscosity [26,27]. When processing by injection molding, this leads to underfilling and defective products, and, when processed by 3D printing (FDM), to poor adhesion of the filaments in the sample and, as a result, to low mechanical properties [28]. Moreover, 3D printing demands that the materials have higher rheological properties than injection molding does. The fact is that the melts of polymer materials, being non-Newtonian fluids, change viscosity depending on shear rates—as a rule, the higher the shear rate, the lower the melt viscosity. It is known that 3D printing takes place at very low shear rates [29,30], when the melt has the highest viscosity; therefore, materials for 3D printing should have very high rheological properties.

It is possible to achieve high melt flow in two ways—by increasing the processing temperature or by introducing modification additives into the composition of the material. Printing of polymeric materials by the FDM method takes place at temperatures significantly higher than injection molding temperatures; therefore, a decrease in the melt viscosity of fiber-filled composites by increasing the already high printing temperatures will lead to the destruction of the material. The most optimal way to achieve the necessary rheological properties is the introduction of modification additives into the polymer material.

The existing industrial plasticizers and lubricants on the market have low thermal properties and are mainly used for standard plastics with low processing temperatures [31–33] and are therefore unsuitable for plasticizing high-performance polymers. This served as the basis for the study of the effectiveness of modifiers for fiber-filled composite materials based on PEI intended for 3D printing.

2. Materials and Methods

The object of the study was PEI Ultem 1010 manufactured company Sabic (Saudi Arabia) as a matrix polymer. An oligomer of polyphenylene sulphone (OPSU) with the following characteristics was used as a modifier: a molecular weight (MW) of about 19,000, an onset of destruction temperature of 470 °C and a glass transition temperature of 190 °C. It was synthesized according to the method described in [34] using 4,4′-dihydroxy diphenyl and 4,4′-dichloro diphenyl sulfone. The polycarbonate

(PC) of the K-20 MRA brand of the Carbotex company (Japan) was used as modifier either. It has a temperature of the onset of destruction at about 430 °C and a glass transition temperature of 145 °C. As a filler, milled carbon fibers with a length of 0.2 mm and diameter of 7 micrometers from R&G (Germany) were used.

First, starting components were dried under vacuum at temperatures of 120–170 °C and mixed on a high-speed mixer. The dry mixture was processed at temperatures of 310–360 °C on a TwinTech twin-screw microextruder (UK) with L/D = 30. Test samples were obtained by injection molding on an SZS-20 machine from Haitai Machinery (China) at a material cylinder temperature of 340–380 °C and a mold temperature of 180 °C. Standard test samples were printed on a Stratasys Fortus 400mc 3D printer (Figure 1a) at 416 °C.

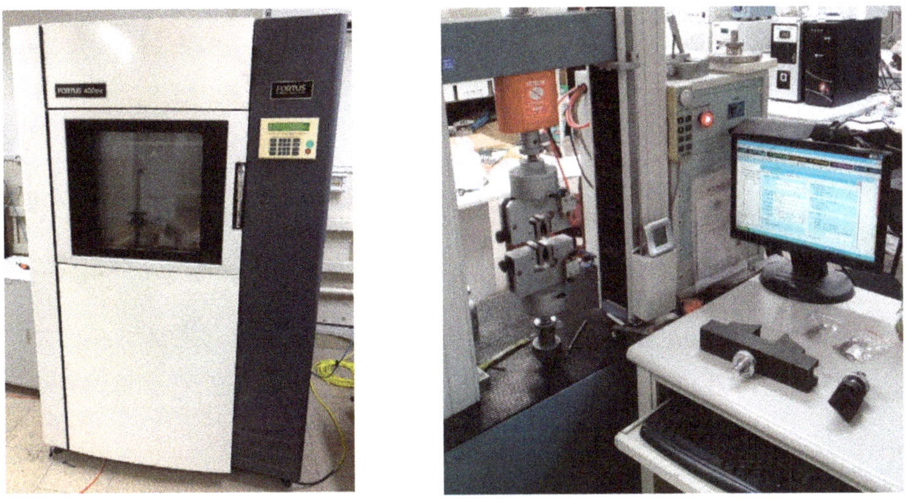

(a) (b)

Figure 1. 3D printer Fortus 400mc (**a**) and universal testing machine GT-TCS 2000 (**b**).

Rheological properties were determined by melt flow rate (MFR) on an IIRT-5 device (Russia) at a temperature of 350 °C and a load of 5 kg, as well as using a LCR 7001 capillary rheometer from Dynisco (USA) at a temperature of 380 °C. Heat resistance was determined by the method of thermogravimetric analysis (TGA) on a TGA 4000 device from PerkinElmer (USA), at a heating rate of 10 °C/min in air. Glass transition temperature was determined by the method of differential scanning calorimetry (DSC) on a DSC 4000 device from the same company at a heating rate of 10 °C/min in air. Microscopic studies were carried out on a Vega 3 scanning electron microscope (SEM) from Tescan (Czech Republic).

The uniaxial tensile mechanical tests were carried out on dog-bone samples with dimensions according to GOST 112 62-80 type 5. The tests were carried out on a universal testing machine Gotech Testing Machine GT-TCS 2000 (Taiwan) at a temperature of 23 °C (Figure 1b). Impact tests were performed with and without notch, according to the Izod method according to GOST 19109-84 on a Gotech Testing Machine, model GT-7045-MD (Taiwan) with a pendulum energy of 11 J. The average fiber lengths in composite materials were determined according to GOST R 57730-2017.

3. Results and Discussion

At the first stage, the effect of OPSU and PC on the main properties of unfilled PEI was studied. OPSU was introduced into PEI in the amount of 10, 15, 20, 30 wt.%, and PC 10, 15, 20 wt.%. The study of the rheological properties of PEI after the introduction of modifiers shows (Figure 2) that both OPSU and PC lead to a uniform increase in MFR, with a slight advantage for the values for composites with PC.

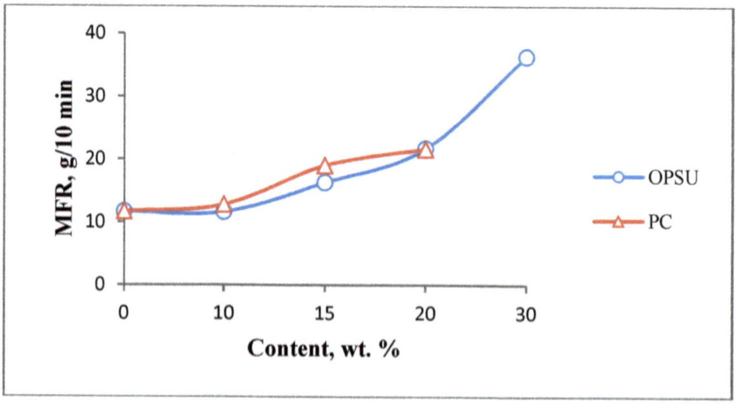

Figure 2. Dependence of melt flow rate polyetherimide (MFR PEI) on the content of oligophenylene sulfone (OPSU) and polycarbonate (PC).

The plasticization of PEI is also observed on flow curves, where a decrease in the viscosity of the PEI melt with an increase in the content of OPSU and PC is seen (Figure 3a,b). It is noteworthy that the flow curves of composites with an equal content of OPSU and PC are similar, i.e., both materials provide effective plasticization of PEI, and equally.

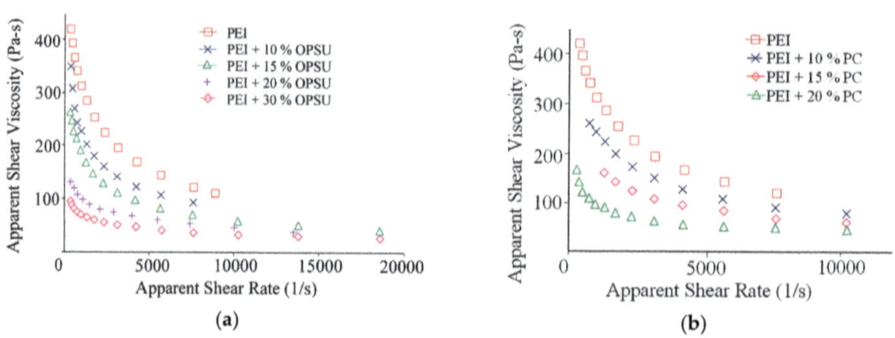

Figure 3. The dependence of the PEI melt viscosity on the content of OPSU (**a**) and PC (**b**).

Table 1 shows that the introduction of OPSU leads to a significant decrease in impact strength, both for notched and unnotched specimens. An increase in the content of OPSU to 30% led to a slight decrease in the strength and elastic modulus of PEI. In this case, the elastic modulus, both in bending and in tension, remains at a sufficiently high level, regardless of the content of OPSU. Upon reaching 20% content, there is a lack of tensile yield strength, i.e., brittle destruction begins.

For composites with PC, a different picture is observed. The introduction of up to 20% PC into PEI leads to a slight increase in toughness and plasticity. The elastic modulus and strength also decrease slightly and remain at a high level.

Table 1. Mechanical properties of the blends PEI/OPSU and PEI/PC.

Content	Impact Strength, kJ/m^2		E_{fl},	σ_{fl},	E_{ten},	σ_{ten},	σ_{yield},	ε,
	Unnotched	Notched	GPa	MPa	GPa	MPa	MPa	%
PEI	75.5 ± 7.6	6.1 ± 0.6	3.63 ± 0.16	112.4 ± 4.5	2.85 ± 0.08	88.0 ± 5	111.0 ± 5	22.7 ± 11.2
				OPSU				
10%	39.2 ± 3.9	1.5 ± 0.2	3.29 ± 0,08	104.0 ± 6.7	2.80 ± 0.1	84.0 ± 1.7	106.9 ± 2.8	13.7 ± 2.6
15%	29.7 ± 3.0	1.5 ± 0.2	3.26 ± 0.14	104.1 ± 7.3	2.82 ± 0.08	78.0 ± 0.5	105.2 ± 0.4	12.0 ± 3.7
20%	27.9 ± 2.8	1.2 ± 0.1	3.18 ± 0.1	103.3 ± 9.8	2.81 ± 0.06	103.8 ± 3.2	-	8.5 ± 1.4
30%	7.3 ± 0.7	1.3 ± 0.1	3.27 ± 0.5	65.1 ± 5.3	2.69 ± 0.08	71.4 ± 9.3	-	4.1 ± 2.5
				PC				
10%	55.7 ± 5.6	4.2 ± 0.4	3.40 ± 0.38	104.2 ± 4.4	2.79 ± 0.09	80.0 ± 0.3	102.5 ± 0.8	15.0 ± 4.6
15%	79.5 ± 8.0	4.6 ± 0.5	3.36 ± 0.13	106.7 ± 8.3	2.77 ± 0.05	81.0 ± 0.5	100.3 ± 0.5	25.0 ± 13.5
20%	84.7 ± 8.5	4.1 ± 0.4	3.28 ± 0.08	100.8 ± 7.7	2.7 ± 0.01	78.0 ± 1.3	96.7 ± 2.9	30.0 ± 5.6

E_{fl}: flexural modulus, σ_{fl}: flexural strength, E_{ten}: tensile modulus, σ_{ten}: tensile strength, σ_{yield}: yield point, ε: elongation at break.

DSC studies of the obtained blends showed that the introduction of OPSU leads to a uniform decrease in the glass transition temperature (T_g) of PEI (Figure 4a; Table 2) and only single thermal effects corresponding to the glass transition temperature of PEI are observed. In the case of introducing 10% PC into PEI, a slight decrease in the T_g of PEI is also observed; however, a further increase in its content practically does not affect the value of T_g (Table 2). On the DSC curve of the composite with 15 % PC content, a second barely noticeable peak is observed in the T_g region of the PC (Figure 4b, see arrows), which indicates a forming biphasic structure. Similar patterns were obtained in [21,35]. At 20% content, the peak becomes more pronounced.

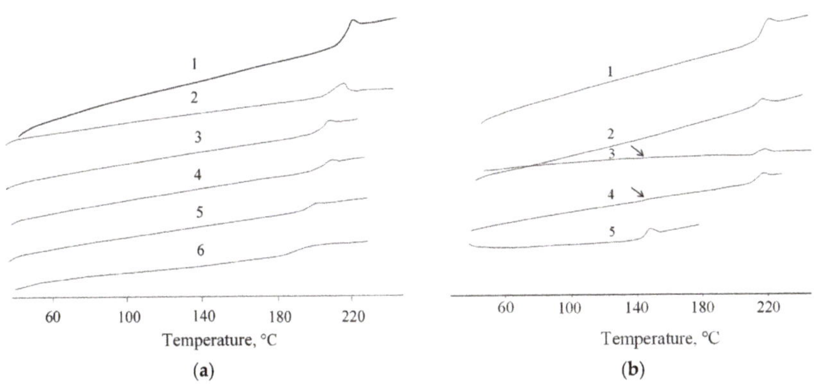

Figure 4. Differential scanning calorimetry (DSC) curves of blends PEI/OPSU (**a**): 1—PEI; 2-5—PEI+10, 15, 20, 30% OPSU, respectively; 6—OPSU and PEI/PC (**b**): 1—PEI; 2-4—PEI+10, 15, 20% PC, respectively; 5—PC.

Table 2. Dependence of the glass transition temperature of PEI on the content of OPSU and PC.

Content	T_g, °C	
	Peak 1	Peak 2
PEI	216.8	-
OPSU	190.0	-
PEI +10% OPSU	211.8	-
PEI +15% OPSU	205.9	-
PEI +20% OPSU	206.4	-
PEI +30% OPSU	198.3	-
PC	145.6	-
PEI +10% PC	214.2	-
PEI +15% PC	214.2	143.0
PEI +20% PC	214.2	142.7

The immiscibility of PEI and PC is also confirmed by the fact that PEI/PC samples are opaque (Figure 5e–g), while the samples of PEI/OPSU mixtures remain optically transparent over the entire content range (Figure 5a–d), which highlights the good compatibility of these materials.

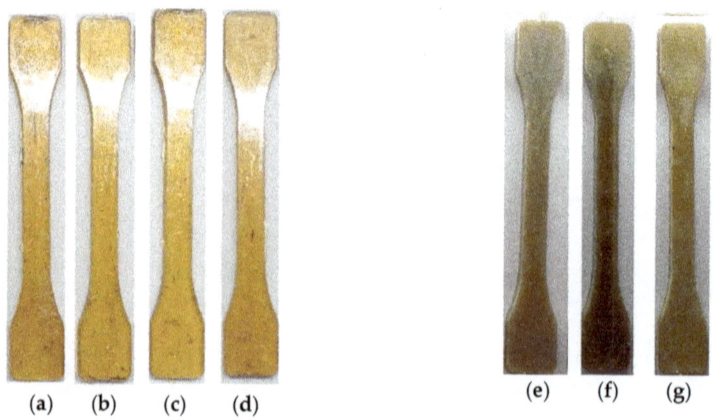

Figure 5. Samples from composites PEI/OPSU (**a**–**d**) and PEI/PC (**e**–**g**).

(TGA) shows that the introduction of both OPSU and PC into PEI leads to a slight decrease in its heat resistance (Table 3). In the case of using OPSU, a decrease in thermal properties is insignificant since the oligomer itself is characterized by high heat resistance. More noticeably, the thermal stability of PEI decreases with the introduction of PC; however, even in this case, the blends have sufficient resistance to thermal degradation.

Table 3. Thermogravimetric analysis (TGA) data of PEI with the various content of OPSU and PC.

Content	$T_{2\%}$	$T_{5\%}$	$T_{10\%}$
PEI	513.3	530.7	539.3
OPSU	472.0	500.0	523.0
PEI +10% OPSU	507.8	521.8	530.9
PEI +15% OPSU	503.2	521.0	532.0
PEI +20% OPSU	496.1	518.4	528.7
PEI +30% OPSU	498.0	518.6	529.5
PC	432.0	459.0	472.0
PEI +10% PC	492.0	507.0	519.7
PEI +15% PC	483.2	498.0	512.3
PEI +20% PC	480.8	499.5	510.1

Therefore, comparative studies of OPSU and PC as PEI modifiers showed their rather effective and equal plasticizing ability. The high mobility of macromolecules and the fluidity of the OPSU and PC melts at the PEI processing temperature lead to the effect of temporary plasticization, i.e., facilitating the flow of PEI melt. At room temperature, in view of the high glass transition temperatures, both OPSU and PC are in a glassy state, which allows for preserving the elastic modulus and strength properties of PEI. At the same time, PC, due to its high molecular weight and plasticity, promotes the elasticization of PEI. OPSU, on the contrary, having a low MW, leads to embrittlement of the samples.

To study the effect of modifiers on the properties of composites, compositions with a content of 30% carbon fibers (PEI-CF) were used, since, at this content, the MFR significantly decreases, and, at the same time, the composite has rather high mechanical properties [36]. Modifiers were introduced in amounts of 10, 15, 20 wt.% of the polymer content in the composite.

Figure 6 shows the dependences of the MFR of a composite with 30% CF on the content of OPSU and PC. The graph shows that both modifiers equally and quite effectively increase the melt flow of the composite. At a 20% content of OPSU, the MFR increases by 212%, and at a similar content of PC, by 192%.

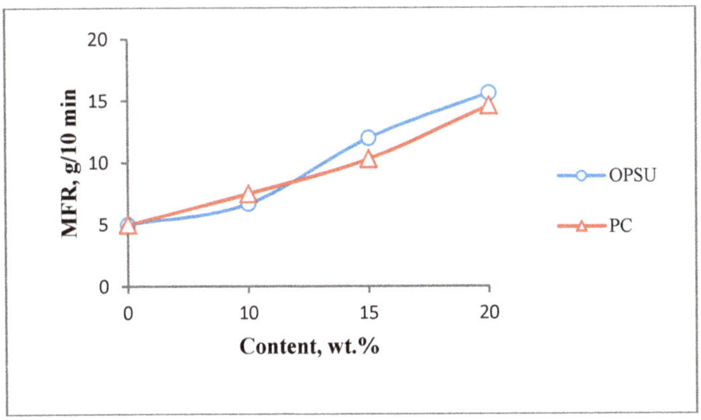

Figure 6. MFR PEI-carbon fibers (CF) with different content of OPSU and PC.

The study of rheological properties using a capillary rheometer confirmed a decrease in the viscosity of carbon-filled PEI with the introduction of OPSU and PC (Figure 7). The flow curves are similar in nature and viscosity values for equal modifiers content. Apparently, the mechanism of the plasticization of the composite is similar to unfilled PEI.

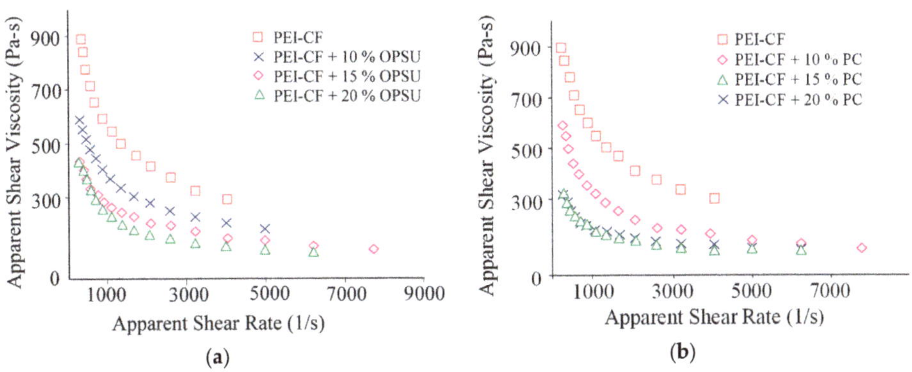

Figure 7. Dependence of the PEI-CF melt viscosity on the content of OPSU (**a**) and PC (**b**).

The mechanical properties of composites containing various quantities of OPSU and PC were also studied to determine their effect on them. Table 4 shows that the introduction of OPSU affects the impact strength of the composite slightly; there is a slight decrease in the impact strength of the composite during tests of unnotched samples and an increase during the testing of notched samples. In this case, a very pronounced increase in the elastic modulus and strength is observed, both in bending and in tension. The introduction of PC noticeably caused a more significant decrease in toughness and elongation, while there is a significant increase in the elastic modulus. However, unlike composites with OPSU, the bending and tensile strength remains practically unchanged.

Table 4. Mechanical properties of carbon-filled PEI with different content of OPSU and PC.

Content	Impact Strength, kJ/m^2		E_{fl},	σ_{fl},	E_{ten},	σ_{ten},	ε,
	Unnotched	Notched	GPa	MPa	GPa	MPa	%
PEI-CF (PEI+30% CF)	40.0 ± 4.0	5.6 ± 0.6	12.1 ± 0.4	257.2 ± 5.3	8.2 ± 0.3	154.0 ± 8.2	4.2 ± 0.3
OPSU							
10%	38.1 ± 3.8	6.5 ± 0.7	14.4 ± 0.4	256.3 ± 4.6	8.6 ± 0.6	178.5 ± 5.1	4.3 ± 0.5
15%	36.6 ± 3.7	6.3 ± 0.6	16.5 ± 0.4	302.3 ± 6.8	9.7 ± 0.4	191.6 ± 12.1	4.1 ± 0.4
20%	36.1 ± 3.6	6.7 ± 0.7	17.8 ± 0.5	306.3 ± 7.5	8.9 ± 0.3	191.0 ± 7.2	3.8 ± 0.6
PC							
10%	20.6 ± 2.1	5.4 ± 0.5	18.6 ± 1.6	264.0 ± 8.1	10.6 ± 0.9	151.0 ± 19.3	2.8 ± 0.5
15%	24.2 ± 2.4	6.0 ± 0.6	18.7 ± 0.3	239.8 ± 4.9	9.7 ± 1.2	157.5 ± 10.3	2.8 ± 0.5
20%	22.7 ± 2.3	6.0 ± 0.6	20.5 ± 0.6	253.0 ± 9.3	11.0 ± 0.6	150.2 ± 15.4	2.6 ± 0.5

E_{fl}: flexural modulus, σ_{fl}: flexural strength, E_{ten}: tensile modulus, σ_{ten}: tensile strength, ε: elongation at break.

The observed changes in the mechanical properties of the carbon-filled composites differ from the effect of these modifiers on unfilled PEI (Table 1), where a greater decrease in toughness and elongation was observed when using OPSU. Apparently, the compatibility of PEI and OPSU in the case of a carbon-filled composite plays a more noticeable positive role, which leads to the production of a composite with higher impact strength and bending and tensile strength, while the heterophasic structure in the case of using PC leads to a lower impact strength.

It is known that, in the process of processing composite materials containing fibrous fillers, there is a significant destruction of the fibers and a reduction in their length, and, consequently, the reinforcing ability [37]. The determination of the residual linear particle sizes of carbon fibers showed that, after extrusion without the addition of a modifier, the particle sizes are reduced by more than 2 times (Table 5). However, it is seen that the introduction of OPSU leads to a greater preservation of fiber lengths; the residual length of CF in composites with the content of OPSU significantly exceed the particle sizes in the composite without modifier. In this case, there is a tendency to produce a slight increase in particle sizes with an increase in the content of OPSU. The introduction of a PC also leads to an increase in size, but to a lesser extent.

Table 5. The dependence of the residual linear particle size of CF on the content of the OPSU and PC.

Content	Arithmetic Average Fiber Length, μm	Weighted Average Fiber Length, μm
CF	220 ± 32	283 ± 26
PEI-CF	87 ± 11	102 ± 14
OPSU		
10%	118 ± 16	133 ± 11
15%	108 ± 14	126 ± 14
20%	125 ± 12	142 ± 17
PC		
10%	96 ± 10	113 ± 18
15%	106 ± 12	124 ± 15
20%	93 ± 11	108 ± 15

Thus, a decrease in the melt viscosity leads to a decrease in shear stresses during processing, which reduces the degree of destruction of the filler particles. Apparently, this leads to a significant increase in mechanical properties with the introduction of OPSU. It is also possible to increase the wettability of the filler particles by a less viscous polymer melt and to improve their adhesive interaction due to the compatibility of OPSU and the polymer matrix, as confirmed by electron microscopy studies.

As can be seen in microphotographs, the number of holes from the pulled-out fibers on the surface of the fracture of the specimen with the content of OPSU is noticeably smaller; moreover, a denser fixing of the polymer matrix on the CF surface is observed (Figure 8a,b). In the case of using a PC, on the contrary, there are obvious gaps between the fibers and the polymer matrix (Figure 8c). In this case, one can notice the inhomogeneous structure of the matrix with PC (Figure 9b), in contrast to the mixture with OPSU (Figure 9a), which once again confirms the biphasic structure of PEI and PC mixture and, on the contrary, the miscibility of PEI and OPSU. Apparently, these reasons are associated with a decrease in toughness with the introduction of PC and the absence of an increase in strength properties.

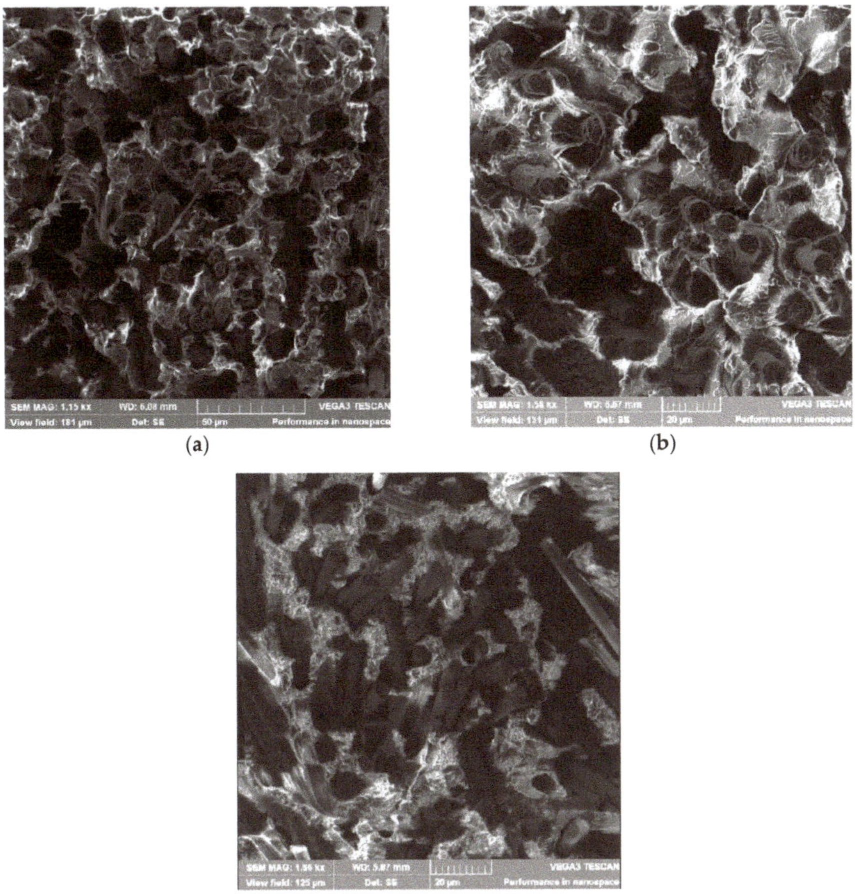

Figure 8. Scanning electron microscope (SEM) images of the broken surface of carbon-filled samples: (**a**)—PEI-CF; (**b**)—PEI-CF + 15% OPSU; (**c**)—PEI-CF + 10% PC.

The thermal stability of carbon-filled composites with different contents of modifiers was studied by the TGA method. Heat resistance was evaluated by temperature loss of 2%, 5% and 10% of the mass. Table 6 shows that OPSU practically does not affect the temperature of the destruction of composites, with the exception of the temperature of loss of 2% of the mass at a 10% content of modifier. It should be noted that the temperature of the loss of 2% of the mass can be affected by various random factors, and they are not always regular, in contrast to the temperatures of loss of 5 and 10% of the mass, which are more correct indicators of destruction. In these cases, it is seen that the temperatures are equal to the values of the initial composite—the difference is within the error of the method.

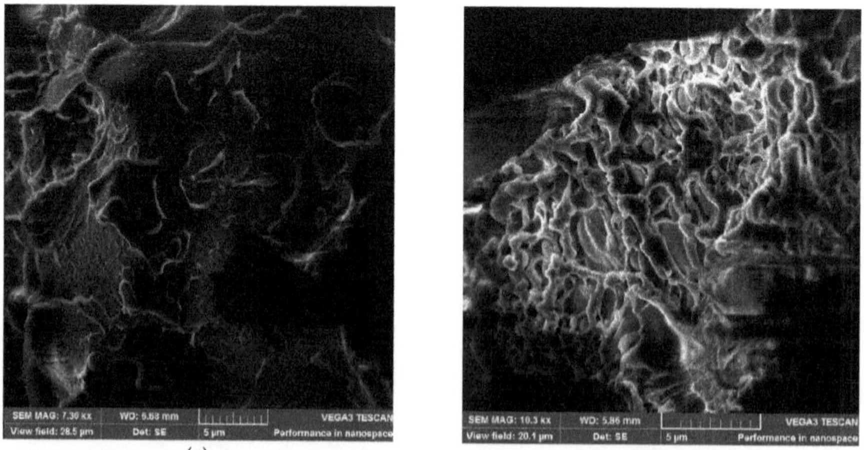

(a) (b)

Figure 9. SEM images of the polymer matrix of carbon-filled samples: (**a**)—PEI-CF + 15% OPSU; (**b**)—PEI-CF + 10% PC.

Table 6. Heat resistance of carbon-filled PEI with different content of OPSU and PC.

Content	$T_{2\%}$	$T_{5\%}$	$T_{10\%}$
PEI-CF	516	529	537
	OPSU		
10%	500	528	539
15%	512	530	540
20%	510	529	539
	PC		
10%	470	495	510
15%	474	508	518
20%	496	510	519

However, it can be noted that PC clearly reduces the temperature of destruction of the material, which is probably due to its lower heat resistance compared to PEI and OPSU. Despite this, the heat resistance of composites with PC content remains quite high.

The thermo-oxidative stability of composites with modifiers was also studied by capillary viscometry. The essence of the study was to measure the melt flow rate of composites at various holding times in the viscometer cylinder at a temperature of 350 °C. As can be seen from Table 7, the starting polymer has a high thermal stability of the melt. The MFR value decreases insignificantly—after holding for 60 min, the MFR decrease is only 10%, which fits into the framework of the method error. The melt stability of the composite with 30% CF also remains quite high—there is no significant change in the MFR.

At 10% OPSU concentration, the MFR practically does not change from the exposure time—the melt is quite stable. However, at higher modifier contents, a noticeable decrease in the MFR with time is observed (by 54% and 63% at 15% and 20% content and exposure time of 60 min). The data obtained indicate a most likely degree of crosslinking with the participation of low molecular weight OPSU, which leads to an increase in melt viscosity.

Table 7. Thermo-oxidative stability of carbon-filled PEI with different content of OPSU and PC.

Content	MFR at Various Exposure Times			
	4 min	5 min	30 min	60 min
PEI	13.2	12.7	12.6	11.3
PEI-CF	5.0	5.4	5.8	6.1
		OPSU		
10%	6.7	7.0	6.4	6.4
15%	12.0	10.9	7.4	5.5
20%	15.6	14.4	8.7	5.8
		PC		
10%	7.5	6.1	2.4	0.1
15%	10.3	10.0	3.7	1.2
20%	14.6	13.0	8.1	4.2

The introduction of PC leads to a noticeable decrease in the thermal stability of the melt. This temperature is high enough for PC; therefore, as can be seen from Table 7, prolonged aging leads to a significant reduction in the MFR—a reduction that is greater than that of composites with OPSU.

Based on these results, a filaments were obtained for 3D printing from composites contenting 25% of CF and 20% OPSU and the same composition but with PC instead OPSU (Figure 10a,b). Standard samples were printed by the FDM method with the longitudinal laying of filaments (filament orientation angle 0 °) and a negative gap between them.

(**a**)

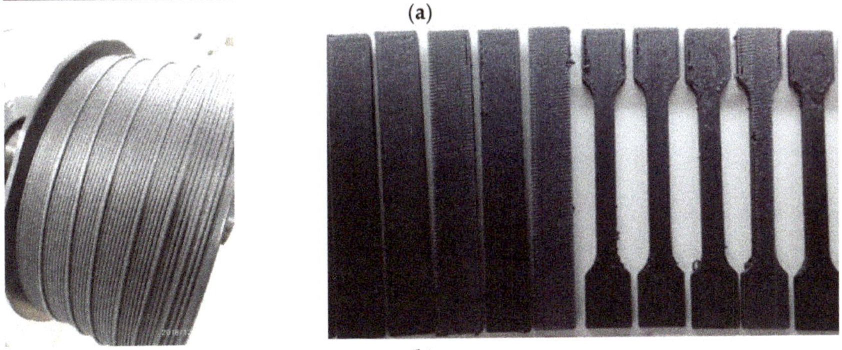

(**b**)

Figure 10. Composite filament for printing and printed test specimens: PEI/CF/OPSU (**a**); PEI/CF/PC (**b**).

As can be seen from the test results (Table 8), the printed samples have high mechanical properties. However, the composite with OPSU significantly exceeds the similar composite with PC in flexural and tension elastic modulus by 58% and 39%, respectively, and in flexural strength by 117%.

Table 8. Mechanical properties of the printed composites based on PEI/CF with OPSU and PC.

Properties	PEI/CF/OPSU	PEI/CF/PC
Flexural modulus, GPa	13.3 ± 1.2	8.4 ± 0.8
Flexural strength, MPa	227.6 ± 7.3	104.8 ± 6.4
Tensile modulus, GPa	5.4 ± 0.5	3.9 ± 0.2
Tensile strength, MPa	73.2 ± 13.2	73.0 ± 10.4
Elongation at break,%	2.2 ± 0.5	3.4 ± 1.6

4. Conclusions

Thus, it was shown that the introduction of OPSU and PC leads to a decrease in the melt viscosity of PEI. At the same time, the elastic modulus and the strength of PEI are preserved; however, in the case of using OPSU, the toughness is significantly reduced. It was found that PEI and PC form a biphasic structure, while PEI and OPSU are thermodynamically compatible, which is confirmed by DSC, electron microscopy, and the appearance of the samples.

It was revealed that PC and OPSU also significantly reduce the melt viscosity of a carbon-filled composite based on PEI, acting as a temporary plasticizer. In the case of using OPSU, there is a significant increase in the elastic modulus and strength, as well as the conservation of impact strength. The introduction of PC, on the contrary, leads to a decrease in toughness, while the elastic modulus and strength remain at the level of the initial composite.

It is shown that the introduction of PC and OPSU into a carbon-filled composite leads to an increase in carbon fiber length in comparison with a composite without modifiers. Moreover, in the composites with OPSU, a higher adhesive interaction of the polymer matrix with the surface of the carbon fiber is observed. The thermal stability and heat resistance of the melts of composites with OPSU significantly exceed composites with PC.

The printed samples of carbon-filled PEI with OPSU and PC have high mechanical properties; however, the composite with OPSU significantly exceeds the properties of the composite with PC in elastic modulus, both in bending and in tension, as well as in flexural strength.

Author Contributions: Investigation, A.S., I.M., A.Z., E.R. and D.K.; Project administration, S.K. All authors have read and agreed to the published version of the manuscript.

Funding: This work was supported by the Ministry of Education and Science of the Russian Federation under agreement No. 14.577.21.0278 of September 26, 2017. Project ID: RFMEFI57717X0278.

Conflicts of Interest: The authors declare no conflicts of interest.

References

1. Tkachuk, A.I.; Grebeneva, T.A.; Chursova, L.V.; Panin, N.N. Thermoplastic binders. Present and Future. *Trudy VIAM* **2013**, *11*, 7.
2. Petrova, G.N.; Bader, E.Y. Development and research of sizing compositions for thermoplastic carbon plastics. *Trudy VIAM* **2016**, *12*. [CrossRef]
3. Vyavahare, S.; Teraiya, S.; Panghal, D.; Kumar, S. Fused deposition modelling: A review. *Rapid Prototyp. J.* **2019**, *26*, 176–201. [CrossRef]
4. Mahmood, S.; Qureshi, A.J.; Talamona, D. Taguchi based process optimization for dimension and tolerance control for fused deposition modeling. *Addit. Manuf.* **2018**, *21*, 183–190. [CrossRef]
5. Gardan, J.; Makke, A.; Recho, N. Improving the fracture toughness of 3D printed thermoplastic polymers by fused deposition modeling. *Int. J. Fract.* **2017**, *210*, 1–15. [CrossRef]
6. Santana, L.; Ahrens, C.H.; da Costa Sabino Netto, A.; Bonin, C. Evaluating the deposition quality of parts produced by an open-source 3D printer. *Rapid Prototyp. J.* **2017**, *23*, 796–803. [CrossRef]
7. Zhang, B.; Seong, B.; Nguyen, V.; Byun, D. 3D printing of high-resolution PLA-based structures by hybrid electrohydrodynamic and fused deposition modeling techniques. *J. Micromech. Microeng.* **2016**, *26*. [CrossRef]
8. Tanikella, N.G.; Wittbrodt, B.; Pearce, J.M. Tensile strength of commercial polymer materials for fused filament fabrication 3D printing. *Addit. Manuf.* **2017**, *15*, 40–47. [CrossRef]

9. Salazar-Martin, A.G.; Perez, M.A.; García-Granada, A.A.; Reyes, G.; Puigoriol-Forcada, J.M. A study of creep in polycarbonate fused deposition modelling parts. *Mater. Des.* **2018**, *141*, 414–425. [CrossRef]
10. Rohde, S.; Cantrell, J.; Jerez, A.; Kroese, C.; Damiani, D.; Gurnani, R.; Di Sandro, L.; Anton, J.; Young, A.; Steinbach, D.; et al. Experimental characterization of the shear properties of 3D-printed ABS and polycarbonate parts. *Exp. Mech.* **2018**, *58*, 871–884. [CrossRef]
11. Mohan, N.; Senthil, P.; Vinodh, S.; Jayanth, N. A review on composite materials and process parameters optimisation for the fused deposition modelling process. *Virtual Phys. Prototyp.* **2017**, *12*, 47–59. [CrossRef]
12. Mohamed, O.A.; Masood, S.H.; Bhowmik, J.L. Characterization and dynamic mechanical analysis of PC-ABS material processed by fused deposition modelling: An investigation through I-optimal response surface methodology. *Measurement* **2017**, *107*, 128–141. [CrossRef]
13. Baca Lopez, D.M.; Rafiq, A. Tensile Mechanical Behavior of Multi-Polymer Sandwich Structures via Fused Deposition Modelling. *Polymers* **2020**, *12*, 651. [CrossRef] [PubMed]
14. Arif, M.F.; Kumar, S.; Varadarajan, K.M.; Cantwell, W.J. Performance of biocompatible PEEK processed by fused deposition additive manufacturing. *Mater. Des.* **2018**, *146*, 249–259. [CrossRef]
15. Wang, Y.; Müller, W.-D.; Rumjahn, A.; Schwitalla, A. Parameters influencing the outcome of additive manufacturing of tiny medical devices based on PEEK. *Materials* **2020**, *13*, 466. [CrossRef]
16. Slonov, A.L.; Khashirov, A.A.; Zhansitov, A.A.; Rzhevskaya, E.V.; Khashirova, S.Y. The influence of the 3D-printing technology on the physical and mechanical properties of polyphenylene sulfone. *Rapid Prototyp. J.* **2018**, *24*, 1124–1130. [CrossRef]
17. Ramiro, J.; Eguiazabal, J.I.; Nazabal, J. Phase structure and mechanical properties of blends of poly(ether imide) and bisphenol A polycarbonate. *Polym. J.* **2004**, *36*, 705–715. [CrossRef]
18. Jiang, S.; Liao, G.; Xu, D.; Liu, F.; Li, W.; Cheng, Y.; Li, Z.; Xu, G. Mechanical properties analysis of polyetherimide parts fabricated by fused deposition modeling. *High Perform. Polym.* **2019**, *31*, 97–106. [CrossRef]
19. Bagsik, A.; Schöppner, V.; Klemp, E. FDM part quality manufactured with Ultem*9085. In Proceedings of the 14th International Scientific Conference on Polymeric Materials, Halle (Saale), Germany, 15–17 September 2010; Volume 15, pp. 307–315.
20. Gebisa, A.; Lemu, H. Investigating effects of fused-deposition modeling (FDM) processing parameters on flexural properties of ULTEM 9085 using designed experiment. *Materials* **2018**, *11*, 500. [CrossRef]
21. Cicala, G.; Ognibene, G.; Portuesi, S.; Blanco, I.; Rapisarda, M.; Pergolizzi, E.; Recca, G. Comparison of Ultem 9085 Used in Fused Deposition Modeling (FDM) with Polytherimide Blends. *Materials* **2018**, *11*, 285. [CrossRef]
22. Mikhailin, Y.A. *Heat-Resistant Polymers and Polymeric Materials*; Professiya: St. Petersburg, Russia, 2006; 624p.
23. PEI with Carbon Fiber (Ultem CF). Available online: https://filament2print.com/gb/advanced/893-pei-carbon-fiber.html (accessed on 5 March 2020).
24. CarbonX™ CF-PEI 3D Printing Filament. Available online: https://www.3dxtech.com/carbonx-carbon-fiber-pei-3d-filament-made-using-ultem-pei/ (accessed on 6 March 2020).
25. Chuang, K.C.; Grady, J.E.; Draper, R.D. Additive manufacturing and characterization of Ultem polymers and composites. In Proceedings of the CAMX Conference, Dallas, TX, USA, 26–29 October 2015.
26. Xanthos, M. *Functional Fillers for Plastics*; Nauchnyye osnovy i tekhnologii: St. Petersburg, Russia, 2010; 462p.
27. Slonov, A.L.; Zhansitov, A.A.; Rzhevskaya, E.V.; Khakulova, D.M.; Khakyasheva, E.V.; Khashirova, S.Y. Study of the Geometric Characteristics of Carbon Fiber Fillers on the Properties of Polyphenylene Sulfone. *Mater. Sci. Forum* **2018**, *935*, 5–10. [CrossRef]
28. Slonov, A.L.; Zhansitov, A.A.; Musov, I.V.; Rzhevskaya, E.V.; Khakulova, D.M.; Khashirov, A.A.; Khashirova, S.Y. Research of influence of fillers different nature on the properties of polysulfone and the definition of possibility of applying the composites on their base in the 3D printing. *Plasticheskie Massy* **2018**, *7–8*, 34–37. [CrossRef]
29. Kishore, V.; Chen, X.; Ajinjeru, C.; Hassen, A.A.; Lindah, J.; Failla, J.; Kunc, V.; Duty, C. Additive manufacturing of high performance semicrystalline thermoplastics and their composites solid freeform fabrication. In Proceedings of the Additive Manufacturing of High Performance Semicrystalline Thermoplastics and Their Composites, Austin, TX, USA, 8–10 August 2016.
30. Turner, N.B.; Strong, R.; Gold, A.S. A review of melt extrusion additive manufacturing processes: I. Process design and modeling. *Rapid Prototyp. J.* **2014**, *20*, 192–204. [CrossRef]

31. Barstein, R.S.; Kirilovich, V.I.; Nosovsky, Y.E. *Plasticizers for Polymers*; Khimiya: Moscow, Russia, 1982; 200p.
32. Kozlov, P.V.; Pankov, S.P. *Physicochemical Principles of Plasticization of Polymers*; Khimiya: Moscow, Russia, 1982; 224p.
33. Kalinchev, E.L.; Sakovtseva, M.B. Effective injection molding of polymeric materials with lubricants. *Polym. Mater.* **2014**, *7*, 14–26.
34. Shakhmurzova, K.T.; Zhansitov, A.A.; Kurdanova, Z.I.; Baykaziev, A.E.; Guchinov, V.A.; Khashirova, S.Y. Study of influence of molecular weight on the physical and chemical properties of polyphenylene sulfones. *Proc. Kabardino-Balkarian State Univ.* **2016**, *6*, 64–66.
35. Yong, S.C.; Heon, S.L.; Woo, W.K. Thermal Properties and Morphology of Blends of Poly (ether imide) and Polycarbonate. *Polym. Eng. Sci.* **1996**, *36*. [CrossRef]
36. Slonov, A.L.; Musov, I.V.; Zhansitov, A.A.; Khakulova, D.M.; Rzhevskaya, E.V.; Khashirova, S.Y. Investigation of the Influence of Linear Dimensions and Concentration of Carbon and Glass Fibers on the Properties of Polyetherimide. *Key Eng. Mater.* **2019**, *816*, 48–54. [CrossRef]
37. Paul, D.R.; Bucknell, K.B. *Polymer Mixtures. Volume II: Functional Properties*; Kuleznev, V.N., Ed.; Nauchnyye osnovy i tekhnologii: St. Petersburg, Russia, 2009; 606p.

polymers

MDPI

Article

Surface Modification of Staple Carbon Fiber by Dopamine to Reinforce Natural Latex Composite

Xiaolong Tian [1], Shuang Han [1], Qianxiao Zhuang [1], Huiguang Bian [1], Shaoming Li [1,2], Changquan Zhang [1], Chuansheng Wang [1,2,*] and Wenwen Han [1,2,3,*]

1 College of Electromechanical Engineering, Qingdao University of Science and Technology, Qingdao 266061, Shandong Province, China; 15165268516@163.com (X.T.); hanshuang258@163.com (S.H.); zqx_112@163.com (Q.Z.); bhg@qust.edu.cn (H.B.); jdwz@qust.edu.cn (S.L.); qw18561929710@163.com (C.Z.)

2 Academic Division of Engineering, Qingdao University of Science & Technology, Qingdao 266061, Shandong Province, China

3 National Engineering Laboratory for Advanced Tire Equipment and Key Materials, Qingdao University of Science and Technology, Qingdao 266061, Shandong Province, China

* Correspondence: wangchuanshengtxl@163.com (C.W.); hbhanwenwen@qust.edu.cn (W.H.)

Received: 30 March 2020; Accepted: 22 April 2020; Published: 24 April 2020

Abstract: Carbon fiber significantly enhances the mechanical, thermal and electrical properties of rubber composites, which are widely used in aerospace, military, national defense and other cutting-edge fields. The preparation of a high-performance carbon fiber rubber composite has been a research hotspot, because the surface of carbon fiber is smooth, reactive inert and has a poor adhesion with rubber. In this paper, a high-performance rubber composite is prepared by mixing dopamine-modified staple carbon fiber with natural latex, and the mechanisms of modified carbon fiber-reinforced natural latex composite are explored. The experimental results show that the surface-modified staple carbon fiber forms uniform and widely covered polydopamine coatings, which significantly improve the interface adhesion between the carbon fiber and the rubber matrix. Meanwhile, when the concentration of dopamine is 1.5 g/L and the staple carbon fiber is modified for 6h, the carbon fiber rubber composite shows excellent conductivity, thermal conductivity, and dynamic mechanical properties, and its tensile strength is 10.6% higher than that of the unmodified sample.

Keywords: surface modification of staple carbon fiber; natural rubber latex; reinforcement mechanism; dopamine; rubber composite

1. Introduction

Carbon fiber (CF) is formed by the thermal conversion of organic fibers, and has a high strength, high modulus, high thermal conductivity, chemical corrosion resistance and other excellent properties [1,2]. At present, CF-reinforced rubber composites (CFRC) have been widely used in aerospace, military, national defense, automobile and other fields due to their excellent properties [3–6]. A good interface between the CF and the rubber matrix is significantly difficult to obtain, and this plays a crucial role in preparing high-performance CFRC. This has also resulted in extensive research by scholars at home and abroad.

The surface modification of CF has become the main method to enhance the interfacial adhesion, which can be divided into physical modification and chemical modification [7–10]. The main function is to improve the surface roughness of the CF and increase the meshing effect between the CF and the matrix, on the other hand, to introduce active groups on the surface of carbon fiber to enhance the interaction between the CF and the rubber matrix [11–14]. As a new surface modification material, the application of dopamine is gradually expanding [15,16]. Kim et al. [17] found that, under the condition of high oxygen concentration, dopamine can be evenly deposited in a short time, and a smooth layer of dopamine can be obtained. Win et al. [18] proved that the addition of oxidants can greatly shorten the self-polymerization time of dopamine. Du et al. [19] used ultraviolet radiation to control the self-polymerization of dopamine.

The processing method of CFRC is also an important factor affecting the interfacial adhesion. In the traditional processing, the natural rubber (NR) subdivision, CF incorporation, agglomeration of CF and uniform distribution of CF successively occurs, which mainly depends on the strong shear force provided by the internal mixer in order to realize the uniform dispersion of CF in the rubber matrix, as shown in Figure 1. CF is easy to agglomerate and is broken in the traditional processing method, which decreases the properties of CFRC. To prepare high-performance CFRC, natural rubber latex (NRL), which has a good film-forming performance and can evenly cover CF to form a film, replaces NR to reinforce the interfacial adhesion between the CF and the rubber matrix.

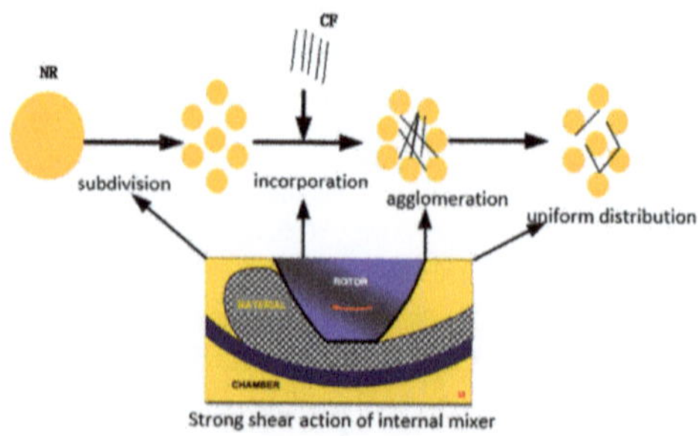

Figure 1. The processing of carbon fiber-reinforced rubber composites (CFRCs).

The present work investigates the difference in the properties of CFRC prepared by NRL or NR and verifies the advantages of wet mixing NRL and CF to prepare NRL/CF composites. Consequently, the mechanisms of dopamine surface-modified carbon fiber-reinforced natural rubber latex composites are systematically studied. Furthermore, the effects of dopamine concentration and modification time on the properties of composites are studied.

2. Materials and Methods

2.1. Experimental Method

Table 1 shows the experimental scheme, exploring the influence of dopamine surface-modified CF on the properties of CFRC.

Table 1. Experimental scheme.

Sample Number	Rubber	Dopamine [1] Modified CF Process	
		Dopamine Concentration (g/L)	Processing Time (h)
A	NR	0	0
B	NRL	0	0
C	NRL	0.5	2
D	NRL	1	2
E	NRL	1.5	2
F	NRL	2	2
G	NRL	1.5	4
H	NRL	1.5	6
I	NRL	1.5	8

[1] Dopamine purchased in Shanghai Aladdin Biochemical Technology Co., Ltd., Shanghai, China.

2.2. Materials

Table 2 shows the composites' formulations and suppliers.

Table 2. Compound formulation and suppliers.

Component	Formulation (phr)	Suppliers
NR/NRL	100	Von Bundit Co. Ltd., Phuket, Thailand
CF [1]	Variable	Toray Co., Ltd., Tokyo, Japan, Tyle: T700s, Diameter: 7 μm, Length: 3 mm
Zin Oxide	5	Hebei Shijiazhuang Zinc Oxide Factory, Shijiazhuang, China
Adhesive RA65	1.5	Wuxi Huasheng Rubber New Material Technology Co., Ltd., Wuxi, China
Carbon black N326	40	Jiangxi Black Cat Carbon Black Co., Ltd., Jiangxi, China
Stearic acid	2	Fengyi Grease Technology (Tianjin, China) Co., Ltd., China
Anti-aging agent 4020	2	Shandong shangshun Chemical Co., Ltd., Weifang, China
Resin SL3020	1	Sino Legend (China) Chemical Company Ltd., Suzhou, China
Accelerator CZ	1.5	Shandong shangshun Chemical Co., Ltd., Weifang, China
Sulfur	1.5	ChaoyangTianming Industry and Trade Co., Ltd., Beijing, China

[1] CF is purchased from Toray Co., Ltd., Tokyo, Japan, without laboratory sizing and surface treatment.

2.3. Adhesion Mechanism of CF and NRL

The interface behavior between the CF and the rubber matrix is a key factor affecting the performance of composite materials. A polydopamine layer was grown on the surface of CF, after being modified by dopamine, which was covered by a latex film formed by natural latex, and then bonded with the rubber matrix to produce good interfacial interaction. Figure 2 shows the adhesion interface between the CF and the rubber matrix.

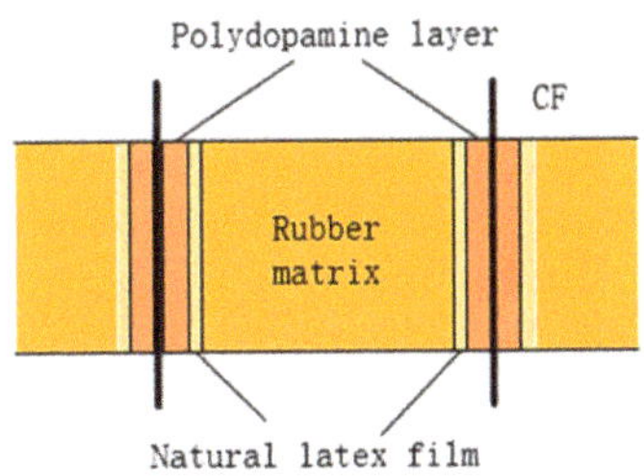

Figure 2. The adhesion interface between the CF and rubber matrix.

Dopamine surface-modified carbon fiber can enhance the bonding effect between the CF and the rubber matrix. Dopamine can be oxidized and self-polymerized in alkaline aqueous solution, forming a polydopamine layer and a covalent bond with double bond groups in rubber molecules, while binding with CF through intermolecular interaction, which involves a Van der Waals interaction, hydrogen bond, and other non-covalent interactions, thus forming a good interface effect. The formation process of the polydopamine layer is shown in Figure 3.

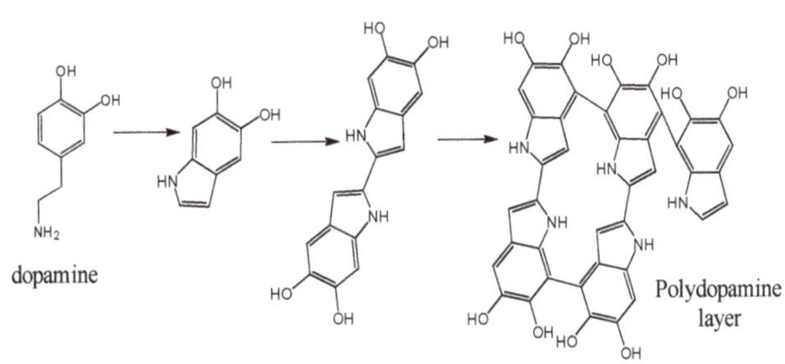

Figure 3. The formation process of the polydopamine layer.

2.4. Preparation of Composite Materials

2.4.1. Pretreatment of the Material

(1) Impregnate CF with NRL:

Use ultrasound (VCY-1500, Shanghai Yanyong Chaosheng Equipment Co., Ltd., Shanghai, China; experimental parameters: ultrasonic power, 1000 W; ultrasonic time, 5 min) to vibrate the NRL to destroy the protein layer and phospholipid layer of the latex particles. Clean the CF with deionized water and add it to the treated NRL. Use the high-speed disperser (T 25 easy clean digital, German IKA company, Staufen im Breisgau, Germany; rotation rate: 400 rpm) to mix the composite for 10 min, and then pour it into a large tray and tile it for drying in order to prepare the NRL/CF masterbatch.

(2) Modify CF with different concentrations of dopamine:

Prepare dopamine solutions with concentrations of 0.5, 1, 1.5 and 2 g/L. Add the Tris (Trihydroxymethyl aminomethane, purchased from Shanghai Aladdin Biochemical Technology Co., Ltd., Shanghai, China) to the dopamine solution and adjust the pH to around 8.5. Clean the CF with deionized water, and add it into different concentrations of dopamine, heating for 2 h in a water bath. Then, clean the modified CF by deionized water again, in order to mix it with the NFL treated by ultrasound.

(3) Modify CF with different processing times of dopamine:

Clean the CF with deionized water and add the CF into four dopamine solutions with a concentration of 1.5 g/L, heating for 2 h in a water bath. Then, clean the modified CF by deionized water again, in order to mix it with the NFL treated by ultrasound.

2.4.2. Mixing Processing

(1) NR/CF composite:

CF, NR and other fillers were incorporated into the mixer for mixing. The mixing process was as follows: the rotating speed was 60 rpm, the filling coefficient of the mixer was 0.7, and the cooling water temperature was 60 °C.

Sulfur and accelerant were incorporated into the open mill to prepare the compound of NR/CF composites.

(2) NRL/CF composite:

(a) The mixture of modified CF and NRL treated by ultrasound was mixed by the high-speed disperser for 10 min, and then drying to prepare NRL/CF masterbatch;

(b) The masterbatch and other fillers were incorporated into the mixer according to the mixing process of the compound of NRL/CF composites.

Sulfur and accelerant were incorporated into the open mill to prepare the compound of NRL/CF composites.

2.4.3. Curing Process

The rubber compounds were cured at 150 °C at a pressure of 10 MPa for an optimum cure time (t90) + 3 min.

2.5. Characterization

Curing Characterization. The curing characterization of rubber compounds was evaluated using a moving-die rheometer (M-2000-AN) from GOTECH TESTING MACHINES CO., Ltd. The specimens were tested according to ISO 6502-2: 2018. The Mooney viscosity values of the rubber compounds were evaluated using a Mooney viscometer (UM-2050, GOTECH TESTING MACHINES CO., Ltd., Taichung, Taiwan) according to ISO 289-2: 2016.

Physical and Mechanical Properties. The hardness of the rubber vulcanizate was evaluated using Shore Hardmeter (LX-A, Shanghai Liuling Instrument Factory, Shanghai, China) according to ISO 7619-2: 2004, three points were measured for each sample and the final result produced the median [6]. The tensile and tear properties of the vulcanized rubber were tested using a universal testing machine (TS 2005 b, GOTECH TESTING MACHINES CO., Ltd., Taichung, Taiwan) at a drawing rate of 500 mm/min according to the standards ISO 37: 2005 and ISO 34-1: 2004, respectively [20]. Five specimens were tested for each sample type and the final result produced the median. The abrasion of the rubber vulcanizates were evaluated using a DIN wear machine (GT-2012-D, GOTECH TESTING MACHINES CO., Ltd., Taichung, Taiwan) according to GB/T 1689-1998, three specimens were tested and the final result produced the median.

Dynamic Mechanical Thermal Analysis. The viscoelastic mechanical properties of the vulcanizates were evaluated using a dynamic thermomechanical analyzer (EPLEXOR-150N, Gabo Qualimeter Testanlagen GmbH, Ahlden, Germany), in which the test temperature range was −65 to 65 °C, the heating rate was 2 K/min, the vibration frequency was 10 Hz, the static strain was 5%, the static force was 70 N, the dynamic strain was 0.25%, and the dynamic stress was 60 N.

Morphology Analysis. The cross-section of the sample after tensile fracture, which adhered to the conductive adhesive and was fixed on the sample table for gold spraying, was observed under a scanning electron microscope (JSM-7500F, Japan Electronics Corporation).

Payne effect. The Payne effect of the composites was tested by the rubber processing analyzer, in which the scanning conditions were set at 60 °C, the scanning frequency was 1 Hz, and the scanning range was 0.28%–40%.

3. Results and Discussion

3.1. Dopamine Surface-Modified CF on the Microstructure of the Composite

The cohesion interaction between the CF and the rubber matrix can be seen clearly from the electron microscope image, which is important for the evaluation of surface-modified CF by dopamine.

The CF is impregnated with NRL to form an adhesive film on its surface, which strengthens the adhesion between the CF and NRL from Figure 4. There is a large gap between the CF and the rubber matrix in the composite of NR/CF, which weakens the interface adhesion between the CF and the rubber matrix.

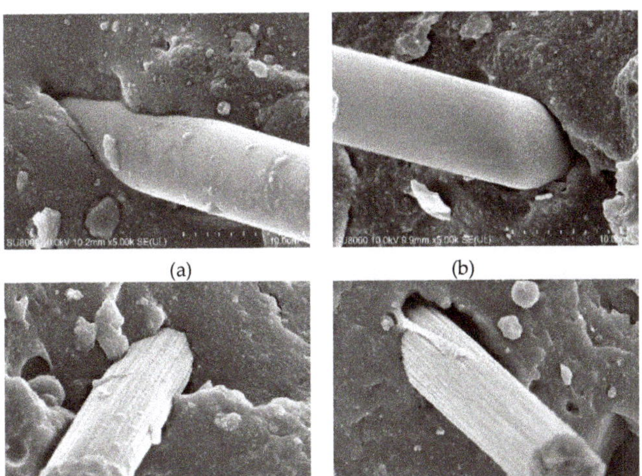

(a) (b) (c) (d)

Figure 4. SEM pictures of CF in natural rubber (NR) and natural rubber latex (NRL): (**a**) SEM pictures of NRL/CF; (**b**) SEM pictures of NR/CF; (**c**) SEM pictures of NRL/CF; (**d**) SEM pictures of NR/CF.

The CF modified by dopamine adheres closely to the rubber matrix, and there is no difference from Figure 5. Different concentrations of dopamine solution can make the surface of the carbon fiber uneven, which makes the bonding effect between the carbon fiber and the rubber matrix different. At lower treatment concentrations (for B), a polydopamine covering could not be formed on the surface of carbon fiber completely, and stress concentration point was easily formed when the composite was stressed; with the increase in the treatment concentration, a uniform covering (for c, d, e) was formed on the surface of the carbon fiber. The interface adhesion between the carbon fiber and the rubber matrix was improved. CF was modified in the higher treatment concentration (for F), in which a non-uniform dopamine coating, which was not conducive to the adhesion between the CF and the rubber matrix, was formed.

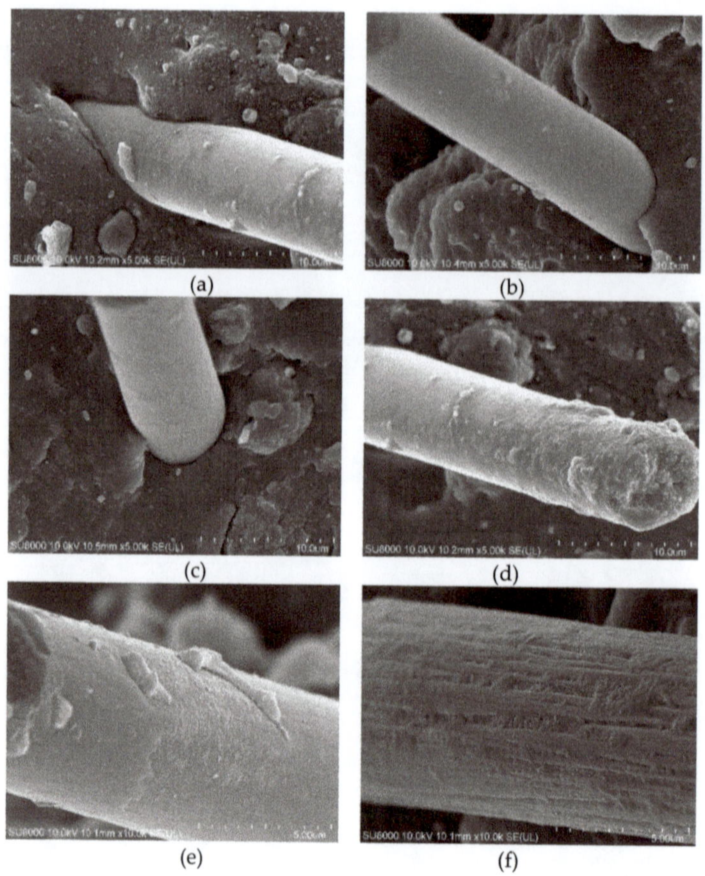

Figure 5. SEM picture of composites of CF modified by different concentration of dopamine. (**a**) The concentration of dopamine was 0 g/L; (**b**) the concentration of dopamine was 0.5 g/L; (**c**) the concentration of dopamine was 1 g/L; (**d**) the concentration of dopamine was 1.5 g/L; (**e**) the concentration of dopamine was 2 g/L; (**f**) the concentration of dopamine was 2.5 g/L.

When CF is separated from the rubber matrix, the surface morphology of NRL/CF composite prepared under different modification times is different from Figure 6. With the increase in modification time, the inhomogeneity of the dopamine coating on the surface of the carbon fiber increases, which is not conducive to the adhesion between the CF and the rubber matrix, thus reducing the adhesion performance of the composite. Dopamine coating evenly distributed, when CF was modified by dopamine for 6 h.

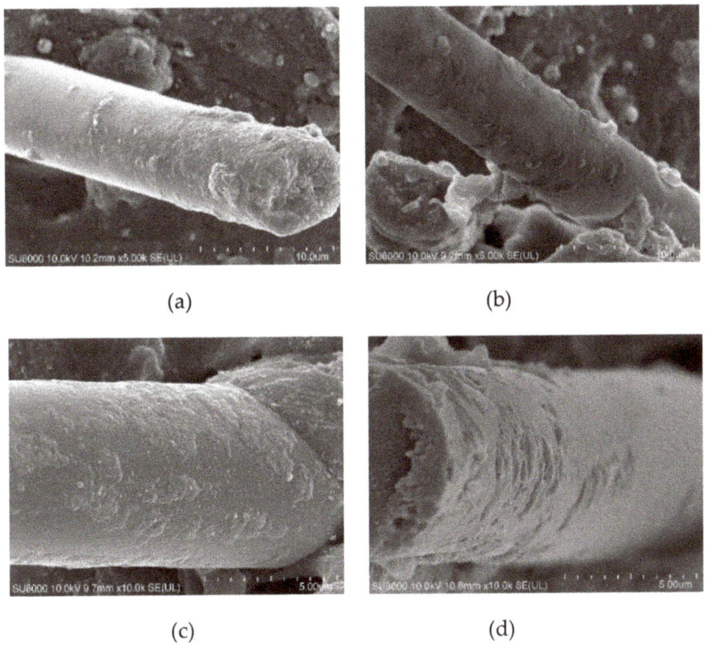

Figure 6. SEM picture of composites of CF modified by different processing times of dopamine. (a) The processing time of dopamine was 2 h; (b) the processing time of dopamine was 4 h; (c) the processing time of dopamine was 6 h; (d) the processing time of dopamine was 8 h;

3.2. Dopamine Surface-Modified CF on the Processability of Composites

The processing properties of the NRL/CF composition were improved by CF impregnated by NFL, meaning that the value of scorching time (tc10), curing time (tc90), minimum torque (M_L), maximum torque (M_H), degree of crosslinking ($M_H - M_L$) decreased, and the Mooney viscosity (ML(1 + 4)) increased from group A to B in Table 3.

Table 3. Processing properties of composite materials.

Test Item	A	B	C	D	E	F	G	H	I
tc10/min	3:54	1:58	2:33	2:29	2:32	2:27	3:29	3:23	3:46
tc90/min	8:26	7:47	6:10	5:54	5:59	5:51	8:35	7:35	8:06
M_L/(dN·m)	2.04	1.95	2.01	2.46	2.48	2.32	2.40	2.57	2.43
M_H/(dN·m)	17.28	16.03	16.23	16.63	16.98	16.22	16.78	17.24	17.05
$M_H - M_L$/(dN·m)	15.24	14.08	14.22	14.23	14.50	13.90	14.38	14.67	14.26
ML(1 + 4)100 °C	44.5	47.3	47.2	46.9	47.0	47.2	48.1	48.0	47.7

The composite of NRL/CF was reinforced by dopamine-modified CF, while the values of scorching time and the degree of crosslinking increased and the curing time and Mooney viscosity decreased from groups B, C, D, E and to group F in Table 3. Meanwhile, as the curing time is shortened, the degree of crosslinking tends to rise at first, and then fall, while the Mooney viscosity tends to rise at first, then fall, then rise again with the increase in the dopamine concentration.

A layer of dopamine with a certain thickness, forming on the surface of dopamine-treated CF, reinforces the cohesion action between the CF and the rubber matrix. Thus, the non-uniform dopamine layer weakens the cohesion action, which is caused by the increasing concentrations.

The scorching time and curing time of the NRL/CF composite tend to increase due to the faintly acidic nature of dopamine, making the curing speed slow and causing the degree of crosslinking to increase slightly alongside the extension of the dopamine modification time in groups E, G, H and I in Table 3. This is mainly because the generation of a uniform dopamine layer requires a certain amount of time, which is the optimum modification point. If the dopamine modification time is insufficient, the polydopamine coating distributes unevenly on the surface of CF. However, an excessive modification time leads to thickness variations in the polydopamine layer. These non-uniform layers weaken the cohesion action.

3.3. Dopamine Surface-Modified CF on the Comprehensive Properties of Composite

The composite of NRL/CF has similar properties in terms of its hardness, tensile stress at 100% elongation (TS 100%), tensile stress at 300% elongation (TS 300%), tensile strength, elongation at break, resilience and DIN abrasion compared with the NR/CF composite, whereas volume resistivity is prominently improved from A to B in Table 4. Natural rubber is made from natural latex by the processes of drying and acidifying (and other procedures), in which some conductive impurities may be incorporated, which can promote its conductivity to some extent. Thus, the conductivity of the composite of NR/CF is inferior to that of NRL/CF.

Table 4. The physical and mechanical properties of composites.

Test Item	A	B	C	D	E	F	G	H	I
Hardness/°	64	65	65	66	66	66	66	66	67
TS 100% [1]/MPa	3.21	2.97	3.01	2.93	3.15	2.97	2.64	2.68	2.35
TS 300% [2] /MPa	12.80	12.5	12.38	12.47	12.72	12.43	13.23	13.81	11.90
TS [3]/MPa	24.78	24.8	24.42	25.75	26.15	25.44	26.42	27.45	26.13
Elongation at break/%	510.36	490.04	552.96	519.68	544.44	550.52	490.84	509.91	524.16
Resilience/%	71.98	71.84	70.87	71.24	71.98	71.46	72.03	71.99	72.14
Abrasion/cm^{-3}	0.139	0.140	0.151	0.148	0.140	0.142	0.141	0.139	0.140
Volume resistivity/Ω·cm	1.04×10^6	4.71×10^7	2.52×10^7	2.62×10^6	1.02×10^7	1.61×10^7	1.01×10^7	1.54×10^7	2.8×10^8

[1] Tensile stress at 100% elongation (TS 100%); [2] tensile stress at 300% elongation (TS 300%); [3] tensile strength (TS).

The hardness of the composite of NRL/CF is basically unchanged, TS 100% and TS 300% tend to grow with the increase in dopamine concentrations from groups B, C, D, E and to group F in Table 4. When the dopamine concentrations are 1.5 g/L, the tensile strength of the composite of NRL/CF reaches the peak, which is 5.4% higher than the unmodified one (A) and the DIN abrasion decreases. This is mainly due to the even distribution of the polydopamine coating on the surface of the CF, which can reinforce the cohesion interaction between the carbon fiber and the rubber matrix, thus improving the tensile strength and the abrasion of the composite. Meanwhile, the volume resistivity of NRL/CF composites decreased first and then increased; however, overall, it is lower than that of the unmodified one (B), which correlates with the increase in the dopamine concentration. The polydopamine coating on the CF aggregated uniformly, making the interfacial interaction between the rubber matrix and the CF enhanced and the conductivity of the composite improved. At the same time, the polydopamine coating on the surface of CF will be unevenly distributed at a higher treatment concentration, meaning that the volume resistivity of the composite increases.

The hardness, resilience and abrasion of NRL/CF composites were less affected by the processing time of dopamine-modified carbon fiber. The comprehensive properties of NRL/CF composites were the best when carbon fibers were modified by the dopamine solution for 6 h. The main reason is that, before the optimal modification time is reached, the longer the time is, the larger the coverage area of the polydopamine coating on the surface of the CF is and the more uniform the thickness is, which is conducive to the interfacial adhesion between the CF and the rubber matrix. When the modification time exceeds the optimal value, the thickness of the polydopamine coating on the surface of the carbon fiber will be too thick or uneven, which will lead to a decline in the interfacial adhesion between the carbon fiber and the rubber matrix, thus reducing the mechanical properties of the rubber composite. The volume resistivity of the NRL/CF composite increases with the increase in modification time, indicating that the conductivity of the rubber sample decreases. Because of the poor conductivity of dopamine, the formation of polydopamine growing on the surface of carbon fiber will hinder the conduction pathway of the carbon fiber, thus reducing the conductivity of the NRL/CF composite.

3.4. Dopamine Surface-Modified CF on Thermal Conductivity of Composites

CF has good thermal conductivity, which is dispersed evenly in the rubber matrix and overlaps to form a thermal conduction channel, enabling the NRL/CF composite to have a strong thermal conductivity. The thermal conductivity coefficient of the composite presents the tendency of rising first and then decreasing at temperatures of 60, 90 and 120 °C with the increase in the dopamine concentration from Figure 7. The main factors are that the polydopamine layer gradually changes alongside the increase in the concentration, and cannot be distributed evenly on either lower or higher modification concentrations, resulting in the formation of a clearance at the bonding point of the rubber matrix and the CF. This decelerates the conduction of heat through the composite. Thus, the dopamine layer on the surface of the carbon fiber was evenly distributed and the thermal conductivity coefficient of the composite was high at concentrations of 1 and 1.5 g/L.

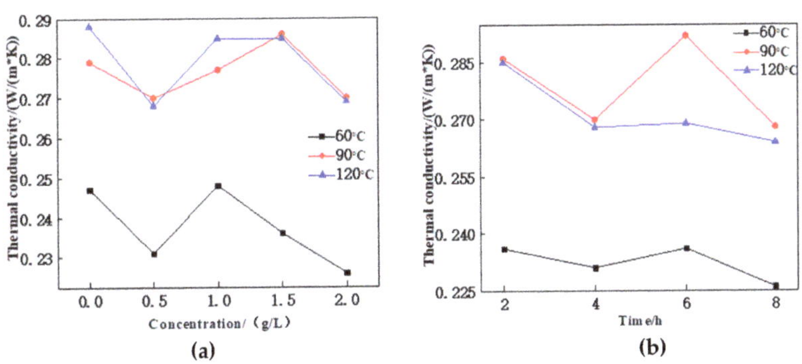

Figure 7. Thermal conductivity of composites: (**a**) effect of CF modified with different concentrations of dopamine on thermal conductivity of composites; (**b**) effect of processing time of dopamine-modified CF on thermal conductivity of composites.

The effect of CF modification time on NRL/CF composite is the same as that of modification concentrations. This is mainly because the modification time also affects the uniformity of the dopamine layer on the surface of the carbon fiber. The dopamine layer cannot be distributed evenly under either a shorter or longer modification time. When the modification time is 6 h, the dopamine layer distributes evenly on the surface of CF, and the thermal conductivity coefficient of the composite is high.

3.5. Dopamine Surface-Modified CF on Payne Effect of Composites

The Payne effect [21] can evaluate the interaction between composite fillers, which is usually quantified by $\Delta G'$. The Payne effect was calculated as follows:

$$\Delta G' = G'(0.28\%) - G'(40\%)$$

Generally speaking, the better the dispersion of the filler is, the lower the Payne effect is. Table 5 shows the test results of the Payne effect of the composites.

Table 5. The test results of Payne effect of composites.

	A	B	C	D	E	F	G	H	I
$G'(0.28\%)$/KPa	513.25	421.13	851.28	963.64	765.4	759.94	396.5	390.44	385.63
$G'(40\%)$/KPa	241.47	221.9	346.54	382.96	344.6	338.82	238.47	229.98	252.97
$\Delta G'$/KPa	271.78	199.23	504.74	580.68	420.8	421.12	158.03	160.46	132.7

The shear modulus $G'(0.28\%)$ and $\Delta G'$ of the NRL/CF composite is obviously lower than that of the NR/CF composite from Figure 8 and Table 5. The CF impregnated by latex can be distributed uniformly across the rubber matrix, which can weaken the interaction between the fillers, and decrease the Payne effect.

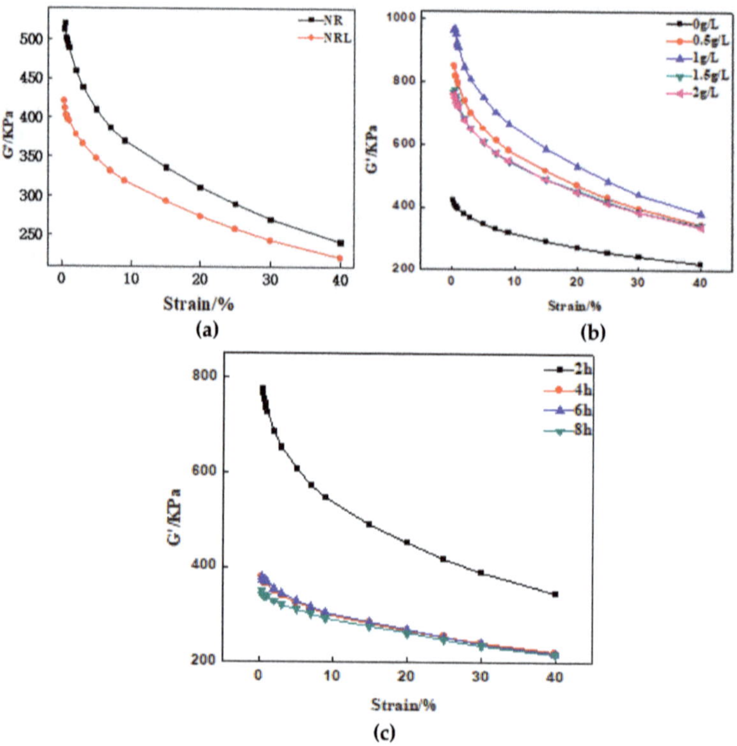

Figure 8. Viscoelasticity curve of composite: (**a**) viscoelasticity curve of CF with natural rubber and natural latex; (**b**) viscoelasticity curve of composites with CF modified by different concentrations of dopamine; (**c**) viscoelasticity curve of composites with CF modified by different processing times of dopamine.

The $\Delta G'$ of the dopamine-modified NRL/CF composite rises significantly from Figure 8, which indicates that dopamine-modified CF can restrict the rubber matrix deformation to improve the interfacial adhesion between the CF and the rubber matrix. The value of $\Delta G'$ in terms of the dopamine-modified NRL/CF composite increases significantly from Table 5. The polydopamine layer not only reinforces the adhesion between the CF and the rubber matrix, but also causes the adhesion between CFs, so the CF aggregates and the dispersion of CF in the rubber matrix gets worse. The $\Delta G'$ of the composite is relatively low at the dopamine concentrations of 1.5 and 2 g/L from Table 5, which indicates that the dispersion of CF in the rubber matrix and the adhesion between the CF and the rubber matrix are preferable.

The value of $\Delta G'$ tends to decrease with the modification time, making it clear that the agglomeration of the carbon fiber is weakened and the dispersion of the carbon fiber in the rubber matrix is improved with an increase in the modification time of CF.

3.6. Dopamine Surface Modification CF on Dynamic Viscoelasticity of Composites

The dynamic viscoelasticity of vulcanized rubber includes rolling resistance, wear resistance and wet-skid resistance. Rolling resistance reflects fuel economy, wear-resistance reflects the durability and service life of the tire, and the anti-skid performance is directly related to the safety of the tires, which attracts much attention.

The curves of the loss factors with the temperature changes in the NRL/CF composites basically overlap with the curves of the NR/CF composites from Figure 9. This means that the properties of the composites prepared by these two methods are almost the same in terms of wet resistance and rolling resistance.

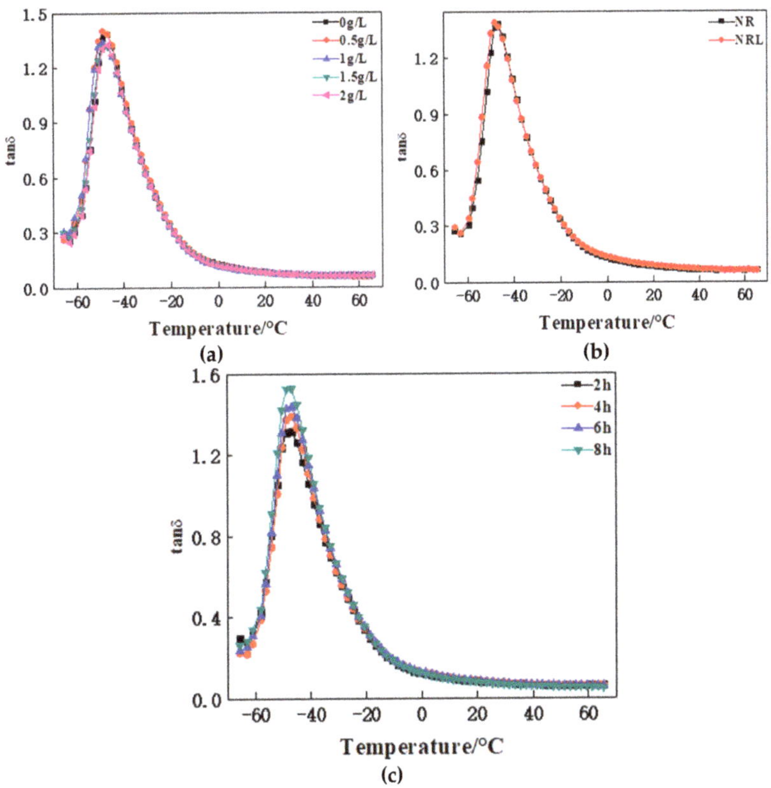

Figure 9. tanδ-T Curve of composites: (**a**) tanδ-T Curve of CF in natural rubber and natural latex; (**b**) tanδ-T Curve composites of CF modified by different concentration of dopamine; (**c**) tanδ-T Curve of composites of CF modified by different processing time of dopamine.

The loss factor of the composite of NRL/CF modified by dopamine of different concentrations is lower than that of the unmodified one at temperatures of 0 and 60 °C (Figure 6). The interfacial interaction between modified CF and the rubber matrix become stronger, and restrict the movement of the rubber molecule chain, making the shear modulus of the composite increase. Meanwhile, the CF becomes the stress point.

The loss factor of the composite of NRL/CF modified by dopamine of different modification times changes slightly at a temperature of 0 °C, and decreases gradually at 60 °C (Figure 6). A polydopamine layer is uniformly generated on the surface of the CF, enlarging the contact area of the CF and the rubber matrix, with the modification time increasing. The interfacial interaction is improved in order to restrict the deformation of the rubber matrix.

4. Conclusions

(1) Impregnated by the natural latex, the adhesion between staple carbon fiber and rubber matrix is reinforced, and the material properties of the composite are better than those prepared by the traditional method;

(2) When the concentration of dopamine was 1.5 g/L and the modification time was 6 h, the properties of the composites modified by dopamine were the best. The main reason is that when the concentration and modification time reaches an optimal value, dopamine forms a uniform polydopamine coating with a wide coverage on the surface of the carbon fiber, which significantly improves the interface adhesion between the carbon fiber and rubber matrix.

Author Contributions: Conceptualization, C.W. and X.T.; writing—original draft preparation, X.T., S.H. and Q.Z.; writing—review and editing, W.H., C.Z. and S.L.; funding acquisition, C.W., H.B. and W.H. All authors have read and agreed to the published version of the manuscript.

Funding: This research was funded by the Shandong Provincial Natural Science Foundation (grant number ZR2016XJ003 and ZR2019BEE056), Qingchuang Technology Plan (No. 2019KJB007) and Key Research and Development Plan of Shandong Province (No. 2019GGX102018).

Conflicts of Interest: The authors declare no conflict of interest.

References

1. Li, J.-W.; Hsu, H.-H.; Chang, C.-J.; Chiu, Y.-J.; Tseng, H.-F.; Chang, K.-C.; Krishna Karapala, V.; Lu, T.-C.; Chen, J.-T. Preparation and Thermal Dissipation of Hollow Carbon Fibers from Electrospun Polystyrene/Poly(amic acid) Carboxylate Salt Core-Shell Fibers. *Eur. Polym. J.* **2020**, *130*, 109648. [CrossRef]
2. Liu, D.; Ouyang, Q.; Jiang, X.; Ma, H.; Chen, Y.; He, L. Thermal properties and thermal stabilization of lignosulfonate-acrylonitrile-itaconic acid terpolymer for preparation of carbon fiber. *Polym. Degrad. Stabil.* **2018**, *150*, 57–66. [CrossRef]
3. Hao, Z.; Shen, J.Q.; Sheng, X.; Shen, Z.; Yang, L.; Lu, X.F.; Luo, Z.; Zheng, Q. Enhancing Performances of Polyamide 66 Short Fiber/Natural Rubber Composites via In Situ Vulcanization Reaction. *Fibers Polym.* **2020**, *21*, 392–398. [CrossRef]
4. Yantaboot, K.; Amornsakchai, T. Effect of preparation methods and carbon black distribution on mechanical properties of short pineapple leaf fiber-carbon black reinforced natural rubber hybrid composites. *Polym. Test.* **2017**, *61*, 223–228. [CrossRef]
5. Wu, C.M.; Cheng, Y.C.; Lai, W.Y.; Chen, P.H.; Way, T.D. Friction and Wear Performance of Staple Carbon Fabric-Reinforced Composites: Effects of Surface Topography. *Polymers* **2020**, *12*, 141. [CrossRef]
6. Zheng, B.; Deng, T.; Li, M.; Huang, Z.; Zhou, H.; Li, D. Flexural Behavior and Fracture Mechanisms of Short Carbon Fiber Reinforced Polyether-Ether-Ketone Composites at Various Ambient Temperatures. *Polymers* **2018**, *11*, 18. [CrossRef]
7. Chen, S.; Feng, J. Epoxy laminated composites reinforced with polyethyleneimine functionalized carbon fiber fabric: Mechanical and thermal properties. *Compos. Sci. Technol.* **2014**, *101*, 145–151. [CrossRef]
8. Dreyer, D.R.; Miller, D.J.; Freeman, B.D.; Paul, D.R.; Bielawski, C.W. Elucidating the Structure of Poly(dopamine). *Langmuir* **2012**, *28*, 6428–6435. [CrossRef] [PubMed]
9. Wu, W.L.; Li, J.K. Study on Carbon Fiber Reinforced Chloroprene Rubber Composites. *Adv. Mater. Res.* **2014**, *1052*, 254–257. [CrossRef]
10. Yi, M.; Sun, H.; Zhang, H.; Deng, X.; Cai, Q.; Yang, X. Flexible fiber-reinforced composites with improved interfacial adhesion by mussel-inspired polydopamine and poly(methyl methacrylate) coating. *Mater. Sci. Eng. C* **2016**, *58*, 742–749. [CrossRef]

11. Huang, J.; Xu, C.; Wu, D.; Lv, Q. Transcrystallization of polypropylene in the presence of polyester/cellulose nanocrystal composite fibers. *Carbohydr. Polym.* **2017**, *167*, 105–114. [CrossRef] [PubMed]
12. Jiang, J.; Yao, X.; Xu, C.; Su, Y.; Zhou, L.; Deng, C. Influence of electrochemical oxidation of carbon fiber on the mechanical properties of carbon fiber/graphene oxide/epoxy composites. *Compos. Part A Appl. Sci. Manuf.* **2017**, *95*, 248–256. [CrossRef]
13. Luo, G.; Li, W.; Liang, W.; Liu, G.; Ma, Y.; Niu, Y.; Li, G. Coupling effects of glass fiber treatment and matrix modification on the interfacial microstructures and the enhanced mechanical properties of glass fiber/polypropylene composites. *Compos. Part B Eng.* **2017**, *111*, 190–199. [CrossRef]
14. Xu, Z.; Huang, Y.; Zhang, C.; Liu, L.; Zhang, Y.; Wang, L. Effect of γ-ray irradiation grafting on the carbon fibers and interfacial adhesion of epoxy composites. *Compos. Sci. Technol.* **2007**, *67*, 3261–3270. [CrossRef]
15. Li, M.; Gu, Y.; Liu, Y.; Li, Y.; Zhang, Z. Interfacial improvement of carbon fiber/epoxy composites using a simple process for depositing commercially functionalized carbon nanotubes on the fibers. *Carbon* **2013**, *52*, 109–121. [CrossRef]
16. Newcomb, B.A. Processing, structure, and properties of carbon fibers. *Compos. Part A Appl. Sci. Manuf.* **2016**, *91*, 262–282. [CrossRef]
17. Kim, H.W.; McCloskey, B.D.; Choi, T.H.; Lee, C.; Kim, M.-J.; Freeman, B.D.; Park, H.B. Oxygen Concentration Control of Dopamine-Induced High Uniformity Surface Coating Chemistry. *ACS Appl. Mater. Int.* **2013**, *5*, 233–238. [CrossRef]
18. Wei, Q.; Zhang, F.; Li, J.; Li, B.; Zhao, C. Oxidant-induced dopamine polymerization for multifunctional coatings. *Polym. Chem.* **2010**, *1*. [CrossRef]
19. Du, X.; Li, L.; Lia, J.; Yang, C.; Frenkel, N.; Welle, A.; Heissler, S.; Nefedov, A.; Grunze, M.; Levkin, P.A. UV-Triggered Dopamine Polymerization: Control of Polymerization, Surface Coating, and Photopatterning. *Adv. Mater.* **2014**, *26*, 8029–8033. [CrossRef]
20. Li, Z.; Wan, J.; Li, Y.; Li, Y.; Zhao, F.; Zhao, S. Effects of coupling agents on the properties of an NR/SBR matrix and its adhesion to continuous basalt fiber cords. *J. Appl. Polym. Sci.* **2018**, *136*. [CrossRef]
21. Payne, A. Effect of Dispersion on Dynamic Properties of Filler-Loaded Rubbers. *J. Appl. Polym. Sci.* **1965**, *9*, 2273–2284. [CrossRef]

polymers

MDPI

Article

PEEK Composites as Self-Lubricating Bush Materials for Articulating Revolute Pin Joints

Juanjuan Zhu [1,*], Fang Xie [2,*] and R S Dwyer-Joyce [1]

1 The Leonardo Centre for Tribology, Department of Mechanical Engineering, University of Sheffield, Mappin Street, Sheffield S1 3JD, UK; r.dwyer-joyce@sheffield.ac.uk

2 School of Mechanical & Automotive Engineering, Nanyang Institute of Technology, Nanyang 473004, China

* Correspondence: juan.zhu@sheffield.ac.uk (J.Z.); xiefang@nyist.edu.cn (F.X.)

Received: 10 February 2020; Accepted: 13 March 2020; Published: 17 March 2020

Abstract: In this study, bearing bushes made of polyetheretherketone (PEEK), 30 wt % carbon fibre reinforced PEEK, 30 wt % glass fibre reinforced PEEK, each 10 wt % of PTFE, graphite and carbon fibre modified PEEK were investigated on a purpose built pin joint test rig. The unlubricated friction and wear behaviour was assessed in sliding contact with a 300M shaft, subjected to a nominal pressure of 93 MPa, articulating sliding speed of 45 °/s. The worn surface and the subsurface layer were studied using optical profilometry and scanning electron microscopy (SEM). Due to thermal sensitivity of PEEK composites, friction energy and temperature rise were analysed for determining the friction and wear mechanism. The bush made of PTFE, graphite and carbon fibre (each 10 wt %) modified PEEK presented the best performance for friction coefficient, wear loss, friction energy and temperature rise. Current work demonstrated that reinforcement modified PEEK composite possesses desirable properties to perform as a load bearing bush in certain tribological applications.

Keywords: PEEK composites; reinforcements; self-lubricating bush; friction and wear; pin joints

1. Introduction

Compared with metals, polymers possess certain desired properties for engineering use, i.e., lightweight (low density), low cost, ease of manufacturing, self-lubricating and corrosion resistance [1–4]. They can provide significant weight savings while maintaining structural performance, and therefore offering improved fuel efficiency for aerospace and other transport applications. In addition, polymers are increasingly used in tribological applications, especially for harsh lubrication conditions, such as bearings, gears, piston rings and seals in aerospace machines and ocean engineering machines or other mechanical components used in high temperature and corrosive environment [5–8].

Among the speciality polymers, polyetheretherketone (PEEK) is one of the most promising engineering materials for tribological applications. Studies have been conducted on the friction and wear of pure PEEK in comparison with other polymers [9,10]. However, there are limitations of pure PEEK, such as low thermal stability, heat conductivity and dissipativity. In order to minimise these disadvantages and to further improve the friction and wear property, PEEK based composites have been tailored with variety of reinforcements, fillers and solid lubricants [11]. In the past twenty years, researchers have made great efforts to develop PEEK-based composites. Mechanical strength, friction and wear properties were studied for carbon fibre reinforced PEEK composites [12,13]. Sumer et al. [14] reported that the glass fibre in the composite improved friction and wear under dry sliding contact. Wang et al. reported that the composite with 7.5 wt % ZrO_2 particles produced a low wear rate and friction coefficient through the block-on-ring tests (PEEK composite block against sliding steel ring) due to the formation of a thin, uniform and tenacious transfer film at the interface [15]. The influence of Polytetrafluoroethylene (PTFE) on the mechanical and tribological properties were studied by Zhang et al. [16] and Bijwe et al. [17].

Tribological behaviour of PEEK composites is also affected by the operating environment, i.e., gas, temperature, lubricant, load, etc. [18–21]. Theiler and Gradt evaluated the tribological behaviour of PEEK composites in air, vacuum and hydrogen environments [18] from pin-on-disc (PEEK composite pin against steel disc) contact. It was found that PEEK composites presented lower environmental sensitivity compared with pure PEEK [18]. Varying lubricants, i.e., water [6,14], sea water [1], mineral oil [5,21,22] were used in the study of friction and wear for PEEK composites. Zhang et al. observed enhanced lubricity under boundary and mixed lubrication regime for the PEEK composites reinforced with graphitic carbon nitride nanosheets when lubricated by PAO4 oil through the plate-on-ring (PEEK composite plate against sliding steel ring) tests [5].

Tribological characteristics of PEEK composites are highly dependent on the tribo-system. The friction energy dissipated in the sliding contact usually causes a consistent temperature rise in the two contacting bodies [7,23]. The temperature variation in service plays an important role in affecting the mechanical, physical and thermal properties, resulting in structural changes of polymer components. Most of the work relating to PEEK composites has been conducted in the lab using standard tribo-meters. For engineering use, some research has been carried out where PEEK composites form the tribological component, including ball bearings [8,24], thrust bearing [25], orthopaedic device [26] and crank shaft bush on the robot joint [27]. There is no work conducted towards journal bearing bushes made of PEEK composites. The aim of the current work is to investigate the tribological performance of PEEK composite used as bearing bushes through a purpose build pin joint test rig subjected to a contact pressure of 93 MPa without lubrication. A thorough assessment was conducted on tested bushes, including friction coefficient, bush wall deformation, wear rate, friction energy and temperature increase. The wear tracks and the subsurface layer were examined to assess the tribological behaviour of PEEK composites used as load bearing bush material.

2. Experimental Methods

2.1. Specimen

In this work, pure PEEK and three PEEK composites produced by injection moulding (Ensinger Ltd., Manchester, UK) were studied. The as-bought materials had the same shape and size (bar with outer diameter of 25 mm). The melting temperature was 334 °C from the manufacture's data sheet. The three PEEK composites were: 30 wt % carbon fibre (~6 μm diameter) reinforced PEEK; 30 wt % glass fibre (~15 μm diameter) reinforced PEEK and each 10 wt % of PTFE, graphite and carbon fibre modified PEEK. Unfilled PEEK was tested for comparison. Figure 1 shows the SEM images of the fracture cross-section for PEEK and PEEK composites, in which how the reinforced fibres distribute and orientate in the matrix are indicated. These reinforced and unreinforced PEEKs were thereafter referred to Bush A, B, C and D respectively, listed in Table 1, including their mechanical and thermal properties.

Table 1. Composition, mechanical and thermal properties of PEEK and PEEK composites [28].

Specimen	PEEK Composite Reinforcements	Density, g/cm^3	Elastic Modulus, GPa	Compression Strength @ 10% Strain, MPa	Rockwell Hardness, M Scale	Elongation at Break @22.8 °C, %	Thermal Conductivity, Wm^{-1}°C^{-1}
Bush A	30 wt % carbon fibre	1.41	6.34	165	107	7	0.92
Bush B	30 wt % glass fibre	1.53	6.89	172	103	2.2	0.3
Bush C	10 wt % each, carbon fibre, graphite, PTFE	1.46	5.52	114	95	2.5	0.82
Bush D	None	1.31	4.48	121	99	40	0.29

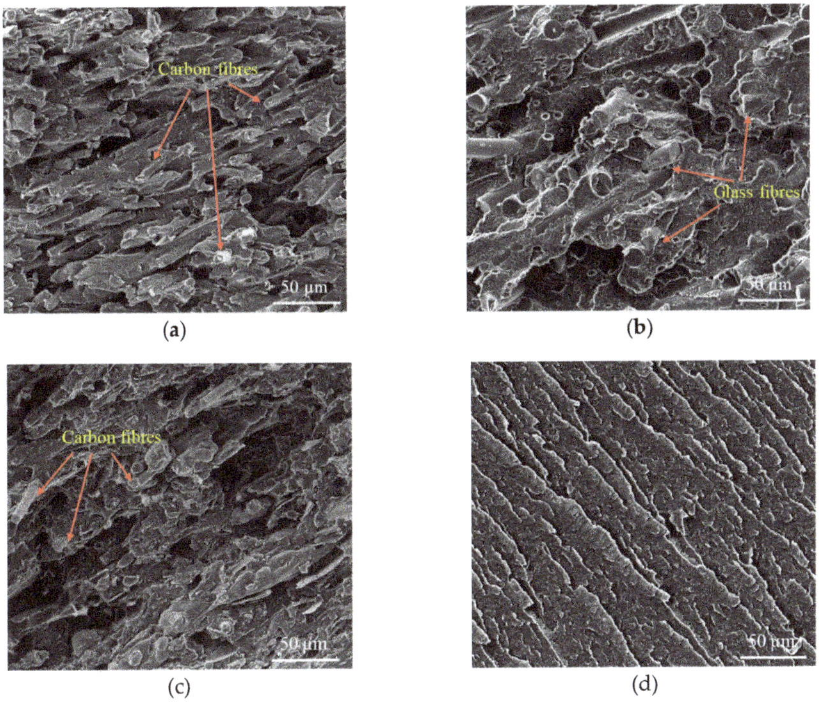

Figure 1. SEM images of fracture cross-section for polyetheretherketone (PEEK) and PEEK composites, (**a**) 30 wt % carbon fibre reinforced PEEK, (**b**) 30 wt % glass fibre reinforced PEEK, (**c**) each 10 wt % of PTFE, graphite and carbon fibre modified PEEK, (**d**) neat PEEK.

PEEK and PEEK composite bars were mechanically machined to bush halves for testing. The machining process involved turning the outer diameter to 15 and 20 mm, bored inner hole and precise reaming to the final inner diameter 10 mm, which left the roughness Ra = 0.7–1.1 μm for the bearing surface. Bushes were slit into halves to accommodate the loading design on the pin joint test rig, shown in Figure 2a. Figure 2b,c shows bush samples, bush holder and bush/pin contact configuration.

2.2. Wear Test

Wear tests were performed on a pin joint rig, shown in Figure 2a. To apply the normal load, the loading platform (shown in Figure 2b) with fitted lower bush holder was raised using an Enerpac RSM200 manual hydraulic cylinder, while the upper bush holder was kept static. The shaft was driven by an AKM42H (120 V) motor attached with a Micron XTRUE 160 planetary gearhead. In this study, the shaft performed an oscillating motion from −60° to +60° at a speed of 45 °/s (3.9 mm/s). A C-FW compression load cell (capacity of 100 kN) was located under the platform for measuring the normal load. A plunger dial indicator was attached to the loading platform to record its vertical displacement, which was the radial deformation occurring in the bush wall. A FUTEK FSH02059 torque transducer (200 Nm capacity, Irvine, CA, USA) was used to measure the frictional torque between the bush and shaft. The overall monitoring, recording and control of the rig was via a PC using a software program written in LabVIEW (National Instruments, Austin, TX, USA). More details of the test rig have been reported in [29].

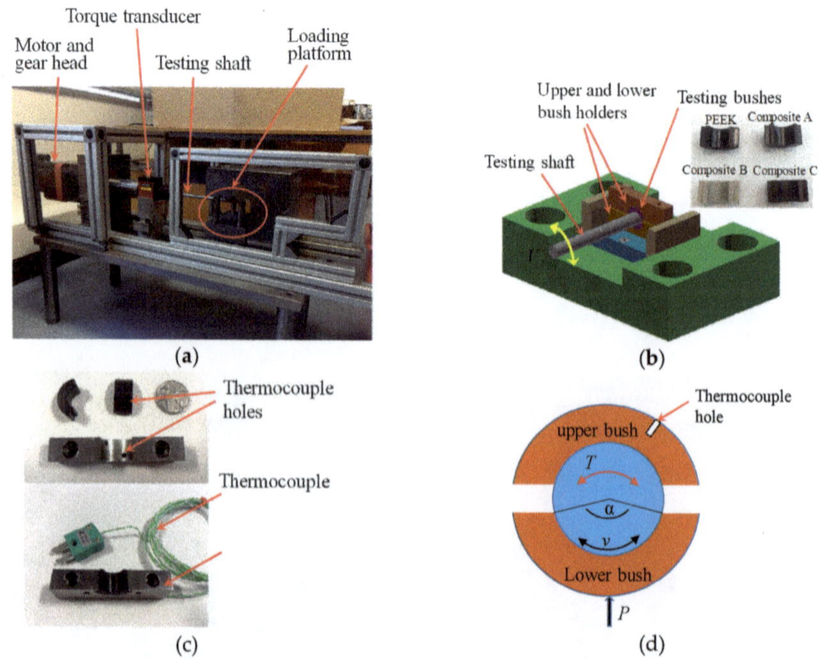

Figure 2. Pin joint test rig and bush specimen arrangement, (**a**) photo of the pin joint test rig; (**b**) loading platform with pin/bush assembly, bush halves located in two separate holders above and below the shaft, inset: bush specimens made of PEEK and PEEK composites; (**c**) thermocouple location; (**d**) contact geometry between shaft and bush halves subjected to normal load P applied from the lower bush, shaft oscillating speed v and required frictional torque T.

A thermocouple hole was drilled in the bush wall, 1.2 and 2.5 mm deep against the inner surface for wall thickness 2.5 and 5 mm respectively, shown in Figure 2d. The temperature change in the wall material was measured using a K-type thermocouple. Figure 2c shows the assembly of bush, bush holder and thermocouple. A fresh bush pair was used for each test. Prior to test, bushes were cleaned with isopropanol in an ultrasonic bath. Bush wall thickness (WT) and mass were measured before and after each test. Mass loss (Δm) and wall thickness change were recorded for assessing the wear resistance.

A 300M steel shaft with a diameter of 10 mm and surface roughness Ra = 0.5 μm was adopted to contact with the bush specimen. It was cleaned with isopropanol prior to each test and reused. Tests were carried out without lubricant and at room temperature and humidity. Three repeats for each test were conducted using fresh bushes and a newly cleaned shaft.

Based on the contact geometry shown in Figure 2d, the nominal contact pressure between the rotating pin and the bush half, p, and the friction coefficient, μ, are calculated from the following equations,

$$p = \frac{P}{2RL\sin(\alpha/2)} \tag{1}$$

$$\mu = \frac{T}{2PR} \tag{2}$$

where P is the normal load, 9 kN, R is the radius of the shaft, 10 mm, L is the contact width, and α is the arc angle of the bush halve. It should be noted that even though every effort has been made in sample preparation to reduce the variance between bush halves, it was impossible to have exactly identical samples. In this work, the contact mechanisms from the lower and upper bush halves were assumed to be the same, i.e., same friction force/torque occurred from each bush half.

Table 2 shows the testing conditions including the shaft and bush dimensions. Due to the varying reinforcements, differences in mechanical and tribological properties were expected. In order to fully understand their tribological capacity, varying test durations were applied. A defined radial deformation of the bush wall was used as an indicator to end the test. In testing, the reading from the plunger dial indicator was used to calculate the deformation in the bush wall. The test was manually stopped when the wall thickness change was 10% of its original thickness, which was defined as a failure in this study. The corresponding maximum articulating cycles were then compared among tested bushes. There was an exception for the Bush C (WT = 5 mm) caused by excessive lower deformation. In this case, the test was ended after 6 h running.

Table 2. Oscillating test conditions.

Nominal Contact Pressure	Articulating Displacement	Articulating Speed	Pin Radius	Bush Arc Angle	Bush Width	Bush Wall Thickness	Oscillating Cycles
p = 93 MPa	−60° to +60°	45 °/s (3.9 mm/s)	R = 5 mm	120°	L = 10 mm	WT = 2.5/5 mm	Vary

Friction coefficient and wear coefficient (referred to as mass loss) were used to analyse the contact mechanism between the shaft and bush. Wear was measured by mass loss, Δm. Wear coefficient of the material W, in mm^3/Nm, was calculated using the following equation,

$$W = \frac{\Delta m}{\rho PS} \tag{3}$$

where ρ is the density of the specimen listed in Table 1, P is the normal load, and S is the total sliding distance.

During the test, frictional heating occurred [30] at the contact between the shaft and bush halves due to combined normal and tangential loading. As polymers are more sensitive to mechanical stresses and temperature [31], it is necessary to take into account frictional energy in the investigation of the friction and wear properties. For the current contact configuration, the frictional energy equals the work required to enable the shaft to rotate inside the bush halves. It is calculated by the following equation,

$$E = Pv \int_{t_s}^{t_e} \mu(t)dt \tag{4}$$

where v is the sliding speed, 3.9 mm/s, μ(t) is the coefficient of friction (CoF), the shaft starts articulating at t_s and ends at t_e.

Specific wear energy [23] that combines friction coefficient and wear was used to assess the friction and wear properties. It is the ratio of the frictional work divided by the bush mass loss in the wear process, shown as the following equation,

$$E_W = \frac{E}{\Delta m} = \frac{Pv \int_{t_s}^{t_e} \mu(t)dt}{\Delta m} \tag{5}$$

2.3. Characterization

In this study, an Inspect F FEG-SEM (FEI, Eindhoven, Netherlands) was used to characterize worn surfaces of the tested samples. Wear debris were assessed using an Alicona InfiniteFocusSL microscope (Alicona Imaging GmbH, Graz, Austria). The contact zone on the shaft after each test was examined by using an optical microscope (Zeiss Optical Microscope, Cambridge, UK). The bush mass was measured using a Sartorius Electronic Analytical Balance BP210D (accuracy 0.01 mg).

3. Results and Discussions

3.1. Friction and Wear

Figure 3 presents three repeats of the wear test for Bush A, characterised by CoF and temperature in the bush wall. It can be seen that good repeatability was seen among the three repeats. The small difference of CoF and temperature curves may arise from the variance of specimen surface texture and roughness produced in the process of mechanical machining.

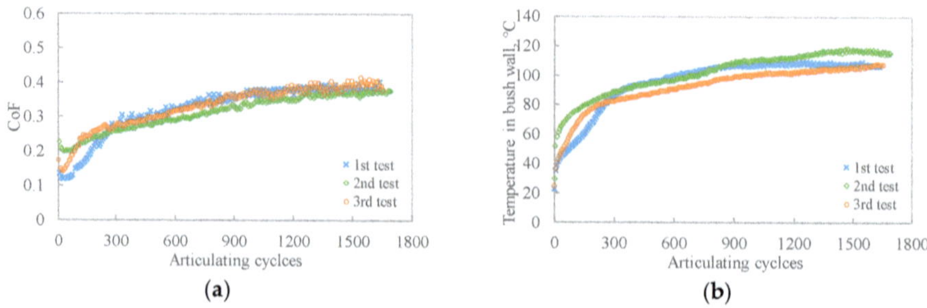

Figure 3. Three repeat wear tests for Bush A: (**a**) CoF varying with articulating cycles; (**b**) temperature rise in the bush wall varying with articulating cycles.

The comparison of CoF and temperature increase among four composite bushes are shown in Figures 4 and 5. The tests for Bush A, B and D were stopped when the bush wall thickness reached a 10% change compared with the original size. Overall, the CoF increased over the testing period for all bushes while the composite C presented the lowest CoF values and temperature increase. This indicated that the incorporation of PTFE, graphite and carbon fibre significantly reduced both the friction and temperature rise.

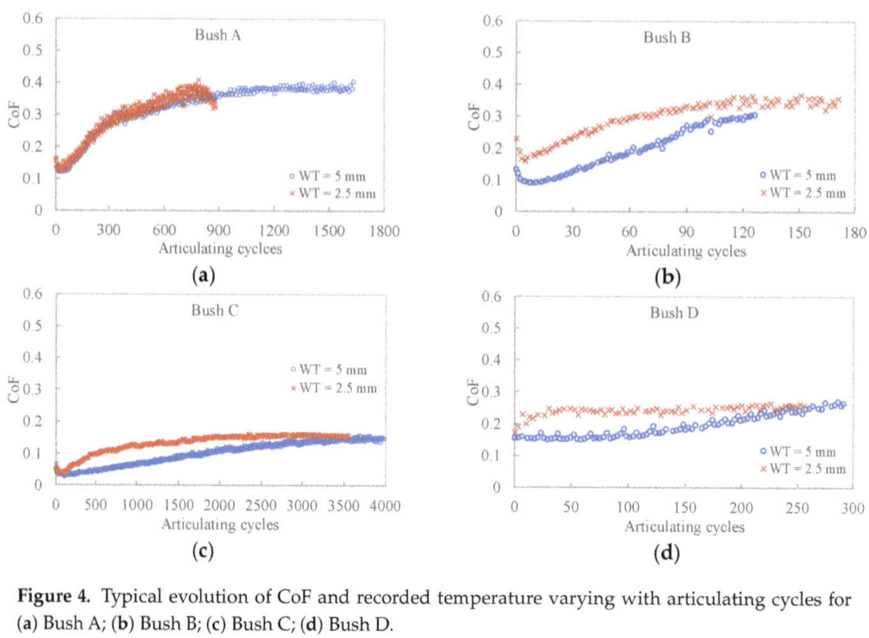

Figure 4. Typical evolution of CoF and recorded temperature varying with articulating cycles for (**a**) Bush A; (**b**) Bush B; (**c**) Bush C; (**d**) Bush D.

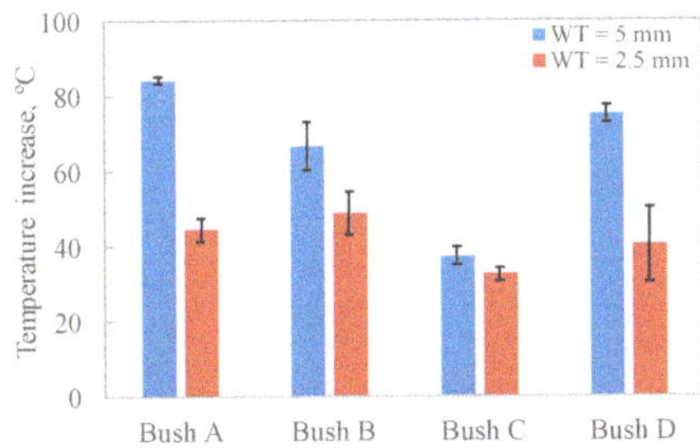

Figure 5. Temperature increase in the middle of the bush wall thickness.

Comparing the two bush wall thicknesses, the thinner ones had slightly higher CoF and lower temperature increase. As they were subjected to the same testing conditions, the difference in CoF could only be caused by the energy dissipation efficiency. In other words, the contact temperature played as an influential factor for the contact mechanism. Apparently, thinner wall bushes reduced the accumulation of the friction heat by dissipating heat to the adjacent metal parts. While for thick wall bushes, the increase in contact temperature decreases the stiffness of the matrix, the shear strength, and therefore resulted in a lower CoF [32].

For the first 300 cycles in Figure 4a,d, the carbon fibre reinforcement in Bush A did not seem to reduce the friction as unfilled PEEK shows a constant and relatively lower CoF. Compared with Bush B (glass fibre reinforced) and Bush D (unfilled PEEK), Bush A did show improved bearing capacity (higher articulating cycles), which is in agreement with the findings of [1].

After each test, the bush wall thickness was measured to determine maximum wall reduction, as shown in Table 3. Under the same load, the thinner bushes presented more deformation indicating lower load bearing capacity. For Bush A, the highest radial deformation, 14.09% reduction, was observed at WT = 2.5 mm. Unsurprisingly, Bush C showed the lowest deformation for both wall sizes. During the wear test, there was no wear debris observed for Bush B and D. The bush mass loss was also too low to be measured. The wear coefficients for Bushes A and C were calculated from their mass loss and are shown in Table 3. Both CoF and wear loss for Bush A were found to be significantly higher than that of Bush C. In other words, Bush C exhibited notably superior bearing properties among the four tested composites. From this study, it is clear that no correlation between mass loss and bearing capacity can be concluded. The friction coefficient and wear loss did not provide enough information to disclose the contact mechanism either. As there was no wear loss occurred on Bush B and D, in order to compare the contact mechanism among the tested bushes, it is necessary to study the interface at a microscopic scale.

Table 3. Maximum radial change of the bush wall and wear coefficient for tested bushes.

	Bush A		**Bush C**	
WT, mm	**2.5**	**5**	**2.5**	**5**
Wear coefficient, $\times 10^{-6}$ mm^3/Nm	4.33 ± 0.78	3.76 ± 0.65	0.73 ± 0.04	0.13 ± 0.04

3.2. Wear Debris and Worn Surfaces

In order to understand the wear mechanism, wear debris from Bush A and C were collected and assessed using the Alicona InfiniteFocusSL, shown in Figure 6. No wear debris were observed from Bush B (glass fibre reinforced) and D (unfilled PEEK) from the bush wear test. For carbon fibre reinforced Bush A, large fragments of debris, up to 3–5mm in length, were observed, while the 2.5 mm wall bush produced similar but thicker flakes, shown in Figure 6a,b. The highest wear loss and wear coefficient were presented by Bush A when WT = 2.5 mm. Figure 6c,d shows much finer wear debris from Bush C. Slightly coarse wear particles were found for thicker wall bush. The presence of graphite and PTFE in the matrix reduced the formation of larger debris chips.

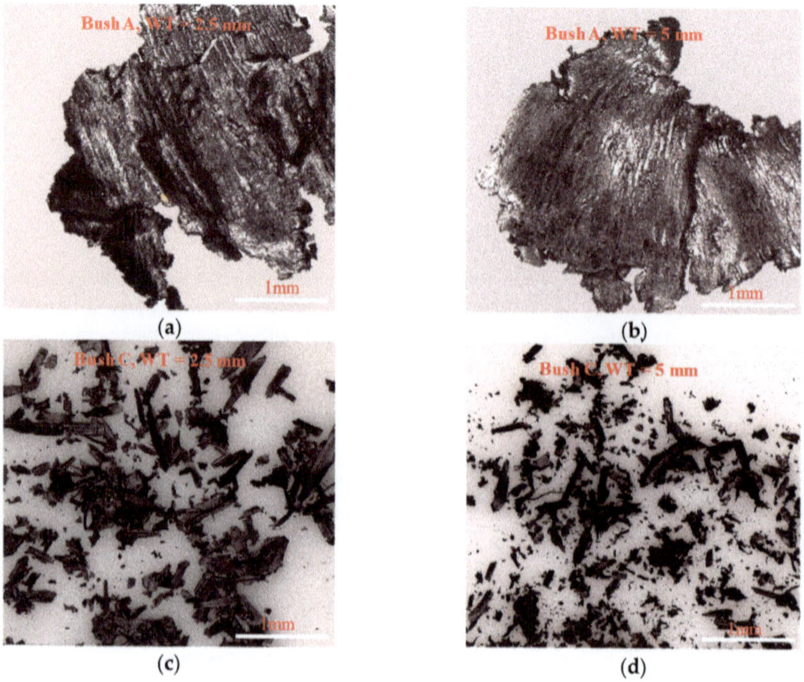

Figure 6. Typical debris observed from the wear test for (**a**) and (**b**) Bush A; (**c**) and (**d**) Bush C for wall thickness 2.5 and 5 mm, respectively. No wear debris observed from Bush B (glass fibre reinforced PEEK) and Bush D (unfilled PEEK).

Figure 7 gives lower and higher magnification SEM images of the top-view worn surfaces from tested bushes. It is clear that under the combined action of compression, shearing and frictional heating, PEEK and PEEK composites displayed diverse patterns on the surface layer, caused by different wear mechanisms. Due to repeated stressing, cracks were produced at the surface and/or just sub-surface in the composite. These cracks gradually grew and joined each other until wear debris, including spalls, were detached after a certain number of stressing cycles. Therefore, adhesion and fatigue were the main wear mechanisms occurring at the interfaces. Bushes with thinner walls showed patches of overlapping platelets on the surface, demonstrating a severe deformation and shearing of the surface materials.

Bush B showed the overall worst case, with large blocky particles over 1 mm in length. For fibre reinforced matrixes, shown in the higher magnification images in Figure 7a,c,e, fibres were pulled out, broken and crushed, either exposed on the surface or pressed in the deformed layer. Through block-on-ring test, Zhang et al. found that carbon fibre thinning (fibre wear) dominate the wear mechanism at low pressure of 1MPa [33]. This phenomenon was not observed on tested bushes. It implies that under higher contact pressure (93 MPa), the contact mechanism mainly fell in severe deformation and tearing of surface material. Glass fibres or carbon fibres were broken into short pieces, remaining the same diameter, rather than gradually being thinned by shear stress caused fatigue.

For the thicker wall bushes, a smoother surface layer (Figure 7b,d,f,h) was observed under the same loading and sliding conditions. Continuous micro ploughing along the sliding direction associated with platelet patches was exhibited by Bush A at WT = 5mm. The wear mechanism falls into a combination of abrasion and adhesion. These slightly 'smoother' surface topographies agreed with higher articulating cycles in Figure 4, inferring greater bearing capacities for thick wall bushes. Compared with other bushes, Bush C showed smoother surfaces without gaps between platelet patches. This was due to the existence of self-lubricating agents, graphite and PTFE, reducing the shear stress on the interface. Obvious matrix shear failure was rarely observed for Bush D (unfilled PEEK) while the worn surface showed some patches detached on the surface. Plastic flow was observed at WT = 5 mm. Bush B and D showed comparable articulating cycles and radial deformation which were much worse than Bush A and C. Glass fibres did not bring any enhancement to the mechanical strength or friction and wear resistance of the matrix.

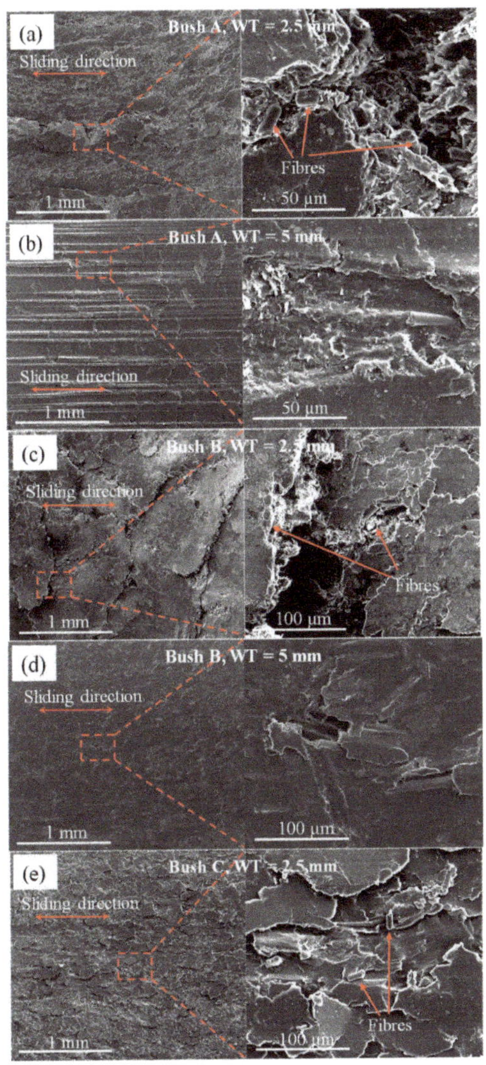

Figure 7. *Cont.*

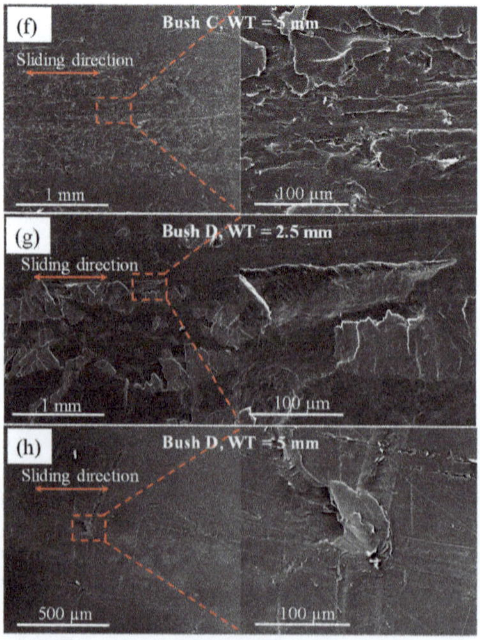

Figure 7. SEM images of worn bush surfaces, (**a**) and (**b**) Bush A; (**c**) and (**d**) Bush B; (**e**) and (**f**) Bush C; (**g**) and (**h**) Bush D for wall thickness of 2.5 and 5 mm.

3.3. Cross Section of Worn Surfaces

Worn surface morphology is useful in the analysis of friction and wear behaviour, while subsurface material change provides important information regarding the bulk properties such as load carrying capacity, resistance to compression, cracking and fatigue. After testing, bush halves were quenched in liquid nitrogen and fractured to expose the cross-section. Figure 8 shows fractography for tested bushes after cyclic loading in normal and tangential directions. The direction of sliding is into the page. An extensively deformed subsurface layer was observed for all bushes except Bush C. For example, Figure 8a for the 2.5 mm thick Bush A shows a 200 µm thick layer composed of deformed material on top of the substrate matrix. The surface layer material appears compressed by the high normal load. The inset image shows the deformed matrix with broken carbon fibres and the PEEK. The surface layer in Figure 8b for WT = 5 mm is slightly thicker, around 270 µm, but less compressed as delaminated sub-layers can be seen. The inset image in Figure 8b shows an unmodified substrate matrix. It is reasonable to conclude that the deformation occurring in the surface layer is an attribute of the wall thickness reduction. Chen et al. claimed that the exposed carbon fibre on the sliding surface carried most of the normal load and therefore improved the matrix load bearing capacity [1]. While in this study, this thick surface layer stacked on top of the substrate matrix was presumed to support the normal load and dissipate frictional heat into bulk material underneath.

Bush B showed a similar deformed depth after 171 articulating cycles (Figure 8c) compared with Bush A of 882 articulating cycles (Figure 8a). A layer was detached from the substrate matrix, shown in Figure 8d. Again, in this layer, glass fibres were found to be fractured, crushed and blended in the matrix shown in the inset. Cross sections of the unfilled PEEK bush are shown in Figure 8g,h. Plastic flow has occurred shown by the inset in Figure 8h.

Figure 8. SEM images of cross section of worn surfaces for (**a**) and (**b**) Bush A, (**c**) and (**d**) Bush B, (**e**) and (**f**) Bush C, and (**g**) and (**h**) Bush D for wall thickness of 2.5 and 5 mm, respectively.

Unsurprisingly, the best performance was shown by the Bush C (Figure 8e,f). Only a very thin surface layer was affected by compression and shearing. In the sliding contact under normal load, strain occurs both in the normal and tangential directions [33,34]. For the reinforced matrix, normal and shear stresses transferred at the interface between reinforcements and the matrix material. Due to the very low surface energy of graphite and PTFE, the shear stress is low leading to a low CoF. As the carbon fibres were dispersed uniformly in the matrix with varying orientations, the normal stress was therefore supported effectively by high modulus carbon fibres in varying directions [33].

3.4. Worn Surface on Shafts

Figure 9 shows microscopic images of the pin surface before and after contacting with bush samples (WT = 5 mm). The sliding direction is marked on Figure 9b. Compared with the fresh surface (Figure 9a), a continuous thick layer of transferred material was observed for the Bush A (Figure 9b). It is clear that material transfer from bush surface has occurred. Weakening and debonding between carbon fibres and the matrix were expected due to the shearing and compression. In sliding, when the friction force is greater than the adhesive interaction between polymer matrix and reinforcements, the asperities of the composite material can be removed to form a transfer layer on the counterface [35]. Bely et al. [36] reported that the transfer of polymer is the most important characteristic of adhesive wear in polymers. The adhesion process is normally associated with other wear types (fatigue, abrasion and so on) [35].

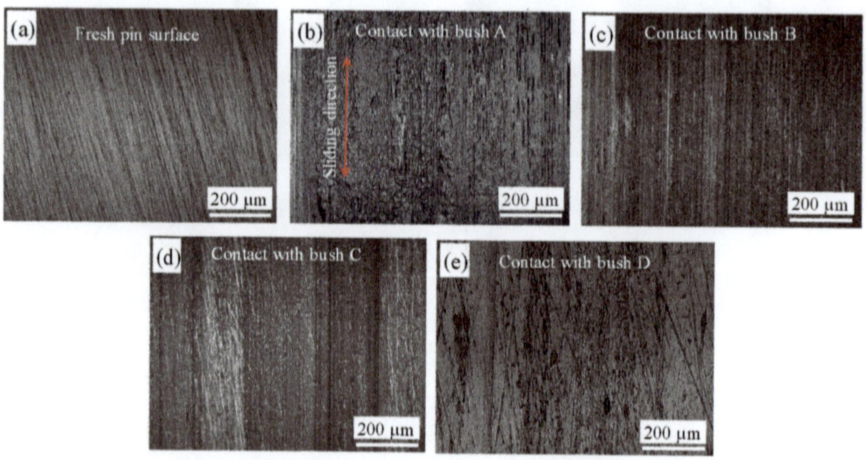

Figure 9. Microscopic images of the shaft surfaces, (**a**) fresh surface; (**b**) contacted with Bush A; (**c**) contacted with Bush B; (**d**) contacted with Bush C; (**e**) contacted with Bush D.

There was no visible wear debris accumulated next to the sample for Bushes B and D after testing. However, due to the extreme hard of glass fibres in the Bush B, fine furrows were observed along the sliding direction, shown in Figure 9c. Without contacting with hard glass fibres, the shaft surface presented limited wear and material transfer (Figure 9e). Even a thin layer of bush material was deposited on the pin surface, no measurable mass loss was produced for neither Bush B or D. This indicated that only the top asperity layer was removed from the bush surfaces while the bulk matrix underwent a plastic deformation.

3.5. Friction Energy

Over the testing duration, the friction energy that is transformed as a consequence of frictional contact is dissipated and converted to heat, vibration, material deformation or stored in the tribo-system. For sliding between the shaft and bush, the frictional energy and specific frictional energy were calculated from Equations (4) and (5), respectively (see Table 4). The table includes the overall temperature increase (temperature difference between the start and end of each test). It can be seen that bush A and C show significantly higher frictional energy compared to Bush B and D. This is because the testing durations (articulating cycles) were much longer than that of Bush B and D. Low thermal conductivity usually leads to the accumulation of the frictional heat and therefore reduces the bearing capability of composites [37]. This explains the performance of Bush B and D, due to their low thermal conductivities, they showed higher temperature increase and lower articulating cycles to failure. This is also evidenced by the formation of a peeling layer shown in Figure 8d,h. The multilayers immediately beneath the wear track in Figure 8h show the plastic deformation due to the combined action of reciprocating shearing and thermal softening. However, Bush A and C have similar thermal conductivities, 0.92 and 0.82 $Wm^{-1}°C^{-1}$, respectively; the articulating cycles of Bush C are more than twice that of Bush A. In this case, CoF played an important role in the temperature increase. In other words, in order to achieve good performance of the bearing bush, both low friction coefficient and high thermal conductivity are required [38].

Table 4. Frictional energy for tested bushes.

	Bush A		Bush B		Bush C		Bush D	
WT, mm	2.5	5	2.5	5	2.5	5	2.5	5
Friction energy, $\times 10^4$ J	4.46 ± 0.48	9.53 ± 0.01	1.08 ± 0.11	0.44 ± 0.01	7.52 ± 1.59	7.15 ± 1.12	0.84 ± 0.16	0.97 ± 0.13
Specific wear energy, $\times 10^4$ J/mg	0.29 ± 0.06	0.46 ± 0.09	-	-	0.79 ± 0.07	4.36 ± 0.64	-	-
Temperature rise, °C	44.5 ± 3.06	84.16 ± 0.9	48.74 ± 5.58	66.54 ± 6.37	32.47 ± 1.8	37.4 ± 2.31	40.33 ± 10.03	75.02 ± 2.4

Due to a significant difference in articulating cycles for each bush test, it is better to compare bushes using friction energy per cycle (Figure 10a) and temperature rise per cycle (Figure 10b). It can be seen that Bush C produced the least friction energy per cycle due to the lowest CoF. Less frictional heat produced at the interface leads to lower temperature increase in the material bulk, shown in Figure 10b. The graphite lamellar structure reduced CoF and hence the heat generation. In addition, during material deformation, PTFE helped to store much of the work, which was used in crystallographic and amorphous chain rearrangement, resulting in less sample heating [39]. Therefore, the thermal softening in Bush C was limited. This agreed with the microscopic morphology displayed in Figure 8e,f where the lest deformation of the material was observed.

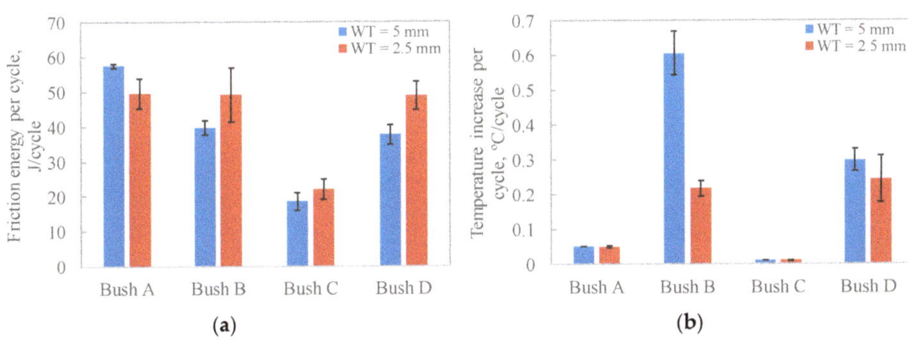

Figure 10. Averaged friction energy and temperature increase per cycle for tested bushes, (**a**) friction energy per cycle; (**b**) temperature increase per cycle.

4. Conclusions

A comparative study of PEEK composite bushes for use in articulating revolute pin joints has been conducted. The friction, wear, friction energy and temperature rise have been studied. The friction and wear mechanisms were assessed by studying the microscopic worn surfaces and deformation layer beneath. The thermal accumulation and dissipation were studied to improve the understanding of the tribological performance for PEEK composite used as bearing bushes.

Due to low thermal conductivity, unfilled PEEK and glass fibre reinforced PEEK presented much lower articulating cycles to failure than that of graphite, PTFE and carbon fibre filled PEEK. The load bearing capacity of the composite is much higher than that of the matrix, and thus, any sub-surface fracture and yielding are diminished due to the presence of the hard and strong reinforcements. Presence of graphite and PTFE in the PEEK matrix not only reduced shear force at the interface but also minimised the temperature increase in the bulk material. In addition, the wear resistance was significantly improved. The wear coefficient of Bush C was found to be 0.13×10^{-6} mm^3/Nm compared with 4.33×10^{-6} mm^3/Nm for Bush A.

Bushes made of PEEK composite formulated with PTFE, graphite and carbon fibre exhibit low friction, self-lubricating, low temperature rise, and therefore present superior bearing properties, including enhanced bearing life and reducing energy consumption in machinery. The findings facilitate the application of this PEEK composite used as self-lubricating bearing bushes.

Author Contributions: Conceptualization, methodology, formal analysis and draft preparation, J.Z.; experimentation, data curation and formal analysis, F.X.; supervision and writing—review and editing, R.S.D.-J. All authors contributed to writing and correcting the manuscript. All authors have read and agreed to the published version of the manuscript.

Funding: This research was supported by the Aerospace Technology Institute (UK) funded project Large Landing Gear of the Future (LLGF) led by Safran Landing Systems UK Ltd. [grant number 113077] and the Engineering and Physical Sciences Research Council through R S Dwyer-Joyce's fellowship on Tribo-Acoustic Sensors [grant number EP/N016483/1].

Acknowledgments: The authors gratefully acknowledge Le Ma (Sorby Centre, Kroto Research Institute, North Campus, University of Sheffield, UK; l.ma@sheffield.ac.uk) for her professional support in SEM sample preparing and imaging.

Conflicts of Interest: The authors declare no conflict of interest.

References

1. Chen, B.; Wang, J.; Yan, F. Comparative investigation on the tribological behaviors of CF/PEEK composites under sea water lubrication. *Tribol. Int.* **2012**, *52*, 170–177. [CrossRef]
2. Aldousiri, B.; Shalwan, A.; Chin, C.W. A review on tribological behaviour of polymeric composites and future reinforcements. *Adv. Mater. Sci. Eng.* **2013**, *2013*, 1–8. [CrossRef]
3. Qi, Y.; Gong, J.; Cao, W.; Wang, H.; Ren, J.; Gao, G. Tribological behavior of PTFE composites filled with peek and nano-Al_2O_3. *Tribol. Trans.* **2018**, *61*, 694–704. [CrossRef]
4. Hufenbach, W.A.; Stelmakh, A.; Kunze, K.; Böhm, R.; Kupfer, R. Tribo-Mechanical properties of glass fibre reinforced polypropylene composites. *Tribol. Int.* **2012**, *49*, 8–16. [CrossRef]
5. Zhang, L.; Li, G.; Guo, Y.; Qi, H.; Che, Q.; Zhang, G. PEEK reinforced with low-loading 2D graphitic carbon nitride nanosheets: High wear resistance under harsh lubrication conditions. *Compos. Part A Appl. Sci. Manuf.* **2018**, *109*, 507–516. [CrossRef]
6. Tang, Q.; Chen, J.; Liu, L. Tribological behaviours of carbon fibre reinforced PEEK sliding on silicon nitride lubricated with water. *Wear* **2010**, *269*, 541–546. [CrossRef]
7. Friedrich, K. Polymer composites for tribological applications. *Adv. Ind. Eng. Polym. Res.* **2018**, *1*, 3–39. [CrossRef]
8. Lewis, S.D.; Rowntree, R.A. Hybrid polymeric bearings for space applications. In Proceedings of the 7th European Space Mechanisms and Tribology Symposium, ESTEC, Noordwijk, The Netherlands, 1–3 October 1997; Kaldeich-Schürmann, B.H., Ed.; ESA SP-140.
9. Laux, K.A.; Jean-Fulcrand, A.; Sue, H.J.; Bremner, T.; Wong, J.S.S. The influence of surface properties on sliding contact temperature and friction for polyetheretherketone (PEEK). *Polymer* **2016**, *103*, 397–404. [CrossRef]
10. Laux, K.A.; Schwartz, C.J. Influence of linear reciprocating and multi-Directional sliding on PEEK wear performance and transfer film formation. *Wear* **2013**, *301*, 727–734. [CrossRef]
11. Omrani, E.; Menezes, P.L.; Rohatgi, P.K. State of the art on tribological behavior of polymer matrix composites reinforced with natural fibers in the green materials world. *Eng. Sci. Technol. Int. J.* **2016**, *19*, 717–736. [CrossRef]

12. Molazemhosseini, A.; Tourani, H.; Khavandi, A.; Yekta, B.E. Tribological performance of PEEK based hybrid composites reinforced with short carbon fibers and nano-Silica. *Wear* **2013**, *303*, 397–404. [CrossRef]
13. Regis, M.; Lanzutti, A.; Bracco, P.; Fedrizzi, L. Wear behavior of medical grade PEEK and CFR PEEK under dry and bovine serum conditions. *Wear* **2018**, *408–409*, 86–95. [CrossRef]
14. Sumer, M.; Unal, H.; Mimaroglu, A. Evaluation of tribological behaviour of PEEK and glass fibre reinforced PEEK composite under dry sliding and water lubricated conditions. *Wear* **2008**, *265*, 1061–1065. [CrossRef]
15. Wang, Q.; Xue, Q.; Liu, H.; Shen, W.; Xu, J. The effect of particle size of nanometer ZrO2 on the tribological behaviour of PEEK. *Wear* **1996**, *198*, 216–219. [CrossRef]
16. Zhang, Z.; Breidt, C.; Chang, L.; Friedrich, K. Wear of PEEK composites related to their mechanical performances. *Tribol. Int.* **2004**, *37*, 271–277. [CrossRef]
17. Bijwe, J.; Sen, S.; Ghosh, A. Influence of PTFE content in PEEK-PTFE blends on mechanical properties and tribo-performance in various wear modes. *Wear* **2005**, *258*, 1536–1542. [CrossRef]
18. Theiler, G.; Gradt, T. Environmental effects on the sliding behaviour of PEEK composites. *Wear* **2016**, *368–369*, 278–286. [CrossRef]
19. Oyamada, T.; Ono, M.; Miura, H.; Kuwano, T. Effect of gas environment on friction behaviour and tribofilm formation of PEEK/Carbon fiber composite. *Tribol. Trans.* **2013**, *56*, 607–614. [CrossRef]
20. Wang, Q.; Zheng, F.; Wang, T. Tribological properties of polymers PI, PTFE and PEEK atcryogenic temperature in vacuum. *Cryogenics* **2016**, *75*, 19–25. [CrossRef]
21. Zhang, G.; Burkhart, T.; Wetzel, B. Tribological behavior of epoxy composites under diesel-Lubricated conditions. *Wear* **2013**, *307*, 174–181. [CrossRef]
22. McCarthy, D.M.C.; Glavatskih, S.B. Assessment of polymer composites for hydrodynamic journal-Bearing applications. *Lubr. Sci.* **2009**, *21*, 331–341. [CrossRef]
23. Conte, M.; Pinedo, B.; Igartua, A. Frictional heating calculation based on tailored experimental measurements. *Tribol. Int.* **2014**, *74*, 1–6. [CrossRef]
24. Mizobe, K.; Honda, T.; Koike, H.; Santos, E.C.; Kida, K.; Kashima, Y. Relationship between load, rotation speed and, strength in all-PEEK and PEEK Race–PTFE retainer hybrid polymer bearings under dry rolling contact fatigue. *Adv. Mater. Res.* **2012**, *567*, 66–70. [CrossRef]
25. Shi, X.C.; Orito, M.; Kashima, Y.; Mizobe, K.; Kida, K. Observation of cracks of PEEK polymer thrust bearings under rolling contact fatigue in water. *Key Eng. Mater.* **2016**, *703*, 172–177. [CrossRef]
26. Brockett, C.L.; Carbone, S.; Abdelgaied, A.; Fisher, J.; Jennings, L.M. Influence of contact pressure, cross-shear and counterface material on the wear of PEEK and CFR-PEEK for orthopaedic applications. *J. Mech. Behav. Biomed. Mater.* **2016**, *63*, 10–16. [CrossRef] [PubMed]
27. Koike, H.; Kanemasu, K.; Itakura, K.; Okazaki, S.; Takamiya, M.; Santos, E.C.; Kida, K. Measurement of fatigue and wear of PEEK bush and A7075 cam plate in humanoid robot joints. *Mater. Res. Innov.* **2014**, *18*, S38–S43. [CrossRef]
28. Ensinger Ltd., UK. Available online: https://www.ensingerplastics.com/en-gb (accessed on 1 February 2020).
29. Zhu, J.; Ma, L.; Dwyer-Joyce, R. Friction and wear behaviours of self-Lubricating peek composites for articulating pin joints. *Tribol. Int.* **2019**. [CrossRef]
30. Hoskins, T.J.; Dearn, K.D.; Chen, Y.K.; Kukureka, S.N. The wear of PEEK in rolling-sliding contact-Simulation of polymer gear applications. *Wear* **2014**, *309*, 35–42. [CrossRef]
31. Myshkin, N.K.; Petrokovets, M.I.; Kovalev, A.V. Tribology of polymers: Adhesion, friction, wear, and mass-Transfer. *Tribol. Int.* **2005**, *38*, 910–921. [CrossRef]
32. Zhang, G.; Liao, H.; Li, H.; Mateus, C.; Bordes, J.M.; Coddet, C. On dry sliding friction and wear behaviour of PEEK and PEEK/SiC-Composite coatings. *Wear* **2006**, *260*, 594–600. [CrossRef]
33. Zhang, G.; Rasheva, Z.; Schlarb, A.K. Friction and wear variations of short carbon fiber (SCF)/PTFE/graphite (10 vol.%) filled PEEK: Effects of fiber orientation and nominal contact pressure. *Wear* **2010**, *268*, 893–899. [CrossRef]
34. Zhang, G.; Zhang, C.; Nardin, P.; Li, W.Y.; Liao, H.; Coddet, C. Effects of sliding velocity and applied load on the tribological mechanism of amorphous polyether-Ether-Ketone (PEEK). *Tribol. Int.* **2008**, *41*, 79–86. [CrossRef]
35. Myshkin, N.; Kovalev, A. Adhesion and surface forces in polymer tribology—A review. *Friction* **2018**, *6*, 143–155. [CrossRef]

36. Bely, V.A.; Sviridenok, A.I.; Petrokovets, M.I.; Savkin, V.G. *Friction and Wear in Polymer-Based Materials*; Pergamon Press: Oxford, UK, 1982; pp. 109–125.
37. Bijwe, J.; Kumar, M.; Gurunath, P.V.; Desplanques, Y.; Degallaix, G. Optimization of brass contents for best combination of tribo-Performance and thermal conductivity of non-Asbestos organic (NAO) friction composites. *Wear* **2008**, *265*, 699–712. [CrossRef]
38. Mu, L.; Shi, Y.; Feng, X.; Zhu, J.; Lu, X. The effect of thermal conductivity and friction coefficient on the contact temperature of polyimide composites: Experimental and finite element simulation. *Tribol. Int.* **2012**, *53*, 45–52. [CrossRef]
39. Rae, P.J.; Dattelbaum, D.M. The properties of poly (tetrafluoroethylene) (PTFE) in compression. *Polymer* **2004**, *45*, 7615–7625. [CrossRef]

polymers

MDPI

Article

Structure and Properties of Polysulfone Filled with Modified Twill Weave Carbon Fabrics

Dilyus I. Chukov *, Sarvarkhodza G. Nematulloev, Victor V. Tcherdyntsev, Valerii G. Torokhov, Andrey A. Stepashkin, Mikhail Y. Zadorozhnyy, Dmitry D. Zherebtsov and Galal Sherif

NUST "MISiS", Center of composite materials, Leninskiy pr. 4, Moscow 119049, Russia; nematulloev.sarvar@yandex.ru (S.G.N.); vvch@misis.ru (V.V.T.); vgtorohov@gmail.com (V.G.T.); a.stepashkin@yandex.ru (A.A.S.); priboy38@mail.ru (M.Y.Z.); dmitry_zherebtsov@bk.ru (D.D.Z.); eng_galal_emad@mu.edu.eg (G.S.)

* Correspondence: dil_chukov@mail.ru; Tel.: +7-495-638-44-13

Received: 15 November 2019; Accepted: 24 December 2019; Published: 30 December 2019

Abstract: Carbon fabrics are widely used in polymer based composites. Nowadays, most of the advanced high-performance composites are based on thermosetting polymer matrices such as epoxy resin. Thermoplastics have received high attention as polymer matrices due to their low curing duration, high chemical resistance, high recyclability, and mass production capability in comparison with thermosetting polymers. In this paper, we suggest thermoplastic based composite materials reinforced with carbon fibers. Composites based on polysulfone reinforced with carbon fabrics using polymer solvent impregnation were studied. It is well known that despite the excellent mechanical properties, carbon fibers possess poor wettability and adhesion to polymers because of the fiber surface chemical inertness and smoothness. Therefore, to improve the fiber–matrix interfacial interaction, the surface modification of the carbon fibers by thermal oxidation was used. It was shown that the surface modification resulted in a noticeable change in the functional composition of the carbon fibers' surface and increased the mechanical properties of the polysulfone based composites. Significant increase in composites mechanical properties and thermal stability as a result of carbon fiber surface modification was observed.

Keywords: polymer-matrix composites; carbon fibers; polysulfone; surface modification

1. Introduction

Among the various types of inorganic and organic fibers, carbon fibers (CFs) show the highest tensile strength and elastic modulus at relatively low density [1,2]. This is why CFs, despite their high cost, are widely investigated and applied. The most promising CF application is as a reinforcement for polymer based composites [1–3]. CF reinforced polymer composites have been widely used as engineering materials in the aeronautic and automotive industries. Specific applications of these composites include body structure for electric vehicles [4], wet clutches to distribute torque in vehicle drive-trains of automatic transmissions and limited slip differentials [5], deployable space structures [6], heat exchangers [7], electromagnetic shielding [8], ballistic protection [2,9], and some other engineering areas. For example, in [10–12], a family of novel high-strength, lightweight structural epoxy/CF composites with self-healing function were proposed for potential use in aerospace and aeronautical structures, sports utilities, etc.

The properties of CF reinforced polymer composites strongly depend on the fiber shape and location in the matrix. Polymer composites filled with short CFs are widely used as materials with good mechanical and tribological performance [3,13–15]. The technology of the formation of such composites is relatively cheap and easy [16,17], moreover, they are suitable for production by additive manufacturing [18]. Short CFs can either be chaotically distributed in a polymer matrix or oriented;

in the latter, composites show anisotropy in mechanical [18], thermal [18], and tribological [19] properties. However, short CF filled polymers do not achieve the mechanical characteristics required for high-performance structural application in the aeronautic and automotive industries. Realization of the excellent mechanical properties of CFs in the polymer matrix can be more effectively achieved by using continuous fibers as reinforcement.

Continuous CF reinforced unidirectional composites are popular objects for modeling and theoretical approaches [20,21]. Unidirectional composites allow for the realization of CF mechanical characteristics better than 2D ones [1]. Unidirectional composites based on thermoplastics can be easily produced by simply using impregnation with polymer melt [22]. Application of unidirectional composites is restricted to very specific areas [6], however, such composites are promising as a semi-product in advanced additive manufacturing [23].

Most of the investigation and structural applications of CF reinforced polymer composites is based on 2D CF fabrics. Among the carbon fabrics, the most commonly used as reinforcement in polymer composites are woven fabrics, whereas knitted and braided fabrics are used much less often [3]. The number of weave structures that can be produced is practically unlimited, but in the investigations and applications, based structures such as plain, twill and satin are most commonly used. The weave structure of reinforcing woven CF fabrics can significantly affect the mechanical behavior of polymer composites. For instance, a comparison of epoxy based composites reinforced with CF fabrics showed that composites with twill weave fabrics show a higher shear modulus, shear strength, and ultimate strength than composites reinforced with satin weave CFs [24]. Similar results were observed in [25], where tensile tests showed that epoxy based composites reinforced with twill weave CF fabrics possessed higher magnitudes of strength, modulus, and strain than composites containing satin weave CFs. Mechanical tests, carried out with epoxy based composites, showed that the tensile properties of composites reinforced with twill weave CFs were more stable with an increase in the strain rate than the properties of composites with plain and satin weave CF fabrics [26]. The investigation of polyetherimide based composites showed that composites containing twill weave CF fabrics showed better tensile and flexural properties than composites filled with plain and satin weave CFs; however, composites with twill weave CFs had good wear resistance properties in adhesive mode only, whereas in abrasive wear mode, better properties were observed for composites with satin weave CFs and in erosive wear mode both satin and plain weave CF filled composites shows better behavior than the twill weave reinforced one [27]. On the other hand, in [28], it was reported than epoxy composites filled with twill weave CFs showed lower tensile characteristics than composites reinforced with plain and satin weave CFs, both in static and dynamic test modes. Furthermore, twill weave CFs are often used as the model object in theoretical investigations [29,30] and can be a suitable material for the elaboration of new polymer based composites.

For critical structures operating under high loads and elevated temperatures, composite materials with matrices based on thermoplastic polymers are actively beginning to be applied. Thermoplastic based composites show a considerably higher static fracture toughness compared to the thermoset (epoxy) composites [3]. Important advantages of thermoplastics are their unlimited shelf life, low curing duration, maintainability (i.e., the ability to correct defects and damage by reheating), the possibility of reforming defective products [31], high environmental resistance, and high chemical resistance including to aviation fuels and oils [1].

In recent years, a number of thermoplastic based CF reinforced composites have been elaborated and investigated [32]. Among the thermoplastics, high performance polymers are of particular interest due to their thermal stability and high mechanical properties. Polysulfone (PSU) has one of the highest service temperatures of all melt-processable high performance polymers. The high temperature nature of the PSU allows them to be used in demanding applications that other polymeric materials cannot satisfy. PSU is highly resistant to acids, alkali, and electrolyte materials, oxidizing agents, surfactants, and hydrocarbon oils. Its resistance to high temperatures allows for the use of PSU as a flame retardant material [33].

PSU is frequently used as a modifier for the epoxy matrix for CF reinforced composites [34–36]. In the 1990s, CF reinforced PSU was widely considered to be promising in medicine for arthoplastic applications, where composites filled with short [37] or unidirectional [38–40] CFs were mainly used in these investigations. In recent years, only few papers related to PSU based CF reinforced composites have appeared. In [41], the laminated gradient PSU composites with the formation of layers by parallel laying of continuous CFs as materials for artificial intervertebral discs were elaborated. The authors in [42] report on the formation of thermal conductive short CF filled PSU composites by injection molding. No data on the PSU composites reinforced with CF fabrics were observed in the literature.

The aim of the present paper was to study the effect of CF surface modification by thermal oxidation on the mechanical and thermal properties of the polysulfone based composites reinforced with carbon fabrics. Whereas polymer melt impregnation, as was shown in [22], is a suitable method for PSU based unidirectional CF composites, in the case of the less permeable structure of CF fabrics, this technique is not effective. Recently, [43,44] suggested using a PSU solution impregnation method to obtain CF fabric reinforced PSU based composites. In the present study, we applied the PSU solution impregnation method to produce composites reinforced with CF twill weave fabrics.

2. Materials and Methods

2.1. Preparation of Carbon Fiber/Polysulfone Composites

Ultrason S2010 (BASF, Ludwigshafen, Germany) PSU and twill weave fabrics 3K-1200-200 (HC Composite, Moscow, Russia) based on high-modulus carbon fibers were used as the raw materials. Obtaining the composites was carried out in several stages including obtaining a polysulfone solution, impregnation of the CF with the solution; and drying, which resulted in prepregs (pre-impregnated fabrics). The prepregs were further cut into the desired shapes and molded by compression molding into the final composites. Figure 1 shows the scheme of the used process.

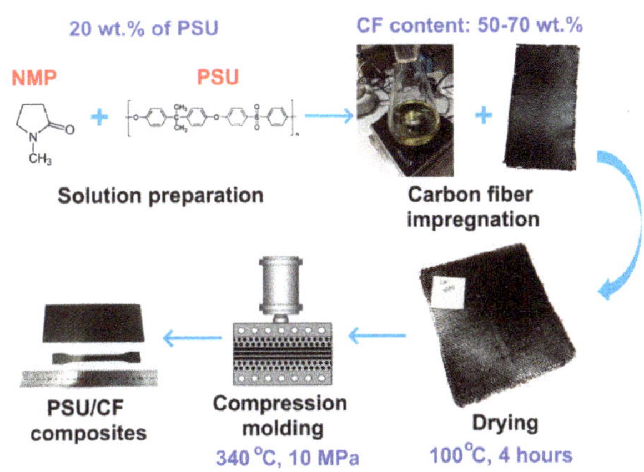

Figure 1. Scheme of the preparation of the carbon fiber reinforced polysulfone based composites.

To obtain the PSU solutions, N-methyl-2-pyrrolidone (Eastchem, Jiangsu, China) was used as a solvent, and a 20 wt % PSU solution was formed using a magnetic stirrer. After impregnation, the carbon fabrics were dried at 100 °C. CF/PSU composites with various fiber to polymer ratios were produced: 50 wt % of the CF and 50 wt % of the PSU (denoted as 50/50); 60 wt % of the CF and 40 wt % of the PSU (60/40); and 70 wt % of the CF and 30 wt % of the PSU (70/30). These prepregs were stacked in the mold carefully to avoid misalignment. The stainless steel mold was previously coated with the mold release agent. During compression molding, the mold was heated to attain the temperature of 340 °C. The prepregs were compression molded at the above-mentioned temperature under the pressure of 10 MPa. Thereafter, samples were cooled under applied pressure. The specimens for investigation were cut from the composites in accordance with the standards for mechanical testing.

2.2. *Carbon Fibers Surface Modification*

To improve the fiber–matrix interfacial interaction, surface modification of the CF by thermal oxidation (TO) in an air atmosphere at 300, 400, and 500 °C was applied and the fibers denoted as TO 300, TO 400, and TO 500 °C, respectively. The muffle furnace was heated up to the desired temperature, then the fibers were loaded into the furnace and held for 30 min. The fiber to polymer ratio of 60/40 was used for all of the modified fiber reinforced composites.

2.3. *Thermogravimetric Analysis*

To study the kinetics of solvent removal, thermogravimetric analysis (TGA) was performed using the TA Instruments Q600 system (TA Instruments, New Castle, DE, USA). During the tests, the polymer solution was heated at a rate of 10 °C/min to various temperatures of isothermal exposure (80, 100, and 120 °C); after reaching the isotherm, air blowing was started. The total measurement time was 3 h. The weight of the samples of the studied materials was chosen as close as possible to each other and varied in the range of 20–25 mg.

2.4. *Fourier-Transform Infrared Spectroscopy Analysis*

The FTIR analysis was carried out using a Nicolet 380 spectrometer (Thermo Scientific, Waltham, MA, USA) with spectral range of 3750–650 cm^{-1} and resolution of 1 cm^{-1}.

2.5. *X-ray Photoelectron Spectroscopy*

The x-ray photoelectron spectroscopy (XPS) studies of the CF were carried out using a ULVAC-PHI VersaProbe II spectrometer (ULVAC-PHI, Inc., Chigasaki, Kanagawa, Japan) with monochromatic Al Kα radiation (hν = 1486.6 eV), a power of 50 W, and a beam diameter of 200 µm. Atomic concentrations were determined from survey spectra using the method of relative elemental sensitivity factors. The binding energies (B.E.) of the photoelectron lines (C1s, O1s, N1s) were determined from high-resolution spectra taken at an analyzer transmittance energy of 23.5 eV and a data acquisition density of 0.2 eV/step. For decomposition of the photo peal, we used a Gaussian Lorentzian mix function and a linear background subtraction. XPS analysis of the modified fabrics was performed immediately after surface modification.

2.6. *In-Plane Shear Strength Tests*

In-plane shear strength of the composites was determined according to ASTM D 3846–02 using 80 × 10 × 4 mm samples at a crosshead speed of 1.3 mm/min. For the tests, a Zwick/Roell Z020 universal test machine (Zwick Roell Group, Ulm, Germany) was used. Compressive load was applied to a notched specimen of uniform width to measure the shear strength. The specimen was loaded edgewise in a supporting jig of the same description as that referenced in ASTM D 695 for testing thin specimens. A failure of the specimen occurs in shear between two centrally located notches machined halfway through its thickness and spaced a fixed distance apart on opposing faces. The distance between the notches was 6.5–8 mm.

2.7. *Mechanical Tests*

A Zwick/Roell Z020 universal test machine equipped with 1 and 20 kN sensors and a contact strain measurement system MultiXtens was used for mechanical tests. Tensile tests were performed in accordance with ISO 527:2009 for 110 × 10 × 2 mm samples and the flexural tests were performed in accordance with ISO 14125:1998 for 110 × 10 × 2 mm samples at a span length of 80 mm. Cross head speed during the tests was 10 mm/min. Five samples were tested for each type of composites in each test. Composite structure was investigated using a VEGA 3 TESCAN scanning electron microscope ((TESCAN ORSAY HOLDING, a.s., Brno–Kohoutovice, Czech Republic)) in a backscattered electron image mode. For the scanning electron microscopy (SEM) test, all specimens were sputter coated with a thin layer of carbon (10–15 nm) to provide the electrical conductivity of the samples.

2.8. Dynamic Mechanical Analysis

A Dynamic mechanical analyzer DMA Q800 (TA Instruments, New Castle, DE, USA) dynamic mechanical analyzer was used in these investigations. Specimens approximately 2 mm wide, 2 mm thick and 45 mm long were used for the DMA tests. The measurements were performed using a double cantilever clamp at a frequency of 1 Hz and deformation of 0.1%, in a temperature range from 30 to 220 °C and the heating rate was 2 °C/min.

3. Results and Discussion

3.1. Polysulfone Solution

Due to the high melt viscosity of the main part of high-performance thermoplastic polymers, the forces exerted on the fibers during melt impregnation are extremely high and may cause fiber damage [45]. The viscosity of the polymer solution is much lower than that of the melt, which allows it to impregnate the fabric uniformly and increase the wettability of the fibers. Another important advantage is that, unlike the melt impregnation, solution impregnation can be carried out at room temperature, followed by removal of the solvent at elevated temperatures. Therefore, solution impregnation can be considered as the preferred method to prepare high performance composites based on a soluble high performance thermoplastic [46,47]. The main disadvantage of this approach is removing the solvent after impregnation. Incomplete removal may produce voids that have a deleterious effect on the mechanical properties of the composite. Obtaining composites with the best set of properties is possible only if the minimum amount of residual solvent is reached. The results of the solvent evaporation at various temperatures obtained by thermogravimetric analysis are presented in Figure 2. It was found that the kinetics of solvent removal strongly depend on the measurement temperature. The results show that the higher the temperature, the faster the TGA curve reaches a plateau, and if at 80 °C, intense mass loss (evaporation of the solvent) lasts about 110 min, then at 100 °C and 120 °C, the time to the plateau decreases to 50 and 25 min, respectively. It was found that drying at 80 °C did not completely remove the solvent, and its residual ratio was about 2 wt %, while, at higher temperatures, the solvent could be completely removed.

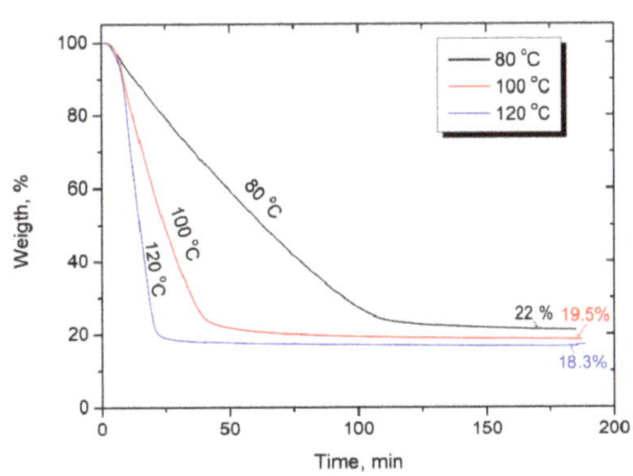

Figure 2. Solvent evaporation from Polysulfone/N-methyl-2-pyrrolidone solutions.

These results were further confirmed by FTIR spectroscopy of the PSU after various treatments. Figure 3 shows the FTIR spectra of pure PSU after compression molding at 340 °C (designated as the neat polymer (340 °C)); the polymer that was dissolved in N-methyl-2-pyrrolidone, dried at 100 °C, and compression molded at 340 °C (solution (dried at 340 °C)); and also the dissolved polymer just after drying at 100 °C (solution (dried at 100 °C)). Among the many observed peaks, a peak at 1690 cm^{-1} was chosen as the reference, which corresponds to C=O bond oscillations [48,49]. This peak can be associated with the presence of residual solvent in the sample or partial oxidation of the polymer during one or another treatment. It should be noted that during compression molding, there is no significant oxidation of the polymer, as evidenced by the low intensity of the peak at 1690 cm^{-1} for the pure PSU after compression molding at 340 °C, therefore in this case, we associated this peak with the presence of residual solvent. The FTIR results for the just dried PSU solution showed that the peak for the C=O bond oscillations had a rather high intensity, but after compression molding of the same sample, a noticeable decrease in the intensity of this peak occurred, which confirms the additional removal of the solvent during the compression molding. Therefore, it can be concluded that the residual solvent content in the final composites is very low and will not affect the composites' mechanical and other performance properties. Any noticeable changes or appearance of new peaks in the FTIR spectra of the PSU samples after various treatments were not found.

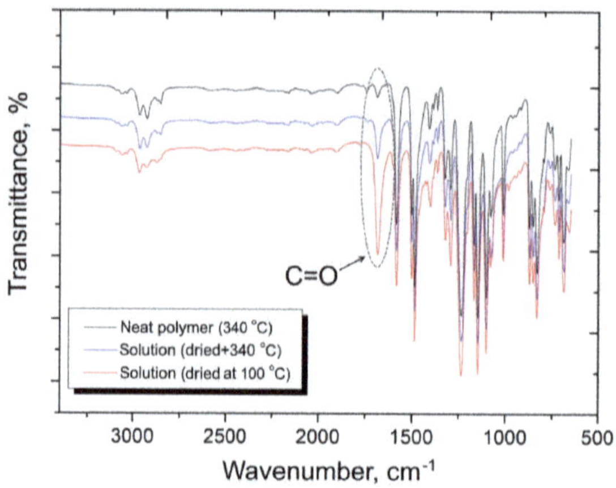

Figure 3. Infrared spectra of polysulfone after various treatments.

3.2. *Surface Modification of CFs*

It is well known, that despite the excellent mechanical properties, CFs show poor wettability and adhesion to matrix polymers, most likely because the CF surface is chemically inert, smooth, and exhibits low surface free energy [50,51]. Thermal oxidation is one of the most widely used methods of CF surface modification, which allows for improvement in the fiber–matrix interaction in polymer based composites. We used various thermal oxidation temperatures (up to 500 °C) to study the effect of surface treatment on the fibers and composite structure and properties.

To study the elemental composition and functional groups on the surface of the initial and modified carbon fibers, the x-ray photoelectron spectroscopy (XPS) method was used, which is the most commonly used method that provides comprehensive information about the CF surface. Elemental compositions and specific contents of various functional groups on the CF surface were investigated. The wide-scan spectra allows for the elemental compositions of the carbon fiber surface and the high resolution spectra of C1s peak to be obtained, which allows for the identification of functional groups on the CF surface, as shown in Figure 4 and the results are summarized in Tables 1 and 2.

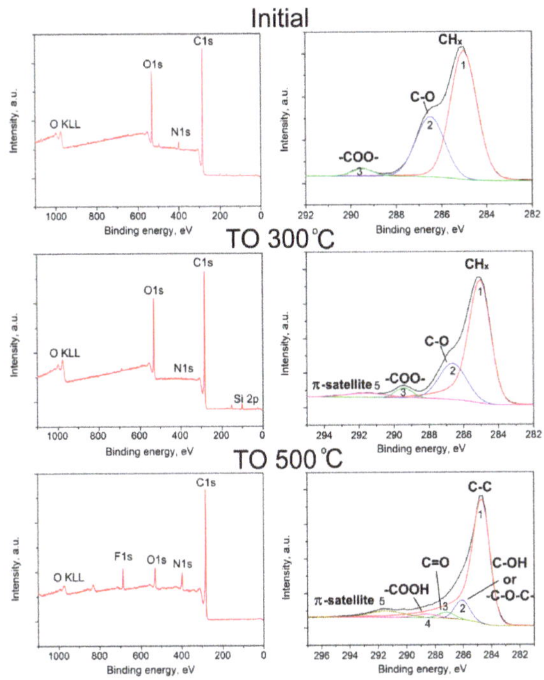

Figure 4. X-ray photoelectron spectroscopy survey spectra (left) and the high resolution spectra of C1s (right) of the initial and modified (TO 300 °C and TO 500 °C) carbon fibers.

Table 1. Concentration of different elements in atomic percent obtained from the x-ray photoelectron spectroscopy results for the carbon fiberssurface before and after thermal oxidation (TO).

Sample	Concentration, at. %							
	C	O	N	Si	S	Na	Cl	F
initial	77.5	18.0	3.0	0.7	0.3	0.3	0.3	-
TO 300 °C	79.0	18.1	0.4	2.3		0.1		0.1
TO 500 °C	82.5	7.5	8.5	0.3	-	-		1.0

Table 2. Binding energies (E_B) obtained from the high resolution spectra of the initial and modified carbon fibers [43].

Sample		C					O		N		
		1	2	3	4	5	1	2	1	2	3
Initial	E_B, eV	285.0	286.5	289.5	-	-	531.7	532.9		400.3	
	%	65	32	3	-	-	15	85		100	
TO 300 °C	E_B, eV	285.0	286.5	289.5	-	291.5	531.3	533.5		400.3	
	%	69	23	4	-	4	10	90		100	
TO 500 °C	E_B, eV	284.7	286.1	287.5	288.6	291.4	531.3	533.4	398.7	400.55	403.5
	%	80	9	3	2	6	50	50	40	50	10

It was found that the used regimes of surface modifications resulted in a noticeable change in the elemental composition of the fiber surface. It is known that the sizing applied to the CF surface is usually an organic polymer such as polyetherimide, epoxy resin, etc., with a high oxygen content [52,53]. Additionally, amine hardeners are often used to cure epoxy resins. As a result, a sufficiently high content of oxygen and nitrogen was found in the initial fibers (18.0 and 3.0 at. %, respectively). The concentration of carbon on the surface of the initial fibers was 77.5 at. %. Surface modification of carbon fibers increased the concentration of carbon to 79.0 at. % and 82.5 at. % after thermal oxidation (TO) at 300 °C (denoted as TO 300 °C) and 500 °C (denoted as TO 300 °C), respectively. At the same time, if after thermal oxidation at a temperature of 300 °C the oxygen concentration did not significantly change, caused by incomplete removal of the sizing at this temperature, the modification at 500 °C decreased the oxygen concentration to 7.5 at. %. After treatment at 300 °C, the nitrogen concentration decreased, which is due to the partial removal of the sizing agent, while the clean surface of the fibers was depleted in nitrogen, which may be due to the fact that it is covered with a graphitized residue that is formed during the production of the fiber. After surface modification at 500 °C, the nitrogen content increased to 8.5%.

The high-resolution spectrum shown in Figure 4 (right) was used to analyze the C1s region. Each spectrum was decomposed into a set of subspectra that allowed for the detection of the specific contents of various functional groups on the CF surface. The initial CF surface C1s spectrum corresponds to the polymer spectrum, which is typical for sized CF. For the sample after treatment at 300 °C, it was found that there is no obvious asymmetry of the C1s spectrum, which means that removing the sizing from the CF was not completed in this treatment. Treatment at 500 °C resulted in a sharp change in the shape of the C1s spectrum and it acquired the typical peaks for clean CF surface features: the main asymmetric peak 1 is from graphitized carbon, and peaks 2, 3, and 4 correspond to functional groups.

The binding energies (E_B) obtained from the C1s spectra are summarized in Table 2. Typical features of the C1s spectrum of the original fiber are the symmetric peak 1 (E_B = 285.0 eV) from CHx groups, peak 2 (E_B = 286.5 eV) from C–O bonds and peak 3 (E_B = 289.5 eV), which can be attributed to the N–COO– or –COO– group. The center of the oxygen spectrum is located at E_B = 532.8 eV, which can be attributed to several forms of oxygen. A high-resolution spectrum of nitrogen in the sample was not obtained; E_B was determined from a survey spectrum, 400.3 eV, and corresponded to the N–COO– group.

For the CF after treatment at 300 °C, the most satisfactory approximation of the C1s spectrum was made by adding a small asymmetry to the shape of peak 1, but the asymmetry parameters did not correspond to a pure carbon surface. Typical features of the C1s spectrum after this treatment are symmetric peak 1 (E_B = 285.0 eV) from the CHx groups, peak 2 (E_B = 286.5 eV) from the CO bonds, peak 3 (E_B = 289.5 eV) from the groups N–COO– or –COO–, and peak 5 (E_B = 291.5 eV)-π satellite [54,55]. TO 500 °C resulted in a significant change in the functional composition of the fibers, and several peaks related to the various functional groups were observed. The following peaks were detected in these spectra: peak 2 (E_B = 286.1 eV) from hydroxyl –C–OH, ethereal –C–O–C– or epoxy groups, and C–N– (it should be noted that the chemical shift for C–N is slightly less than C–O, however, it is almost impossible to distinguish them reliably), peak 3 (E_B = 287.5 eV) corresponded to double bonds C=O and –C=N, and peak 4 (E_B = 288.6 eV) carboxyl groups COOH– [56]. The presence of double bonds N=O and C=O was additionally confirmed by the O1s and N1s spectra, from which two components were distinguished. In the O1s spectrum, a peak at 531.3 eV corresponded to a double bond, and a peak at 533.4 eV to a single bond. Similarly, for the N1s spectrum, the peak at 398.7 eV corresponded to double bonds, and the peak at 400.5 eV to single bonds. Hence, the used regimes of modification allowed for a change in the CF surface chemical composition, and a greater number of functional groups on the surface appeared as a result of thermal oxidation. An increase in the activity of the CF surface affected the fiber–matrix interaction in the composites.

3.3. *Interface Analysis of Carbon Fibers/Polysulfone Composites*

In-plane shear strength (IPSS) is an important indicator of the interfacial adhesion between the fiber and matrix [57,58]. Here, we used shear tests to evaluate the IPSS of the PSU based composites reinforced with the initial and surface modified carbon fibers. Figure 5 shows the outlook of the investigated samples before and after the shear test. It can be seen that destruction of samples proceeds along the layers of composites, thus the results of the shear strength tests can be unambiguously attributed to the fiber–matrix interfacial interaction. The results of the in-plane shear strength of the CF/PSU composites with various fiber to polymer ratios are shown in Figure 6a.

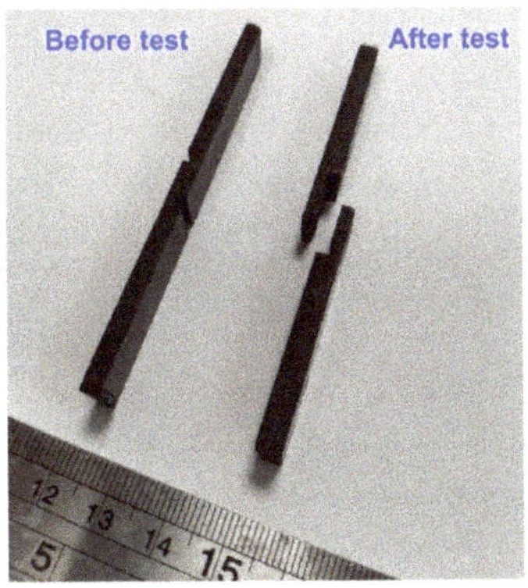

Figure 5. The experimental samples for the in-plane shear strength tests.

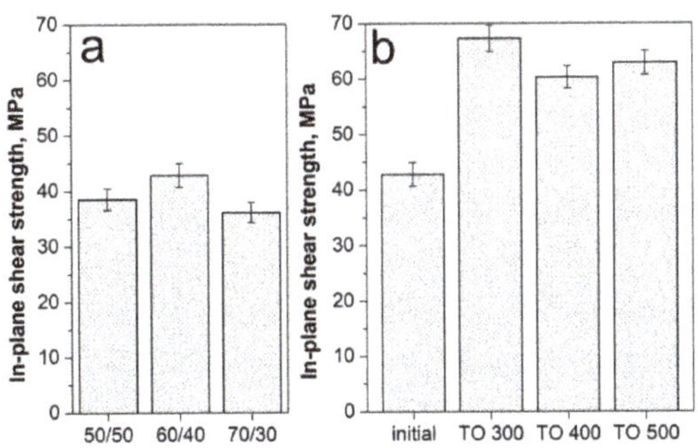

Figure 6. The in-plane shear strength of the CF/PSU composites depending on fiber content (**a**) and the fiber surface modification regime (**b**).

It was established that the shear strength of the composites with a 50/50 fiber to polymer ratio and reinforced with the initial fibers was 38.5 ± 1.4 MPa, for the 60/40 composites it was 42.8 ± 1.9 MPa, and for the 70/30 composites, it was 36.1 ± 1.2 MPa. The reason for the lowest shear strength values of the 70/30 composites was the low polymer content, which resulted in poor binding of the fiber filaments, and the polymer layer was too thin so it did not ensure the integrity of the composite. As a result, the formation of a weak interface was observed. Since the 60/40 composites showed the highest IPSS values, this composition was used to determine the influence of the CF surface modification on ILSS values. Additionally, the fiber to polymer ratio of 60/40 was used for all the modified fiber reinforced composites.

The IPSS results of the modified CF reinforced composites showed that the selected regimes of surface modification could significantly increase the shear strength of PSU-based composites. We found that thermal oxidation of the CF in the temperature range of 300–500 °C for 30 min allowed for an increase in shear strength up to 60–67.3 MPa (Figure 6b) (TO 300, 400, 500 designations mean the PSU based composites reinforced with CF after thermal oxidation at 300, 400, 500 °C, respectively). The increase in shear strength of the composites was due to the fact that, as shown by XPS analysis, the modification of the CF surface resulted in the formation of new functional groups on the fiber surface. Increase in the number of hydroxyl (–OH), carbonyl (–C=O), and carboxyl (–COOH) functional groups resulted in the chemical bonds between the filler and the polymer matrix formation, which resulted in the formation of a strong interfacial interaction in the composites. Favorable chemical and physical interaction between modified CF and PSU during the preparation of the composites https://www.sciencedirect.com/science/article/pii/S1359835X18303543-b0180 leads to the interfacial adhesion improvement [59,60]. As will be shown later, the observed improvement in interfacial interaction also resulted in a noticeable improvement in the mechanical properties of the composites reinforced with modified fibers.

Further demonstration of the CF surface modification positive effect on the interfacial interaction in the composites can be realized by SEM investigation of the fracture morphologies of the composites after tensile tests. Structural analysis of the composites reinforced with initial CF and after surface modification showed that the selected regimes could have a significant impact on the state of the interfaces, and, therefore, on the fracture mechanism.

It was shown that the fiber–matrix interaction was significantly improved in the case of using CF after TO. As can be seen from Figure 7, the method applied for composites allowed us to achieve good impregnation of the fabrics with polysulfone due to the low-viscosity polymer solution used. The polymer penetrated not only between the individual fabric threads, but also impregnated the threads, penetrating between the CF individual filaments (Figure 7a). However, the composites reinforced with the initial CFs showed poor fiber–matrix interaction due to the inertness of the CF surface. In this case, mechanical interlocking between the PSU and CF was mainly realized. In this case, the CF surface remains clean and smooth; almost no polymer traces were observed on the CF surface, which means that during the mechanical tests, an intense destruction of the fiber–polymer interface occurs. Figure 7b shows that a number of fibers were pulled out from the PSU matrix, which was caused by the interface failure and suggests poor interfacial adhesion.

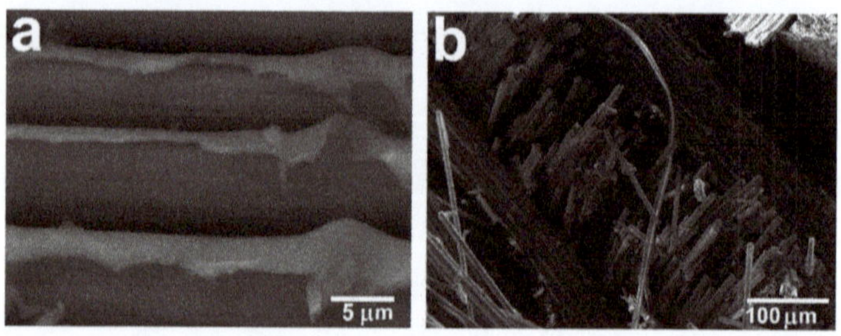

Figure 7. The structure of the CF/PSU composites (60/40 wt %) reinforced with the initial carbon fibers after tensile tests: (**a**) high-magnification image of the interfiber space; (**b**) low-magnification image of the fractured surface [43].

Comparatively, the CF surface modification significantly changed the composites' fracture surface morphology. Figure 8 shows good fiber–matrix adhesion in the case of the composite containing thermally modified CFs rather than for composites reinforced with untreated CFs. Even after the mechanical tests, a significant amount of fragmented PSU adhered to the CF surface still remained. There was almost no CF pulled out from the PSU matrix and the interface debonding was reduced. In this case, the composites' destruction proceeded with an equal probability both through the polymer matrix and the fibers, which is shown by the large quantity of destroyed CF ends located in the same level with the broken PSU surface of the polymer as well as presented in Figure 8 on high resolution image cracks directly in the CF. Formation of such a strong interface provides a good transfer of stress from the matrix material onto the CF, which is accompanied by a significant increase in the general mechanical properties of the composite.

Figure 8. The structure of the carbon fiber/polysulfone composites (60/40 wt %) reinforced with surface modified carbon fibers (TO 500 °C, 30 min) after the tensile tests.

3.4. Mechanical Tests of CF/PSU Composites

To further evaluate the mechanical properties of the CF/PSU composites, tensile and three-point flexural tests were performed using a universal testing machine. As known, reinforcement content is one of the main factors that affects the mechanical properties of fiber reinforced polymer composites. Fiber content strongly affect the stiffness, strength, and conductivity as well as other performance composite properties. Figure 9 shows typical stress–strain curves for the initial fiber reinforced PSU based composites during tensile and flexural tests depending on fiber content. The obtained curves were nearly linear, and the strain rate mainly depended on the fiber content. It can be seen that higher fiber content resulted in a lower strain rate that was caused by the fact that the carbon fibers had a much higher stiffness and lower plasticity compared with the polymer matrix.

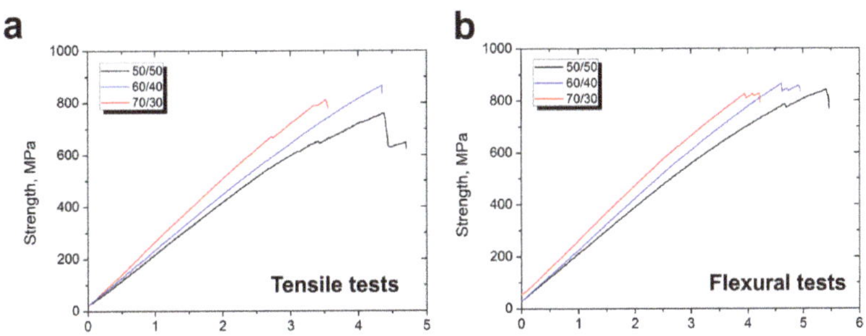

Figure 9. Typical stress–strain curves of the CF/PSU composites (**a**) during tensile and (**b**) flexural tests depending on the fiber content (initial fibers).

Figure 10a presents the tensile test results for composites with various fiber content. The minimum values of 748 ± 32 MPa and 55 ± 2.1 GPa for the ultimate tensile strength and Young's modulus, respectively, was found for the composites with a fiber to polymer ratio of 50/50. Increase in the fiber content up to 60 wt % resulted in the increase in the tensile strength of the composites up to 880 ± 44 MPa and Young's modulus up to 57.5 ± 2.2 GPa. An increase in the fiber content to 70% resulted in a decrease in the composite mechanical properties, the tensile strength decreased to 775 ± 3 6 MPa, and the Young's modulus decreased to 53.7 ± 2.3 GPa.

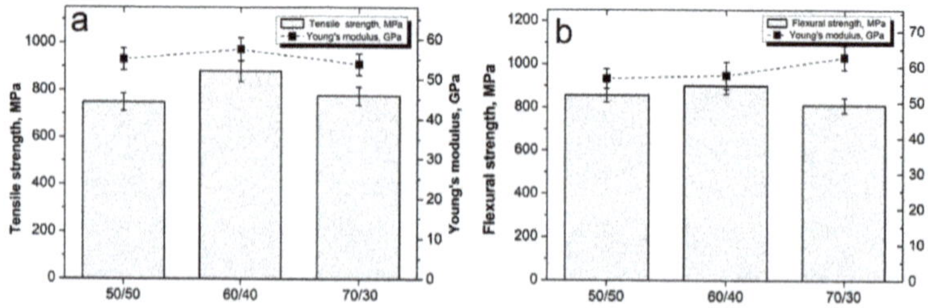

Figure 10. (**a**) Tensile and (**b**) flexural properties of the CF/PSU composites with various fiber to polymer ratios.

Figure 10b shows the results of the flexural tests. It can be seen that the minimum value of the flexural strength (855 ± 31 MPa) was observed for the 50/50 composites. Increase in the fiber content up to 60% resulted in an increase in the strength of the composites up to 899 ± 27 MPa; a further increase in the CF content up to 70% resulted in a decrease in strength to 811 ± 28 MPa. The Young's modulus in this case tended to increase from a value of 56.7 ± 1.7 GPa for a 50/50 composite up to 62.7 ± 2.4 GPa for a 70/30 composite.

We can summarize that at low CF content, the strength of the composites was rather low because of the significant content of the "weak" PSU phase. Increase in the CF content resulted in an increase in fiber packing density, while a decrease in the polymer interlayer between the individual fibers occurred. A decrease in the strength with a further increase in the CF content proceeded because the matrix content was too low to bind the CF and distribute the applied load to CF, in this case.

As above-mentioned, the mechanical properties of composites mainly depend on the fiber–matrix interfacial interaction. Tensile and flexural tests of the composites reinforced with CFs after TO were performed. In this case, composites with an optimum fiber content of 60 wt % that showed the maximum values of the strength-elastic properties were used. Figure 11 shows that the CF surface modification significantly affected the composites' mechanical characteristics.

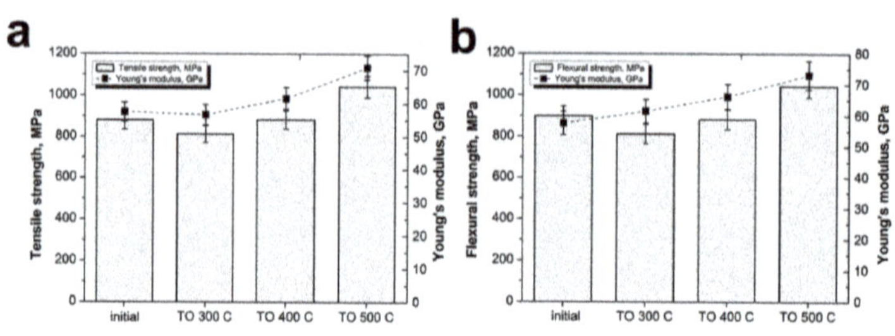

Figure 11. (**a**) Tensile and (**b**) flexural properties of the CF/PSU composites (60/40 wt %) reinforced with fiber after various surface modifications.

The maximum ultimate strength and Young's modulus values were obtained at the TO temperature of 500 °C. In this case, the tensile strength and Young's modulus of the composites increased from the initial 880 ± 44 MPa and 57.5 ± 2.2 GPa up to 1047 ± 2.8 MPa and 70.9 ± 2.6 GPa, respectively. The same behavior after the flexural tests was observed: from the initial 899 ± 27 MPa and 57.6 ± 2.8 GPa, the flexural strength and flexural modulus increase up to 1042 ±3.2 MPa and 73.1 ± 3.5 GPa, respectively. The observed increase in mechanical properties, due to the good CF/PSU interfacial interaction, and a stronger interface for the composites reinforced with modified fibers in comparison with initial fiber reinforced composites, allows for an improvement in the mechanical properties of the composites.

3.5. Dynamic Mechanical Analysis

The effect of the TO of CF on the composites' dynamic mechanical response was studied by DMA. This method of thermal analysis, which allows the mechanical properties of a material to be measured during its periodic deformation, makes it possible to determine the thermal behavior of a material under cyclic loads. Figure 12 shows the storage modulus and tan δ of the CF/PSU composites, which was plotted as a function of temperature. Storage modulus indicates a measure of the elastic response of the composite and tan δ (=E″/E′) is equal to the ratio of the loss modulus (E″) to the storage modulus (E′), thus representing the damping capacity of a material. It was shown that for all samples, the values of E′ decreased gradually with increasing temperature, followed by a sharp drop after the Tg region (Figure 12a).

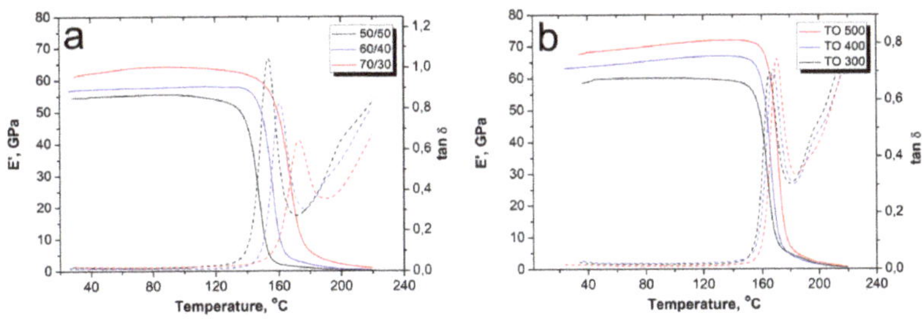

Figure 12. The DMA results of the CF/PSU composites (60/40 wt %) depending on the (**a**) initial fiber content and the (**b**) fibers' surface modification regime.

The behavior of the storage modulus values at room temperature of the obtained composites with different fiber to polymer ratios was similar to the behavior during the flexural tests (see Figure 9). The minimum values of the E′ were observed for the 50/50 composites and the highest values were observed for the 60/40 composites. Up to temperatures of 130–140 °C, the storage modulus of the composites underwent only slight changes, however, above this temperature, the E′ values began to decrease. It can be seen that the temperature of the beginning of the E′ decrease and the temperature to which the samples retain their stability depend on the fiber content. The lowest thermal stability was demonstrated by the 50/50 composites, which lost their stability at temperatures more than 10 °C less than that for the 60/40. The 70/30 composites showed the best thermal stability. The observed decrease in storage modulus was due to the softening of the polymer matrix due to glass transition. The softening of the PSU can be judged by the behavior of tan δ, as shown in Figure 12a. Usually, the maximum of the tan δ peak relates to the relaxation processes of the polymer matrix such as glass transition (α-transition), melting, or other. In our case, the tan δ peaks were affected by the fiber to polymer ratio, and were 155, 161, and 172 °C, for 50/50, 60/40, and 70/30 composites, respectively. Consequently, an increase in the fiber content resulted in a slight increase in the thermal stability of the composites, which was caused both by a decrease in the mobility of the macromolecular segments of the polymer matrix and by a decrease in the volume of the polymer matrix, which is prone to softening at temperatures close to the glass transition temperature. As a result of softening of the polymer matrix, it was no longer able to transfer an external load to the reinforcing fibers, which was accompanied by a sharp drop in the mechanical properties of the composites.

It was shown that the CF surface modification allowed for an increase in the storage modulus of the composites and shifted the temperature of the tan δ peak toward higher temperatures. In Figure 12b, it can be seen that the higher TO temperature resulted in the higher tan δ peak temperature, which corresponded to temperatures of 165, 167, and 170 °C for the composites reinforced with fibers after TO at 300, 400, and 500 °C, respectively. This is due to the fact that reinforcement with modified fibers reduces the free volume and decreases the polymer chains' mobility because of strong interfacial interaction between the CF and PSU. An increase in the interfacial interaction between the fibers and the matrix means that the mobility of the chains decreases even more, which results in higher thermal stability temperatures of the tan δ peak.

4. Conclusions

The solution impregnation method was used to form PSU based composites reinforced with twill wave CF fabrics. The elaborated technique allowed us to achieve homogeneous impregnation of CFs with the polymer and to further remove the solvent from the composite almost completely. Low viscosity of the impregnation solution resulted in polymer penetration even between the individual CF filaments. To improve the adhesion between CF and PSU, the surface of the fabrics was modified by thermal oxidation in an air atmosphere. It was shown that the surface modification by heating at a temperature to 500 °C for 30 min significantly changes the functional composition of the CF surface. Hydroxyl –C–OH, ethereal –C–O–C–, carboxyl COOH–, carbonyl –C=O, and epoxy groups form on the CF surface. Shear strength tests and SEM analysis of the fracture surface confirmed that surface modification resulted in a significant increase in the adhesion between the CF and polymer matrix. The in-plane shear strength magnitude for composites reinforced with modified CF fabrics was found to be more than 1.5 times higher than for composites containing unmodified CFs. Fracture analysis showed that in the case of unmodified CF, destruction proceeded on the boundary between the CFs and PSU. In the case of modified CF, the destruction of composites occurred both through the polymer matrix and CF body.

Mechanical tests in the tensile and flexural modes were carried out for the obtained composites. Carbon fiber content affects the composites' mechanical properties. At low CF content, the composite strength was not that high due to the significant content of the "weak" polymer phase. Increase in the fiber content provided an increase in the fiber packing density, while a decrease in the polymer interlayer between the individual fibers occurred. Decrease in the strength at a further increase in the CF content proceeded because the matrix content was too small to bind the CF and distribute the applied load to the CFs in this case. Surface modification of CF also led to an increase in the tensile and flexural properties of the composites. The maximum ultimate strength and Young's modulus values were found in the case of the CF TO temperature value of 500 °C. In this case, the tensile strength and Young's modulus of the composites increased from the initial 880 ± 44 MPa and 57.5 ± 2.2 GPa up to 1047 ± 28 MPa and 70.9 ± 2.6 GPa, respectively. The same behavior after the flexural tests was observed: from the initial 899 ± 27 MPa and 57.6 ± 2.8 GPa, the flexural strength and flexural modulus increased up to 1042 ± 32 MPa and 73.1 ± 3.5 GPa, respectively. Dynamic mechanical analysis showed that the investigated composites were stable up to the temperatures of 130–140 °C, and that an increase in the CF content was accompanied by an increase in the thermal stability of the composites. Surface modification of CFs also resulted in an increase in the composites' thermal stability due to the decrease in the polymer chains mobility, which was caused by good interfacial interaction between the CFs and PSU.

Author Contributions: Conceptualization, D.I.C. and A.A.S.; Investigation, S.G.N., V.G.T., M.Y.Z., G.S., and D.D.Z.; Methodology, V.V.T.; Writing—original draft preparation, D.I.C.; Writing—review and editing, V.V.T.; Supervision, D.I.C.; Project administration, V.V.T. All authors have read and agreed to the published version of the manuscript.

Funding: The reported study was funded by the Russian Science Foundation, grant No. 18-19-00744.

Conflicts of Interest: The authors declare no conflicts of interest.

References

1. Chung, D.D.L. Processing-structure-property relationships of continuous carbon fiber polymer-matrix composites. *Mater. Sci. Eng. R Rep.* **2017**, *113*, 1–29. [CrossRef]
2. Santulli, C. Mechanical and Impact Damage Analysis on Carbon/Natural Fibers Hybrid Composites: A Review. *Materials* **2019**, *12*, 517. [CrossRef]

3. Alam, P.; Mamalis, D.; Robert, C.; Floreani, C.; Ó Brádaigh, C.M. The fatigue of carbon fibre reinforced plastics-A review. *Compos. Part B Eng.* **2019**, *166*, 555–579. [CrossRef]
4. Liu, Q.; Lin, Y.; Zong, Z.; Sun, G.; Li, Q. Lightweight design of carbon twill weave fabric composite body structure for electric vehicle. *Compos. Struct.* **2013**, *97*, 231–238. [CrossRef]
5. Wenbin, L.; Jianfeng, H.; Jie, F.; Liyun, C.; Chunyan, Y. Mechanical and wet tribological properties of carbon fabric/phenolic composites with different weave filaments counts. *Appl. Surf. Sci.* **2015**, *353*, 1223–1233. [CrossRef]
6. Gao, J.; Chen, W.; Yu, B.; Fan, P.; Zhao, B.; Hu, J.; Zhang, D.; Fang, G.; Peng, F. Effect of temperature on the mechanical behaviours of a single-ply weave-reinforced shape memory polymer composite. *Compos. Part B Eng.* **2019**, *159*, 336–345. [CrossRef]
7. Viana, S.T.; Scariot, V.K.; Provensi, A.; Barra, G.M.O.; Barbosa, J.R. Fabrication and thermal analysis of epoxy resin-carbon fiber fabric composite plate-coil heat exchangers. *Appl. Therm. Eng.* **2017**, *127*, 1451–1460. [CrossRef]
8. Munalli, D.; Dimitrakis, G.; Chronopoulos, D.; Greedy, S.; Long, A. Electromagnetic shielding effectiveness of carbon fibre reinforced composites. *Compos. Part B Eng.* **2019**, *173*, 106906. [CrossRef]
9. Zulkifli, F.; Stolk, J.; Heisserer, U.; Yong, A.T.-M.; Li, Z.; Hu, X.M. Strategic positioning of carbon fiber layers in an UHMwPE ballistic hybrid composite panel. *Int. J. Impact Eng.* **2019**, *129*, 119–127. [CrossRef]
10. Neisiany, R.E.; Lee, J.K.Y.; Khorasani, S.N.; Ramakrishna, S. Towards the development of self-healing carbon/epoxy composites with improved potential provided by efficient encapsulation of healing agents in core-shell nanofibers. *Polym. Test.* **2017**, *62*, 79–87. [CrossRef]
11. Wu, X.-F.; Rahman, A.; Zhou, Z.; Pelot, D.D.; Sinha-Ray, S.; Chen, B.; Payne, S.; Yarin, A.L. Electrospinning core-shell nanofibers for interfacial toughening and self-healing of carbon-fiber/epoxy composites. *J. Appl. Polym. Sci.* **2013**, *129*, 1383–1393. [CrossRef]
12. Neisiany, R.E.; Khorasani, S.N.; Naeimirad, M.; Lee, J.K.Y.; Ramakrishna, S. Improving Mechanical Properties of Carbon/Epoxy Composite by Incorporating Functionalized Electrospun Polyacrylonitrile Nanofibers. *Macromol. Mater. Eng.* **2017**, *302*, 1600551. [CrossRef]
13. Capela, C.; Oliveira, S.E.; Ferreira, J.A.M. Fatigue behavior of short carbon fiber reinforced epoxy composites. *Compos. Part B Eng.* **2019**, *164*, 191–197. [CrossRef]
14. Chukov, D.I.; Stepashkin, A.A.; Tcherdyntsev, V.V.; Kaloshkin, S.D.; Danilov, V.D. Strength and thermophysical properties of composite polymer materials filled with discrete carbon fiber. *Inorg. Mater. Appl. Res.* **2014**, *5*, 386–391. [CrossRef]
15. Chukov, D.I.; Stepashkin, A.A.; Salimon, A.I.; Kaloshkin, S.D. Highly filled elastomeric matrix composites: Structure and property evolution at low temperature carbonization. *Mater. Des.* **2018**, *156*, 22–31. [CrossRef]
16. Huang, Z.-M.; Zhang, C.-C.; Xue, Y.-D. Stiffness prediction of short fiber reinforced composites. *Int. J. Mech. Sci.* **2019**, *161–162*, 105068. [CrossRef]
17. Dorigato, A.; Fredi, G.; Pegoretti, A. Application of the thermal energy storage concept to novel epoxy–short carbon fiber composites. *J. Appl. Polym. Sci.* **2019**, *136*, 47434. [CrossRef]
18. Spoerk, M.; Savandaiah, C.; Arbeiter, F.; Traxler, G.; Cardon, L.; Holzer, C.; Sapkota, J. Anisotropic properties of oriented short carbon fibre filled polypropylene parts fabricated by extrusion-based additive manufacturing. *Compos. Part A Appl. Sci. Manuf.* **2018**, *113*, 95–104. [CrossRef]
19. Zhang, G.; Rasheva, Z.; Schlarb, A.K. Friction and wear variations of short carbon fiber (SCF)/PTFE/graphite (10 vol.%) filled PEEK: Effects of fiber orientation and nominal contact pressure. *Wear* **2010**, *268*, 893–899. [CrossRef]
20. Swolfs, Y.; Verpoest, I.; Gorbatikh, L. A review of input data and modelling assumptions in longitudinal strength models for unidirectional fibre-reinforced composites. *Compos. Struct.* **2016**, *150*, 153–172. [CrossRef]
21. Ma, Y.; Yang, Y.; Sugahara, T.; Hamada, H. A study on the failure behavior and mechanical properties of unidirectional fiber reinforced thermosetting and thermoplastic composites. *Compos. Part B Eng.* **2016**, *99*, 162–172. [CrossRef]
22. Chukov, D.; Nematulloev, S.; Zadorozhnyy, M.; Tcherdyntsev, V.; Stepashkin, A.; Zherebtsov, D. Structure, Mechanical and Thermal Properties of Polyphenylene Sulfide and Polysulfone Impregnated Carbon Fiber Composites. *Polymers* **2019**, *11*, 684. [CrossRef]

23. Stepashkin, A.A.; Chukov, D.I.; Senatov, F.S.; Salimon, A.I.; Korsunsky, A.M.; Kaloshkin, S.D. 3D-printed PEEK-carbon fiber (CF) composites: Structure and thermal properties. *Compos. Sci. Technol.* **2018**, *164*, 319–326. [CrossRef]
24. Liang, Y.; Wang, H.; Gu, X. In-plane shear response of unidirectional fiber reinforced and fabric reinforced carbon/epoxy composites. *Polym. Test.* **2013**, *32*, 594–601. [CrossRef]
25. Bilisik, K.; Karaduman, N.S.; Sapanci, E. Tensile properties of nanoprepreg/nanostitched 3D carbon/epoxy MWCNTs composites. *Mech. Mater.* **2019**, *128*, 11–23. [CrossRef]
26. Bergmann, T.; Heimbs, S.; Maier, M. Mechanical properties and energy absorption capability of woven fabric composites under ±45° off-axis tension. *Compos. Struct.* **2015**, *125*, 362–373. [CrossRef]
27. Bijwe, J.; Rattan, R. Influence of weave of carbon fabric in polyetherimide composites in various wear situations. *Wear* **2007**, *263*, 984–991. [CrossRef]
28. Foroutan, R.; Nemes, J.; Ghiasi, H.; Hubert, P. Experimental investigation of high strain-rate behaviour of fabric composites. *Compos. Struct.* **2013**, *106*, 264–269. [CrossRef]
29. Lundström, F.; Frogner, K.; Andersson, M. A method for inductive measurement of equivalent electrical conductivity in thin non-consolidated multilayer carbon fibre fabrics. *Compos. Part B Eng.* **2018**, *140*, 204–213. [CrossRef]
30. Thompson, A.J.; El Said, B.; Ivanov, D.; Belnoue, J.P.-H.; Hallett, S.R. High fidelity modelling of the compression behaviour of 2D woven fabrics. *Int. J. Solids Struct.* **2018**, *154*, 104–113. [CrossRef]
31. Gaurav, A.; Singh, K.K. Fatigue behavior of FRP composites and CNT-Embedded FRP composites: A review. *Polym. Compos.* **2018**, *39*, 1785–1808. [CrossRef]
32. Yao, S.-S.; Jin, F.-L.; Rhee, K.Y.; Hui, D.; Park, S.-J. Recent advances in carbon-fiber-reinforced thermoplastic composites: A review. *Compos. Part B Eng.* **2018**, *142*, 241–250. [CrossRef]
33. de Leon, A.C.; Chen, Q.; Palaganas, N.B.; Palaganas, J.O.; Manapat, J.; Advincula, R.C. High performance polymer nanocomposites for additive manufacturing applications. *React. Funct. Polym.* **2016**, *103*, 141–155. [CrossRef]
34. Mujika, F.; Benito, A.D.; Fernández, B.; Vázquez, A.; Llano-Ponte, R.; Mondragon, I. Mechanical properties of carbon woven reinforced epoxy matrix composites. A study on the influence of matrix modification with polysulfone. *Polym. Compos.* **2002**, *23*, 372–382. [CrossRef]
35. Solodilov, V.I.; Korokhin, R.A.; Gorbatkina, Y.A.; Kuperman, A.M. Comparison of Fracture Energies of Epoxy-polysulfone Matrices and Unidirectional Composites Based on Them. *Mech. Compos. Mater.* **2015**, *51*, 177–190. [CrossRef]
36. Cauich-Cupul, J.I.; Herrera-Franco, P.J.; García-Hernández, E.; Moreno-Chulim, V.; Valadez-González, A. Factorial design approach to assess the effect of fiber–matrix adhesion on the IFSS and work of adhesion of carbon fiber/polysulfone-modified epoxy composites. *Carbon Lett.* **2019**, *29*, 345–358.
37. Brown, S.A.; Hastings, R.S.; Mason, J.J.; Moet, A. Characterization of short-fibre reinforced thermoplastics for fracture fixation devices. *Biomaterials* **1990**, *11*, 541–547. [CrossRef]
38. Zimmerman, M.C.; Scalzo, H.L.; Parsons, J.R.; Torop, A.H.; Lin, T.S. Effects of environmental exposure on carbon polysulphone composites. *Biomaterials* **1991**, *12*, 424–430. [CrossRef]
39. Mai, K.; Li, J.; Zeng, H. Mechanical property and fracture morphology of fiber-reinforced polysulfone plasticized with acetylene-terminated sulfone. *J. Appl. Polym. Sci.* **1994**, *52*, 1279–1291. [CrossRef]
40. Antonov, A.V.; Zelenskii, E.S.; Kuperman, A.M.; Lebedeva, O.V.; Rybin, A.A. Behavior of reinforced plastics based on polysulfone matrix under impact loading. *Mech. Compos. Mater.* **1998**, *34*, 12–19. [CrossRef]
41. Migacz, K.; Chłopek, J.; Morawska-Chochół, A.; Ambroziak, M. Gradient composite materials for artificial intervertebral discs. *Acta Bioeng. Biomech.* **2014**, *16*, 3–12.
42. Sun, D.; Lin, G.; Sui, G.; Hao, Y.; Yang, R.; Zhang, K. Multi-gating injection molding to enhance the thermal conductivity of carbon fiber/polysulfone composite. *Polym. Compos.* **2017**, *38*, 185–191. [CrossRef]
43. Chukov, D.; Nematulloev, S.; Torokhov, V.; Stepashkin, A.; Sherif, G.; Tcherdyntsev, V. Effect of carbon fiber surface modification on their interfacial interaction with polysulfone. *Results Phys.* **2019**, *15*, 102634. [CrossRef]
44. Chukov, D.; Nematulloev, S.; Stepashkin, A.; Maksimkin, A.; Zherebtsov, D.; Tcherdyntsev, V. Novel carbon fibers reinforced composites based on polysulfone matrix. *MATEC Web Conf.* **2018**, *242*, 01004. [CrossRef]
45. Iyer, S.R.; Drzal, L.T. Manufacture of Powder-Impregnated Thermoplastic Composites. *J. Thermoplast. Compos. Mater.* **1990**, *3*, 325–355. [CrossRef]

46. Zheng, L.; Liao, G.X.; Jian, X.G. Preparation of Solution Impregnated Continuous Carbon Fibre Reinforced Poly (Phthalazinone Ether Sulfone Ketone) Composites. *Adv. Compos. Lett.* **2009**, *18*, 096369350901800101. [CrossRef]
47. Wu, G.M.; Schultz, J.M. Processing and properties of solution impregnated carbon fiber reinforced polyethersulfone composites. *Polym. Compos.* **2000**, *21*, 223–230. [CrossRef]
48. Masuelli, M.A. Synthesis Polysulfone-Acetylethanol Ultrafiltration Membranes. Application to Oily Wastewater Treatment. *J. Mater. Phys. Chem.* **2013**, *1*, 37–44.
49. Kumar, M.; McGlade, D.; Ulbricht, M.; Lawler, J. Quaternized polysulfone and graphene oxide nanosheet derived low fouling novel positively charged hybrid ultrafiltration membranes for protein separation. *RSC Adv.* **2015**, *5*, 51208–51219. [CrossRef]
50. Duchoslav, J.; Unterweger, C.; Steinberger, R.; Fürst, C.; Stifter, D. Investigation on the thermo-oxidative stability of carbon fiber sizings for application in thermoplastic composites. *Polym. Degrad. Stab.* **2016**, *125*, 33–42. [CrossRef]
51. Karsli, N.G.; Aytac, A. Effects of maleated polypropylene on the morphology, thermal and mechanical properties of short carbon fiber reinforced polypropylene composites. *Mater. Des.* **2011**, *32*, 4069–4073. [CrossRef]
52. Dilsiz, N.; Wightman, J.P. Surface analysis of unsized and sized carbon fibers. *Carbon* **1999**, *37*, 1105–1114. [CrossRef]
53. Dai, Z.; Zhang, B.; Shi, F.; Li, M.; Zhang, Z.; Gu, Y. Effect of heat treatment on carbon fiber surface properties and fibers/epoxy interfacial adhesion. *Appl. Surf. Sci.* **2011**, *257*, 8457–8461. [CrossRef]
54. Vautard, F.; Grappe, H.; Ozcan, S. Stability of carbon fiber surface functionality at elevated temperatures and its influence on interfacial adhesion. *Appl. Surf. Sci.* **2013**, *268*, 61–72. [CrossRef]
55. Kundu, S.; Wang, Y.; Xia, W.; Muhler, M. Thermal Stability and Reducibility of Oxygen-Containing Functional Groups on Multiwalled Carbon Nanotube Surfaces: A Quantitative High-Resolution XPS and TPD/TPR Study. *J. Phys. Chem. C* **2008**, *112*, 16869–16878. [CrossRef]
56. Dilsiz, N.; Wightman, J.P. Effect of acid–base properties of unsized and sized carbon fibers on fiber/epoxy matrix adhesion. *Colloids Surf. A Physicochem. Eng. Asp.* **2000**, *164*, 325–336. [CrossRef]
57. Huang, S.-Y.; Wu, G.-P.; Chen, C.-M.; Yang, Y.; Zhang, S.-C.; Lu, C.-X. Electrophoretic deposition and thermal annealing of a graphene oxide thin film on carbon fiber surfaces. *Carbon* **2013**, *52*, 613–616. [CrossRef]
58. Yuan, H.; Zhang, S.; Lu, C. Surface modification of carbon fibers by a polyether sulfone emulsion sizing for increased interfacial adhesion with polyether sulfone. *Appl. Surf. Sci.* **2014**, *317*, 737–744. [CrossRef]
59. Li, N.; Wu, Z.; Huo, L.; Zong, L.; Guo, Y.; Wang, J.; Jian, X. One-step functionalization of carbon fiber using in situ generated aromatic diazonium salts to enhance adhesion with PPBES resins. *RSC Adv.* **2016**, *6*, 70704–70714. [CrossRef]
60. Liu, L.; Yan, F.; Li, M.; Zhang, M.; Xiao, L.; Shang, L.; Ao, Y. A novel thermoplastic sizing containing graphene oxide functionalized with structural analogs of matrix for improving interfacial adhesion of CF/PES composites. *Compos. Part A Appl. Sci. Manuf.* **2018**, *114*, 418–428. [CrossRef]

polymers

MDPI

Article

Effect of Formation Route on the Mechanical Properties of the Polyethersulfone Composites Reinforced with Glass Fibers

Galal Sherif *, Dilyus Chukov, Victor Tcherdyntsev and Valerii Torokhov

Center of composite materials, National University of Science and Technology "MISIS", Leninskii prosp, 119049 Moscow, Russia

* Correspondence: eng_galal_emad@mu.edu.eg; Tel.: +7-9267878736

Received: 4 July 2019; Accepted: 16 August 2019; Published: 19 August 2019

Abstract: Interfacial interaction is one of the most important factors that affect the mechanical properties of the fiber reinforced composites. The effect of fabrics' sizing removal from glass fibers' surface by thermal treatment on the mechanical characteristics of polyethersulfone based composites at different fiber to polymer weight ratios was investigated. Three fiber to polymer weight ratios of 50/50, 60/40, and 70/30 were studied. Flexural and shear tests were carried out to illustrate the mechanical properties of the composites; the structure was studied using Fourier-transform infrared spectroscopy and scanning electron microscopy. It was shown that solution impregnation of glass fabrics with polyethersulfone before compression molding allows to achieve good mechanical properties of composites. The thermal treatment of glass fabrics before impregnation results in an increase in flexural and shear strength for all the composites due to the improvement of fiber–matrix interaction.

Keywords: glass fibers; surface modification; polyethersulfone; impregnation

1. Introduction

Growing attention has been given to the improvement of polymeric composite properties, especially in regards to their high strength-to-weight ratio [1–3]. Polyethersulfone (PES) is a superior performance engineering plastic with a high glass transition temperature T_g of 225 °C and operating temperature up to 180 °C. Due to several advantages such as high toughness; ease to produce and form complex shapes; good tribological properties; high modulus and strength; perfect fatigue resistance and dimensional stability; as well as rich fire, chemical and radiation resistance; PES is exceedingly eligible as a high-temperature tribo-material to substitute metals or ceramics [4]. Nevertheless, the tribological and mechanical properties of PES have to be evolved to gain the requirements necessary for several sophisticated applications, such as aerospace, automotive, and microelectronics [5].

Since enhancing the mechanical properties of polymer composites is the major task, glass fibers (GF) are widely used as reinforcing material [4]. GF supply beneficial adhesion to the polymer matrix, perfect aesthetic quality, and revised strength of the resulting composites [6]. Because GF possess high mechanical properties, low weight ratios, suitable heat resistance, and have a very cheap cost, they are attractive as reinforcers for polymers [7]. Thermoplastic composites reinforced with glass fibers were paid great attention because of their wide applications in automotive, aerospace, and many engineering applications [8]. Several studies have been carried out to investigate the mechanical properties of thermoplastics reinforced with glass fibers [9–20]. It was found that the properties of the composites are affected by the fibers' geometry, orientation, concentration, and by the nature of the fibers [13,15–17,21,22] in addition to the interfacial adhesion between the fiber and polymer [23–25], which plays a crucial role in the composites' strength.

The interfacial adhesion is affected both by the raw fibers' surface coating (sizing) and by any GF surface treatment or modification [26–33]. Sizing prevents the damage of fibers after producing, makes their use in manufacturing easier, protects fibers from the environmental impact [28], and enhances the composites' properties in the case of using sizing compatible with the matrix material [28,34]. Sizing could be provided by various materials, such as silanes, epoxies, paraffin and other coupling agents; and it could be one or more from these materials, but the actual and accurate sizing formula remains the secret of manufacturers [28]. However, most of the composites are made from epoxy resin, and most of the commercial sizings are often of the same nature. These sizings have relatively low degradation temperatures (about 250 °C) which are much lower than the processing temperatures of the engineering plastics such as PES. This would inevitably cause the problem of the sizing being degraded under high processing temperature, leading to weakened interfacial adhesion. A lot of methods (electrochemical, chemical, thermal, grafting, coating, and discharge plasma treatments, etc.) have been elaborated to enhance the interfacial adhesion between GF and polymers [35–37]. In the present work, we investigate the effect of formation route in addition to fabrics' sizing removal from GF surface on the structure and mechanical properties of PES based composites.

2. Materials and Methods

Preparation of PES Composites

Woven glass fabrics (NPO "Stekloplastic" Russia) (T-23 "260 ± 10 g/m^2, 12 + 1 warp, 8 + 1 weft yarn/cm, 0.27 ± 0.03 thickness") and PES Ultrason E2010 (Basf, Germany) powder were used as raw materials. A solvent of N-Methyl-2-pyrrolidone was used to produce PES solution to provide a good impregnation of fabrics with polymer. Solution formation was carried out with 20/80 polymer to solvent weight ratio for 24 h using a magnetic stirrer. Four routes were used for sample preparation: (1) compression molding of PES powder together with as-received glass fabrics; (2) compression molding of PES powder together with preheated glass fabrics; (3) compression molding of impregnated as-received glass fabrics with PES solution, and (4) compression molding of impregnated preheated glass fabrics with PES solution. The samples of PES solution impregnated glass fabrics were dried at 150 °C for 5 h to remove the solvent before compression molding. The bulk samples were prepared using the compression molding technique as shown in Figure 1. All samples were produced using the compression molding method at 350 °C and pressure of 10 MPa. The fiber to polymer weight ratio was varied as follows: 50/50, 60/40, and 70/30 wt %.

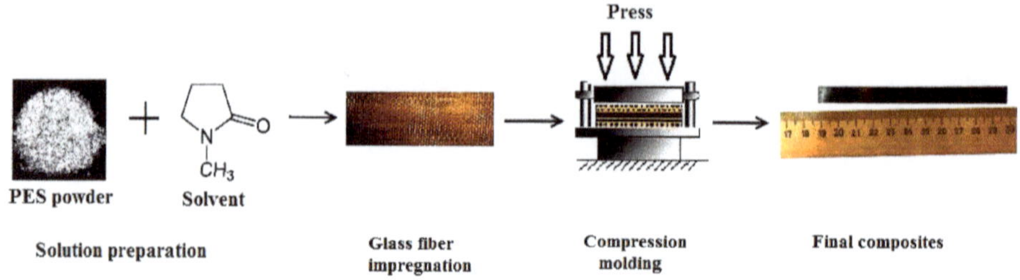

Figure 1. Preparation steps of polyethersulfone (PES) composites with compression molding.

The glass fabrics were preheated using the furnace in air atmosphere at three temperatures (300, 350, and 400 °C) for 1 h and they were used to produce bulk samples which were tested to choose the best-preheated temperature.

Flexural and shear tests were performed to examine the PES-based composites using a Zwick/Roell Z020 universal test machine (Boston, MA, USA) equipped with 1 and 20 kN sensors and a contact strain measurement system MultiXtens. For flexural tests, samples of 110 mm × 10 mm × 2 mm and a span length of 80 mm (according to ISO 14125:1998) were prepared. Samples of 110 mm × 10 mm

× 4 mm and a gauge length of 80 mm (according to ASTM D 3846) were prepared for shear tests. The flexural and shear tests were carried out at speeds of 10 and 1.3 mm/min, respectively, at room temperature. At least six samples were tested at each condition.

FTIR spectroscopy of the samples after various treatments were obtained using a Nicolet 380 FT-IR spectrometer (spectral range of 4000–450 cm^{-1}, resolution of 1 cm^{-1}). A scanning electron microscope (VEGA 3 TESCAN) (Brno - Kohoutovice, Czech Republic) in a backscattered electron image mode was used to study the structure of the fracture surface of the composites. The samples were coated with a thin layer (10–15 nm) of carbon in a sputter coater. For the studies the composites' fracture surface after flexural tests were used.

3. Results and Discussion

Elaboration of the advanced methods to improve the mechanical properties of the polymeric composites has become one of biggest challenges facing the industry recently. The effect of the preheating temperatures on the composites' mechanical properties were studied. Figure 2 shows the effect of the fabrics which were preheated at different temperatures on the flexural strength of the composites. The composites were prepared by the compression molding of impregnated fabrics with 50/50 wt %. It can be noticed that the deflections were affected by the temperature, and were higher (3.5 cm) in case of 350 °C compared with 3 and 2.7 cm for 300 and 400 °C preheating temperature, respectively. In the case of 300 °C it seems that it was not enough to remove all the sizing coating which affected the adhesion between fibers and polymer, so the failure strain was less than in the 350 °C samples, while in the 400 °C samples the fiber strength decreased because of the heating and made the samples weaker [28]. It is seen that flexural strength was higher in case of GF preheating temperature of 350 °C (501 MPa) compared with 300 °C (450 MPa), and 400 °C (440 MPa). This trend was similar to that observed by many researchers [38–41] who studied the influence of heating the glass fiber. The results showed that if the fibers are heated above 400 °C, their strength drops rapidly. Taking into account these results, we used GF heated at 350 °C in further study.

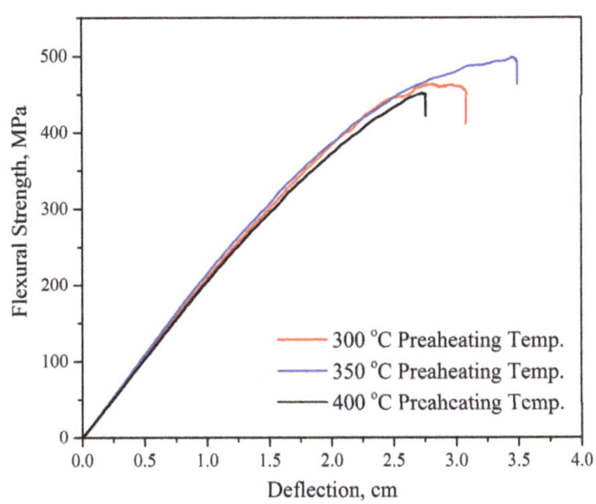

Figure 2. The strain–stress curves of 50/50 wt % composites prepared by route 4 using glass fibers (GF) preheated at various temperatures.

Four types of composites (PES powder with untreated glass fibers (as-received) (route 1), PES powder with preheated glass fibers (route 2), solution impregnated PES with untreated glass fibers (route 3), and PES solution with preheated glass fibers (route 4) were examined to choose the optimum conditions for producing the samples. The composites prepared with 50/50 fiber to polymer weight ratio and the GF preheating temperature of 350 °C were studied. Figure 3 shows flexural strength magnitude depending on the composite's formation route. As it is seen, the highest value of flexural strength was observed for sample obtained by PES solution impregnated preheated glass fabrics. The flexural strength of this composite was 501 MPa, which was significantly higher than values of 103, 148, 417 MPa for PES powder with as-received glass fabrics, PES powder with preheated glass fabrics, and PES solution with as-received glass fabrics samples, respectively. It can be concluded that PES in the solution form provides a good impregnation with the fabrics, and it was better with preheated fabrics. The presence of the solution enhanced the wettability and adsorption, which led to a good interfacial force that results in increased strength [42].

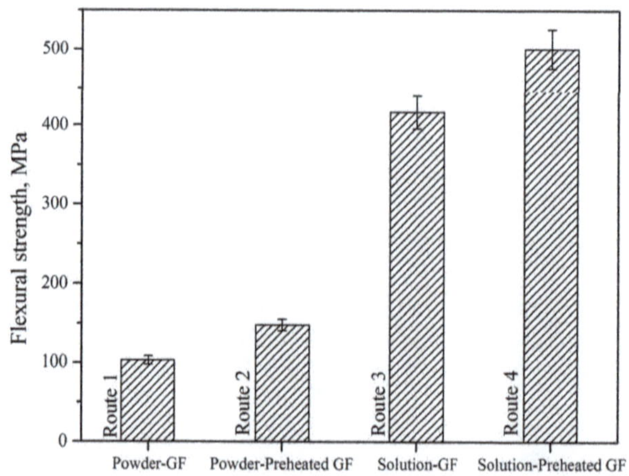

Figure 3. Flexural strength of different types of PES composites (50/50%, 350 °C preheating temperature).

FTIR spectra were carried out for characterization of the composites during the different stages of production. Figure 4a illustrates the characterization of the PES powder before any processing, PES samples prepared from PES powder by compression molding, and PES sample prepared from the PES solution after removing the solvent by heating at 150 °C for 5 h. From the results, it can be noticed that the major change is the amplitude of the C=O peaks (1678–1683) cm^{-1}. It was found that the amplitude increased after compression molding due to the oxidation occurred during the heating, and in the case of the sample of PES in solution form, the presence of the small amount of solvent which contains C=O band makes the amplitude increases in the sample. Figure 4b shows the FTIR spectra of the PES composites with different fiber to polymer weight ratios (50/50, 60/40, and 70/30%). The 50/50 sample shows a higher C=O peak amplitude because the amount of the solvent is more than those in the other samples as a result of higher amount of polymer in 50/50 samples compared with other samples. The characterization of the as-received and preheated glass fabrics using FTIR is shown in Figure 4c. Since the silicone oxide band (1100–900 cm^{-1}) in the glass fiber is a strong band, any band below 1200 cm^{-1} will not appear because of the strong absorption of the silicone oxide band. In addition, the as-received glass fabrics spectra shows noticeable peaks around 2969–2831 cm^{-1} which referred to stretching in C–H of CH, CH_2, and CH_3; while the spectra of preheated glass fabrics did not show any noticeable peaks before 2200 cm^{-1} which means that the sizing coating after heating was below the detection limit [29]. So, we can expect that almost of the sizing coating was removed.

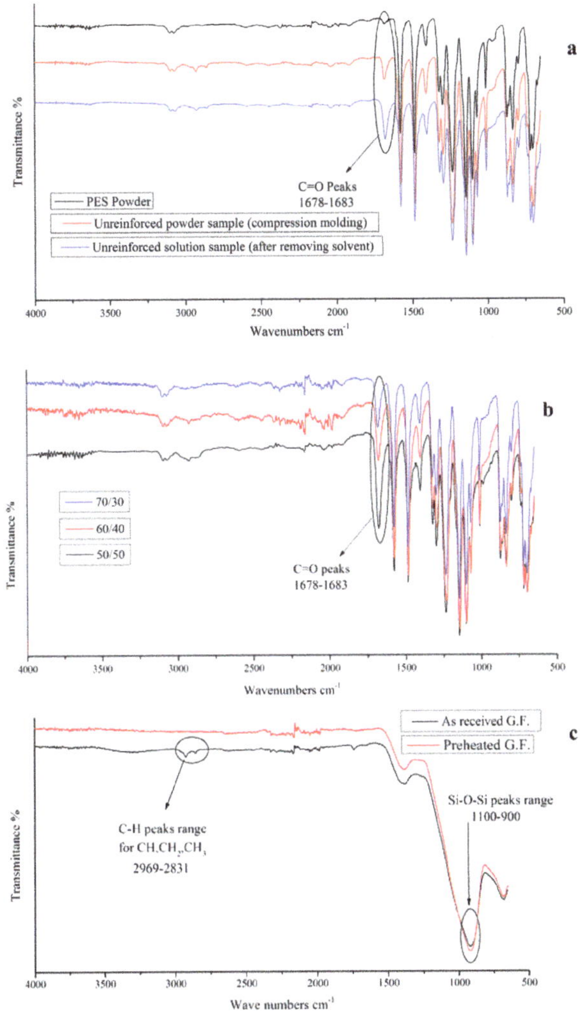

Figure 4. FTIR spectra for (**a**) initial PES powder, unreinforced PES powder sample after compression molding, and unreinforced solution sample after removing solvent. (**b**) PES impregnated composites (route 4) with different fiber to polymer weight ratios. (**c**) As-received GF and GF preheated at 350 °C.

The comparison of the flexural strength and Young's modulus of the PES composites reinforced with as-received and preheated GF is shown in Figure 5. For samples reinforced with as-received GF (Figure 5a) the highest value of flexural strength was observed for samples with the fiber to polymer ratio of 60/40, whereas for samples filled with preheated GF (Figure 5b) flexural strength nearly not depend on sample composition. Young's modulus value gradually increases with an increase in the GF content for PES filled both with as-received and preheated GF. It can be proposed that in case of PES filled with as-received GF, an increase in the fiber amount affected the interface between the fiber and the polymer in the case of 70/30 samples which leads to flexural strength decrease and the fiber being pulled out from the matrix, as it will be shown in SEM images (Figure 7e). The problem of the poor interface was solved by preheating the fiber before using, which leads to removal of the sizing coating and enhanced adhesion between the fibers and polymer. It can be concluded that in addition to the increase in the properties of the composites due to the increase of the fiber content, the interfacial interaction between the fiber and the polymer was improved. This improvement of the adhesion between fiber and matrix provides a good stress transfer, which controls the strength of the composites [42,43].

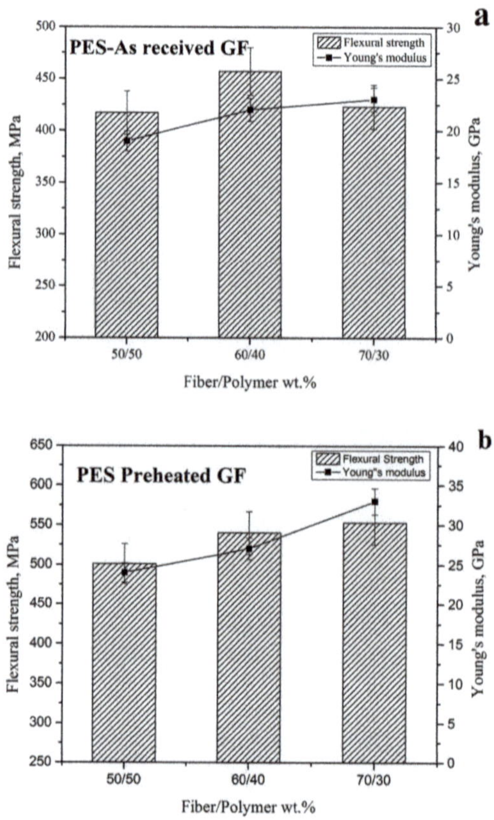

Figure 5. Flexural strength and Young's modulus for composites reinforced with (**a**) as-received GF (route 3) and (**b**) preheated at 350 °C GF (route 4).

To clarify this interfacing behavior, a shear test was carried out for the as-received and preheated GF reinforced composites to study the effect of preheating on the interfacial interaction. Figure 6 shows the shear strength for as-received and preheated GF reinforced composites. As shown in the figure the shear stress for the as-received glass fabrics composites increases from 48 MPa (for 50/50 samples) to 60 MPa (for 60/40 samples), while the shear strength in 70/30 samples (47 MPa) was affected by the decrease in polymer's content which bond the fabrics layers together and directly affect the shear stress. On the other hand, a clear enhancement was noted in the preheated samples (56, 59, and 57 MPa for 50/50, 60/40, and 70/30 samples, respectively) compared with the as-received samples. It can be concluded that the shear strength increases in the case of preheated GF reinforced composites, which was an indication of the improvement of the interfacial interaction which leads to the increase of the flexural strength of the preheated composites compared with the as-received composites.

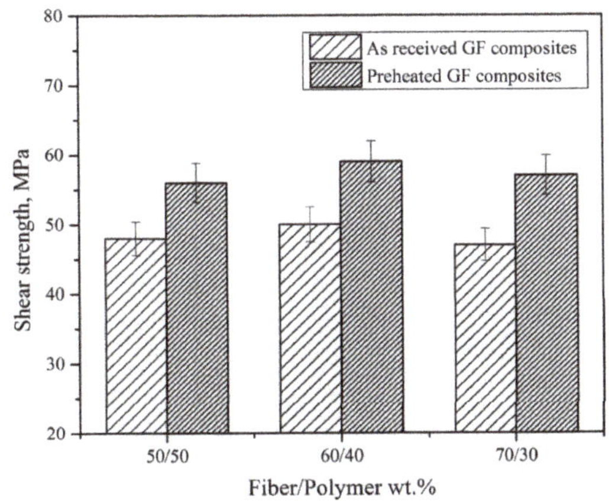

Figure 6. Shear strength of composites reinforced with as-received GF (route 3) and preheated at 350 °C GF (route 4).

The improvement of the flexural strength and Young's modulus for as-received and preheated glass fibers can be concluded from Table 1. It can be noticed that the flexural strength and Young's modulus increased when the fiber percentage increased, which behaved like most of the thermoplastic composites [2,4,44]. The thermal treatment showed an additional improvement of all the samples [36], 20.1, 18.2, and 30.7% were the improvement percentages in flexural strength for the preheated composites for 50/50, 60/40, and 70/30, respectively, compared with the as-received composites and the improvement percentages in Young's modulus were 26.3, 22.7, and 43.5%, respectively.

Table 1. The flexural strength and Young's modulus for composites reinforced with as-received GF (route 3) and preheated at 350 °C GF (route 4).

Fiber/Polymer	50/50		60/40		70/30	
Property	Flexural Strength, MPa	Young's Modulus, GPa	Flexural Strength, MPa	Young's Modulus, GPa	Flexural Strength, MPa	Young's Modulus, GPa
As-received GF composites	417	19	457	22	423	23
Preheated GF composites	501	24	540	27	553	33
Increase, %	20.1	26.3	18.2	22.7	30.7	43.5

SEM images of the fracture surfaces of the PES filled with 50/50 as-received GF, 50/50 preheated GF, 70/30 preheated GF, and 70/30 as-received GF are shown in Figure 7. The preheating of fibers allowed a good impregnation in the 50/50 sample as shown in Figure 7b compared with the sample 50/50 as-received GF in Figure 7a. The large amount of PES particles in preheated samples referred to the good interfacing between the fiber and polymer while a small amount of PES particles on the surface of the filaments of the as-received sample indicates bad adhesion. This leakage of adhesion led to fiber pull-out that appears in as-received samples. The 60/40 as-received composites shown in Figure 7c show a good distribution of PES particles on the glass fibers' surface, these particles increased in amount and size in the case of 60/40 preheated composites shown in Figure 7d. This improvement was because of the good interface due to removing the sizing coating (in case of using preheated GF). However, the amount of PES particles decreased in the case of 70/30 preheated samples shown in Figure 7f but still provided a good impregnation which appears on the fibers' surface. On the contrary, a poor adhesion occurred in the 70/30 as-received glass fibers reinforced composites (Figure 7e), which was evident in the form of fiber pull-out phenomenon.

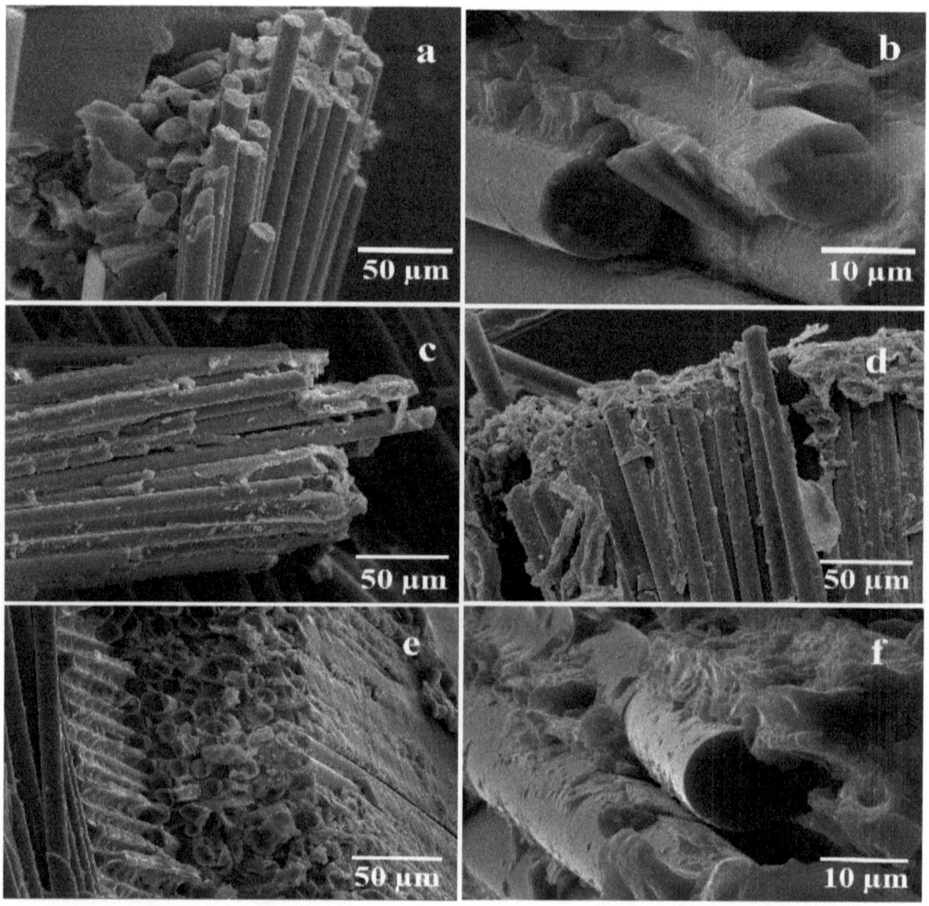

Figure 7. The structure of the fracture surface of the composites: (**a**) 50/50 as-received GF; (**b**) 50/50 preheated GF; (**c**) 60/40 as-received GF; (**d**) 60/40 preheated GF; (**e**) 70/30 as-received GF, and (**f**) 70/30 preheated GF composites.

4. Conclusions

A new method to improve the mechanical properties of PES composites by preheating the glass fibers to remove the sizing coating was suggested. Firstly, it was found that the method used to produce the samples has a great effect on the properties of the prepared composites. Using the PES in the solution form provides a good impregnation between the fibers and the polymer which improves the interaction between matrix and reinforcement resulting composite with better properties. The influence of the fiber to polymer ratio in addition to the effect of the thermal treatment of the fiber on the mechanical behavior of the composites was investigated. The results show that the flexural strength of the composites significantly increased when using the PES solution to impregnate the GF and preheated glass fabrics as reinforcements. The composites reinforced with as-received glass fabrics showed insufficient interfacial interaction between fibers and polymer, which resulted in lower mechanical properties compared with preheated composites. In the as-received composites 60/60 fiber to polymer ratio showed good results. The results reveal that the mechanical properties increased with increasing the fiber to polymer ratio, and the 70/30 samples were the best composition in case of using the preheated GF. Improvement was noticed in the mechanical properties of the PES-based composites due to the heat treatment of the glass fibers before using, which led to remove most of the sizing coating of fabrics (according to the FTIR of the fibers after preheating) and enhance the adhesion between the fabrics and polymer. According to SEM images, a good interface occurred in the case of using the preheated glass fabrics compared with those in as-received composites.

Author Contributions: Conceptualization, D.C.; investigation, G.S., and V.T. (Valerii Torokhov); writing—original draft preparation, G.S.; writing—review and editing, V.T. (Victor Tcherdyntsev); supervision, D.C.; project administration, V.T. (Victor Tcherdyntsev).

Funding: The reported study was funded by Russian Science Foundation grant No. 18-19-00744.

Acknowledgments: G.S. gratefully acknowledges the financial support of the Ministry of Science and Higher Education of the Russian Federation in the framework of Increase Competitiveness Program of MISiS (support project for young research engineers).

Conflicts of Interest: The authors declare no conflict of interest.

References

1. Gupta, M.K.; Srivastava, R.K. Mechanical properties of hybrid fibers-reinforced polymer composite: A review. *Polym. Plast. Technol. Eng.* **2016**, *55*, 626–642. [CrossRef]
2. Kumar, V.V.; Balaganesan, G.; Lee, J.K.Y.; Neisiany, R.E.; Surendran, S.; Ramakrishna, S. A review of recent advances in nanoengineered polymer composites. *Polymers* **2019**, *11*, 644. [CrossRef] [PubMed]
3. Dikshit, V.; Bhudolia, S.K.; Joshi, S.C. Multiscale polymer composites: A review of the interlaminar fracture toughness improvement. *Fibers* **2017**, *5*, 38. [CrossRef]
4. Munirathnamma, L.M.; Ravikumar, H.B. Microstructural characterization of short glass fibre reinforced polyethersulfone composites: A positron lifetime study. *J. Appl. Polym. Sci.* **2016**, *133*, 43647. [CrossRef]
5. Sen Du, S.; Li, F.; Xiao, H.M.; Li, Y.Q.; Hu, N.; Fu, S.Y. Tensile and flexural properties of graphene oxide coated-short glass fiber reinforced polyethersulfone composites. *Compos. Part B Eng.* **2016**, *99*, 407–415.
6. Caixeta, R.W.; Guiraldo, R.D.; Berger, S.B.; Kaneshima, E.N.; Faria Júnior, E.M.; Drumond, A.C.; Gonini Júnior, A.; Lopes, M.B. Influence of glass-fiber reinforcement on the flexural strength of different resin composites. *Appl. Adhes. Sci.* **2015**, *3*, 24. [CrossRef]
7. Jing, M.; Che, J.; Xu, S.; Liu, Z.; Fu, Q. The effect of surface modification of glass fiber on the performance of poly (lactic acid) composites: Graphene oxide vs. silane coupling agents. *Appl. Surf. Sci.* **2018**, *435*, 1046–1056. [CrossRef]
8. Thomason, J.L. The influence of fibre length, diameter and concentration on the modulus of glass fibre reinforced polyamide 6, 6. *Compos. Part A Appl. Sci. Manuf.* **2008**, *39*, 1732–1738. [CrossRef]
9. Sarasini, F.; Tirillò, J.; Sergi, C.; Seghini, M.C.; Cozzarini, L.; Graupner, N. Effect of basalt fibre hybridisation and sizing removal on mechanical and thermal properties of hemp fibre reinforced HDPE composites. *Compos. Struct.* **2018**, *188*, 394–406. [CrossRef]
10. Nassir, N.A.; Birch, R.S.; Cantwell, W.J.; Wang, Q.Y.; Liu, L.Q.; Guan, Z.W. The perforation resistance of glass fibre reinforced PEKK composites. *Polym. Test.* **2018**, *72*, 423–431. [CrossRef]
11. Yang, L.; Thomason, J.L. Interface strength in glass fibre-polypropylene measured using the fibre pull-out and microbond methods. *Compos. Part A Appl. Sci. Manuf.* **2010**, *41*, 1077–1083. [CrossRef]
12. Cech, V.; Prikryl, R.; Balkova, R.; Grycova, A.; Vanek, J. Plasma surface treatment and modification of glass fibers. *Compos. Part A Appl. Sci. Manuf.* **2002**, *33*, 1367–1372. [CrossRef]
13. Zhang, S.; Caprani, C.; Heidarpour, A. Influence of fibre orientation on pultruded GFRP material properties. *Compos. Struct.* **2018**, *204*, 368–377. [CrossRef]
14. Larson, B.K.; Drzal, L.T. Glass fibre sizing/matrix interphase formation in liquid composite moulding: Effects on fibre/matrix adhesion and mechanical properties. *Composites* **1994**, *25*, 711–721. [CrossRef]
15. Zheng, L.; Jian, X. Enhanced continuous glass fibre-reinforced poly (phthalazinone ether sulfone ketone) composites by blending polyetherimide and polyethersulfone. *Polym. Polym. Compos.* **2011**, *19*, 445–450. [CrossRef]

16. Boufaida, Z.; Farge, L.; André, S.; Meshaka, Y. Influence of the fiber/matrix strength on the mechanical properties of a glass fiber/thermoplastic-matrix plain weave fabric composite. *Compos. Part A Appl. Sci. Manuf.* **2015**, *75*, 28–38. [CrossRef]
17. Feih, S.; Wei, J.; Kingshott, P.; Sørensen, B.F. The influence of fibre sizing on the strength and fracture toughness of glass fibre composites. *Compos. Part A Appl. Sci. Manuf.* **2005**, *36*, 245–255. [CrossRef]
18. Zhong, Y.; Cheng, M.; Zhang, X.; Hu, H.; Cao, D.; Li, S. Hygrothermal durability of glass and carbon fiber reinforced composites–A comparative study. *Compos. Struct.* **2018**, *211*, 134–143. [CrossRef]
19. Jafari, A.; Ashrafi, H.; Bazli, M.; Ozbakkaloglu, T. Effect of thermal cycles on mechanical response of pultruded glass fiber reinforced polymer profiles of different geometries. *Compos. Struct.* **2019**, *223*, 110959. [CrossRef]
20. Lee, N.; Jang, J. The effect of fibre content on the mechanical properties of glass fibre mat/polypropylene composites. *Compos. Part A Appl. Sci. Manuf.* **1999**, *30*, 815–822. [CrossRef]
21. Bajerova, M.; Krejcova, K.; Rabiskova, M.; Gajdziok, J.; Masteikova, R. Oxycellulose: Significant characteristics in relation to its pharmaceutical and medical applications. *Adv. Polym. Technol.* **2009**, *28*, 199–208.
22. Thomason, J.L.; Vlug, M.A.; Schipper, G.; Krikor, H.G.L.T. Influence of fibre length and concentration on the properties of glass fibre-reinforced polypropylene: Part 3. Strength and strain at failure. *Compos. Part A Appl. Sci. Manuf.* **1996**, *27*, 1075–1084. [CrossRef]
23. Thomason, J.L.; Schoolenberg, G.E. An investigation of glass fibre/polypropylene interface strength and its effect on composite properties. *Composites* **1994**, *25*, 197–203. [CrossRef]
24. Nygård, P.; Gustafson, C.G. Interface and impregnation relevant tests for continuous glass fibre-polypropylene composites. *Compos. Part A Appl. Sci. Manuf.* **2003**, *34*, 995–1006. [CrossRef]
25. Wongpajan, R.; Mathurosemontri, S.; Takematsu, R.; Xu, H.Y.; Uawongsuwan, P.; Thumsorn, S.; Hamada, H. Interfacial Shear Strength of Glass Fiber Reinforced Polymer Composites by the Modified Rule of Mixture and Kelly-Tyson Model. *Energy Procedia* **2016**, *89*, 328–334. [CrossRef]
26. Harinath, P.N.V.; Radadia, N.; Bramhe, S.; Surendran, D.; Sabane, V.; Singh, A.; Kim, T.N.; Sharbidre, R.; Cheol, B.J.; Peters, L.; et al. Analysis of microstructural, thermal and mechanical properties of unidirectional glass fiber fabrics exhibiting sizing migration. *Compos. Part B Eng.* **2019**, *164*, 570–575. [CrossRef]
27. Yang, L.; Thomason, J.L.; Zhu, W. The influence of thermo-oxidative degradation on the measured interface strength of glass fibre-polypropylene. *Compos. Part A Appl. Sci. Manuf.* **2011**, *42*, 1293–1300. [CrossRef]
28. Palmese, G.R.; Andersen, O.A.; Karbhari, V.M. Effect of glass fiber sizing on the cure kinetics of vinyl-ester resins. *Compos. Part A Appl. Sci. Manuf.* **1999**, *30*, 11–18. [CrossRef]
29. Petersen, H.; Kusano, Y.; Brøndsted, P.; Almdal, K. Preliminary characterization of glass fiber sizing. *Proc. 34th Risø Int. Symp. Mater. Sci.* **2013**, *34*, 333–340.
30. Laura, D.M.; Keskkula, H.; Barlow, J.W.; Paul, D.R. Effect of glass fiber surface chemistry on the mechanical properties of glass fiber reinforced, rubber-toughened nylon 6. *Polymer* **2002**, *43*, 4673–4687. [CrossRef]
31. Thomason, J.L.; Nagel, U.; Yang, L.; Bryce, D. A study of the thermal degradation of glass fibre sizings at composite processing temperatures. *Compos. Part A Appl. Sci. Manuf.* **2019**, *121*, 56–63. [CrossRef]
32. Thomason, J.L.; Dwight, D.W. Use of XPS for characterization of glass fibre coatings. *Compos. Part A Appl. Sci. Manuf.* **1999**, *30*, 1401–1413. [CrossRef]
33. Qin, Y.J.; Xu, Y.H.; Zhang, L.Y.; Zheng, G.Q.; Yan, X.R.; Dai, K.; Liu, C.T.; Shen, C.Y.; Guo, Z.H. Interfacial interaction enhancement by shear-induced β-cylindrite in isotactic polypropylene/glass fiber composites. *Polymer* **2016**, *100*, 111–118. [CrossRef]
34. Dey, M.; Deitzel, J.M.; Gillespie, J.W.; Schweiger, S. Influence of sizing formulations on glass/epoxy interphase properties. *Compos. Part A Appl. Sci. Manuf.* **2014**, *63*, 59–67. [CrossRef]
35. Li, F.; Liu, Y.; Qu, C.B.; Xiao, H.M.; Hua, Y.; Sui, G.X.; Fu, S.Y. Enhanced mechanical properties of short carbon fiber reinforced polyethersulfone composites by graphene oxide coating. *Polymer* **2015**, *59*, 155–165. [CrossRef]
36. Kim, N.; Kim, D.Y.; Kim, Y.J.; Jeong, K.U. Enhanced thermomechanical properties of long and short glass fiber-reinforced polyamide 6,6/polypropylene mixtures by tuning the processing procedures. *J. Mater. Sci.* **2014**, *49*, 6333–6342. [CrossRef]
37. Sheng, Y.; Wen, B.; Li, X.; Hu, X. Effect of surface treatment on the mechanical properties of BF/PLA composite. *Fuhe Cailiao Xuebao/Acta Mater. Compos. Sin.* **2012**, *29*, 6.

38. Feih, S.; Boiocchi, E.; Mathys, G.; Mathys, Z.; Gibson, A.G.; Mouritz, A.P. Mechanical properties of thermally-treated and recycled glass fibres. *Compos. Part B Eng.* **2011**, *42*, 350–358. [CrossRef]
39. Manikandan, V.; Winowlin Jappes, J.T.; Suresh Kumar, S.M.; Amuthakkannan, P. Investigation of the effect of surface modifications on the mechanical properties of basalt fibre reinforced polymer composites. *Compos. Part B Eng.* **2012**, *43*, 812–818. [CrossRef]
40. Thomason, J.L.; Yang, L.; Meier, R. The properties of glass fibres after conditioning at composite recycling temperatures. *Compos. Part A Appl. Sci. Manuf.* **2014**, *61*, 201–208. [CrossRef]
41. Karger-Kocsis, J.; Mahmood, H.; Pegoretti, A. Recent advances in fiber/matrix interphase engineering for polymer composites. *Prog. Mater. Sci.* **2015**, *73*, 1–43. [CrossRef]
42. Chen, J.; Zhao, D.; Jin, X.; Wang, C.; Wang, D.; Ge, H. Modifying glass fibers with graphene oxide: Towards high-performance polymer composites. *Compos. Sci. Technol.* **2014**, *97*, 41–45. [CrossRef]
43. Cech, V.; Palesch, E.; Lukes, J. The glass fiber-polymer matrix interface/interphase characterized by nanoscale imaging techniques. *Compos. Sci. Technol.* **2013**, *83*, 22–26. [CrossRef]
44. Zhou, X.; Lin, Q.; Dai, G. Studies on mechanical properties of discontinuous glass fiber/continuous glass mat/polypropylene composite. *Polym. Polym. Compos.* **2002**, *10*, 299–306. [CrossRef]

polymers

MDPI

Article

Effect of Glass Fibers Thermal Treatment on the Mechanical and Thermal Behavior of Polysulfone Based Composites

Galal Sherif [1,2,*], Dilyus I. Chukov [1], Victor V. Tcherdyntsev [1], Valerii G. Torokhov [1] and Dmitry D. Zherebtsov [1]

1 Center of composite materials, National University of Science and Technology "MISIS", Leninskiy prosp. 4, Moscow 119049, Russia; dil_chukov@mail.ru (D.I.C.); vvch@misis.ru (V.V.T.); vgtorohov@gmail.com (V.G.T.); dmitry_zherebtsov@bk.ru (D.D.Z.)

2 Production and Design Dept., Faculty of Engineering, Minia University, Minia 61111, Egypt

* Correspondence: eng_galal_emad@mu.edu.eg; Tel.: +7-9267878736

Received: 10 March 2020; Accepted: 10 April 2020; Published: 13 April 2020

Abstract: The effect of thermal treatment of glass fibers (GF) on the mechanical and thermo-mechanical properties of polysulfone (PSU) based composites reinforced with GF was investigated. Flexural and shear tests were used to study the composites' mechanical properties. A dynamic mechanical analysis (DMA) and a heat deflection temperature (HDT) test were used to study the thermo-mechanical properties of composites. The chemical structure of the composites was studied using IR-spectroscopy, and scanning electron microscopy (SEM) was used to illustrate the microstructure of the fracture surface. Three fiber to polymer ratios of initial and preheated GF composites (50/50, 60/40, 70/30 (wt.%)) were studied. The results showed that the mechanical and thermo-mechanical properties improved with an increase in the fiber to polymer ratio. The interfacial adhesion in the preheated composites enhanced as a result of removing the sizing coating during the thermal treatment of GF, which improved the properties of the preheated composites compared with the composites reinforced with initial untreated fibers. The SEM images showed a good distribution of the polymer on the GF surface in the preheated GF composites.

Keywords: glass fibers; heat treatment; polysulfone; mechanical properties

1. Introduction

Polysulfone (PSU) is a high-performance amorphous thermoplastic with excellent mechanical properties, high service temperature due to its high glass transition temperature (T_g) 185 °C, flexibility, and excellent thermal stability. These superior properties make PSU the most appropriate choice for wide applications such as medicine, food, processing equipment, and relatively high-temperature components [1–4]. PSU is broadly used because of increasing demands for high-temperature polymers in many industries, such as automotive, aerospace, and microelectronics.

Many materials and types of fillers are used to reinforce polymer matrix composites [5–8]. Nowadays, most of the high-performance polymer-based composite materials are produced using fibrous fillers [4–9]. GF are one of these materials that offer high specific strength and stiffness, low cost, and suitable heat resistance [6,10,11]. The mechanical properties of composites mainly depend on reinforcing fiber/matrix properties, fibers' surface morphology, and the interfacial bonding between the fiber and the matrix [8,10]. It is also recognized that the bonding strength at the fiber–matrix interface has a significant effect on composite materials' mechanical properties. Therefore, so far, many efforts have been made to propose an appropriate engineered fiber/matrix interface to significantly increase the composite's strength, toughness, and environmental stability [11,12]. As it is known, the bonding strength mainly depends on physical absorption, chemical reaction, and bonding between the fiber surface layer and the matrix polymer. The bonding strength is strongly affected by a fiber surface modification such as surface treatment or chemical sizing [10].

Several studies dealing with thermoplastic composites reveal that temperature has a noticeable influence on mechanical properties [12–18]. It has been found that tensile strength and Young's modulus of thermoplastic composites decrease with increasing temperature and drop sharply close to the T_g [19,20]. On the other hand, above the T_g, the strain increases because of the intensive motion of polymers' molecular chains. This effect in unfilled polymers is greater than that of in reinforced polymers [21,22]. The effect of temperature and different environmental conditions on the thermoplastic reinforced with GF was studied, and it was observed that a bi-linear reduction in strength and stiffness occurred in the T_g range [23,24].

In addition to traditional methods of polymer composites mechanical properties investigations, such as tensile, compression, flexural tests, etc., thermal and dynamical tests methods are widely used nowadays. One of these methods is dynamic mechanical analysis (DMA), which is widely and successfully used to study the dynamic mechanical response of composites. The data used as a function of temperature, time, frequency, and stress can also be an indicator of the interface, morphology, and presence of an internal defect in the composite structure. It is an excellent technique to study the effect of temperature on the mechanical properties of composite materials. Since polymeric composites in many applications exposed to different types of dynamic stressing during service, studying the viscoelastic behavior of these materials have become critical [25–27]. The heat deflection temperature (HDT) test is another effective tool to evaluate the physical performance of a polymer under load and elevated temperature. The HDT data represent the maximum service temperature without a large deflection [28–31].

Recently [32], we investigated the effect of the formation route on the GF reinforced polyethersulfone based composites. It was observed that composite formation via compression molding of a polymer powder together with GF does not allow samples to be obtained with high mechanical properties, whereas the formation of composites by GF impregnation with polyethersulfone solution results in the formation of composites with high flexural strength. Therefore, in the present study, we applied the polymer solution route to obtain PSU based composites. The current study aims to illustrate the thermal treatment effect of the removal of GF sizing coating on the mechanical and thermal properties of PSU composites. According to our results, the mechanical and thermal properties increase with an addition of a preheated GF, which leads to an expansion of high-temperature applications of these composites. Additionally, the current study seeks to increase the knowledge base of thermoplastic composites especially in terms of understanding the effect of temperature on the performance of PSU/GF composites. Additionally, the comparison of the present study results with the data obtained in [32] allows the effect of the polymer nature on the interaction between the matrix and GF in the composites to be revealed.

2. Materials and Methods

2.1. Material and Sample Preparation

Woven glass fabrics (NPO "Stekloplastic", Moscow, Russia) (T-23/1 "260 ± 10 g/m^2) and PSU Ultrason S2010 (Basf, Ludwigshafen, Germany) powder were used as raw materials. A polysulfone solution was obtained by dissolving the PSU powder in N-methyl-2-pyrrolidone (Eastchem, Jiangsu, China). Bulk composite samples were formed in accordance with the method described in [32]. The solution was prepared in a 20/80 polymer to solvent weight ratio for 24 h using a magnetic stirrer. The samples were dried at a temperature of 150 °C for 5 h, and then they were compression molded at 340 °C and 10 MPa. Figure 1 shows a scheme of the preparation process of the PSU solution and composites. Three fiber to polymer weight ratios were prepared (50/50, 60/40, and 70/30 (wt.%)). There are many sizing compositions used in commercial GF, which can be completely wiped out by using a thermal treatment in a range of 200 to 400–500 °C [33–38]. The method of removing the sizing coating from the GF surface was elaborated previously [32]. Notably, the investigation was carried out using the same type of GF; it was shown that the optimal preheating conditions for the type of GF used is annealing in an air-atmosphere furnace at 350 °C for 1 h, according to Fourier-transform infrared (FTIR) spectra; preheated GF used in this study was prepared using the above-mentioned conditions. The composites were reinforced using initial and preheated GF.

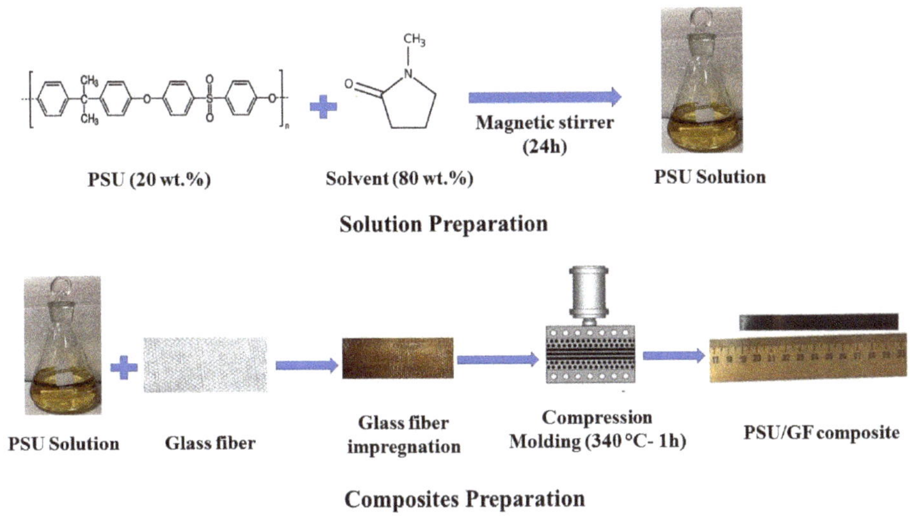

Figure 1. Scheme of the preparation of the polysulfone (PSU) solution and composites.

2.2. Characterization of the Samples' Structures

An FTIR spectrometer Nicolet 380 (Thermo Scientific, Waltham, MA, USA) (spectral range of 4000–450 cm^{-1}, resolution of 1 cm^{-1}) was used to study the chemical structures of the samples. The microstructure, fracture, interfacial bonding, and fiber pulling out were studied using a scanning electron microscope (VEGA 3 TESCAN) (TESCAN ORSAY HOLDING, a.s., Brno–Kohoutovice, Czech Republic) in backscattered electron image mode. Before the SEM examination, the samples were coated with a thin layer (10–15 nm) of carbon in a sputter coater.

2.3. Mechanical Tests

Flexural and shear properties were measured using a Zwick/Roell Z020 universal test machine (Zwick Roell Group, Ulm, Germany) provided with 1 and 20 kN sensors and a MultiXtens contact strain measurement system. Conforming with ISO 14125:1998 standards, the samples for the flexural tests were prepared in a dimension of 110 mm × 10 mm × 2 mm and 80 mm span. For shear tests (according to ASTM D 3846), 110 mm × 10 mm × 4 mm samples were used with a gauge length of 80 mm. According to this method, the shear strength was measured by applying a compressive load to a notched specimen of uniform width. The specimen was loaded edgewise in a supporting jig of the same description in ASTM D 695 for testing thin specimens. A failure of the specimen occurred in shear between two centrally located notches machined halfway through its thickness and spaced a fixed distance apart on opposing faces. The distance between the notches was 6.5–8 mm. The test speeds were 10 and 1.3 mm/min for the flexural and shear tests, respectively. At least five samples were examined at room temperature in each condition.

2.4. *Thermo-Mechanical Tests*

A DMA Q800 (TA Instruments, New Castle, DE, USA) dynamic mechanical analyzer was used to study the dynamic mechanical properties. The specimens sized 2 mm × 2 mm × 45 mm were used for the DMA tests. The measurements were realized using a double cantilever clamp at a frequency of 1 Hz and a deformation of 0.1%, in a temperature range from 30 to 220 °C; the heating rate was of 2 °C/min. The HDT tests were carried out using an Instron CEAST 6910 HDT/Vicat tester. The samples sized 80 mm × 10 mm × 4 mm were used in the HDT test at a load of 1.8 MPa and a span length of 64 mm (ISO 75). The deflection in the HDT test was set up to 1 mm as a maximum deflection. The DMA and HDT were performed for both PSU composites reinforced with initial and preheated GF. In each condition, three fiber to polymer weight ratios (50/50, 60/40, and 70/30 (wt.%)) were used.

3. Results and Discussion

3.1. *FTIR*

Figure 2 shows the FTIR spectra for the initial PSU and PSU reinforced with both initial and preheated GF 50/50 composites. For the initial PSU spectra, the C–H band for the aryl group was noticed in a range of 3000–3100 cm^{-1}. The peaks at 2800 and 3000 cm^{-1} related to symmetric and asymmetric bands of CH_3 and CH_2. The C–C in-ring bands were revealed by 1401, 1501, and 1586 cm^{-1} peaks. The stretching vibration of the asymmetric O=S=O band occurred at 1292 and 1325 cm^{-1}, while the peak at 1232 cm^{-1} referred to the stretching vibration of the C–O band. The stretching of the symmetric O=S=O bands appeared at 1140 and 1168 cm^{-1}, and the aryl group was indicated by 1019 cm^{-1} peak [1,23,39,40].

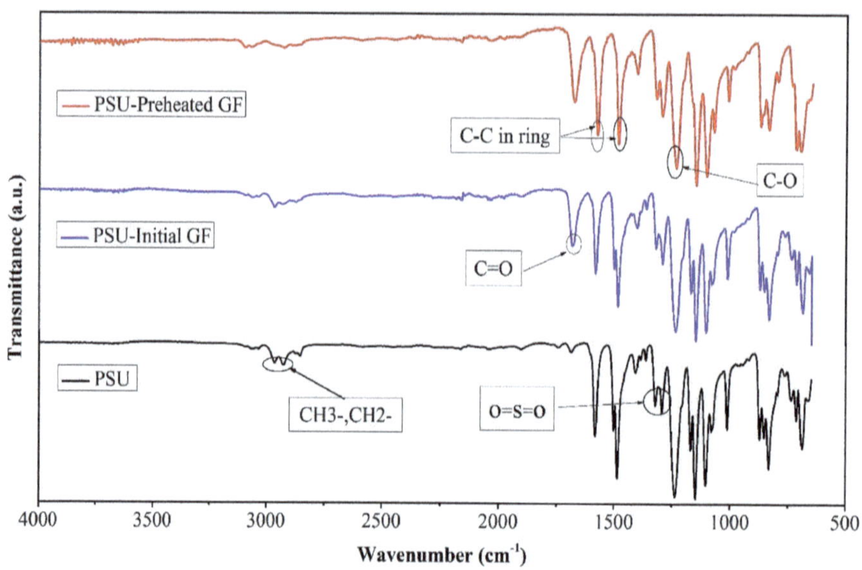

Figure 2. FTIR spectra for the initial PSU and 50/50 PSU composites reinforced with initial and preheated GF.

Few differences could be distinguished in the spectral attribution of PSU composites. In the spectra of the composites, the C=O band appeared clearly due to the presence of some residual solvent [32]. The amplitude of this peak was stronger in the spectra related to the PSU reinforced with preheated GF due to the oxidation during the preheating process. In the case of the composites containing preheated GF, the peaks between 2800 and 3000 cm^{-1} reduced because of the removal of GF sizing [32]. Based on the data observed from FTIR, it could be noted that the spectrum characteristic for the composite was very similar with that of PSU except some new peaks because of the presence of some residual solvent and the effect of the removal of GF coating.

3.2. Mechanical Tests

The flexural and shear tests were implemented to study the mechanical properties of the composites. The comparison of flexural strength and Young's modulus values for the initial GF reinforced composites are shown in Figure 3a. The curve showed a trend of increasing flexural strength and Young's modulus with increasing the GF ratio. The composites with a GF to PSU ratio of 70/30 recorded the maximum value of flexural strength (460 MPa) and Young's modulus (26 GPa) compared with the 340 MPa and 18 GPa for the 50/50 composites. The comparison of these data with those observed previously for polyethersulfone based composites [32] shows that in case of PSU matrix composites, no decrease in Young's modulus at the increase of the GF content from 60/40 to 70/30 was observed.

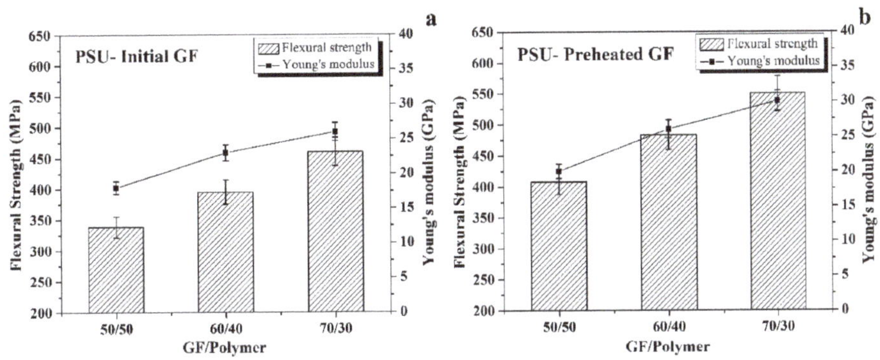

Figure 3. Flexural strength and Young's modulus for the initial GF (**a**,**b**) the preheated GF reinforced composites.

The sizing coating prevents good adhesion between the fiber and the polymer, which mainly affects the composite's strength. A thermal treatment was carried out to remove the sizing coating of the fiber to enhance the interface bonding between the polymer and the fibers [32,41]. Figure 3b illustrates the values of flexural strength and Young's modulus for the preheated GF reinforced composites. It can be noted that the mechanical properties increased with increasing the GF content; the flexural strength increased from 408 MPa for 50/50 composites and 483 MPa for the 60/40 composites to 550 MPa in the case of the 70/30 composites, whereas Young's modulus increased from 20 GPa for the 50/50 composites to 26 and 30 GPa for the 60/40 and 70/30 composites, respectively. A remarkable enhancement occurred in the preheated GF composite properties compared with the initial GF composites at the same ratio. It was considered that heating the GF removed the GF sizing, which contributed to the improvement of interface bonding between the fiber and the matrix.

Shear strength, which is affected mainly by the interface bonding, is illustrated in Figure 4. The effect of the removal of the GF sizing on the interface between GF and the matrix was clearly demonstrated by an increase of shear strength values of the preheated GF composites compared with those of the initial GF composites at the same ratios shown in the figure. Shear strength increased from 43 MPa for the initial GF to 45 MPa for the preheated GF 50/50 composites and from 45/46 MPa for the initial GF 60/40 and the 70/30 composites to 47/49.5 MPa for the preheated GF 60/40 and 70/30 composites, respectively. The comparison of these data with those observed previously for polyethersulfone based composites [32] shows that in case of the PSU matrix composites, shear strength tended to increase with an increase in the GF content from 60/40 to 70/30 both for the composites containing initial and preheated GF, whereas for polyethersulfone based composites such an increase in the GF content resulted in a decrease in shear strength even for the composites reinforced with the preheated GF. It is additional evidence of the important role of the chemical nature of the matrix polymer on the interaction between the matrix and the reinforcers.

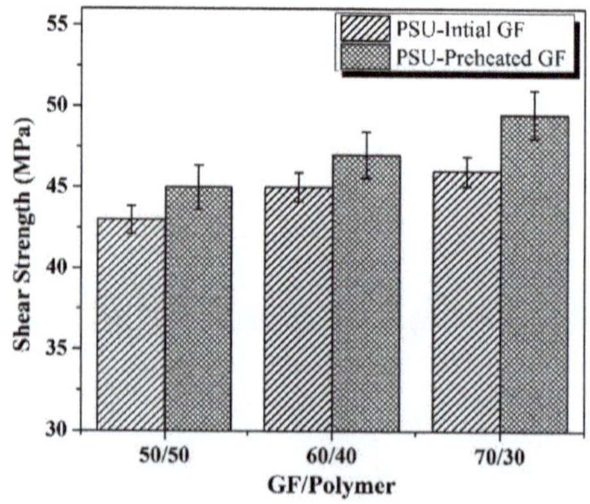

Figure 4. Shear strength for the initial GF and the preheated GF reinforced composites.

3.3. Thermo-Mechanical Tests

Figure 5a shows the temperature dependences of the storage modulus for the initial GF reinforced composites. The values of the storage modulus remained at a plateau in a temperature range below the T_g, while it started to fall around the T_g, which is the region of the transformation from glassy to rubbery state. It can be noted that the storage modulus increased with an increase in the GF content as a result of an increase in the stiffness and the thermo-resistance of the composites with increases in the GF ratio. The results recorded that the storage modulus of 22 GPa was found for the 70/30 composites, while the storage modulus values for the 50/50 and 60/40 composites were 15.5 and 20.5 GPa, respectively. It is considered that the thermo-mechanical characteristic of the composite was improved with increasing the GF/PSU ratio due to the enhancement of thermal stability of the composite as a result of an increase of the composite's stiffness and the interfacial interaction, which increased the thermodynamic compatibility between GF and the polymer [21,25].

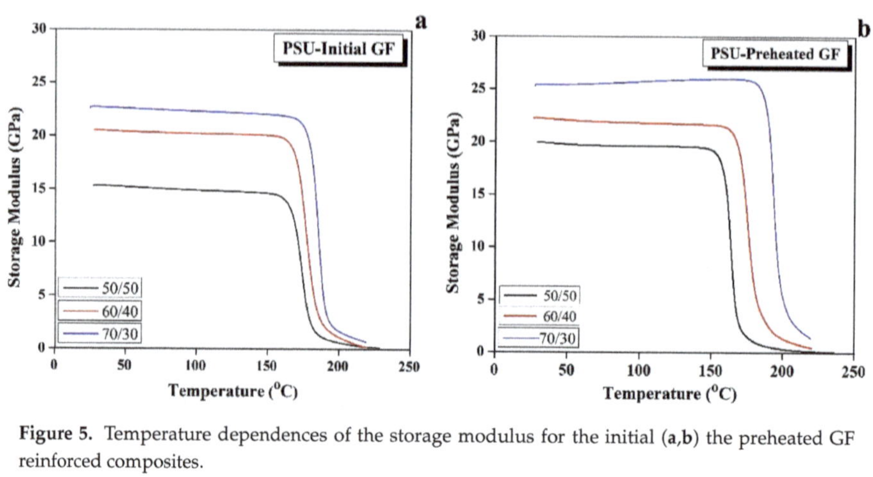

Figure 5. Temperature dependences of the storage modulus for the initial (**a**,**b**) the preheated GF reinforced composites.

The effect of using the preheated GF on the storage modulus of the composites is investigated in Figure 5b. The data showed a noticeable enhancement of the storage modulus values compared with the initial GF composites at the same ratio. This increase was attributable to the improvement of the stiffness of the preheated GF composites due to an increase of the interface between the fiber and the polymer after removing the GF coating, which affected directly on the storage modulus and the ability of the material to store energy. The storage moduli of the 50/50, 60/40, and 70/30 preheated GF composites were 20, 22, and 26 GPa, respectively.

Another way to evaluate thermal stability of the composites' mechanical properties is tangent delta (tan δ) measurement. Tan δ refers to the ratio between loss and storage modulus, and the peak on the Tan δ curve refers to the T_g, which differentiates between the glassy and rubbery region of the thermo-mechanical behavior of the composite. Tan δ of a different initial GF to polymer ratio is shown in Figure 6a. It can be seen that the tan δ maximum decreased with an increase in the GF to PSU ratio, i.e., it decreased from 0.75 for 50/50 composites to 0.65 and 0.53 for 60/40 and 70/30 composites, respectively. On the other hand, the T_g increased from 163 °C for the 50/50 composites to 180 and 192 °C for the 60/40 and 70/30 composites, respectively, due to an increase of thermal stability of the composites. The reduction behavior of tan δ was due to a decrease of the molecular chain's mobility as a result of increasing the fiber/polymer interface bonding [2]. With using the preheated GF, the composites became stiffer so that the values of tan δ decreased in the preheated composites, as shown in Figure 6b, compared with the same ratio of the initial GF reinforced composites. The improvement achieved from using the preheated GF raised thermal stability, which in turn enhanced the T_g of the composites. The results showed an increase from 170 °C for the 50/50 composites to 187 and 198 °C for the 60/40 and 70/30 composites, respectively.

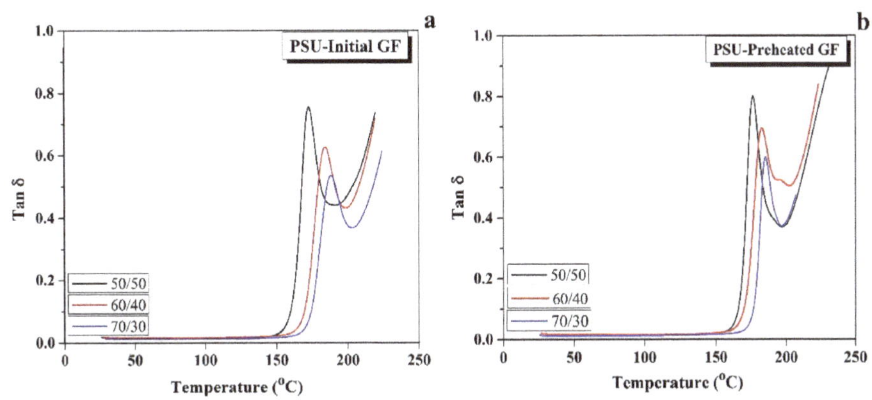

Figure 6. Temperature dependences of Tan δ for the initial (**a**,**b**) the preheated GF reinforced composites.

The HDT tests for the initial and preheated GF reinforced composites were carried out to study the deformation behavior of the composites at the evaluated temperature. The maximum deflection was set to be 1 mm. The HDT results of the initial GF reinforced composites, shown in Figure 7a, indicated that the deflection remained approximately zero up to a temperature near the T_g of the polymer matrix and drastically increased above the T_g. The HDT for the initial GF reinforced composites enhanced from 168 °C for the 50/50 composites to 197 and 209 °C for the 60/40 and 70/30 composites, respectively. It can be proposed that the HDT increases as a result of stiffness and thermal stability enhancement with increasing the GF/PSU ratio [30,42]. The value of deflection works as an indicator of thermal stability. As it is shown in Figure 7, the deflection of the composites was near to zero upon the T_g, whereas above the T_g the deflection increased rapidly. The same behavior was observed in the case of the preheated GF composites, as seen in Figure 7b. The HDT for the composites containing the preheated GF were 181, 202, and 214 °C for the 50/50, 60/40, and 70/30 composites, respectively. The HDT values were found to be higher in the preheated GF composites than those in the initial GF composites. This can be explained by an increase in thermal stability of the composite along the improvement in the interface bonding between the fiber and the matrix.

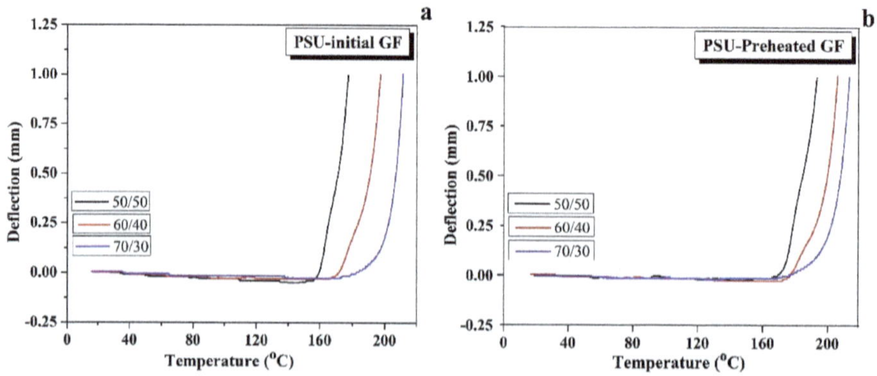

Figure 7. Heat deflection curves for (**a**) the initial and (**b**) the preheated GF reinforced composites.

The values of the T_g, tan δ, and HDT for the initial and preheated GF composites are given in Table 1. It can be noticed that the thermal properties improved with use of the preheated GF instead of the initial GF in polysulfone composites.

Table 1. The T_g, tan δ, and HDT values for the composites reinforced with the initial and preheated GF.

Fiber/Polymer	50/50			60/40			70/30		
Property	T_g (°C)	Tan δ	HDT (°C)	T_g (°C)	Tan δ	HDT (°C)	T_g (°C)	Tan δ	HDT (°C)
Initial GF composites	163	0.75	168	180	0.65	197	192	0.53	209
Preheated GF composites	170	0.81	181	187	0.7	202	198	0.56	214

Figure 8 shows the microstructure of the flexural fracture surface of the PSU based composites reinforced with the initial and preheated GF. Pull-out phenomena appeared in the case of the 50/50 initial fiber-reinforced composites (Figure 8a), which means that the adhesion on the PSU/GF interface, in this case, was not sufficient. Preheating resulted in the improvement in the boundary adhesion (Figure 8b), which was confirmed by the formation of a large amount of PSU particles adherent to the fiber surface. Some pores appeared in the 50/50 preheated fiber-reinforced composites (Figure 8b) because of the presence of some solvent that was not removed during the drying process before compression molding. The evaporation of the solvent resulted in the formation of pores during compression molding. The preheated fiber-reinforced composites of 60/40, as seen in Figure 8d, also showed better interface bonding between the fiber and the polymer than those in the 60/40 initial GF composite shown in Figure 8c. Good interface bonding between the fiber and the matrix occurred in the 70/30 initial GF reinforced composites, as shown in Figure 8e. As seen in Figure 8f, the 70/30 preheated GF reinforced composites, the fracture of fiber was in a brittle form, which was due to sufficient interface bonding between the fiber and the polymer. Moreover, the distribution of the polymer was improved. Thus, an increase in the interfacial interaction due to GF preheating resulted in higher mechanical properties of the composites.

Figure 8. SEM images of the flexural fracture surfaces of the 50/50 (**a**,**b**), 60/40 (**c**,**d**), and 70/30 (**e**,**f**) PSU based composites reinforced with the initial (**a**,**c**,**e**) and the preheated (**b**,**d**,**f**) GF.

4. Conclusions

Mechanical and thermo-mechanical properties of the PSU composites reinforced with initial and preheated GF for a different fiber to polymer weight ratio were studied. The flexural test showed that the composite stiffness and Young's modulus enhanced with increasing the fiber ratio in the initial GF reinforced composites. A remarkable improvement was achieved by using a preheated GF to reinforce PSU. Additionally, shear strength increased in the cases of using a preheated GF. The storage modulus, tangent delta, and T_g values obtained from the DMA test and HDT obtained from the HDT test were increased in the preheated reinforced GF composites compared with those in the initial GF reinforced composites. The fiber to polymer ratio of 70/30 recorded the best properties for the initial and preheated GF reinforced composites. The 70/30 initial GF composites recorded 460 MPa, 26 GPa, and 22 GPa for flexural strength, Young's modulus, and storage modulus, respectively. Due to the improvement of the interfacial adhesion, these magnitudes were increased in the case of the 70/30 preheated GF composites to record 550 MPa, 30 GPa, and 26 GPa for flexural strength, Young's modulus, and storage modulus, respectively. FTIR of the PSU composites showed the main peaks of PSU and GF for the initial and preheated composites. Additionally, the FTIR spectra showed that the sizing coating was removed by heating the GF. It revealed that some of the solvent was not disposed of during the drying process.

The SEM images showed a good distribution of the polymer on the GF surface, which improved with using the preheated GF that led to an increase in the interface bonding between the polymer and GF.

Author Contributions: Conceptualization, D.I.C.; investigation, G.S., V.G.T., and D.D.Z.; writing—original draft preparation, G.S.; writing—review and editing, V.V.T.; supervision, D.I.C.; project administration, V.V.T. All authors have read and agreed to the published version of the manuscript.

Funding: The reported study was funded by Russian Science Foundation grant No. 18-19-00744.

Acknowledgments: G.S. gratefully acknowledges the financial support of the Ministry of Science and Higher Education of the Russian Federation in the framework of Increase Competitiveness Program of MISiS (Support project for young research engineers).

Conflicts of Interest: The authors declare no conflict of interest.

References

1. Jose, A.J.; Wilson, R.; Jacob, G.; Alagar, M. Studies on thermo mechanical and surface properties of polysulfone/poly(ether imide ester) blends. *Mater. Today Proc.* **2019**, *9*, 279–294. [CrossRef]
2. Zhao, J.; He, G.; Liu, G.; Pan, F.; Wu, H.; Jin, W.; Jiang, Z. Manipulation of interactions at membrane interfaces for energy and environmental applications. *Prog. Polym. Sci.* **2018**, *80*, 125–152. [CrossRef]
3. Ou, Y.; Zhu, D.; Zhang, H.; Huang, L.; Yao, Y.; Li, G.; Mobasher, B. Mechanical Characterization of the Tensile Properties of Glass Fiber and Its Reinforced Polymer (GFRP) Composite under Varying Strain Rates and Temperatures. *Polymers* **2016**, *8*, 196. [CrossRef] [PubMed]
4. Chukov, D.; Nematulloev, S.; Zadorozhnyy, M.; Tcherdyntsev, V.; Stepashkin, A.; Zherebtsov, D. Structure, Mechanical and Thermal Properties of Polyphenylene Sulfide and Polysulfone Impregnated Carbon Fiber Composites. *Polymers* **2019**, *11*, 684. [CrossRef] [PubMed]
5. Moriana, R.; Vilaplana, F.; Karlsson, S.; Ribes-Greusa, A. Improved thermo-mechanical properties by the addition of natural fibres in starch-based sustainable biocomposites. *Compos. Part A Appl. Sci. Manuf.* **2011**, *42*, 30–40. [CrossRef]
6. Di Landro, R.; Lorenzi, W. Mechanical Properties and Dynamic Mechanical Analysis of Thermoplastic-Natural Fiber/Glass Reinforced Composites. *Macromol. Symp.* **2009**, *286*, 145–155. [CrossRef]
7. Kim, N.; Kim, D.-Y.; Kim, Y.-J.; Jeong, K.-U. Enhanced thermomechanical properties of long and short glass fiber-reinforced polyamide 6,6/polypropylene mixtures by tuning the processing procedures. *J. Mater. Sci.* **2014**, *49*, 6333–6342. [CrossRef]
8. Chukov, D.; Nematulloev, S.; Torokhov, V.; Stepashkin, A.; Sherif, G.; Tcherdyntsev, V.V. Effect of carbon fiber surface modification on their interfacial interaction with polysulfone. *Results Phys.* **2019**, *15*, 102634. [CrossRef]
9. Stepashkin, A.A.; Chukov, D.I.; Gorshenkov, M.V.; Tcherdyntsev, V.V.; Kaloshkin, S.D. Electron microscopy investigation of interface between carbon fiber and ultra high molecular weight polyethylene. *J. Alloys Compd.* **2014**, *586*, S168–S172. [CrossRef]
10. Obande, W.; Mamalis, D.; Ray, D.; Yang, L.; Brádaigh, C.M.O. Mechanical and thermomechanical characterisation of vacuum-infused thermoplastic- and thermoset-based composites. *Mater. Des.* **2019**, *175*, 107828. [CrossRef]
11. Zhao, Z.K.; Du, S.S.; Li, F.; Xiao, H.M.; Li, Y.Q.; Zhang, W.G.; Hu, N.; Fu, S.Y. Mechanical and tribological properties of short glass fiber and short carbon fiber reinforced polyethersulfone composites: A comparative study. *Compos. Commun.* **2018**, *8*, 1–6. [CrossRef]
12. Thomason, J.L.; Vlug, M.A.; Schipper, G.; Krikor, H.G.L.T. Influence of fibre length and concentration on the properties of glass fibre-reinforced polypropylene: Part 3. Strength and strain at failure. *Compos. Part A Appl. Sci. Manuf.* **1996**, *27*, 1075–1084. [CrossRef]
13. Laura, D.M.; Keskkula, H.; Barlow, J.W.; Paul, D.R. Effect of glass fiber surface chemistry on the mechanical properties of glass fiber reinforced, rubber-toughened nylon 6. *Polymer* **2002**, *43*, 4673–4687. [CrossRef]
14. Chen, J.; Xu, H.; Liu, C.; Mi, L.; Shen, C. The effect of double grafted interface layer on the properties of carbon fiber reinforced polyamide 66 composites. *Compos. Sci. Technol.* **2018**, *168*, 20–27. [CrossRef]

15. Qu, M.; Nilsson, F.; Qin, Y.; Yang, G.; Pan, Y.; Liu, X.; Rodriguez, G.H.; Chen, J.; Zhang, C.; Schubert, D.H. Electrical conductivity and mechanical properties of melt-spun ternary composites comprising PMMA, carbon fibers and carbon black. *Compos. Sci. Technol.* **2017**, *150*, 24–31. [CrossRef]
16. Lv, G.; Zhang, N.; Huang, M.; Shen, C.; Castro, J.; Tan, K.; Liu, X.; Liu, C. The remarkably enhanced particle erosion resistance and toughness properties of glass fiber/epoxy composites via thermoplastic polyurethane nonwoven fabric. *Polym. Test.* **2018**, *69*, 470–477. [CrossRef]
17. Liao, M.; Yang, Y.; Hamada, H. Mechanical performance of glass woven fabric composite: Effect of different surface treatment agents. *Compos. Part B Eng.* **2016**, *86*, 17–26. [CrossRef]
18. Hwang, D.; Cho, D. Fiber aspect ratio effect on mechanical and thermal properties of carbon fiber/ABS composites via extrusion and long fiber thermoplastic processes. *J. Ind. Eng. Chem.* **2019**, *80*, 335–344. [CrossRef]
19. Nikforooz, M.; Montesano, J.; Golzar, M.; Shokrieh, M.M. Assessment of the thermomechanical performance of continuous glass fi ber-reinforced thermoplastic laminates. *Polym. Test.* **2018**, *67*, 457–467. [CrossRef]
20. Cen-Puc, M.; Pool, G.; Oliva-Avilés, A.I.; May-Pat, A.; Avilés, F. Experimental investigation of the thermoresistive response of multiwall carbon nanotube/polysulfone composites under heating-cooling cycles. *Compos. Sci. Technol.* **2017**, *151*, 34–43. [CrossRef]
21. Saba, N.; Jawaid, M. A review on thermomechanical properties of polymers and fibers reinforced polymer composites. *J. Ind. Eng. Chem.* **2018**, *67*, 1–11. [CrossRef]
22. Karsli, N.G.; Demirkol, S.; Yilmaz, T. Thermal aging and reinforcement type effects on the tribological, thermal, thermomechanical, physical and morphological properties of poly(ether ether ketone) composites. *Compos. Part B Eng.* **2016**, *88*, 253–263. [CrossRef]
23. Ficai, D.; Ficai, A.; Trusca, R.; Vasile, B.S.; Voicu, G.; Guran, C.; Andronescu, E. Synthesis and Characterization of Magnetite-Polysulfone Micro- and Nanobeads with Improved Chemical Stability in Acidic Media. *Curr. Nanosci.* **2013**, *9*, 271–277. [CrossRef]
24. Mhatre, A.M.; Chappa, S.; Ojha, S.; Pandey, A.K. Functionalized glass fiber membrane for extraction of iodine species. *Sep. Sci. Technol.* **2019**, *54*, 1469–1477. [CrossRef]
25. Nair, K.C.M.; Thomas, S.; Groeninckx, G. Thermal and dynamic mechanical analysis of polystyrene composites reinforced with short sisal fibres. *Compos. Sci. Technol.* **2001**, *61*, 2519–2529. [CrossRef]
26. Mouhmid, B.; Imad, A.; Benseddiq, N.; Benmedakhene, S.; Maazouz, A. A study of the mechanical behaviour of a glass fibre reinforced polyamide 6,6: Experimental investigation. *Polym. Test.* **2006**, *25*, 544–552. [CrossRef]
27. Eftekhari, M.; Fatemi, A. Tensile behavior of thermoplastic composites including temperature, moisture, and hygrothermal effects. *Polym. Test.* **2016**, *51*, 151–164. [CrossRef]
28. Senturk, O.; Senturk, A.E.; Palabiyik, M. Evaluation of hybrid effect on the thermomechanical and mechanical properties of calcite/SGF/PP hybrid composites. *Compos. Part B Eng.* **2018**, *140*, 68–77. [CrossRef]
29. Tábi, T.; Hajba, S.; Kovács, J.G. Effect of crystalline forms (α' and α) of poly(lactic acid) on its mechanical, thermo-mechanical, heat deflection temperature and creep properties. *Eur. Polym. J.* **2016**, *82*, 232–243. [CrossRef]
30. Daghigh, V.; Lacy, T.E., Jr.; Daghigh, H.; Gu, G.; Baghaei, K.T.; Horstemeyer, M.F.; Pittman, C.U., Jr. Heat deflection temperatures of bio-nano-composites using experiments and machine learning predictions. *Mater. Today Commun.* **2020**, *22*, 100789. [CrossRef]
31. Bledzki, A.K.; Mamun, A.A.; Feldmann, M. Polyoxymethylene composites with natural and cellulose fibres: Toughness and heat deflection temperature. *Compos. Sci. Technol.* **2012**, *72*, 1870–1874. [CrossRef]
32. Sherif, G.; Chukov, D.; Tcherdyntsev, V.; Torokhov, V. Effect of Formation Route on the Mechanical Properties of the Polyethersulfone Composites Reinforced with Glass Fibers. *Polymers* **2019**, *11*, 1364. [CrossRef] [PubMed]
33. Thomason, J. A review of the analysis and characterisation of polymeric glass fibre sizings. *Polym. Test.* **2020**, *85*, 106421. [CrossRef]
34. Jenkins, P.G. Investigation of the Strength Loss of Heat Treated Glass Fibre. Ph.D. Thesis, University of Strathclyde, Glasgow, Scotland, 2016.
35. Mäder, E. Study of fibre surface treatments for control of interphase properties incomposites. *Compos. Sci. Technol.* **1997**, *57*, 1077–1088.

36. Romanenko, K.V.; Lapina, O.B.; Simonova, L.G.; Fraissard, J. 1H and 29Si-MAS NMR characterization of silicate fiberglass supports. *Phys. Chem. Chem. Phys.* **2003**, *5*, 2686–2691. [CrossRef]
37. Thomason, J.L.; Nagel, U.; Yang, L.; Bryce, D. A study of the thermal degradation of glass fibre sizings at composite processing temperatures. *Compos. Part A Appl. Sci. Manuf.* **2019**, *121*, 56–63. [CrossRef]
38. Zhuravlev, L.T. Concentration of hydroxyl groups on the surface of amorphous silicas. *Langmuir* **1987**, *3*, 316–318. [CrossRef]
39. Dizman, C.; Ates, S.; Torun, L.; Yagci, Y. Synthesis, characterization and photoinduced curing of polysulfones with (meth)acrylate functionalities. *Beilstein J. Org. Chem.* **2010**, *6*, 56. [CrossRef]
40. Wei, X.; Wang, Z.; Wang, J.; Wang, S. A novel method of surface modification to polysulfone ultrafiltration membrane by preadsorption of citric acid or sodium bisulfite. *Membr. Water Treat.* **2012**, *3*, 35–49. [CrossRef]
41. Thomason, J.L. Glass fibre sizing: A review. *Compos. Part A Appl. Sci. Manuf.* **2019**, *127*, 105619. [CrossRef]
42. Saha, N.; Basu, D.; Banerjee, A.N. Heat-distortion temperature of unidirectional polyethylene-glass fiber-PMMA hybrid composite laminates. *J. Appl. Polym. Sci.* **1999**, *71*, 541–545. [CrossRef]

polymers

MDPI

Article

Bi-Functional Composting the Sulfonic Acid Based Proton Exchange Membrane for High Temperature Fuel Cell Application

Guoxiao Xu [1], Juan Zou [1], Zhu Guo [1], Jing Li [1,*], Liying Ma [2,*], Ying Li [3] and Weiwei Cai [1,4]

1 Sustainable Energy Laboratory, Faculty of Materials Science and Chemistry, China University of Geosciences, Wuhan 430074, China; guoxiao_xu123@163.com (G.X.); Ascin123@163.com (J.Z.); zj15090728708@163.com (Z.G.); willcai1985@gmail.com (W.C.)

2 School of Chemistry and Materials Science, Guizhou Normal University, 116 Baoshan North Road, Guiyang 550001, China

3 Research Institute for New Materials Technology, Chongqing University of Arts and Sciences, Chongqing 402160, China; leoyingchem@163.com

4 Zhejiang Institute, China University of Geosciences, Hangzhou 311305, China

* Correspondence: kinijing1984@gmail.com (J.L.); maliying2018@163.com (L.M.)

Received: 5 December 2019; Accepted: 11 January 2020; Published: 26 April 2020

Abstract: Although sulfonic acid (SA)-based proton-exchange membranes (PEMs) dominate fuel cell applications at low temperature, while sulfonation on polymers would strongly decay the mechanical stability limit the applicable at elevated temperatures due to the strong dependence of proton conduction of SA on water. For the purpose of bifunctionally improving mechanical property and high-temperature performance, Nafion membrane, which is a commercial SA-based PEM, is composited with fabricated silica nanofibers with a three-dimensional network structure via electrospinning by considering the excellent water retention capacity of silica. The proton conductivity of the silica nanofiber–Nafion composite membrane at 110 °C is therefore almost doubled compared with that of a pristine Nafion membrane, while the mechanical stability of the composite Nafion membrane is enhanced by 44%. As a result, the fuel cell performance of the silica nanofiber-Nafion composite membrane measured at high temperature and low humidity is improved by 38%.

Keywords: bifunctionally composite; sulfonic acid based proton exchange membrane; silica nanofiber; mechanical stability; high temperature fuel cell

1. Introduction

The excessive dependence of human on fossil fuels causes not only serious environment pollution but also global energy crises [1–3]. In response to these crises, new generations of sustainable energy have attracted widespread attention, such as wind energy and solar energy [4]. In order to efficiently utilize sustainable energy sources, secondary energy technologies, represented by lithium-ion battery and hydrogen energy, have to be applied as energy media [5–7]. As a primary segment in the hydrogen energy chain, proton-exchange membrane fuel cells (PEMFCs) have also been regard as one of the most promising green energies due to high efficiency and pollution-free properties [8–10]. As the core component in a PEMFC, a proton-exchange membrane (PEM) acts as a proton-transporting medium and can directly affect the performance of a PEMFC [11,12]. For numerous developed PEMs, sulfonic acid (SA) groups are the most widely used functional groups due to the great proton conductive efficiency of SA groups with the assistance of water molecules at low temperature [13–15]. However, enhancing the operating temperature of PEMFCs over 100 °C is a convenient strategy to improve the fuel cell performance and reduce the Pt dependence of PEMFCs [16–18]. Unfortunately, proton conduction performance of SA-based PEMs, represented by commercial Nafion membranes, would be strongly decayed by elevating temperature over 100 °C due to the poor water capacity of SA-based PEMs [19–21]. Therefore, various inorganic materials with excellent water retention capacity, including silica [22–24], titania [25,26], and zirconium phosphate [27], were employed to cooperate with SA groups to improve the proton conductivity of SA-based PEMs at elevated temperatures. As a result, SA-based PEMs with particulate additives usually have good high-temperature proton conductivity due to improved water retention capacity [28–30]. Nevertheless, poor mechanical stability of composite membranes caused by particle agglomeration or excessive swelling and bed compatibility with matrices greatly affect fuel cell performance [31]. Over the past decade, various nanofiber materials prepared by electrospinning were widely used in modification of SA-based PEMs, such as poly(vinylidene fluoride) (PVDF) [32], sulfonated polyether ether ketone (SPEEK) [33,34], and polyvinyl alcohol (PVA) [35]. Nanofibers with three-dimensional foam-like structures can not only avoid the agglomeration of particulate fillers, but also prevent excessive swelling of membranes.

Inspired by this PEM-reinforcing strategy, silica nanofibers was fabricated to bifunctionally composite a Nafion membrane, which was used as a model of SA-based PEMs, for stable application at elevated temperatures. The silica nanofibers with a porous three-dimensional network structure not only prevented excessive swelling of the membrane, but also facilitated the impregnation of a sulfonic-based conductor into the nanofibers. More importantly, the silica nanofibers without calcination treatment had abundant hydroxyl on the surface, and the nanofibers exhibited good compatibility with the SA-based proton conductor. The silica nanofibers therefore acted as a bifunctional modifier to simultaneously improve mechanical stability and high-temperature proton conductivity of the SA-based composite PEM. As a result, the proton conductivity of the silica network–Nafion composite membrane was enhanced to 0.045 S/cm at 110 °C and 60% RH with the mechanical stability improved by 44%. As a result, high-temperature fuel cell performance was strongly improved as desired.

2. Experimental

2.1. Materials

The tetrathylorthosilicate (TEOS), dimethyl sulfoxide (DMSO), hydrochioric acid and ethanol were purchased from Sinopharm Chemical Reagent Co., Ltd. (Chengdu, China). All reagents were not further treated. The Nafion 212 was supplied by Dupont Co. (Wilmington, DE, USA).

2.2. Preparation of Silica Nanofiber

Prior to electrospinning, a silica sol precursor solution was prepared by the following steps. Firstly, a solution of TEOS, water, ethanol, and HCl at molar ratios of 1:2:2:0.1 was prepared. Secondly, the solution was magnetically stirred for 5 h. Finally, the solution was heated on a hot plate at 80 °C, until the volume of solution was reduced to 3/8 of the original volume. The silica sol was then electrospun under the following conditions: (i) an applied voltage of 28 kV, (ii) a tip-to-collector distance of 15 cm, (iii) a flow rate of 10 μL/min.

2.3. Preparation of Composite Membranes

A 2 wt % Nafion membrane/dimethyl sulfoxide (DMSO) solution was prepared by the following steps. A pretreated Nafion 212 membrane was cut into small pieces. The sheared Nafion 212 membrane was dissolved in a DMSO solution, and the mixture was stirred at 150 °C under Ar atmosphere for 2 h to form a Nafion solution. A solution-casting method was used to prepare the composite membrane. Ten milliliters of the Nafion membrane/DMSO solution was added to a glass plate with a diameter of 7 cm. The plate acted as a collector, and the nanofiber was directly immersed in the Nafion solution. Then, another 10 mL of the Nafion solution was added. The mixed solution was dried in an oven at 80 °C for 24 h. Finally, the composite membrane was hot-pressed. By controlling the time of electrospinning, the contents of the silica nanofiber in the composite were 1%, 3%, and 5%, and the composite membranes were denoted as SiNF-Nafion-1%, SiNF-Nafion-3%, and SiNF-Nafion-5%, respectively. For a comparison, a pristine Nafion membrane was also prepared by the same solution-casting method.

2.4. Characterizations

An electrochemical workstation (Interface 1000 Gamry and CHI) was used to evaluate proton conductive performances of the membranes. With the help of a self-made test mold, all proton conduction measurements were performed in an in-plane direction by a two-probe method under high-temperature conditions and a four-probe method at full humidity. The result was calculated according to the following equation:

$$\sigma = \frac{d}{RA} \; (S/cm), \tag{1}$$

where σ represents the proton conductivity of the membrane, d is the distance between the electrodes, A is the area of the membrane and R is the resistance (Ω) associated with the proton conductivity of the membrane obtained from the electro-chemical impedance spectroscopy (EIS) data.

Water uptake (WU) and volume swelling (VS) ratios of the membranes were measured at 25 °C. The membrane was dried under a vacuum oven at 80 °C for 24 h, and dry weight (W_{dry}) and dry volume (V_{dry}) were immediately recorded. Then, the membrane sample was immersed in deionized water for 24 h before the weight (W_{wet}) and wet volume (V_{wet}) recording. The WU and VS ratios of the membrane were calculated as shown in Equations (3) and (4):

$$\text{WU} = (W_{wet} - W_{dry})/W_{dry}, \tag{2}$$

$$\text{VS} = (V_{wet} - V_{dry})/V_{dry}. \tag{3}$$

The pretreated membrane was sheared into pieces, and the dry weigh was accurately weighted. The dry membrane was immersed in Fenton's reagent (3 ppm $FeSO_4$ in 3% H_2O_2) at 80 °C. The oxidative stability performance was finally evaluated by the percentage of retained weights, after the membrane was treated in Fenton's reagent for 1 and 24 h.

A scanning electron microscope (SU8010, Hitachi, Tokyo, Japan) with an energy-dispersive X-ray spectrometer (EDS) was used to study the cross-sectional morphology of the modified Nafion membrane. Small-angle X-ray scattering (SAXS) experiments were carried out on NanoSTAR with a q range from 0.007 to 0.228 $Å^{-1}$. A NetzschSTA 409 PC TG-DTA instrument was applied for the thermal stability study for the membranes from 30 to 800 °C under nitrogen atmosphere, and the content of silica was calculated by the thermogravimetric (TG) result. Mechanical properties of all the modified Nafion membranes were evaluated by a universal tensile machine (Labthink XLW) at 25 °C. The membrane samples were cut into rectangles (size: 0.5 cm × 3 cm).

The high-temperature fuel cell was measured as follows. The electrode had a catalyst loading of 0.5 mg/cm^2 (Pt loading: 0.2 mg/cm^2) on both the anode and the cathode. The catalyst-coated membrane (active area: 2 cm × 2 cm) was then sandwiched between two sheets of carbon papers (TGP-H-030, Toray) after being activated in a sulfuric acid solution (0.5 M) for 12 h. PEMFC testing was carried out on a MiniTest3000 Fuel Cell Test System (TOYO Corporation) with flow rates of 0.2 and 0.5 L/min for hydrogen and oxygen, respectively.

3. Results and Discussion

Based on the electrospinning method, silica nanofibers were prepared, and the structure of THE pure thin silica-nanofiber membrane was characterized. Figure 1a is a photograph of a pristine silica-nanofiber membrane prepared by electrospinning. The nanofiber membrane was bouffant and flexible. Surface SEM images of the silica-nanofiber membrane are shown in Figure 1b,c. It can be found that the nanofibers formed a porous three-dimensional network with a porous structure. The large voids inside the three-dimensional network contributed to the impregnation of an SA-based proton-transporting conductor. According to the nanofiber diameter distribution result in Figure 1d, the silica nanofiber had an average diameter of around 390 nm. The functional groups in nanofibers were confirmed by the FT-IR spectrum of the silica-nanofiber membrane in Figure 1e, and the peaks at 1056 and 945 cm^{-1} were attributed to Si–O–Si and Si–OH vibrations, respectively. Therefore, the abundant hydroxyl on the surface of the silica nanofibers can achieve excellent interface compatibility between the nanofiber and the SA-based matrix. The silica nanofibers with a porous three-dimensional structure and abundant hydroxyl groups were successfully prepared. Then, with a Nafion membrane as a representative of the SA-based transport conductor, the silica nanofiber (SiNF)-Nafion membrane was prepared by a solution-casting method. It can be found from the photograph of a SiNF-Nafion composite membrane (SiNF-Nafion-3%) in Figure 1f that the membrane turned transparent after the composition.

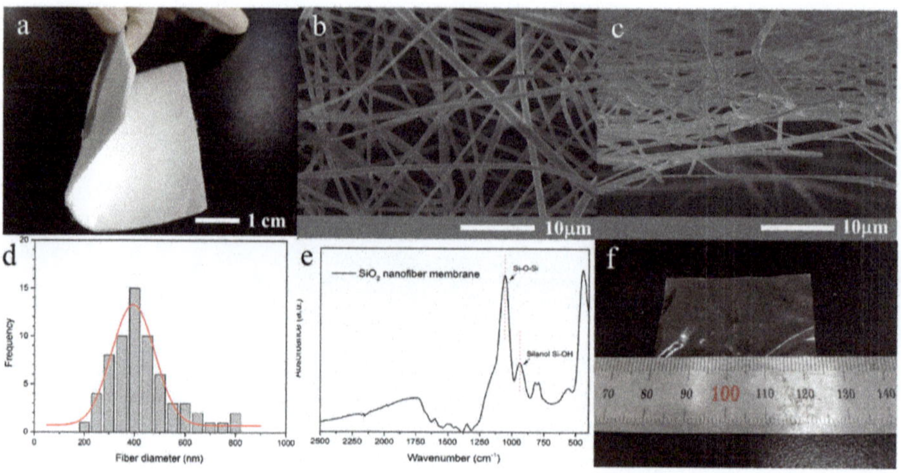

Figure 1. (**a**) Photograph of a pristine silica-nanofiber membrane; (**b**) surface SEM image of the pristine silica-nanofiber membrane; (**c**) cross-sectional surface SEM image of the pristine silica-nanofiber membrane; (**d**) fiber diameter distribution of the pristine silica-nanofiber membrane; (**e**) FT-IR spectrum of the pristine silica-nanofiber membrane; (**f**) photograph of the SiNF-Nafion-3% composite membrane.

Figure 2 displays the cross-sectional SEM images of the SiNF-Nafion membranes and the corresponding elemental maps (Si) by an EDS. The cross-sections of all three SiNF-Nafion membranes remained smooth, and there was no obvious phase separation after the impregnation of Nafion polymer into the silica nanofiber. This result also demonstrated excellent compatibility between the silica nanofiber and the Nafion matrix. As shown in the elemental maps of Si, the silica nanofibers were uniformly distributed in the SiNF-Nafion membrane due to the facile impregnation of Nafion polymer.

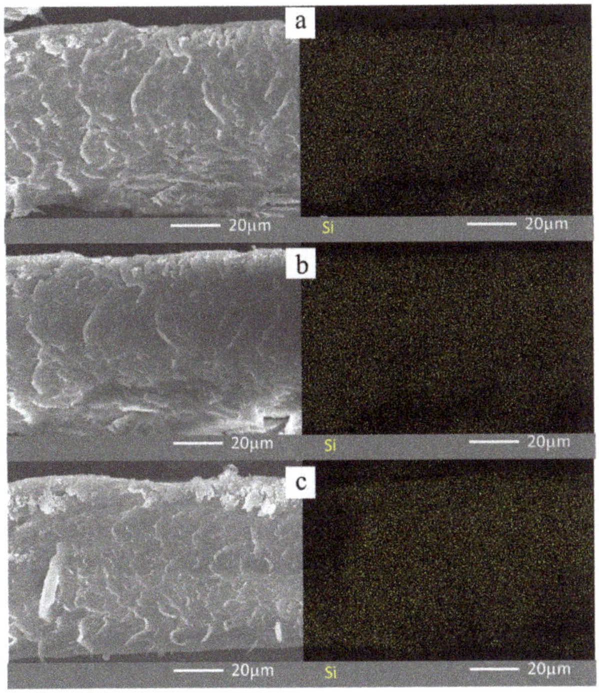

Figure 2. Cross-sectional SEM images and corresponding elemental maps of Si by an energy-dispersive X-ray spectrometer (EDS): (**a**) SiNF-Nafion-1% membrane; (**b**) SiNF-Nafion-3% membrane; and (**c**) SiNF-Nafion-5% membrane.

Since the silica nanofibers were confirmed to be uniformly filled into the composite membranes, the effects of nanofibers on the overall performance of the composite membranes were subsequently studied. Figure 3 compares the stress–strain and TG curves of the pristine Nafion and the SiNF-Nafion membranes. According to the stress–strain curves (Figure 3a), the tensile stresses of the three SiNF-Nafion-3% membranes were measured to be 10.2 MPa, which was 45.7% higher than that of the pristine Nafion membrane. At the same time, elongation at break of all the three SiNF-Nafion membranes was also significantly enhanced compared with that of the pristine Nafion membrane. The improved mechanical stability of the SiNF-Nafion membranes was ascribed to the uniform distribution of silica nanofibers in the composite membranes and the hydrogen bonding interactions between the hydroxyl groups on the silica nanofibers and the SA groups on the Nafion chain. With the silica nanofiber content increased to 5%, the elongation at break of the SiNF-Nafion-5% membrane was significantly reduced due to the fact that excessive silica nanofibers broke the continuity of the polymeric matrix. The mechanical stability improvement of the SiNF-Nafion membranes was also reflected by the change on the VS of the composite membranes. As shown in Table 1, the WU of the SiNF-Nafion-5% membrane was almost doubled compared with that of the pristine Nafion membrane, while the VS ratio was even lowered. The reason is that the silica nanofibers can not only efficiently retain water molecules, but also effectively limit the physical swelling of a polymeric–inorganic composite membrane. The TG curves of the pristine Nafion and SiNF-Nafion membranes shown in Figure 3b indicate that the thermal stabilities of the SiNF-Nafion membranes were not faded after the addition of silica nanofibers. At around 330 °C, the pristine Nafion membrane showed a significant mass loss, which was attributed to the decomposition of SA groups. However, for the SiNF-Nafion membranes, the decomposition temperature of the SA groups was increased to about 380 °C. This result indicated that the nanofibers had a protective effect on the SA groups. At the same time, higher residual weights of the composite membranes confirmed the successful composition of silica nanofibers with the Nafion membrane. As an important factor in evaluating the practicability of a PEM in fuel cell devices, the oxidative stability of the SiNF-Nafion membranes were found to be as good as that of the pristine Nafion membrane due to the property of silica (Table 1).

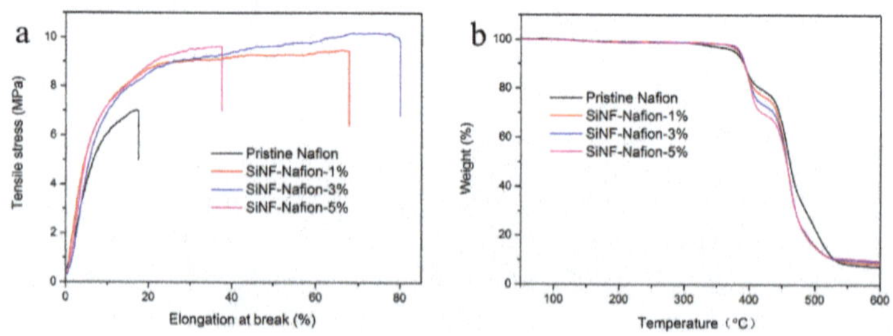

Figure 3. Stress-strain curves (**a**) and thermogravimetric (TG) curves (**b**) of the pristine Nafion and SiNF-Nafion membranes.

Table 1. Water uptake, volume swelling, and oxidative stability of the pristine Nafion and SiNF-Nafion membranes.

Membrane	Water Uptake	Volume Swelling	Oxidative Stability	
			RW-1 [1]	RW-24 [2]
Pristine nafion	15.9%	29.3%	98.6%	98.4%
SiNF-Nafion-1%	20.4%	31.8%	98.9%	98.0%
SiNF-Nafion-3%	25.1%	28.6%	98.5%	98.1%
SiNF-Nafion-5%	28.3%	28.0%	98.8%	98.6%

[1] Retained weight after 1 h Fenton's reagent treatment at 80 °C. [2] Retained weight after 24 h Fenton's reagent treatment at 80 °C.

In addition to systematical stability, proton conductivity at high temperature and low humidity conditions is a great concern for applications of PEMs in high-temperature fuel cells. It can be found from Figure 4a that the SiNF-Nafion-3% membrane exhibited the highest proton conductivity at 20%–60% RH among the four membranes. At 110 °C and 20% RH, the proton conductivity of the SiNF-Nafion-3% membrane was measured to be 0.015 S/cm, which was 25% higher than that of the pristine Nafion membrane. With the humidity increased to 60% RH, the proton conductivity of the SiNF-Nafion-3% membrane was rapidly increased to 0.046 S/cm, which was 1.84 times that of the pristine Nafion membrane. This obvious increase in proton conductivity was ascribed to the great water retention capacity of the silica network uniformly composited in the membrane. However, with the nanofiber content increased to 5%, the proton conductivity of the SiNF-Nafion-5% membrane was reduced, although the increment of proton conductivity by raising the humidity was higher compared with that of the SiNF-Nafion-3% membrane. The reason is that Nafion chains impregnated into voids inside silica nanofibers still acted as a primary proton exchange medium in SiNF–Nafion composite membranes and excess silica broke the continuity of original proton conduction channels of Nafion membranes. As a result, the proton conduction efficiency was decreased, even if the water retention capacity of the membrane was enhanced for the SiNF-Nafion-5% membrane with the highest silica content. The optimum silica nanofiber content was therefore considered to be 3% for the SiNF-Nafion composite membranes. Besides, the structure changes in SA clusters may also affect proton conductivity. The SAXS spectra of the pristine Nafion and SiNF-Nafion membranes are shown in Figure 4b. Generally speaking, the SAXS peak emerging at around 0.5 nm^{-1} represents the presence of ordered SA clusters, which was the main place for proton transport in membranes. From the SAXS result, it can be concluded that no ordered SA clusters appeared in the recast Nafion membrane. With the addition of silica nanofibers, there was an obvious SAXS peak around 0.5 nm^{-1} in the SiNF-Nafion membranes. This results indicated that the ordered SA clusters were formed in the SiNF-Nafion membranes. The SA groups tended to form a cluster structure along the surface of the nanofiber due to the interaction of SA groups with hydroxyl groups on the nanofiber surface. Therefore, the synergistic effect of the excellent water retention capacity of silica nanofibers and the rearrangement of SA clusters led to an increase in the proton conductivity of the SiNF-Nafion membranes. The results were also confirmed by the 100% humidity proton conductivity of the pristine Nafion and SiNF-Nafion membranes, shown in Figure 4c. In the entire temperature range (20–60 °C), the proton conductivities of the SiNF-Nafion membranes were higher than that of the pristine Nafion membrane due to the higher proton transport efficiency of the SiNF-Nafion membranes.

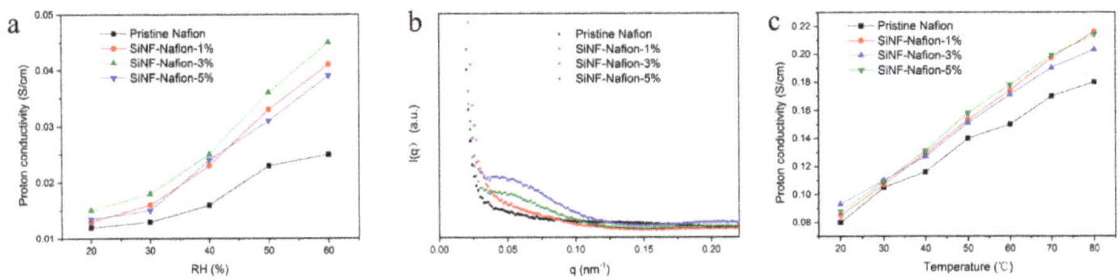

Figure 4. (**a**) High-temperature proton conductivity at low humidity of the pristine Nafion and SiNF-Nafion membranes; (**b**) small-angle X-ray scattering SAXS spectra of the pristine Nafion and SiNF-Nafion membranes; and (**c**) low-temperature proton conductivity at 100% relative humidity of the pristine Nafion and SiNF-Nafion membranes.

Considering the excellent overall performance of the SiNF-Nafion-3% membrane, a fuel cell with the SiNF-Nafion-3% membrane was carried out at 110 °C and 20% RH. The polarization and power density curves of the pristine Nafion and SiNF-Nafion-3% membranes were compared in Figure 5. As expected, the SiNF-Nafion-3% membrane had a higher power density than the pristine Nafion membrane. Moreover, at the middle current range, the slope of the polarization curve of the SiNF-Nafion-3% membrane was decreased compared with that of the pristine Nafion membrane, which indicates that the internal resistance of the fuel cell was reduced. As a result, the maximum power density of the SiNF-Nafion-3% membrane was 118 mW/cm^2, which was 38 % higher than that of the pristine Nafion membrane.

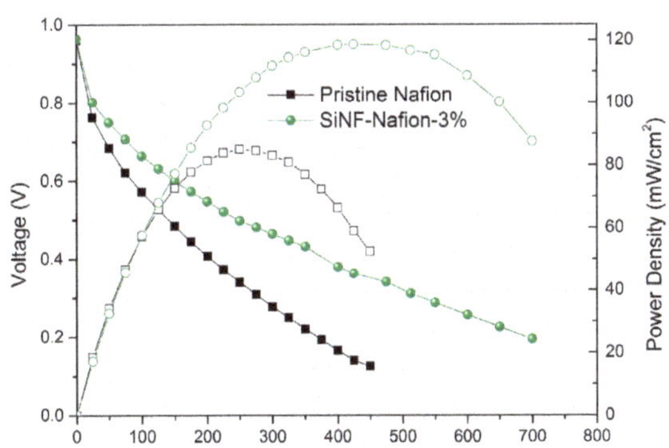

Figure 5. Polarization and power density curves of the pristine Nafion and SiNF-Nafion-3% membranes under 20% RH and humidified under H_2 and O_2 atmospheres at 110 °C.

4. Conclusions

In summary, an SA-based proton conductor can be bifunctionally composited by a silica network. A silica nanofiber membrane with a three-dimensional porous structure can effectively retain water molecules and limit excessive water swelling of a composite membrane at the same time, because an abundant hydrogen bonding between the nanofiber and the SA-based matrix improves the mechanical strength of the composite membrane. Meanwhile, the synergistic effect of excellent water retention capacity of silica nanofiber and rearrangement of SA clusters contributes to the improvement of high-temperature proton conductivity. At 110 °C and 60% RH, the proton conductivity of a reinforced SiNF-Nafion-3% membrane is 1.84 times as high as that of the pristine Nafion membrane. As a result, the SiNF-Nafion-3% membrane exhibits great high-temperature fuel cell performance with a 118 mW/cm^2 power density at low humidity, which is 38 % higher than that of the pristine Nafion membrane.

Author Contributions: Conceptualization, W.C., J.L., L.M., and G.X.; methodology, W.C., J.L., and G.X.; software, W.C., J.L., and G.X.; validation, W.C., J.L., and Y.L.; formal analysis, W.C. and G.X.; investigation, G.X. and J.Z.; resources, W.C. and J.L.; data curation, G.X., J.Z., and Z.G.; writing of the original draft preparation, G.X.; writing of review and editing, W.C., J.L., and Y.L.; visualization, J.L. and G.X.; supervision, W.C. and J.L.; project administration, G.X. and L.M.; funding acquisition, W.C. and J.L. All authors have read and agreed to the published version of the manuscript.

Funding: This research was funded by the National Natural Science Foundation of China, grant numbers 21703211 and 21875224, the Natural Science Foundation of Zhejiang Province grant number: LGG19B030001, the Fundamental Research Funds of Guizhou Normal University, grant number GZNUD [2017]1 and the Science and Technology Project of Guizhou Province grant number [2019]1226.

Acknowledgments: We are grateful for financial support from the National Natural Science Foundation of China (grant numbers: 21703211 and 21875224), the Natural Science Foundation of Zhejiang Province (grant number: LGG19B030001), the Fundamental Research Funds of Guizhou Normal University (grant number: GZNUD [2017]1) and the Science and Technology Project of Guizhou Province (grant number: [2019]1226).

Conflicts of Interest: The authors declare no conflict of interest.

References

1. Kusoglu, A.; Weber, A.Z. New Insights into Perfluorinated Sulfonic-Acid Ionomers. *Chem. Rev.* **2017**, *117*, 987–1104. [CrossRef] [PubMed]
2. Li, J.; Cai, W.; Ma, L.; Zhang, Y.; Chen, Z.; Cheng, H. Towards neat methanol operation of direct methanol fuel cells: A novel self-assembled proton exchange membrane. *Chem. Commun.* **2015**, *51*, 6556–6559. [CrossRef]
3. Thorsten, S.; Su, Z.; Kai, S. Current status of and recent developments in the direct methanol fuel cell. *Chem. Eng. Technol.* **2001**, *12*, 1212–1223. [CrossRef]
4. Liu, C.; Li, F.; Ma, L.; Cheng, H. Advanced materials for energy storage. *Adv. Energy. Mater.* **2010**, *22*, E28–E62. [CrossRef] [PubMed]
5. Li, B.; Xia, D. Anionic redox in rechargeable lithium batteries. *Adv. Mater.* **2017**, *29*, 1701054. [CrossRef] [PubMed]
6. Zhang, Y.; Li, J.; Ma, L.; Cai, W.; Cheng, H. Recent developments on alternative proton exchange membranes: Strategies for systematic performance improvement. *Energy Technol.* **2015**, *3*, 675–691. [CrossRef]
7. Yu, M.; Ma, J.; Xie, M.; Song, H.; Tian, F.; Xu, S.; Zhou, Y.; Li, B.; Wu, D.; Qiu, H.; et al. Freestanding and sandwich-structured electrode material with high areal mass loading for long-life lithium-sulfur batteries. *Adv. Energy Mater.* **2017**, *7*, 1602347. [CrossRef]
8. Ran, J.; Wu, L.; He, Y.; Yang, Z.; Wang, Y.; Jiang, C.; Ge, L.; Bakangura, E.; Xu, T. Ion exchange membranes: New developments and applications. *J. Membr. Sci.* **2017**, *522*, 267–291. [CrossRef]
9. Mosa, J.; Durán, A.; Aparicio, M. Sulfonic acid-functionalized hybrid organic–inorganic proton exchange membranes synthesized by sol–gel using 3-mercaptopropyl trimethoxysilane (MPTMS). *J. Power Sources* **2015**, *297*, 208–216. [CrossRef]
10. Xu, G.; Li, J.; Ma, L.; Xiong, J.; Mansoor, M.; Cai, W.; Cheng, H. Performance dependence of swelling-filling treated Nafion membrane on nano-structure of macromolecular filler. *J. Membr. Sci.* **2017**, *534*, 68–72. [CrossRef]

11. Ma, L.; Cai, W.; Li, J.; Fan, K.; Jiang, Y.; Ma, L.; Cheng, H. A high performance polyamide-based proton exchange membrane fabricated via construction of hierarchical proton conductive channels. *J. Power Sources* **2016**, *302*, 189–194. [CrossRef]
12. Yoon, S.I.; Seo, D.-J.; Kim, G.; Kim, M.; Jung, C.-Y.; Yoon, Y.-G.; Joo, S.H.; Kim, T.-Y.; Shin, H.S. AA′-Stacked trilayer hexagonal boron nitride membrane for proton exchange membrane fuel cells. *ACS Nano* **2018**, *12*, 10764–10771. [CrossRef] [PubMed]
13. Li, J.; Xu, G.; Luo, X.; Xiong, J.; Liu, Z.; Cai, W. Effect of nano-size of functionalized silica on overall performance of swelling-filling modified Nafion membrane for direct methanol fuel cell application. *Appl. Energy* **2018**, *213*, 408–414. [CrossRef]
14. Yin, Y.; Li, Z.; Yang, X.; Cao, L.; Wang, C.; Zhang, B.; Wu, H.; Jiang, Z. Enhanced proton conductivity of Nafion composite membrane by incorporating phosphoric acid-loaded covalent organic framework. *J. Power Sources* **2016**, *332*, 265–273. [CrossRef]
15. Zhao, L.; Li, Y.; Zhang, H.; Wu, W.; Liu, J.; Wang, J. Constructing proton-conductive highways within an ionomer membrane by embedding sulfonated polymer brush modified graphene oxide. *J. Power Sources* **2015**, *286*, 445–457. [CrossRef]
16. Yin, C.; Li, J.; Zhou, Y.; Zhang, H.; Fang, P.; He, C. Enhancement in proton conductivity and thermal stability in nafion membranes induced by incorporation of sulfonated carbon nanotubes. *ACS Appl. Mater. Interfaces* **2018**, *10*, 14026–14035. [CrossRef]
17. Xu, G.; Li, S.; Li, J.; Liu, Z.; Li, Y.; Xiong, J.; Cai, W.; Qu, K.; Cheng, H. Targeted filling of silica in Nafion by a modified in situ sol-gel method for enhanced fuel cell performance at elevated temperatures and low humidity. *Chem. Commun.* **2019**, *55*, 5499–5502. [CrossRef]
18. Wang, K.; Yang, L.; Wei, W.; Zhang, L.; Chang, G. Phosphoric acid-doped poly(ether sulfone benzotriazole) for high-temperature proton exchange membrane fuel cell applications. *J. Membr. Sci.* **2018**, *549*, 23–27. [CrossRef]
19. Kim, K.; Kim, S.-K.; Park, J.O.; Choi, S.-W.; Kim, K.-H.; Ko, T.; Pak, C.; Lee, J.-C. Highly reinforced pore-filling membranes based on sulfonated poly(arylene ether sulfone)s for high-temperature/low-humidity polymer electrolyte membrane fuel cells. *J. Membr. Sci.* **2017**, *537*, 11–21. [CrossRef]
20. Li, J.; Xu, G.; Luo, X.; Cai, W.; Xiong, J.; Ma, L.; Yang, Z.; Huang, Y.; Chen, H. Non-destructive modification on Nafion membrane via in-situ inserting of sheared graphene oxide for direct methanol fuel cell applications. *Electrochim. Acta* **2018**, *282*, 362–368. [CrossRef]
21. Xu, G.; Wei, Z.; Li, S.; Li, J.; Yang, Z.; Grigoriev, S.A. In-Situ sulfonation of targeted silica-filled Nafion for high-temperature PEM fuel cell application. *Int. J. Hydrogen Energy* **2019**, *44*, 29711–29716. [CrossRef]
22. Zhang, J.; Liu, J.; Lu, S.; Zhu, H.; Aili, D.; De Marco, R.; Xiang, Y.; Forsyth, M.; Li, Q.; Jiang, S. Ion-Exchange-Induced selective etching for the synthesis of amino-functionalized hollow mesoporous silica for elevated-high-temperature fuel cells. *ACS Appl. Mater. Interfaces* **2017**, *9*, 31922–31930. [CrossRef] [PubMed]
23. Wang, H.; Li, X.; Zhuang, X.; Cheng, B.; Wang, W.; Kang, W.; Shi, L.; Li, H. Modification of Nafion membrane with biofunctional SiO_2 nanofiber for proton exchange membrane fuel cells. *J. Power Sources* **2017**, *340*, 201–209. [CrossRef]
24. Ke, C.; Li, X.; Qu, S.; Shao, Z.; Yi, B. Preparation and properties of Nafion/SiO_2 composite membrane derived via in situ sol-gel reaction: Size controlling and size effects of SiO_2 nano-particles. *Polym. Adv. Technol.* **2012**, *23*, 92–98. [CrossRef]
25. Bet-moushoul, E.; Mansourpanah, Y.; Farhadi, K.; Tabatabaei, M. TiO_2 nanocomposite based polymeric membranes: A review on performance improvement for various applications in chemical engineering processes. *Chem. Eng. J.* **2016**, *283*, 29–46. [CrossRef]
26. Sayeed, M.D.A.; Kim, H.J.; Park, Y.; Gopalan, A.I.; Kim, Y.H.; Lee, K.-P.; Choi, S.-J. Sulfated titania–silica reinforced Nafion® nanocomposite membranes for proton exchange membrane fuel cells. *J. Nanosci. Nanotechnol.* **2015**, *15*, 7054–7059. [CrossRef]
27. Ni'mah, H.; Chen, W.; Shen, Y.; Kuo, P. Sulfonated nanoplates in proton conducting membranes for fuel cells. *RSC Adv.* **2011**, *1*, 968–972. [CrossRef]
28. Dresch, M.A.; Matos, B.R.; Fonseca, F.C.; Santiago, E.I.; Carmo, M.; Lanfredi, A.J.C.; Balog, S. Small-angle X-ray and neutron scattering study of Nafion-SiO_2 hybrid membranes prepared in different solvent media. *J. Power Sources* **2015**, *274*, 560–567. [CrossRef]

29. Patil, Y.; Kulkarni, S.; Mauritz, K.A. In Situ grown titania composition for optimal performance and durability of Nafion®fuel cell membranes. *J. Appl. Polym. Sci.* **2011**, *121*, 2344–2353. [CrossRef]
30. Kim, A.R.; Vinothkannan, M.; Yoo, D.J. Artificially designed, low humidifying organic–inorganic (SFBC-50/FSiO$_2$) composite membrane for electrolyte applications of fuel cells. *Compos. Part B-Eng.* **2017**, *130*, 103–118. [CrossRef]
31. Lee, J.; Kim, N.; Lee, M.; Lee, S. SiO$_2$-coated polyimide nonwoven/Nafion composite membranes for proton exchange membrane fuel cells. *J. Membr. Sci.* **2011**, *367*, 265–272. [CrossRef]
32. Li, H.; Lee, Y.; Lai, J.; Liu, Y. Composite membranes of Nafion and poly(styrene sulfonic acid)-grafted poly(vinylidene fluoride) electrospun nanofiber mats for fuel cells. *J. Membr. Sci.* **2014**, *466*, 238–245. [CrossRef]
33. Chae, K.; Kim, K.; Choi, M.; Yang, E.; Kim, I.S.; Ren, X.; Lee, M. Sulfonated polyether ether ketone (SPEEK)-based composite proton exchange membrane reinforced with nanofibers for microbial electrolysis cells. *Chem. Eng. J.* **2014**, *254*, 393–398. [CrossRef]
34. Lee, C.; Jo, S.M.; Choi, J.; Baek, K.-Y.; Truong, Y.B.; Kyratzis, I.L.; Shul, Y.G. SiO2/sulfonated poly ether ether ketone (SPEEK) composite nanofiber mat supported proton exchange membranes for fuel cells. *J. Mater. Sci.* **2013**, *48*, 3665–3671. [CrossRef]
35. Mollá, S.; Compañ, V.; Gimenez, E.; Blazquez, A.; Urdanpilleta, I. Novel ultrathin composite membranes of Nafion/PVA for PEMFCs. *Int. J. Hydrog. Energy* **2011**, *36*, 9886–9895. [CrossRef]

polymers

MDPI

Article

Surface Modification of PET Fiber with Hybrid Coating and Its Effect on the Properties of PP Composites

Yapeng Mao, Qiuying Li * and Chifei Wu *

Polymer Processing Laboratory, School of Material Science and Engineering, East China University of Science and Technology, Shanghai 200237, China; maoyp_ecust@163.com
* Correspondence: liqiuying75713@163.com (Q.L.); wucf@ecust.edu.cn (C.W.)

Received: 26 September 2019; Accepted: 18 October 2019; Published: 21 October 2019

Abstract: Surface modification fundamentally influences the morphology of polyethylene terephthalate (PET) fibers produced from abandoned polyester textiles and improve the compatibility between the fiber and the matrix. In this study, PET fiber was modified through solution dip-coating using a novel synthesized tetraethyl orthosilicate (TEOS)/KH550/ polypropylene (PP)-g-MAH (MPP) hybrid (TMPP). The PET fiber with TMPP modifier was exposed to the air. SiO_2 particles would be hydrolyzed from TEOS and become the crystalline cores of MPP. Then, the membrane formed by MPP, SiO_2 and KH550 covered the surface of the PET fiber. TMPP powder was investigated and characterized by fourier transform infrared spectroscopy, scanning electron microscope (SEM) and thermogravimetric analysis (TGA). TMPP-modified PET fiber was researched by X-ray diffraction and SEM. Furthermore, tensile strength of single fiber was also tested. PET fiber/PP composites were studied through dynamic mechanical analysis and SEM. Flexural properties of composites were also measured. The interfacial properties of PET fiber and PP matrix were indirectly represented by contact angle analysis. Results showed that the addition of TEOS is helpful in homogenizing the distribution of PP-g-MAH. Furthermore, TMPP generates an organic-inorganic 'armor' structure on PET fiber, which can make up for the damage areas on the surface of PET fiber and strengthen each single-fiber by 14.4%. Besides, bending strength and modulus of TMPP-modified PET fiber-reinforced PP composite respectively, increase by 10 and 800 MPa. The compatibility between PET fiber and PP was also confirmed to be increased by TMPP. Predictably, this work supplied a new way for PET fiber modification and exploited its potential applications in composites.

Keywords: PET fiber; PP; compatibility; modification

1. Introduction

Polyethylene terephthalate (PET) fiber is a material with poor wettability and weak interfacial bonding with resin matrix, thus, it is difficult to be used to obtain a high-performance fiber-reinforced resin composite [1]. In particular, the ester groups in PET chains make PET fiber inert toward polypropylene (PP) and polyethylene (PE) matrices, which consist of nonpolar molecular chains in the carbon skeletons [2,3]. In order to improve the compatibility between PET fiber and polyolefin resin matrix to produce a composite with better mechanical strength, and completely embody their advantages including high strength and high modulus, some effective and economic methods to modify the surface of PET fiber are highly desirable.

In recent years, great attention on polyester fiber has shifted from performance development to fiber recycling. As the most widely used fiber in the world, the amount of PET waste is still growing with sustained and high speed, and the earth's environment has been seriously damaged. Besides, the accumulation of PET fiber is waste of oil resources. Waste PET could be reused in polymer alloys, toys, soundproof cotton and some other staple fiber products. However, all of these recycling uses only occupy 10% of waste PET, some more effective and economical ways of recycling must be found and industrialized.

PET fiber-reinforced composites may be a potential method because PET fiber is an organic fiber with high strength and modulus, which would be a substitute of glass fiber. However, how to improve the interfacial properties between PET fiber and the matrix is the most important issue. Thus, the study about the surface modification of fiber has attracted the attention of many research laboratories for decades. To date, extensive research efforts have been devoted to the study of the surface modification of PET fiber. Most of these studies are based on surface chemical etching (alkali treatment), grafting modification, and plasma modification, etc. Through alkali weight reduction processing, the antistatic property, permeability and moisture absorption performance of PET fabric were obviously damaged [4]. Moreover, the fracture strength of PET fiber treated with alkali was also seriously damaged [5]. Plasma processing predominated over the amorphous region of PET fiber. Electron bombardment and free radical reaction were also used for the surface modification of fiber or films [6]. However, previous results [7] showed that the severe plasma treatments in low-pressure plasma led to extensive degradation of the PET fiber, resulting in a decrease of tensile strength. Thus, compared to plasma processing, alkali treatment can be easily controlled by reaction time and temperature. Further, the decrease of fiber strength and weight loss could be designed and predicted [8]. Thus, alkali treatment is often chosen as a basic treatment, while grafting and coating treatment are employed for further functionalized treatment.

Currently, coating modification of the fiber surface is gradually becoming a hot research topic. The coating on the fiber surface can not only reduce the thermal stress between the fiber and the resin matrix during processing procedure and repair defects on the surface of fiber and damages caused by alkali treatment, but also improve the molecular structure of the fiber surface and improve the wettability of PET fiber in the resin matrix [9]. However, modification of PET fiber through surface coating has rarely been reported. Kwak et al. [10–12] modified the surface of PET fiber by thermal treatment with a silver-carbamate complex. First, PET fabric was modified by a series of pretreatments with octaethylene glycol monododecyl, sodium hydroxide, distilled water, ethanol and solution of thiopropyl triethoxy silane in acetone. Then, the treated fabric was immersed in silver 2-ethyl-hexylcarbamate solution and squeezed gently. To form silver nanoparticles, the fabric was transferred into a convection oven at 130 °C for 5 min. After thermal reduction, a continuous layer of silver-coated nanoparticles with sizes between 30 and 100 nm was assembled on the PET fiber. Huang et al. [13] modified PET fiber with titania and PE nanoparticles. Results showed that titania and PE particles were effective in controlling agglomeration of each phase and aided in homogenizing the modified layer. According to these studies, inorganic particles and layers have been proven to be helpful for the dispersion of PET fiber.

Moreover, to retain the original shape and properties of PET fiber during the fabrication of PET fiber-reinforced composites, the melting point of the matrix should be lower than that of PET. PE was selected as a matrix in our previous study, and the results showed that the mechanical properties and thermal stability of PE matrix could be obviously enhanced by PET fiber. Tetraethyl orthosilicate (TEOS) and KH550 have been shown to be successful in coating on fiber with an inorganic layer [14,15]. Owing to the slow hydrolysis reaction in air, TEOS generates uniform SiO_2 particles on the modified fiber.

Compared to PE, PP is much better in strength and modulus. As a result, PP was set as the matrix to be strengthened. PP-g-MAH (MPP) was selected as a compatibilizer to improve the compatibility between fillers and the PP matrix [16].

In this study, SiO_2 particles were designed to link with the MPP molecular chain through silane coupling agent KH550, which formed an organic-inorganic film by grain structure on the surface of PET fiber obtained from abandoned polyester textiles. This method, taking both advantages of MPP and SiO_2 particles, is expected to make up for fiber damage caused by alkali treatment and improve the compatibility between the fiber and the PP matrix used in this study.

Moreover, modified PET fiber and untreated PET fiber-filled PP composites were also prepared for comparative analysis. The effects of different surface modification of PET fiber on the interfacial and mechanical properties between the fiber and the matrix were systematically investigated.

2. Experiments

2.1. Materials

PET fibers with 20 μm diameter were obtained by laddering from abandoned polyester textiles (commercially available, 75D/24F/22T). These abandoned polyester textiles are leftovers of the daily production in textile mills. The matrix polymer was polypropylene (PP7633) powder with a density of 0.9 g/cm^3, and MFR = 2 g/10 min (2.16 kg/230 °C, LCY Chemical Corp, Taiwan, China). PP-g-MAH (E516) with a density of 0.93 g/cm^3 and free maleic anhydride content of less than 1.0% was provided by Ningbo MaterChem Technology Co., LTD (Hangzhou, China). Other agents were of analytical grade and all agents were used as received.

2.2. Synthesis of Fiber Modifier

The synthesis of fiber modifier was carried out following the procedure by Chen et al. [17]. Briefly, PP-g-MAH powder was added into boiling xylene solution until it dissolved completely (solution A). TEOS, p-toluenesulfonic acid (PTSA) and xylene were mixed in a mass ratio of 1:0.01:20 (solution B), under stirring for 2 h at room temperature. KH550 and xylene were mixed in a mass ratio of 0.5:20 (solution C). Then, solutions A, B and C were mixed together at 130 °C for 1 h, PET fiber modifier was accomplished, and it was named as TMPP. It is worth noting that the TMPP synthesis process should be carried out without water to avoid the hydrolysis reaction of TEOS.

2.3. Surface Treatment of Polyethylene Terephthalate (PET) Fiber

The PET fiber was immersed in a mixed solution of acetone and water (volume ratio 1:1) for 2 h, and then washed 3 to 4 times with deionized water before being dried for 4 h at 85 °C. Further, the fiber was treated with NaOH solution (0.01 mol L^{-1}) at a ratio of 2% (W/V) at 80 °C for 1 h. Then, they were dried and modified with MPP and TMPP by solution dip-coating, respectively. Finally, the modified fiber was dried at room temperature for 3 days. During the course of the formation of TMPP-modified PET fiber, the vapor in the air induced TEOS into a sol–gel reaction, when the modifier was exposed to the air, and SiO2/MPP composite powders were produced on the surface of the PET fiber. A schematic illustration of the interaction between TEOS and KH550 is presented in Figure 1 [18].

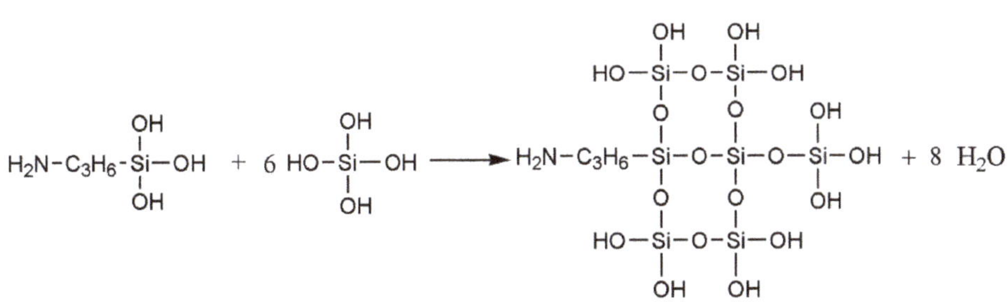

Figure 1. Interaction between TEOS and KH550.

2.4. Preparation of PET/Polypropylene (PP) Composites

The PET fiber and PP were mixed and homogenized in the mass ratio of 15:85 in an internal mixer with the rotor speed 60 rad s^{-1} at 200 °C for 10 min, and then dried at 100 °C for 4 h in an oven after being smashed. According to the modification of PET fiber, these composites were divided into PP, PP filled with untreated fiber, PP filled with MPP-treated fiber and PP filled with TMPP-treated fiber.

2.5. Characterization

2.5.1. Scanning Electron Microscopy

The morphology of TMPP modifier, the modified PET fiber and the bending sections of PET fiber/PP composites were examined by scanning electron microscopy (SEM, S-3400N, Hitachi, Tokyo, Japan). Samples were treated with spray gold craft before observation.

2.5.2. Fourier Transform Infrared Spectroscopy

Fourier transform infrared (FTIR, Nicolet Magna-IR550, Thermo Electron, New York, USA) spectroscopy was used to identify the modification of the surface of the PET fiber. Samples were loaded on potassium bromide tablets.

2.5.3. Thermogravimetric Analysis

The thermal stability of MPP and TMPP modifiers was tested by thermogravimetric analysis (TGA, Pyris6, Perkin-Elmer, company, Waltham, MA, USA). Samples were heated from ambient temperature to 800 °C at a heat rate of 10 °C/min. The TGA was performed in nitrogen atmosphere with the flow rate of 50 mL/min.

2.5.4. X-ray Scattering Analysis

The crystallization of the PET fiber was detected by wide angle X-ray scattering (WAXS, PW1830, Philips) at room temperature. Bragg diffraction angles (2θ) were obtained from a diffraction diagram. The inter-planar spacing (d) and crystallite size (t) of the PET fiber were calculated by using Bragg's law Equation (1) as follows [19]:

$$\lambda = 2dsin\theta \tag{1}$$

where, d is the Bragg spacing (Å), λ is the radiation wavelength (λ = 1.542 Å) and 2θ is the diffraction angle (deg).

Then, t was calculated by using Equation (2):

$$t = \frac{0.9\lambda}{\Delta\omega} cos\theta \tag{2}$$

where, t is the crystallite size (Å), and $\Delta\omega$ is the peak width at half-maximum (rad). The values of $\Delta\omega$ were determined from diffraction diagrams and converted to radians (radians = degrees × π/180).

2.5.5. Measurements of Interfacial Properties

The contact angles of substrate relative to water and glycerol were measured using a contact angle meter (OCA20, Dataphysics Company, Esslingen, Germany). Surface energy, interfacial tension and adhesive energy among these phases were calculated by using the Wu's Equations (3)–(7) as follows:

$$\left(1 + cos\theta_{H_2O}\right)\gamma_{H_2O} = 4\left(\frac{\gamma^{d}_{H_2O}\gamma^{d}}{\gamma^{d}_{H_2O} + \gamma^{d}} + \frac{\gamma^{p}_{H_2O}\gamma^{p}}{\gamma^{p}_{H_2O} + \gamma^{p}}\right) \tag{3}$$

$$\left(1 + cos\theta_{C_3H_8O_3}\right)\gamma_{C_3H_8O_3} = 4\left(\frac{\gamma^{d}_{C_3H_8O_3}\gamma^{d}}{\gamma^{d}_{C_3H_8O_3} + \gamma^{d}} + \frac{\gamma^{p}_{C_3H_8O_3}\gamma^{p}}{\gamma^{p}_{C_3H_8O_3} + \gamma^{p}}\right) \tag{4}$$

$$\gamma = \gamma^{d} + \gamma^{p} \tag{5}$$

$$\gamma_{12} = \gamma_1 + \gamma_2 - 4\left(\frac{\gamma^{d}_1\gamma^{d}_2}{\gamma^{d}_1 + \gamma^{d}_2} + \frac{\gamma^{p}_1\gamma^{p}_2}{\gamma^{p}_1 + \gamma^{p}_2}\right) \tag{6}$$

$$W_{12} = 2\left(\gamma_1^d \gamma_2^d\right)^{1/2} + 2\left(\gamma_1^p \gamma_2^p\right)^{1/2} \quad (7)$$

where, θ is the contact angle, γ is the surface energy (mN m^{-1}), and γ^d and γ^p are the dispersion component and polar component of surface energy, respectively. Subscripts 1 and 2 represent different phases, γ_{12} is the interfacial tension (mN m^{-1}) between two phases, and W_{12} is the adhesive energy of two phases (mN m^{-1}).

2.5.6. Mechanical Properties

The tensile strength of single-fiber was measured by a universal tester (XQ-1, Shanghai) in accordance with GB/T 14337. Bending samples were produced according to standard GB/T 9341-2008 by injection molding (QS-100T, Shanghai Quansheng Plastic Machinery Co., Ltd.). The flexural properties of PP and PET fiber/PP composites were tested using a mechanical property testing machine (WSM-20KN, Changchun).

2.5.7. Dynamic Mechanical Analysis

Dynamic mechanical analysis (DMA) of the composites was carried out using a dynamic mechanical analyzer (Rheogel-E4000, Japan) at a fixed frequency of 11 Hz in bending mode. Results were obtained at a heating rate of 3 °C/min within the temperature range from 70 to 160 °C. A specimen with the dimensions of 40 mm × 5 mm × 2 mm was used in this study.

3. Results and Discussion

3.1. Modifier Powder

3.1.1. Aggregation Morphology and Size of Modifier Powder

Figure 2 shows the SEM images of MPP and TMPP powders. Figure 2a,b exhibit that the average sizes of MPP and TMPP powders are 7.3 and 2.3 μm, respectively. Compared to MPP, the uniformity of TMPP particle size improves significantly. MPP only dissolve in xylene and only when the temperature is high enough; therefore, after heating is stopped, MPP gradually comes out of the solution along with the decrease of solution temperature. Thus, the instability of the MPP modifier limits its practical application in the industry. After adding TEOS, a colloidal solution was formed by sol–gel reaction and the modifier could remain stable for a long time at room temperature. When exposed to air, the anhydrous condition for TMPP/xylene gel was destroyed, and the hydrolysis process of TEOS with water vapor in the air was triggered; thus, homogeneous TMPP powder was produced. Moreover, MPP and TEOS reacted with silane coupling agent KH550 through anhydride and alkoxy groups separately. The schematic illustration of generation of the product is exhibited in Figure 3. In other words, the TMPP modifier could not only form connections between organic polymer and inorganic SiO_2 powder, but also improve the solubility of MPP particles in xylene. Therefore, the stability of the TMPP modifier was improved significantly and the aggregation of SiO_2 particles was prevented, eventually particles acquired the best dispersibility and suitable size.

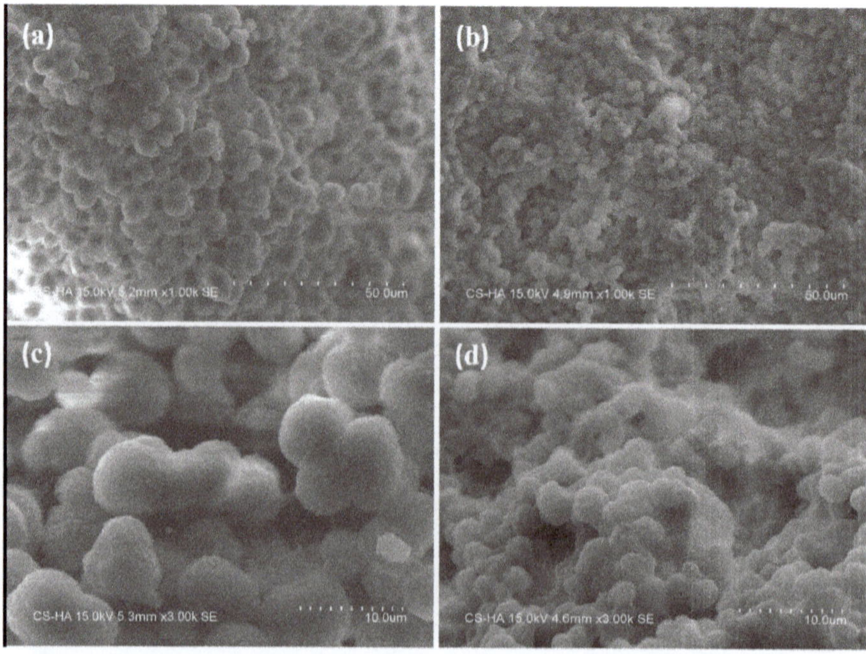

Figure 2. Scanning electron microscope (SEM) images of MPP and TMPP modifiers after hydrolysis. ((**a**) MPP powder, 1000 times, (**b**) TMPP powder, 1000 times, (**c**) MPP powders, 3000 times, (**d**) TMPP powder, 3000 times).

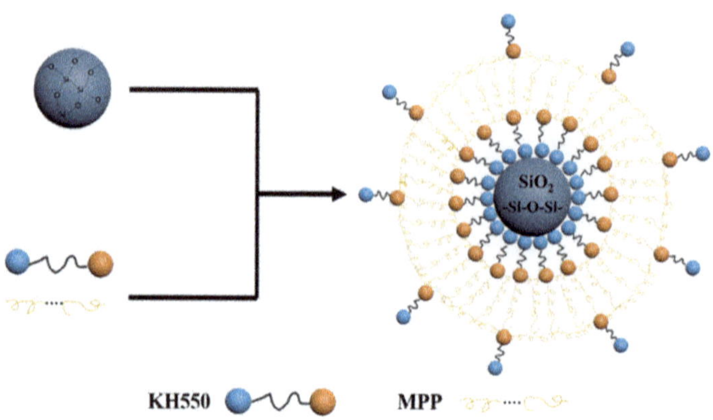

Figure 3. The structural representation of SiO_2/MPP modifier.

3.1.2. Fourier Transform Infrared (FTIR) spectra of MPP and TMPP

FTIR spectra of MPP and TMPP are shown in Figure 4. Peaks at 3411, 2721 and 1627 cm^{-1} are characteristic of amidogens. Only one peak is observed between 3300 and 3500 cm^{-1}, and no peak is present around 2000 cm^{-1}; thus, the amidogens intended to be secondary amidogens [20]. The peak at 1627 cm^{-1} represents that the secondary amidogens are in close proximity with carbonyl groups. The peak at 1097 cm^{-1} corresponds to the Si–O–bond. Taking the structures of KH550, TEOS, and MPP into consideration, the –NH_2 in KH550 reacts with maleic anhydride from MPP molecular chains.

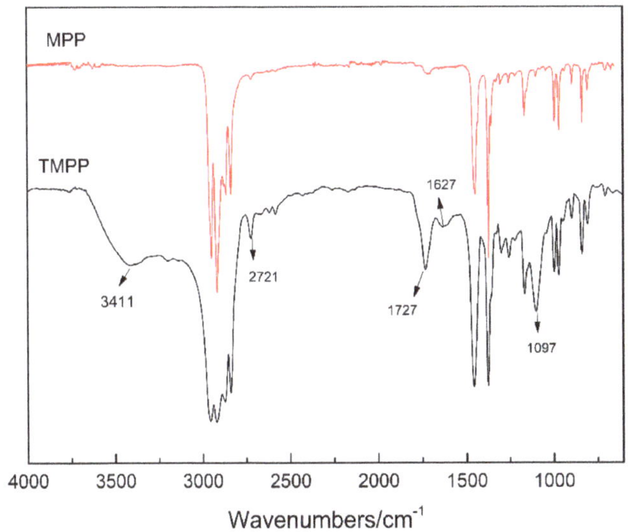

Figure 4. Fourier transform infrared (FTIR) spectra of MPP and TMPP modifiers.

The ideal structures of TMPP modifier and PET fiber are shown in Figure 5. TEOS and KH550 react and form a layer on the surface of the PET fiber, and the sites of $-NH_2$ limit the location of MPP molecules. Then, homogenization of MPP distribution is achieved. However, the reaction of KH550 and TEOS could not be controlled, as shown in Figure 5b, the aggregation of SiO_2 generated a structure as presented in Figure 3, SiO_2 gathered into centers, and MPP linked with them through the reaction between MAH and $-NH_2$.

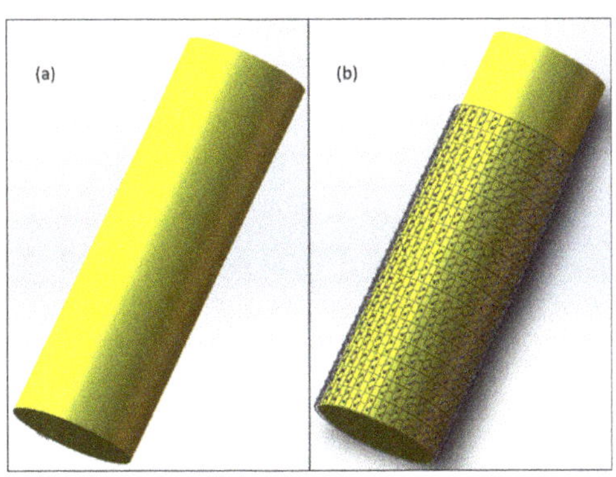

Figure 5. Ideal model of (**a**) PET fiber and (**b**) TMPP-modified PET fiber.

3.1.3. Thermal Stability of MPP and TMPP Powders

Figure 6 displays the thermogravimetry and differential thermogravimetry analysis (TG and DTG) curves of MPP and TMPP modifiers. Figure 6a exhibits that the initial decomposition temperatures of MPP and TMPP modifiers are 300 and 322 °C, respectively. Figure 6b shows that the DTG peak maxima are at 417 and 437 °C for MPP and TMPP modifiers, respectively. The improvement of thermal stability from MPP to TMPP is attributed to the inhibition of thermal decomposition of TMPP because of the introduction of SiO_2 granules. Furthermore, the DTG curve of TMPP shows two peaks, one is at about 417 °C and the other is at 437 °C, which indicates the reaction of only part of MPP molecules with SiO_2. Therefore, the MPP part in the TMPP modifier could be divided into the following two groups: MPP linked with SiO_2, and pure MPP. Besides, with the increase in the temperature to 800 °C, there is no residual of MPP modifier; however, the percentage of TMPP residual mass is 10 wt.% (Figure 6a), which is supposed to be the content of SiO_2 generated by TEOS hydrolysis.

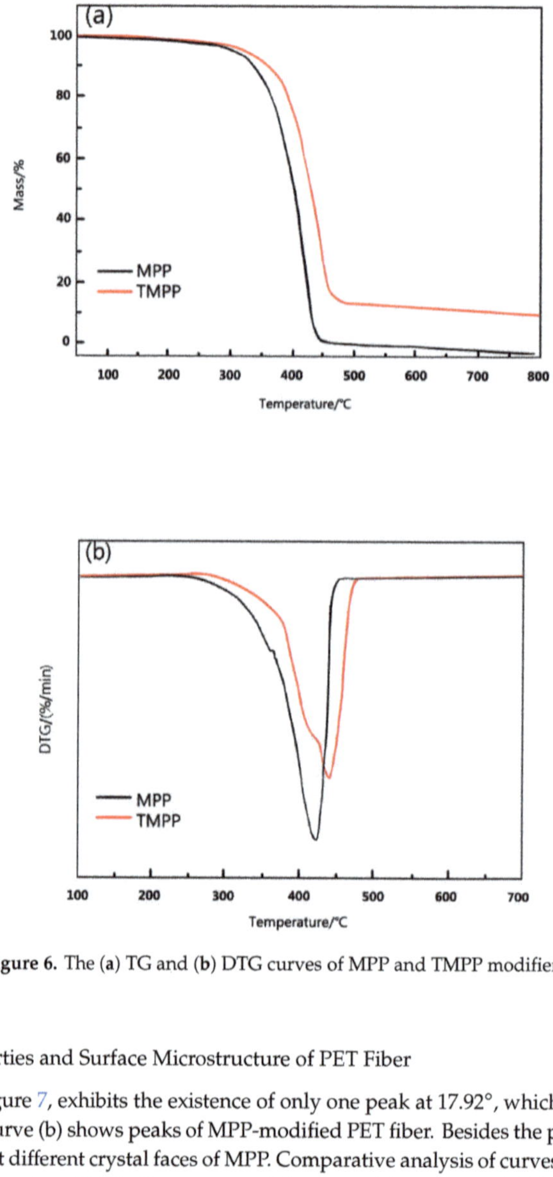

Figure 6. The (**a**) TG and (**b**) DTG curves of MPP and TMPP modifiers.

3.2. *Fiber*

3.2.1. Crystal Properties and Surface Microstructure of PET Fiber

Curve (a) in Figure 7, exhibits the existence of only one peak at 17.92°, which is the characteristic peak of PET fiber. Curve (b) shows peaks of MPP-modified PET fiber. Besides the peak 17.86°, the other three peaks represent different crystal faces of MPP. Comparative analysis of curves (b) and (c) indicates that in curve (c), the peak at 25.20° disappears and the area of the peak at 14.24° apparently increases. The characteristic peak of PET at 17.92° is divided into two peaks. Table 1 lists the characteristic diffraction angles obtained from Figure 7, and the calculation results of Bragg spacing and crystallite size. Table 1 summarizes that the crystallite size of PET decreases from 53.63 Å to 53.28 Å and 52.88 Å when it is modified with MPP and TMPP, respectively. TMPP-modified PET fiber exhibited the smallest

PET crystal size among all the samples. The results of inter-planar spacing were opposite to those of crystal size, and inter-planar spacing increased after the modification. The results indicate that after alkali treatment, PET chains in crystalline regions are partly broken and form smaller microcrystals [21]. TMPP-modified PET fiber exhibited the smallest crystal size among these samples because effective coating of TMPP on PET fiber promoted the interactions among functional groups present on the surface of PET fiber and TMPP. Peaks at 14.24°, 22.12° and 25.20° are characteristic diffraction peaks of MPP on the surface of the fiber. The extra peak at 18.69° of TMPP-modified PET fiber corresponds to PET linked with SiO_2 through KH550 [22]. Shifting of the peak of MPP from 22.12° to 21.98° indicates the reaction between MPP and SiO_2. The reactions between KH550 and TEOS on the surface of TMPP-modified PET fiber lead to the generation of a mesh structure consisting of MPP and Si–O–. Then, the chain segment movement of MPP is restricted, which results in the disappearance of the peak at 25.20°.

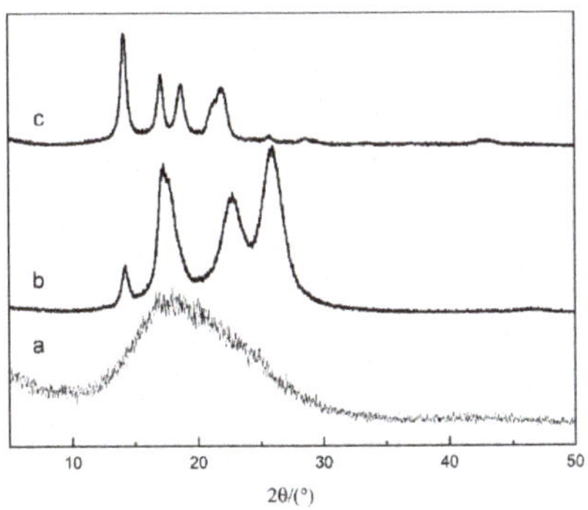

Figure 7. Curves of (**a**) PET fiber, (**b**) MPP-modified PET fiber and (**c**) TMPP-modified PET fiber.

Table 1. Diffraction angles (2θ), inter-planar spacing (d) and crystallite size (t) of PET fiber, and modified PET fiber.

Sample Code	Peak NO.	2θ(°)	d = λ/(2sinθ) (Å)	t = 0.9λ/Δωcosθ (Å)
PET fiber	1	17.92	4.95	53.63
MPP-modified PET fiber	1	14.24	6.22	54.26
	2	17.86	4.97	53.28
	3	22.12	4.02	51.66
	4	25.20	3.53	62.77
TMPP-modified PET fiber	1	14.24	6.22	54.35
	2	17.18	5.16	52.88
	3	18.69	4.75	52.16
	4	21.98	4.04	51.24

These results also show that KH550 and TEOS react and are woven into an armor-like structure on the surface of the PET fiber. They are linked with the Si–O–bond, and the $-NH_2$ groups are left and centered on Si–O–circles. The result of FTIR spectroscopy clearly indicated the reaction of MAH of MPP with $-NH_2$, which then provides the given space in the armor where MPP is located. Reunification of MPP is prevented and a homogeneous MPP layer is placed on the surface of the PET fiber.

3.2.2. Microstructure of MPP- and TMPP-modified PET Fiber

Figure 8 shows the surface morphologies of untreated and modified PET fiber. A columned and smooth fiber is observed in untreated PET fiber. After being dipped into MPP, the smooth surface was partly made up for by severely agglomerated MPP molecules (Figure 8b). Compared to MPP-modified

PET fiber, the surface of TMPP-modified PET fiber is occupied with an integrated compact membrane with embedded and semi-embedded particles (Figure 8c,d), so that the surface roughness and the area of the covered region increased. These results are attributed to the formation of an integrated compact membrane during solvent volatilization and TEOS hydrolysis because of the capillary force when PET fiber soaked with TMPP was exposed to air [23]. The membrane was shaped into a strong cover by SiO_2 from TEOS and molecular chains of MPP. The TMPP modifier cover restrained the reunion of MPP, and then released more functional groups of MPP than the single-MPP modifier-modified layer. As a result, the TMPP modifier cover was more perfect and stronger than the MPP modifier layer. Moreover, the organic–inorganic 'armor' on the PET fiber led to the dramatic increase in the roughness of the PET fiber. The specific surface area of the fiber was enlarged, leading to the generation of the molecular tangles between the fiber and the matrix [24]. All these changes led to the increase in the interfacial bonding strength between the fiber and the matrix and were found to be beneficial for the ultimate performance of PET fiber-reinforced composites.

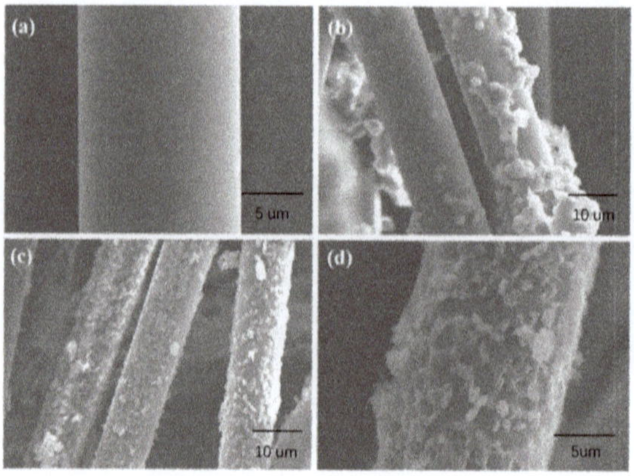

Figure 8. Microstructure of (**a**) untreated PET fiber, (**b**) MPP-modified PET fiber, and (**c**) and (**d**) TMPP-modified PET fiber.

3.2.3. Tensile Test for Single Fiber

Tensile strength of single fiber is an important index to measure the mechanical properties of fiber. Figure 9 shows that the values of tensile strength of untreated PET fiber, MPP-modified PET fiber, and TMPP-modified PET fiber are 1.25, 1.37 and 1.41 GPa, respectively. MPP and TMPP modifiers led to the increase in the tensile strength of PET fiber by 10.4% and 14.4%, respectively. It is well known that PET fiber is composed of highly oriented PET molecular chains. However, numerous terminal hydroxyl groups were generated from the breaking of PET molecular chains on the fiber surface after alkali treatment. The terminal hydroxyls reacted with the epoxy groups from MPP, and MPP was linked to the surface of PET fiber via –C–O–linkages. Therefore, MPP could apparently increase tensile strength of single fiber. Besides the effect of the single-MPP modifier, the TMPP modifier could form a special structure with the MPP-SiO_2 link, through TEOS, MPP, and KH550. The special structure covered the fiber like a crust, and then the coating was tighter and more homogeneous. The structure also made up the hollows on the fiber surface, optimized the surface structure, improved stress transfer ability, and finally, enhanced the strength and carrying capacity of the fiber [25]. TMPP modification could strengthen the PET fiber and improve the compatibility between the fiber and the matrix at the same time.

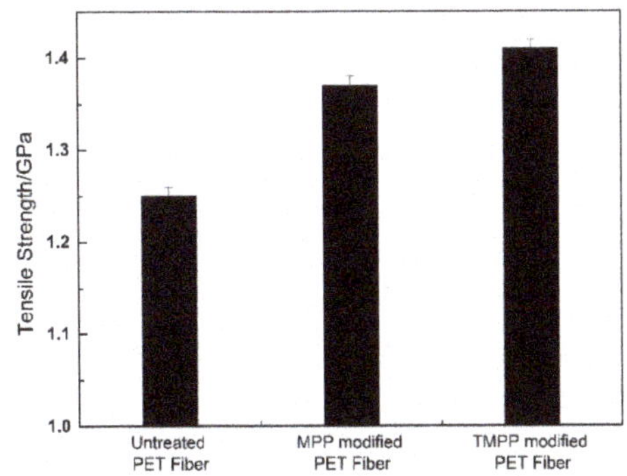

Figure 9. Tensile strength of single fiber.

3.3. Composites

3.3.1. Interfacial Properties

To verify the effect of MPP and TMPP modifiers on surface modification, interfacial properties of the composites were analyzed by using contact angle. Surface energy, interfacial tension and adhesive energy were calculated and are listed in Tables 2 and 3. Table 2 summarizes that the surface energies of PP, MPP, and TMPP modifier are 22.26, 15.59 and 22.75 mN m^{-1}, respectively. Clearly, the surface energy of PP was closer to that of the TMPP modifier than the MPP modifier, which indicates that the TMPP modifier has better compatibility with PP than with the MPP modifier. It is due to the fact that MPP is mostly evenly coated on the surface of the PET fiber when modified with the TMPP modifier, which leads to the surface polarity of the TMPP-modified PET fiber being more similar to that of PP than the MPP-modified PET fiber. Table 3 presents that the interfacial tension of MPP/PP is higher than that of TMPP/PP, and the adhesive energy of MPP/PP is lower than that of TMPP/PP. According to He's work [26], smaller interfacial tension causes tighter bonds between two phases. As a result, the addition of TEOS and KH550 in the TMPP modifier regulates the location and direction of MPP. In a word, the modification of the TMPP modifier was better than that of the MPP modifier.

Table 2. Surface energies between the MPP modifier, the TMPP modifier and the PP matrix.

Sample	Contact Angle θH_2O (°)	Contact Angle $\theta C_3H_8O_3$(°)	Surface Energy γ (mN/m)	Dispersion Component γ^d (mN/m)	Polar Component γ^p (mN/m)
MPP	107.49	97.99	15.59	6.50	9.09
TMPP	97.99	81.99	22.75	14.86	7.89
PP	97.50	85.74	22.26	13.49	8.77
H_2O	-	-	72.80	21.80	51.00
$C_3H_8O_3$	-	-	63.40	37.00	26.40

Table 3. Interfacial tensions and adhesive energies between the MPP modifier, the TMPP modifier and the PP matrix.

Sample	Interfacial Tension γ_{12} (mN/m)	Adhesive Energy W_{12} (mN/m)
MPP/PP	2.45	36.59
TMPP/PP	0.11	44.95

3.3.2. Interface Compatibility of Composite Materials

Figure 10 shows the DMA curves, exhibiting the dynamic properties of PET fiber/PP composites reinforced with different modified PET fibers. Compared to PP, the storage modulus of PET fiber/PP composites increased with the addition of PET fiber, as shown in Figure 10a. The cooperation of modifier and PET fiber leads to the obvious increase in the storage modulus of the PP composite. This is attributed to the fact that compared to untreated PET fiber, the MPP-modified PET fiber exhibits larger specific surface area and better compatibility with PP, which results in higher storage modulus. The TMPP-modified PET fiber leads to the further increase in the storage modulus of the PP composite, because of a more homogeneous modifier layer, larger specific surface area, and higher single-fiber tensile strength than the MPP-modified PET fiber. Figure 10b shows the loss modulus curves of PP and PP composites. Results indicate that the glass-transition temperature (T_g) of composites increased when filling with untreated PET fiber and MPP-modified ones. This is attributed to the fact that after modification, the interfacial compatibility of PET fiber and PP gets promoted, and the interaction force increases; thus, the energy required for releasing molecular chains also increases. As a result, the loss modulus and T_g of composites increase [27,28]. Moreover, compared to the MPP-modified PET fiber-reinforced PP composite, the T_g of the TMPP-modified PET fiber-reinforced PP composite decreases. When PET fiber is modified with the TMPP modifier, the addition of TEOS imports a few SiO_2 particles at the interface between fiber and the matrix, which works as a nucleator for PP. Crystallinity of PP increases for the SiO_2, and the amorphous phase decreases. As a result, in contrast with the PP composite filled with the MPP-modified fiber, the T_g of the PP composite filled with the TMPP-modified PET fiber decreases.

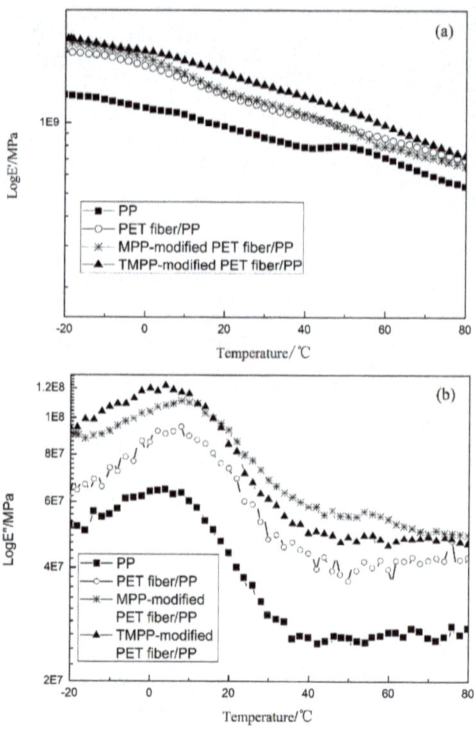

Figure 10. Dynamic properties of different PP composites (**a**): storage modulus, (**b**): loss modulus.

The bending properties of PP and PET fiber/PP composites are shown in Figure 11. Compared to PP, the bending strength and modulus of PET fiber-reinforced composites improved significantly. The properties of the TMPP-modified PET fiber-reinforced PP composite was the best among all the samples and increased by 10 and 800 MPa more than the PP composite filled with untreated PET fiber, with respect to bending strength and modulus respectively, which resulted from the fact that the coatings on PET fiber led to the improvement in the filler-matrix compatibility. In particular, the PET fiber is evenly coated with a layer of membrane in the TPMM-modified PET fiber, similar to a shell to provide the resistance to stress deformation.

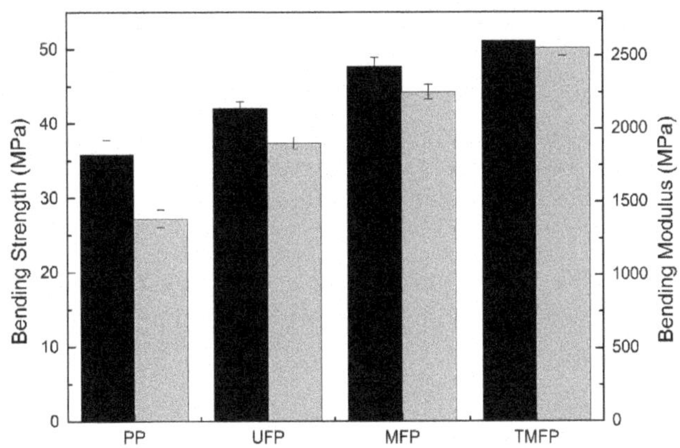

Figure 11. Bending properties of different PP composites ((**a**) PP, (**b**) PP composite filled with untreated PET fiber, (**c**) PP composite filled with the MPP-modified PET fiber, (**d**) PP composite filled with the TMPP-modified PET fiber).

Figure 12 exhibits SEM images, showing the bending sections of different PET fiber/PP composites. Figure 12a demonstrates that the untreated PET fiber apparently separates from the matrix. The PET fibers still remain smooth and unbroken when they are pulled out from the PP matrix. Different from the untreated PET fiber-filled PP, the number of pulled-out PET fiber decreases, as shown in Figure 12b, and the PET fiber is coated with a layer in MPP-modified PET fiber-filled PP. This is because the fiber is partly coated with non-polar groups in the MPP-modified PET fiber and the interfacial compatibility between the fiber and the PP matrix is improved. Figure 12c exhibits the cross-section of TMPP-modified PET fiber-filled PP composites, showing that the PET fiber is finely dispersed in the matrix and is inlaid into the matrix. Moreover, the fiber and the matrix were combined as a whole through a homogeneous amphiprotic TMPP membrane. In general, the TMPP modifier improved the interfacial compatibility of the composites and optimized the dispersion of fiber in the matrix. As a result, the PP composite filled with the TMPP-modified PET fiber presented the best bending properties.

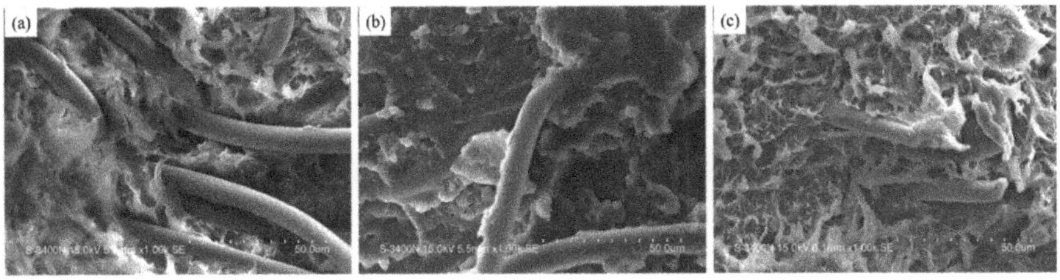

Figure 12. SEM images of PP composites filled with (**a**) untreated PET fiber, (**b**) MPP-modified PET fiber and (**c**) TMPP-modified PET fiber.

4. Conclusions

In summary, the TMPP modifier was firstly obtained in an anhydrous atmosphere, and then a uniform and homogeneous TMPP membrane was synthesized by sol–gel method on the surface of the PET fiber. Finally, a TMPP-modified PET fiber-reinforced PP composite was also produced. FTIR, XRD, and SEM analyses confirmed the formation of a membrane with chemical construction and micro-topography on the PET fiber surface after modification. The TMPP membrane was formed by amid bonds, which came from the reaction between KH550 and MPP. TGA showed the presence of SiO_2 particles and indicated that only partial MPP participated in the reaction with KH550 on the fiber surface. The TMPP membrane increased single-fiber strength by 14.4% and improved the bending strength and modulus of the PET fiber/PP composite by 21% and 34% respectively, compared to the untreated fiber-filled composite. The analysis of contact angles and DMA testified to the increase of compatibility between the TMPP-modified fiber and the PP matrix. The coating modification effectively and conveniently improved the interfacial properties between the fiber and the matrix as well as the mechanical properties of the PET fiber. This method is expected to be an important direction for further research on fiber surface modification and develops the applications of recycled PET fiber in composites.

Author Contributions: Conceptualization, C.W. and Y.M.; methodology, Y.M.; software, Y.M.; validation, C.W., and Q.L.; formal analysis, Q.L.; investigation, Y.M.; resources, C.W.; data curation, Y.M.; writing—original draft preparation, Y.M.; writing—review and editing, Q.L.; visualization, Q.L.; supervision, C.W.; project administration, Q.L.; funding acquisition, C.W.

Funding: This research received no external funding.

Conflicts of Interest: The authors declare no conflict of interest.

References

1. He, H.Z.; Yao, Z.; Wang, K.X.; Xi, H.; Zhou, H.Q.; Yao, Y.D. Modification of sisal fiber by in situ coating steam explosion and electromagnetic interference shielding effectiveness of sisal fiber/pp composites. *Polym. Compos.* **2014**, *35*, 1038–1043. [CrossRef]
2. Yousfi, M.; Soulestin, J.; Vergnes, B.; Lacrampe, M.F.; Krawczak, P. Morphology and mechanical properties of PET/PE blends compatibilized by nanoclays: Effect of thermal stability of nanofiller organic modifier. *J. Appl. Polym. Sci.* **2013**, *128*, 2766–2778. [CrossRef]
3. Zhang, Y.; Zhang, H.; Guo, W.; Wu, C. Effects of different types of polyethylene on the morphology and properties of recycled poly(ethylene terephthalate)/polyethylene compatibilized blends. *Polym. Adv. Technol.* **2011**, *22*, 1851–1858. [CrossRef]
4. Mamun Kabir, S.M.; Koh, J. Alkaline weight reduction and dyeing properties of black dope-dyed poly(ethylene terephthalate) microfibre fabrics. *Color. Technol.* **2017**, *133*, 209–217. [CrossRef]
5. Popescu, V.; Muresan, A.; Constandache, O.; Lisa, G.; Muresan, E.I.; Munteanu, C.; Sandu, I. Tinctorial response of recycled pet fibers to chemical modifications during saponification and aminolysis reactions. *Ind. Eng. Chem. Res.* **2014**, *53*, 16652–16663. [CrossRef]
6. Hwang, Y.J.; McCord, M.G.; Kang, B.C. Helium/Oxygen Atmospheric Pressure Plasma Treatment on Poly(ethylene terephthalate) and Poly(trimethylene terephthalate) Knitted Fabrics: Comparison of Low-stress Mechanical/Surface Chemical Properties. *Fibers Polym.* **2005**, *6*, 113–120. [CrossRef]
7. Cioffi, M.O.H.; Voorwald, H.J.C.; Ambrogi, V.; Monetta, T.; Bellucci, F.; Nicolais, L. Mechanical strength of PET fibers treated in cold plasma and thermal exposed. *J. Mater. Eng. Perform.* **2003**, *12*, 279–287. [CrossRef]
8. Fang, Y.; Liu, X.; Tao, X. Intumescent flame retardant and anti-dripping of PET fabrics through layer-by-layer assembly of chitosan and ammonium polyphosphate. *Prog. Org. Coatings* **2019**, *134*, 162–168. [CrossRef]
9. Munirah Abdullah, N.; Ahmad, I. Effect of Chemical Treatment on Mechanical and Water-Sorption Properties Coconut Fiber-Unsaturated Polyester from Recycled PET. *ISRN Mater. Sci.* **2012**, *2012*, 1–8. [CrossRef]
10. Kwak, W.G.; Oh, M.H.; Son, S.Y.; Gong, M.S. Silver loading on poly(ethylene terephthalate) fabrics using silver carbamate via thermal reduction. *Macromol. Res.* **2015**, *23*, 509–517. [CrossRef]

11. Kwak, W.G.; Oh, M.H.; Gong, M.S. Preparation of silver-coated cotton fabrics using silver carbamate via thermal reduction and their properties. *Carbohydr. Polym.* **2015**, *115*, 317–324. [CrossRef] [PubMed]
12. Kwak, W.G.; Cha, J.R.; Gong, M.S. Surface modification of polyester fibers by thermal reduction with silver carbamate complexes. *Fibers Polym.* **2016**, *17*, 1146–1153. [CrossRef]
13. Jiang, X.; Fan, Y.; Li, F. Preparation and properties of dynamically cured polypropylene (PP)/maleic anhydride-grafted polypropylene (MAH-g-PP)/calcium carbonate ($CaCO_3$)/epoxy composites. *J. Thermoplast. Compos. Mater.* **2013**, *26*, 1192–1205. [CrossRef]
14. Novak, B.M. Hybrid Nanocomposite Materials—between inorganic glasses and organic polymers. *Adv. Mater.* **1993**, *5*, 422–433. [CrossRef]
15. Wang, B.; Wilkes, G.L. Novel hybrid inorganic-organic abrasion-resistant coatings prepared by a sol-gel process. *J. Macromol. Sci. Part A* **1994**, *31*, 249–260. [CrossRef]
16. Dikobe, D.G.; Luyt, A.S. Investigation of the morphology and properties of the polypropylene/low-density polyethylene/wood powder and the maleic anhydride grafted polypropylene/low-density polyethylene/wood powder polymer blend composites. *J. Compos. Mater.* **2017**, *51*, 2045–2059. [CrossRef]
17. Chen, J.; Zhu, Y.; Ni, Q.; Fu, Y.; Fu, X. Surface modification and characterization of aramid fibers with hybrid coating. *Appl. Surf. Sci.* **2014**, *321*, 103–108. [CrossRef]
18. Jeon, H.T.; Jang, M.K.; Kim, B.K.; Kim, K.H. Synthesis and characterizations of waterborne polyurethane-silica hybrids using sol-gel process. *Colloids Surf. A* **2007**, *302*, 559–567. [CrossRef]
19. Bal, S.; Behera, R.C. Structural Investigation of Chemical Treated Polyester Fibers Using Saxs and Other Techniques. *J. Miner. Mater. Charact. Eng.* **2006**, *05*, 179–198. [CrossRef]
20. Jiang, H.L.; Kwon, J.T.; Kim, E.M.; Kim, Y.K.; Arote, R.; Jere, D.; Jeong, H.J.; Jang, M.K.; Nah, J.W.; Xu, C.X.; et al. Galactosylated poly(ethylene glycol)-chitosan-graft-polyethylenimine as a gene carrier for hepatocyte-targeting. *J. Control. Release* **2008**, *131*, 150–157. [CrossRef]
21. Ramesh, M.; Deepa, C.; Aswin, U.S.; Eashwar, H.; Mahadevan, B.; Murugan, D. Effect of Alkalization on Mechanical and Moisture Absorption Properties of Azadirachta indica (Neem Tree) Fiber Reinforced Green Composites. *Trans. Indian Inst. Met.* **2017**, *70*, 187–199. [CrossRef]
22. Han, H.; Li, H.Q.; Liu, M.; Xu, L.; Xu, J.; Wang, S.; Ni, H.; Wang, Z. Effect of "bridge" on the performance of organic-inorganic crosslinked hybrid proton exchange membranes via KH550. *J. Power Sources* **2017**, *340*, 126–138. [CrossRef]
23. Yang, C.H.; Pan, Y.W.; Guo, J.J.; Young, T.H.; Chiu, W.Y.; Hsieh, K.H. Synthesis and application of polyurethane basic organic-inorganic hybrid materials as highly hydrophobic coatings. *J. Polym. Res.* **2016**, *23*, 1–7. [CrossRef]
24. De Lange, P.J.; Mäder, E.; Mai, K.; Young, R.J.; Ahmad, I. Characterization and micromechanical testing of the interphase of aramid-reinforced epoxy composites. *Compos. Part A Appl. Sci. Manuf.* **2001**, *32*, 331–342. [CrossRef]
25. Li, G.; Zhang, C.; Wang, Y.; Li, P.; Yu, Y.; Jia, X.; Liu, H.; Yang, X.; Xue, Z.; Ryu, S. Interface correlation and toughness matching of phosphoric acid functionalized Kevlar fiber and epoxy matrix for filament winding composites. *Compos. Sci. Technol.* **2008**, *68*, 3208–3214. [CrossRef]
26. He, X.; Shi, J.; Wu, L.; Jiang, P.; Liu, B. Study on synergistic toughening of polypropylene with high-density polyethylene and elastomer-olefin block copolymers under ultrasonic application. *Compos. Sci. Technol.* **2018**, *161*, 115–123. [CrossRef]
27. Subasinghe, A.D.L.; Das, R.; Bhattacharyya, D. Fiber dispersion during compounding/injection molding of PP/kenaf composites: Flammability and mechanical properties. *Mater. Des.* **2015**, *86*, 500–507. [CrossRef]
28. Zhang, X.; Liu, Y.; Gao, J.; Huang, F.; Song, Z.; Wei, G.; Qiao, J. Crystallization behavior of nylon-6 confined among ultra-fine full-vulcanized rubber particles. *Polymer* **2004**, *45*, 6959–6965. [CrossRef]

polymers

MDPI

Article

Investigation on the Fiber Orientation Distributions and Their Influence on the Mechanical Property of the Co-Injection Molding Products

Chao-Tsai Huang *, Xuan-Wei Chen and Wei-Wen Fu

Department of Chemical and Materials Engineering, Tamkang University, No. 151, Yingzhuan Rd., Tamsui Dist., New Taipei City 25137, Taiwan; willych1014@gmail.com (X.-W.C.); fu840501@gmail.com (W.-W.F.)
* Correspondence: cthuang@mail.tku.edu.tw or cthuang@moldex3d.com

Received: 10 November 2019; Accepted: 17 December 2019; Published: 20 December 2019

Abstract: In recent years, due to the rapid development of industrial lightweight technology, composite materials based on fiber reinforced plastics (FRP) have been widely used in the industry. However, the environmental impact of the FRPs is higher each year. To overcome this impact, co-injection molding could be one of the good solutions. But how to make the suitable control on the skin/core ratio and how to manage the glass fiber orientation features are still significant challenges. In this study, we have applied both computer-aided engineering (CAE) simulation and experimental methods to investigate the fiber feature in a co-injection system. Specifically, the fiber orientation distributions and their influence on the tensile properties for the single-shot and co-injection molding have been discovered. Results show that based on the 60:40 of skin/core ratio and same materials, the tensile properties of the co-injection system, including tensile stress and modulus, are a little weaker than that of the single-shot system. This is due to the overall fiber orientation tensor at flow direction (A_{11}) of the co-injection system being lower than that of the single-shot system. Moreover, to discover and verify the influence of the fiber orientation features, the fiber orientation distributions (FOD) of both the co-injection and single-shot systems have been observed using micro-computerized tomography (μ-CT) technology to scan the internal structures. The scanned images were further utilizing Avizo software to perform image analyses to rebuild the fiber structure. Specifically, the fiber orientation tensor at flow direction (A_{11}) of the co-injection system is about 89% of that of the single-shot system in the testing conditions. This is because the co-injection part has lower tensile properties. Furthermore, the difference of the fiber orientation tensor at flow direction (A_{11}) between the co-injection and the single-shot systems is further verified based on the fiber morphology of the μ-CT scanned image. The observed result is consistent with that of the FOD estimation using μ-CT scan plus image analysis.

Keywords: co-injection molding; fiber reinforced plastics (FRP); fiber orientation distribution (FOD); micro-computerized tomography (μ-CT) scan technology

1. Introduction

Due to its excellent properties, fiber-reinforced plastics (FRP) material has been applied in industry for years, especially as one of the major lightweight technologies for automotive or aerospace products [1,2]. Specifically, the market of composite products is expected to achieve an estimated $40.2 billion by 2024. The compound annual rate (CAGR) from 2019 to 2024 is about 3.3% [3,4]. Moreover, according to the Composite Industry Market report in 2018, around 1.141 million tons of glass-FRP composite materials were produced in Europe [5]. In the United States, the demand for FRP was 4.3 billion pounds in 2017 [3]. To handle the recycling of FRP composites, there are three primary methods, including mechanical, thermal, and chemical processes [3,6]. However, since the presence of the fiber makes the microstructures inside FRP more complicated than that of general thermoplastics, it causes the recycling of FRP to be very difficult. As higher amounts of FRP are consumed, higher environmental impacts are needed to be addressed now and in future [7]. To overcome this impact, co-injection molding, based on the mechanical recycling process, could be a good solution.

Co-injection molding is commonly used as a daily accessory, and in many other contexts. Basically, the co-injection technology can provide several advantages with integrating materials to reduce cost. This technology also allows for the reuse of materials, offers an upgrade in production efficiency, and can make raw skin/recycled core structures. However, there are some challenges for co-injection processes. For example, how to visualize and control suitable skin/core material distribution is very difficult during the processing. To deal with this complicated process, there is some literature that can be used as guidelines. Seldén [8] monitored different process conditions and found out the skin-to-core material ratio is the main factor causing break-through problems. Moreover, the correlation between internal material distributions, process condition, and material property are discussed in many previous studies [9–11]. The cavity-filling ratio of skin material determines the break-through location. Material viscosity and filling rate affect uniformity of core material distribution. Furthermore, since the geometrical structure of real products is very complicated, the progress of the co-injection processing becomes much more of a challenge. To simulate this situation, Yang and Yokoi [12] proposed a co-injection with a multi-cavity molding system with a fork structure. They found that the core flow pattern in the fork structure is strongly affected by injection flow rate. It is also affected by material property. However, their work is not comprehensive yet because they didn't discuss the flow behavior at the end of filling for the multi-cavity system. Later, Huang [13] and Huang et al. [14] verified that in multi-cavity co-injection systems, the skin-to-core material ratio is still the main factor to dominant the break-through phenomena, while the injection flow rate can be used to adjust the core penetration uniformity. Indeed, the core penetration prediction in the real co-injection system is still very challenging.

On the other hand, as mentioned earlier, due to its excellent properties, the fiber-reinforced plastic (FRP) material has been applied in many industries. The reason that FRP has so much potential is because of the functions of fibers' microstructures, including their orientation, length, and concentration. It is expected that those microstructures will further affect the final shrinkage and warpage of the single-shot and co-injected parts. Moreover, how to manage the glass fiber orientation features and their influence in the recycling of the FRP products are still very complicated. To realize how fiber structures would influence the mechanical properties of the finished parts, Thomason et al. [15,16] studied the phenomenon experimentally and proposed some empirical standards of fiber length to guarantee enough impact strength for automobiles. Moreover, some researchers have studied the important fiber microstructure variables, including fiber orientation, fiber length, and fiber concentration, to determine the level of effectiveness of mechanical property enhancement [17–19]. Cilleruelo et al. further considered the effect of carbon black and nucleating agents on impact properties experimentally [20]. Indeed, the microstructures of fibers were still very difficult to validate in this work. To understand why the fiber orientation is changed, Folgar and Tucker [21], Advani and Tucker [22], and Advani [23] proposed numerical models to predict fiber orientation in short fiber. Using their short fiber model, the fiber orientation distribution (FOD) could be predicted reasonably. As the demand on the impact property increases, the longer fiber length retained in the final injected part is in greater demand. In recent years, some researchers have extended the numerical prediction capability and experimental validation to the long fiber orientation [24–27]. However, no matter how short or long the fibers, to the best of our knowledge, very few researchers have discovered the connection between the fiber microstructures to the mechanical properties of the final injection parts quantitatively. Moreover, knowing how to catch these fiber microstructures in the final parts is still not easy in reality. In general, there are two methods which have been utilized. One is optical section method, and the other is micro-computerized tomography (μ-CT) scan method [28]. Some researchers [28–30] have applied the micro-computerized tomography (μ-CT) scan method to get through finished parts. The μ-CT is a relatively new method to measure the fiber orientation distribution in FRP parts. It is a non-destructive testing method, but the procedure was still not easy to handle. This is especially because there are a huge number of images created after using μ-CT technology, and further image analysis is another key issue to deal with. Overall, this is to say; the ways in which fiber orientation provides the reinforced features in co-injection molding is still not fully understood in reality. In addition, the exact working function of the fibers is not easily visualized and managed.

In this study, we have applied both CAE simulation and experimental methods to investigate the interface (between skin and core) penetration behavior. Furthermore, the connection between the inside FOD variation and the tensile properties of the single-shot and co-injection molding based on the standard tensile bar (ASTM D638 TYPE V) system has been studied. To give better understanding, the main content of this paper is organized as follows. The theoretical background is presented in Section 2. Section 3 describes the model and related information. It will discuss the simulation model and experimental equipment separately. Then, the results and discussion are in Section 4. Finally, the brief conclusion will be addressed in Section 5.

2. Theoretical Background and Numerical Method

2.1. Model for Co-Injection Molding

The numerical simulation was conducted using the Moldex3D R16® software. Both the skin and core materials are considered to be compressible, generalized Newtonian fluid. Surface tension at the melt front is neglected. The governing equations for 3D transient non-isothermal motion are [13]:

$$\frac{\partial \rho}{\partial t} + \nabla \cdot \rho \mathbf{u} = 0 \tag{1}$$

$$\frac{\partial}{\partial t}(\rho \mathbf{u}) + \nabla \cdot (\rho \mathbf{u}\mathbf{u} + \boldsymbol{\tau}) = -\nabla p + \rho \mathbf{g} \tag{2}$$

$$\rho C_P \left(\frac{\partial T}{\partial t} + \mathbf{u} \cdot \nabla T \right) = \nabla \cdot (k \nabla T) + \eta \dot{\gamma}^2 \tag{3}$$

where ρ is density; $\mathbf{u}$ is velocity vector; t is time; $\boldsymbol{\tau}$ is total stress tensor; $\mathbf{u}$ is acceleration vector of gravity; p is pressure; η is viscosity; C_p is specific heat; T is temperature; k is thermal conductivity; $\dot{\gamma}$ is shear rate. For the polymer melt, the stress tensor can be expressed as:

$$\boldsymbol{\tau} = -\eta \left(\nabla \mathbf{u} + \nabla \mathbf{u}^T \right) \tag{4}$$

The modified-Cross model with Arrhenius temperature dependence is employed to describe the viscosity of polymer melt:

$$\eta\left(T, \dot{\gamma}\right) = \frac{\eta_o(T)}{1 + \left(\eta_o \dot{\gamma} / \tau^*\right)^{1-n}} \tag{5}$$

with

$$\eta_o(T) = BExp\left(\frac{T_b}{T}\right) \tag{6}$$

where n is the power law index, η_o is the zero shear viscosity, τ^* is the parameter that describes the transition region between zero shear rate, and the power law region of the viscosity curve.

A volume fraction function, fi, is introduced to specify the evolution of the polymer/air front (i = 1) and skin/core front (i = 2) interfaces. Here, $fi = 0$ is defined as the no-filled region, $f = 1$ as the fully-filled region, and finally the interfacial front is located within cells of an f value between 0 and 1. The advancement of f over time is governed by the following transport equation:

$$\frac{\partial f_i}{\partial t} + \nabla \cdot (\mathbf{u} f_i) = 0 \tag{7}$$

During the polymer melt filling phase, the velocity and temperature are specified at the mold inlet. While the core material is injected, the flow rate setting is specified at the mold inlet. On the mold wall, the non-slip boundary condition is applied, and fixed mold wall temperature is assumed.

2.2. Model for Fiber Orientation Distribution

The fiber orientation is described as follows. A single fiber is regarded as an axisymmetric bond with rigidness. The bond's orientation unit vector **p** along its axis direction can described as the fiber orientation. Orientation state of a group of fibers is given by second moment tensor,

$$\mathbf{A} = \oint \psi(\boldsymbol{p})\boldsymbol{p}\boldsymbol{p}\, d\boldsymbol{p} \tag{8}$$

where $\psi(\boldsymbol{p})$ is the probability density distribution function over orientation space; $\boldsymbol{p}$ is the definition of the orientation vector, as shown in Figure 1. Tensor $\mathbf{A}_4$ is a fourth order orientation tensor, defined as:

$$\mathbf{A}_4 = \oint \psi(\boldsymbol{p})\boldsymbol{p}\boldsymbol{p}\boldsymbol{p}\boldsymbol{p}\, d\boldsymbol{p} \tag{9}$$

where this tensor is also symmetric. The acceptable calculation is obtained through the eigenvalue-based optimal fitting approximation of the orthotropic closure family. To handle this complicated tensor system, Tseng et al. [25,26] developed a new fiber orientation model to couple with Jeffery's hydrodynamic (HD) model, namely, the iARD-RPR model (known as Improved Anisotropic Rotary Diffusion model combined with Retarding Principal Rate model),

$$\dot{\mathbf{A}} = \dot{\mathbf{A}}_{\mathrm{HD}} + \dot{\mathbf{A}}_{\mathrm{iARD}}(C_I, C_M) + \dot{\mathbf{A}}_{\mathrm{RPR}}(\alpha) \tag{10}$$

where $\dot{\mathbf{A}}$ represents the material derivative of A. Parameters C_I and C_M describe the fiber–fiber interaction and fiber–matrix interaction, while parameter α can slow down a response of fiber orientation. Details of the RPR model and the iARD model are available elsewhere [21,22].

$$\dot{\mathbf{A}}_{\mathrm{HD}} = (\mathbf{W}\cdot\mathbf{A} - \mathbf{A}\cdot\mathbf{W}) + \xi(\mathbf{D}\cdot\mathbf{A} + \mathbf{A}\cdot\mathbf{D} - 2\mathbf{A}_4 : \mathbf{D}) \tag{11}$$

where **W** and **D** are the vorticity tensor and rate-of-deformation tensor, respectively. ξ is a shape factor of a particle.

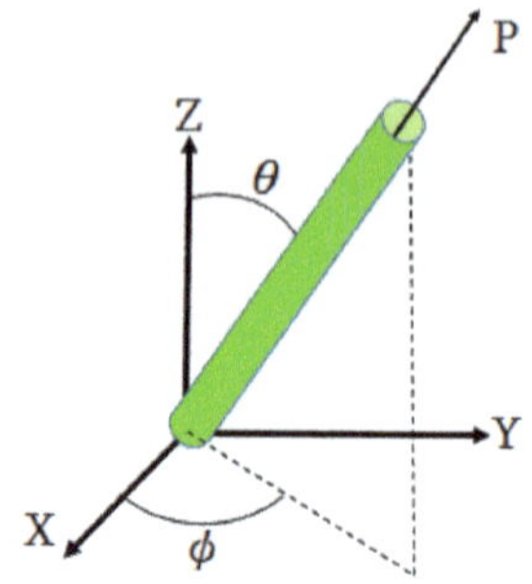

Figure 1. Definition of the orientation vector **p**.

3. Geometrical Model and Related Information

3.1. Simulation Model and Related Information

The geometry model and dimensions of the runner and cavity are shown as in Figure 2. Specifically, it is based on ASTM D638 Type V standard specimen with dimension of 63.5 mm × 9.53 mm × 5.3 mm (see Figure 2a). The moldbase and cooling channel layout is exhibited in Figure 2b. The meshed model is shown in Figure 2c. The associated mesh type and element count are listed in Table 1. In addition, the numerical convergence testing regarding the mesh resolution is shown in Figure 3. Clearly, when it is kept with 20-layer or higher in thickness direction, the sprue pressure history curve is almost unchanged. Hence, in this study, the major mesh type is hexahedron. The selected mesh model is Mesh 4 which has 20-layer in thickness direction and 252,720 total element counts. Moreover, the process conditions for the co-injection process and the counterpart single-shot injection are listed in Tables 2 and 3. Briefly, the filling time is 0.3 s; melt temperature is 230 °C; mold temperature 35 °C; skin-to-core switch over is at 60% by volume. The material used is PP Globalene SF7351 which has 30% short fiber content. Moreover, in order to conduct the interface (between skin and core) penetration behavior and also study the fiber orientation variation dynamically, some measuring nodes have been specified, as shown in Figure 4. Specifically, there are three measuring nodes named A, B, and C. During the co-injection process, when the interface between skin and core arrives at point A, it is specified as time t1 (see Figure 4b). Similarly, when the interface arrives at point B, and then point C, it is specified as t2 and t3, respectively, as displayed in Figure 4c,d.

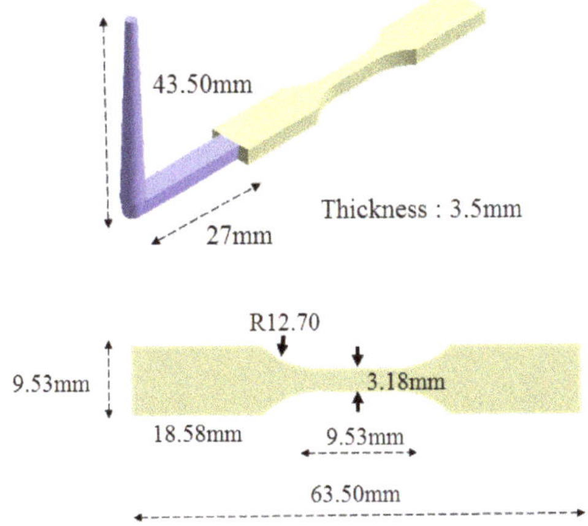

(**a**)

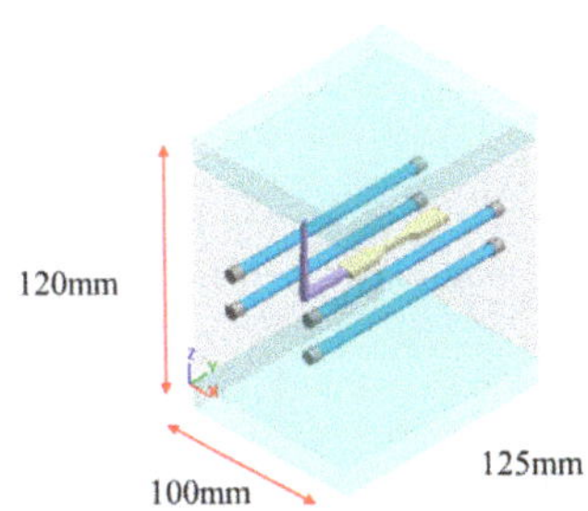

(**b**)

Figure 2. *Cont.*

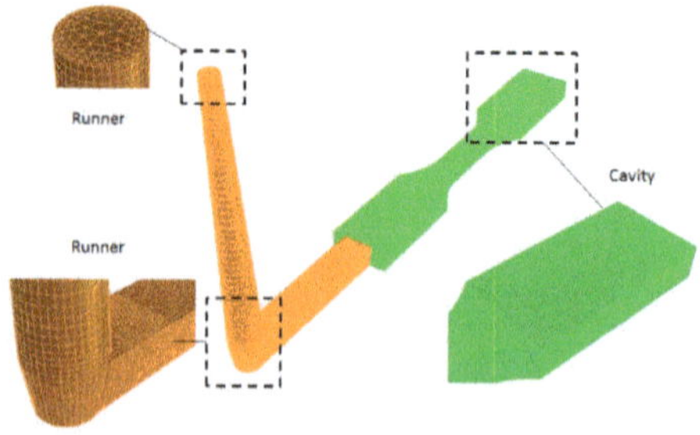

(c)

Figure 2. (**a**) Geometry model and dimensions, (**b**) moldbase and cooling channel layout, (**c**) the meshed model.

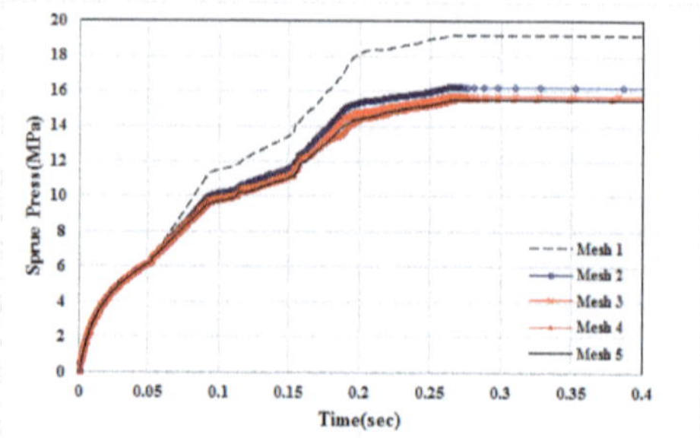

Figure 3. The numerical convergence testing for mesh type and resolution.

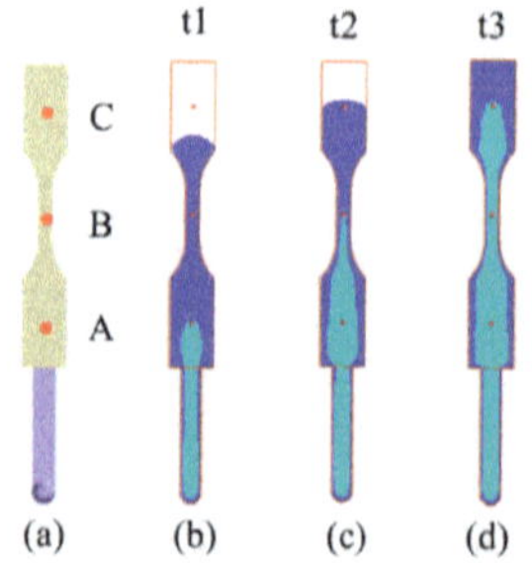

Figure 4. The locations of measuring modes: (**a**) Locations of three measuring nodes, (**b**) t1: The core interface arrives at point A, (**c**) t2: The core interface arrives at point B, (**d**) t3: The core interface arrives at point C.

Table 1. The mesh types and related information.

	Mesh 1	Mesh 2	Mesh 3	Mesh 4	Mesh 5
mesh type (runner)			tetrahedron		
mesh type (cavity)			hexahedron		
layers in thickness	5	10	15	20	25
mesh size (mm)			0.2		
cavity count	51,520	103,040	154,560	206,080	257,600
runner count	22,940	40,560	49,340	46,640	66,900
total element count	74,460	143,600	203,900	252,720	324,500
analysis time (h)	0.25	0.45	0.61	0.75	1

Table 2. Process condition for co-injection.

Material	Skin: PP Globalene SF7351; Core: PP Globalene SF7351
Filling time (s)	0.3
Packing time (s)	—
Flow rate (cm^3/s)	10
Melt temperature (°C)	230
Mold temperature (°C)	35
Injection pressure (MPa)	175
Core switch over (by volume filled) (%)	60

Table 3. Process condition for single shot injection.

Material	PP Globalene SF7351
Filling time (s)	0.3
Packing time (s)	3
Flow rate (cm^3/s)	10
Melt temperature (°C)	230
Mold temperature (°C)	35
Injection pressure (MPa)	175

3.2. Experimental Model and Related Information

In order to realize what happens during the co-injection molding physically, the real co-injection system and the mold were constructed, as shown in Figure 5. The machine model is TA-4.0ST-2ST-80T made by Ta Ai Machinery Co. Ltd. from Taiwan (Figure 5a). The cavity with the same dimension as that of the simulation is listed in Figure 5b. Moreover, to observe the real fiber orientation behavior, micro-computerized tomography (μ-CT) technology was performed using Bruker Skyscan 2211 with 100 kV, 8.3 W, and a resolution of 7 μm, supported by MCL Multiscale X-ray CT laboratory, Industrial Technology Research Institute, Taiwan. Moreover, to realize how the mechanical properties have been affected due to the different processing, the tensile test was performed. The universal tensile testing machine of LS1 high precision testing machine model supplied by Lloyd company was used as shown in Figure 6. The testing procedures are based on [31]. The definition of the dimension parameter and the specific amount are listed in Figure 7 and the Table 4. During each testing, at each time period, the deformation and the associated force were recorded. Then, the deformation and associated force were transferred into stress and strain, based on Equations (12) and (13), described below. The stress and strain obtained in Equations (12) and (13) can be used to make the stress–strain curve. The tensile stress (σ) is obtained by dividing the force (F) by the initial cross-section area of the narrow portion (A).

$$\sigma = \frac{F}{A} = \frac{F}{(W_1 \times H)} \tag{12}$$

The elongation (ε) is obtained by dividing the deformation (ΔL) by the length of narrow portion and stated as a percentage.

$$\varepsilon = \frac{\Delta L}{L_1} \times 100\% \tag{13}$$

The tensile modulus (E) is calculated from the secant between 0.05% and 0.25% strain of the averaged stress–strain curve.

$$\mathrm{E} = \frac{\sigma_{0.25\%} - \sigma_{0.05\%}}{\varepsilon_{0.25\%} - \varepsilon_{0.05\%}} \tag{14}$$

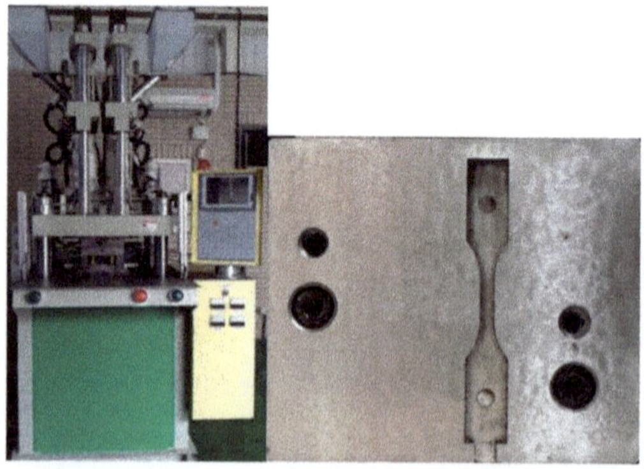

(**a**) (**b**)

Figure 5. (**a**) Co-injection molding system, (**b**) the cavity.

(**a**) (**b**)

Figure 6. (**a**) The universal tensile testing machine, (**b**) the sample holder.

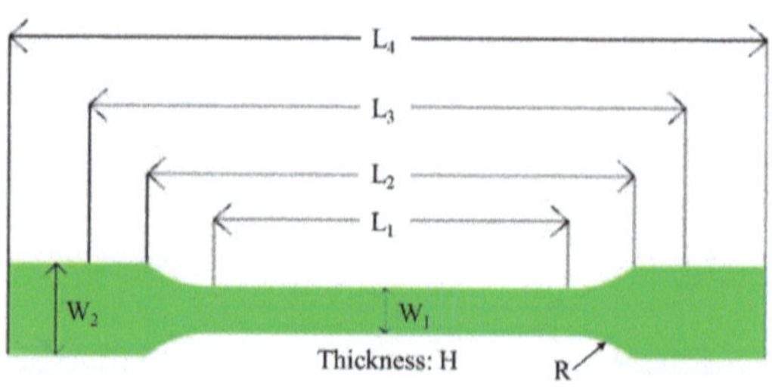

Figure 7. The parameter definition for the tensile test.

Table 4. Specimen Specification.

Symbol	Definition	mm
L_1	Length of narrow portion	
L_2	Distance between broad parallel portions	
L_3	Initial clamping length	
L_4	Overall length	
W_1	Width of narrow portion	
W_2	Width at ends	
H	Thickness	
R	Radius	

4. Results and Discussion

4.1. Skin/Core Ratio Effect

Figure 8 shows the numerical simulation of the movement of the core interface (between skin and core) at various skin/core ratio settings. When the ratio is from 90:10 to 60:40, the higher the ratio, the longer core penetration could be obtained. However, when the ratio is 50:50, the skin break-through phenomenon happened close to the end of the cavity. Furthermore, when the skin/core ratio is increased from 50:50 to 10:90, the skin break-through area was extended. Alongside this, the final core penetration location was be moved from the end of the cavity to the beginning of the cavity, as far as to the runner. This clearly shows how the break-through location is very sensitive to the skin/core ratio in co-injection molding. In addition, in some applications (for example, in plastic recycling), it is expected to that there is more core material covering the inside of the co-injection product to prevent break-through happening. However, to avoid the break-through phenomena happen, a skin/core ratio of 60:40 will be a good choice for our future studies.

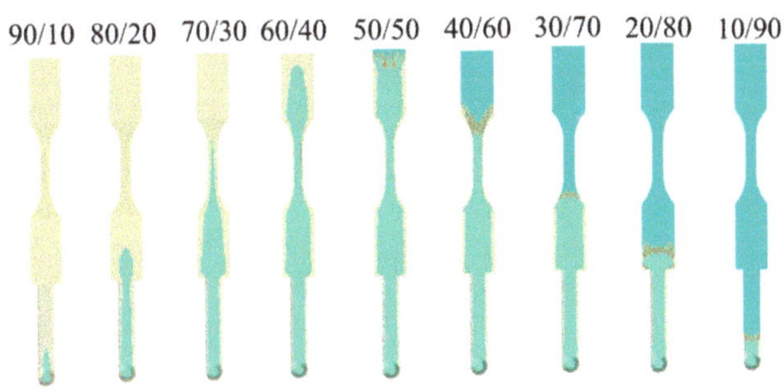

Figure 8. Core penetration behavior at various skin/core ratios, from 90:10 to 10:90.

4.2. Fiber Orientation Distribution (FOD) Prediction

Moreover, to study the fiber effect in FRP products, the fiber orientation distribution (FOD) inside the cavity during single-shot molding has been investigated. For comparison purposes, as the core interface touched point A (time t1), we selected the same melt front time for both single-shot and co-injection systems, as shown in Figure 9a. Meanwhile, Figure 9b presents the fiber orientation tensor components A_{11}, A_{22}, and A_{33} for the single-shot molded specimen. Specifically, from the top surface through the frozen layer to the core layer at the central line of the specimen, the fiber orientation tensor at flow direction (A_{11}) will be slightly increased from 0.75 to 0.85, and then decreased to 0.25. At the same time, the fiber orientation tensor in the cross-flow direction (A_{22}) will be slightly decreased from 0.20 to 0.15, and then increased to 0.5. Furthermore, Figure 9c shows the fiber orientation tensor (A_{11}) at point A through different time periods (from t1 to t3). In Figure 9c, the fiber orientation tensor (A_{11}) is almost the same from t1 to t3 in the single-shot system. It can be inferred that the fiber orientation is strongly affected by the frozen layer and the adjacent shear layer during the conventional injection molding. As long as the frozen layer forms, it influences the shear layer to build a strong A_{11} fiber orientation tensor. Meanwhile, since the FOD calculation in co-injection system is not developed successfully yet, that FOD of the co-injection system will be discovered by using μ-CT plus image analysis technology. The details will be addressed in a later section.

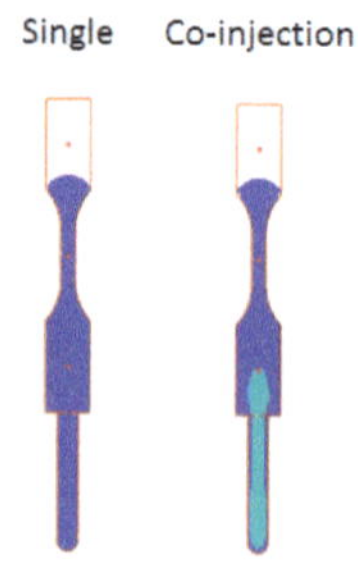

(a)

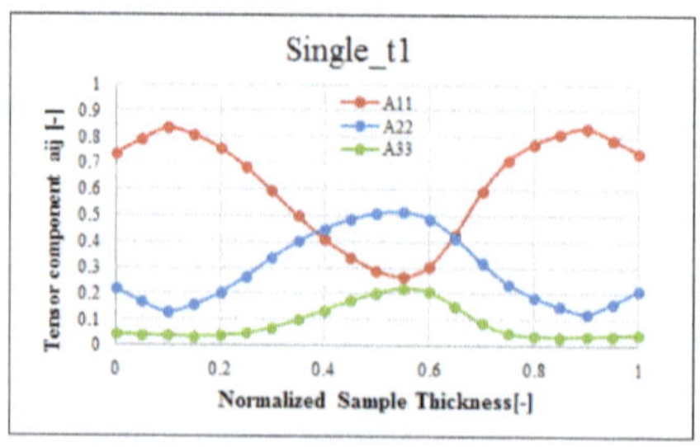

(b)

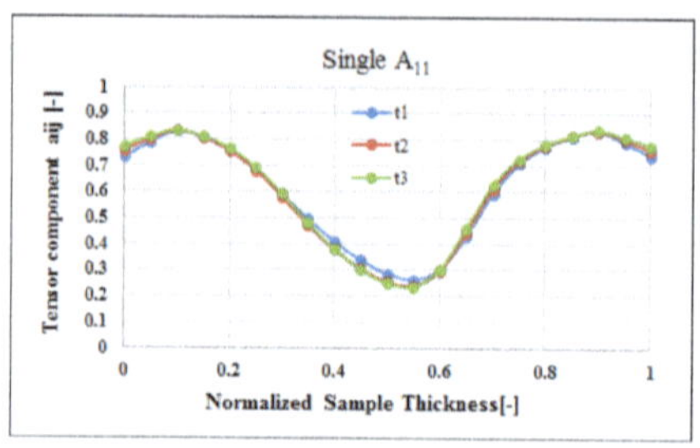

(c)

Figure 9. Fiber orientation features of the single-shot: (**a**) The time period as the flow front is the same for both single and co-injection, (**b**) the FOD for a single shot at point A at time t1, (**c**) the flow direction orientation tensor A_{11} at point A with different time period.

4.3. *Experimental Investigation and Validation*

4.3.1. Short Shot Validation

Figure 10 presents the short shot testing at the skin/core ratio of 60:40 for both the simulation and the experiment. Based on the careful evaluation from 46% to 100% of filling, it is clear that the flow front and interface penetration for the simulation prediction is quite closely matched with the experimental observation.

4.3.2. Break-Through Study and Validation

Moreover, Figure 11 shows the experimental validation of the skin/core ratio effect, specifically for the evaluation of the break-through locations. Basically, from a skin/core ratio of 90:10 to 60:40, no break-through phenomena happened. However, when the skin/core ratio is from 50:50 to 10:90; the more core material injected, the larger the penetration area that is observed. Figure 11b shows the break-through behavior for the skin/core ratio of 40:60. The results show that both simulation and experiment are in close agreement. In addition, the final core penetration location will be moved from the end of the cavity to the beginning of the cavity, even to the runner, when the core material ratio is increased. Overall, this is to say that the observation is matched with the simulation prediction very well.

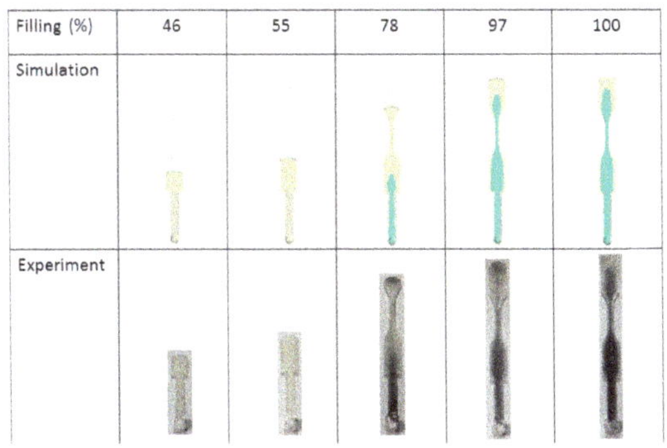

Figure 10. Short shot testing for simulation prediction and experimental study (at skin/core ratio = 40:60).

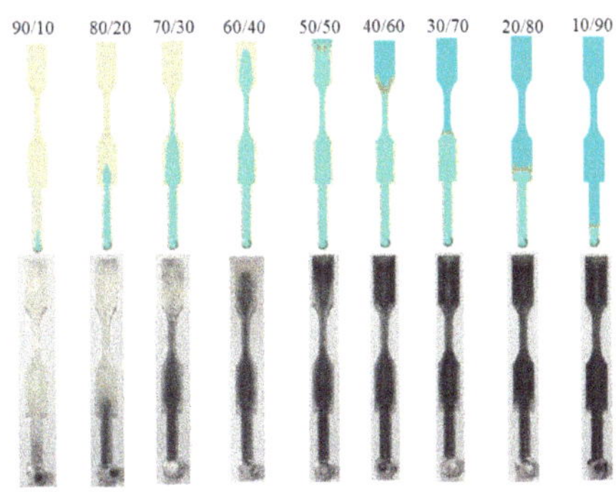

(**a**)

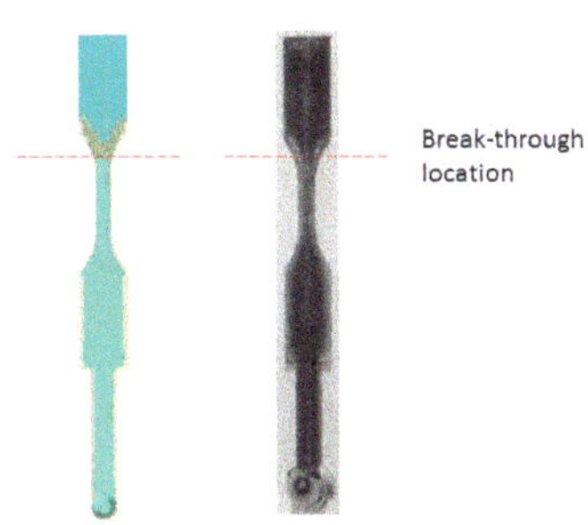

(**b**)

Figure 11. Experimental validation for skin/core ratio effect: (**a**) For various combination from 90:10 to 10:90, (**b**) the observation of the break-through location at skin/core ratio = 40:60.

4.3.3. Fiber Morphology Observation

In order to understand the fiber morphologies and their difference between the single-shot and co-injection systems, the μ-CT scanned images (sliced plane) at different thickness locations have been selected. Figure 12 defines the locations of the selected planes to observe the fiber morphology. Three thickness locations from the top surface have been selected, that is, thickness th = 0.5 mm, 1.0 mm, and 1.75 mm, respectively. Moreover, the fiber morphology inside the co-injected parts can be observed, as shown in Figure 13. Specifically, Figure 13(a1) presents the numerical prediction of the skin/core interface. Since there is no core material observed at th = 0.5 mm, there is no skin/core interface. Figure 13(b1) shows the fiber morphology at thickness th = 0.5 mm experimentally. At this plane, no core material existed. This shows that the fibers are strongly aligned in flow direction near the frozen-layer and shear-layer, but also presents the cross-flow direction in the center core region. Furthermore, when it is moved to thickness th = 1.0 mm, as shown in Figure 13(a2,b2), the skin/core structure appears. In order to trace the skin/core interface and detect the differences across the interface boundary in co-injection molded parts, we have applied numerical simulation (in Figure 13(a2)) to specify the interface first. Then, the marked boundary line can be inserted into the sliced plane of the real scanned image as in Figure 13(b2). When we focused on the areas across the interface boundary from the outer material (skin) to the inner material (core), there was no significant difference across the interface boundary. Similarly, we even moved to the central thickness portion (th = 1.75 mm), as shown in Figure 13(a3,b3). There was no significant difference across the interface boundary from the skin to core materials. Hence, when an FRP co-injection is performed using two of the same materials, the skin/core interface is not significant.

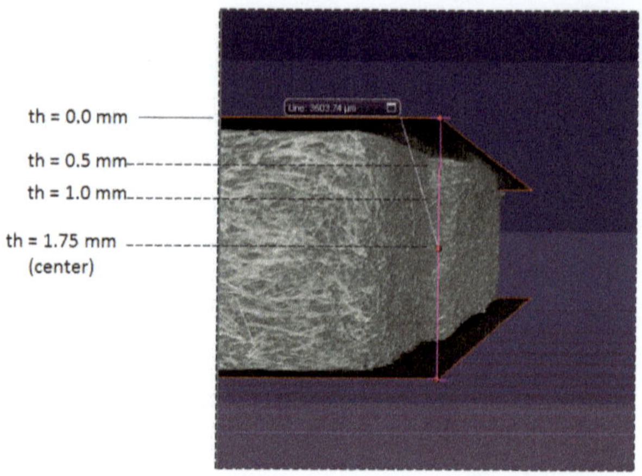

Figure 12. The locations for making the sliced plane to observe the fiber morphology; where th is the distance from the top surface.

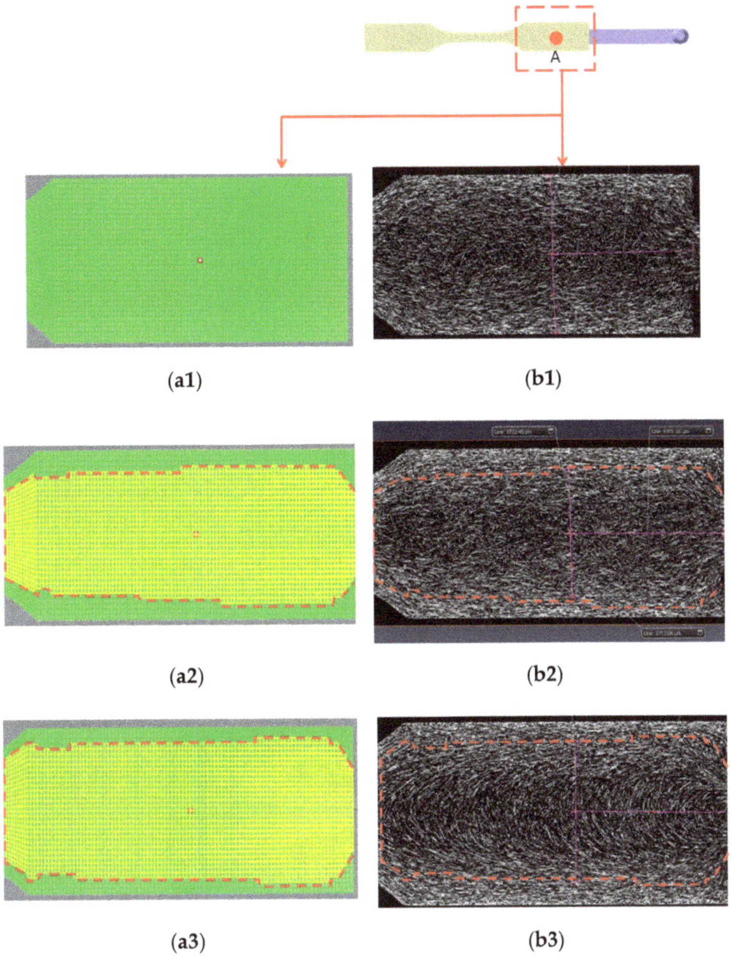

Figure 13. Observation of the fiber morphology based on the sliced plane for co-injected parts at near gate region (NGR) with different thickness locations: (**a1**) Simulation at th = 0.5 mm, (**b1**) sliced image at th = 0.5 mm, (**a2**) simulation at th = 1.0 mm, (**b2**) sliced image at th = 1.0 mm, (**a3**) simulation at th = 1.75 mm (central portion), (**b3**) sliced image at th = 1.75 mm.

Moreover, it is worth discovering what the difference was in fiber morphology between the single-shot and co-injection FRPs parts. Figure 14 shows the observation on the fiber morphology for the single-shot and co-injected parts at different thickness locations. Specifically, one region with (7 mm × 3 mm) has been selected for comparison. In Figure 14(a1,b1), from frozen-layer to central core-layer, no core material existed at th = 0.5 mm. In this region, the strongly aligned fiber area of the single-shot is larger than that of the co-injection system. Similarly, when we focused on the fiber morphology at th = 1.0 mm, as shown in Figure 14(a2,b2), the fibers presented similar behavior, as shown in Figure 14(a1,b1). Finally, the similar trend can be observed in Figure 14(a3,b3). Hence, when it generates one boundary interface via skin/core sequential co-injection molding using two of the same FRPs, the flow field drives the fibers to orientate less in the flow direction. Even the change is small, and this behavior will further affect the mechanical property of the final parts.

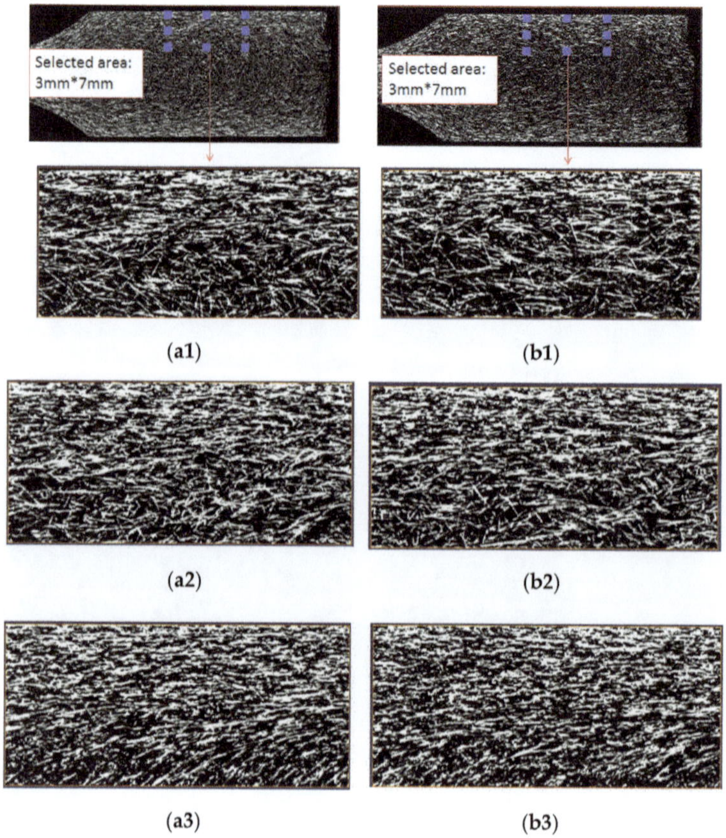

Figure 14. Observation of the fiber morphology based on the sliced plane for the single-shot and co-injected parts at point A with different thickness locations: (**a1**) Single-shot at th = 0.5 mm, (**b1**) co-injected at th = 0.5 mm, (**a2**) single-shot at th = 1.0 mm, (**b2**) co-injected at th = 1.0 mm, (**a3**) single-shot at th = 1.75 mm (central portion), (**b3**) co-injected at th = 1.75 mm.

4.3.4. Fiber Orientation Distribution (FOD) Estimation and Validation

Furthermore, it is worth connecting fiber reinforced features to the product quality. The fiber microstructures can be characterized partially based on fiber orientation distribution (FOD). Figure 15a shows the simulation prediction of the FOD at end of filling (t = t3). From the top surface through the frozen layer to the core layer, at central line of the specimen, the fiber orientation tensor at flow direction (A_{11}) will be increased slightly from 0.75 to 0.85, and then decreased to 0.25. At the same time, the fiber orientation tensor at cross direction (A_{22}) will be decreased slightly from 0.20 to 0.15, and then increased to 0.5. This result is consistent with that observed in Figure 14(a1,a3). Meanwhile, the real specimens are scanned by μ-CT to get the detailed inner microstructure images. Then, those images are further analyzed using AVIZO software to rebuild the structures. Figure 15b presents the FOD of the real measurement, from top surface through frozen layer to core layer at central line of the specimen. The fiber orientation tensor at flow direction (A_{11}) is increased slightly from 0.7 to 0.78, and then decreased to 0.2. At the same time, the fiber orientation tensor at cross direction (A_{22}) will slightly be decreased from 0.30 to 0.2, and then increased to 0.7. Moreover, there is a little difference between the numerical prediction and experimental study from normalized thickness 0 to 0.2 region. This could be due to the shrinkage of the co-injection parts which will further affect the direction of the

FOD along the frozen layer. Overall, the FOD via the real measurement is in a reasonable agreement with that of simulation prediction.

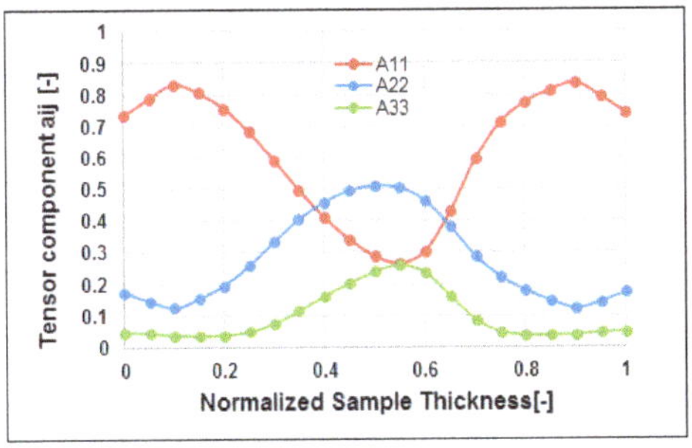

(a)

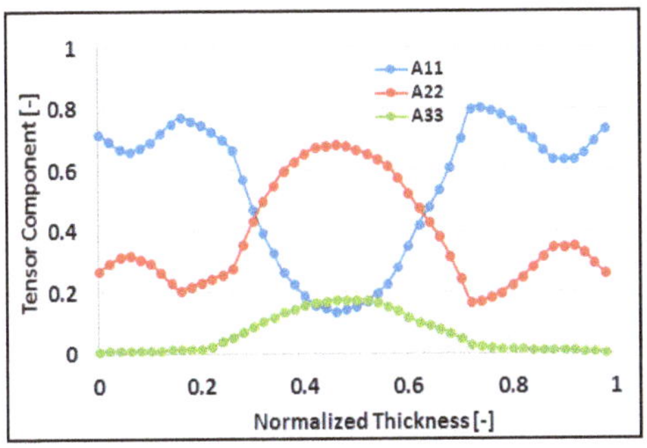

(b)

Figure 15. The comparison of the fiber orientation distribution (FOD) estimation: (**a**) Simulation at point A, (**b**) experiment around point A (NGR).

Moreover, the detailed FOD behavior at flow direction (A_{11}) and cross-flow direction (A_{22}) for both single-shot and co-injection were investigated. Figure 16a shows that the flow direction FOD tensor (A_{11}) of the co-injection is lower than that of the single-shot at the central portion. Furthermore, the overall fiber orientation capability can be described by the integration of the area under the FOD curve. For example, the area below the curve of the co-injection to that of the single-shot is (0.47/0.53 = 0.89). That is to say, the flow direction FOD tensor (A_{11}) of the co-injection is about 89% of that of the single-shot. On the other hand, in Figure 16b, the cross-flow direction FOD tensor (A_{22}) of the co-injection is higher than that of the single-shot at the central portion. Specifically, the area below the curve of the co-injection molding to that of the single-shot is (0.44/0.38 = 1.16). The cross-flow direction FOD tensor (A_{22}) of the co-injection is about 116% of that of the single-shot. From the above results the fiber orientation features of the co-injection molding can be quantified as "the flow direction A_{11} is decreased by 11%, and cross-flow direction A_{22} is increased by 16%." It is inferred that the FOD variation of the co-injection system is due to the presence of the skin layer which can reduce the influence of the solid boundary for changing the fiber alignment.

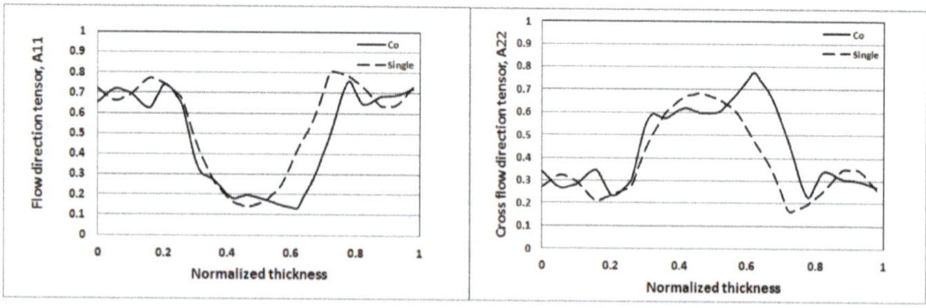

(a)

(b)

Figure 16. The comparison of the fiber orientation distribution (FOD) tensor component at NGR: (**a**) flow direction tensor A_{11}, (**b**) cross-flow direction tensor A_{22}.

4.3.5. Tensile Property Measurement

Moreover, to understand the difference of the fiber reinforced performance between two processes, the mechanical properties based on standard tensile test can be applied. As mentioned earlier, the tensile testing was performed following the standard procedures described in Ref. [31]. Specifically, the specimen is installed into the machine holder (see Figure 6b) under a constant strain at 20 mm/min without pre-tensioning. For each system (single-shot or co-injection), five specimens have been used for the same testing. After finished five tests for each type, the average stress–strain behavior is presented in Figure 16. In Figure 17a, the stress–strain behavior of the single-shot injected part is a little higher than that of co-injection part. Moreover, the associated tensile properties including elongation at break, tensile strength, and tensile modulus are further discussed. Figure 17b shows that the single-shot sample has a larger elongation at break feature than that of the co-injection system, but it is not significant. Furthermore, the detailed tensile stresses and modulus are recorded in Figure 17c and Table 5. Obviously, the average tensile strength of the single-shot is a little higher than that of the co-injection molding by 1.6% (i.e., (86.07 − 84.69)/86.07 × 100% = 1.6%). As per our previous observation and conduction, overall, the fiber orientation tensor at flow direction (A_{11}) of the single-shot is higher than that of the co-injection. Hence, the lower tensile strength and stress modulus of co-injection are consistent with that observation of fiber orientation variation due to the sequential co-injection process. However, the reduction of the tensile properties by co-injection molding is not significant. Hence, using co-injection molding to execute FRP recycling is feasible theoretically. Moreover, some connection between the fiber orientation and the mechanical properties of the final co-injection parts can be obtained.

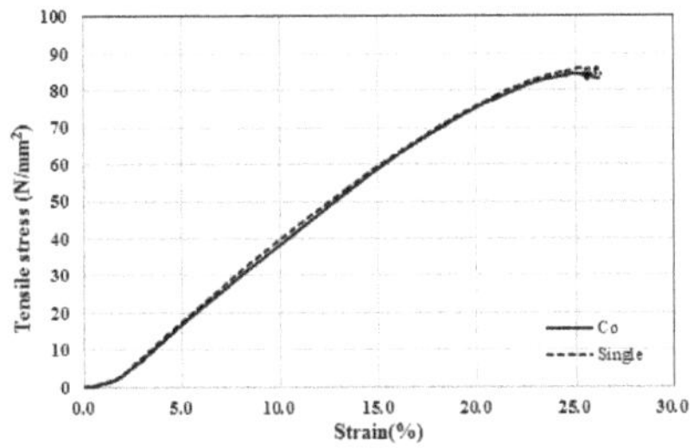

(a)

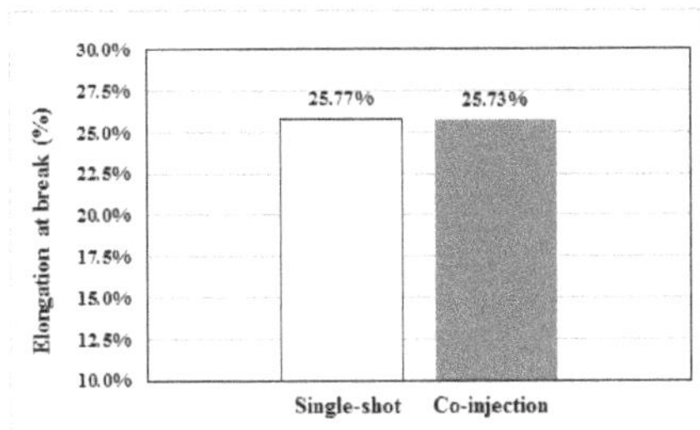

(b)

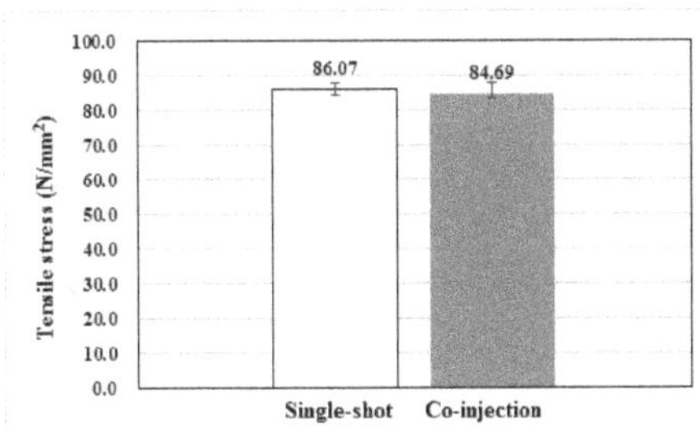

(c)

Figure 17. Tensile property measurement for single-shot and co-injection specimens: (**a**) The average tensile stress-strain behavior, (**b**) the average elongation at break, (**c**) the average tensile strength.

Table 5. The comparison of the tensile properties.

Item	Single-Shot	Co-Injection
Tensile Strength (N/mm^2)	86.07	84.69
Tensile Modulus (N/mm^2)	75.93	39.08
Elongation at Break (%)	25.77%	25.73%

5. Conclusions

In this study, we have applied both CAE simulation and an experimental study to investigate the fiber orientation distributions and their influence on the tensile properties for single-shot and co-injection molding. Specifically, we have obtained some connection between the fiber orientation and the mechanical properties of the final co-injection parts. Some key findings are as follows:

1. The skin/core ratio of 60:40 can provide suitable core-layer penetration without break-through, numerically and experimentally.
2. To discover and verify the influence of the fiber orientation features, the fiber orientation distributions (FOD) of both co-injection and single-shot systems have been observed using μ-CT technology to scan the internal structures, and then software used to perform image analyses for those scanned images. Specifically, the fiber orientation tensor at flow direction (A_{11}) of the co-injection is about 89% of that of the single-shot in the testing conditions. The lower the A_{11}, the lower the tensile property that is expected.
3. The difference of the fiber orientation tensor at flow direction (A_{11}) between the co-injection and the single-shot systems is further verified based on the fiber morphology of the μ-CT scanned image. The observed result is consistent with that of the FOD estimation using the μ-CT scan plus image analysis.
4. To validate the FOD effect on the mechanical properties due to the co-injection, tensile testing was performed. The tensile strength and tensile modulus of the co-injection part is a little weaker than that of the single-shot system. The reason inferred is that the overall fiber orientation tensor at flow direction (A_{11}) of the co-injection system is lower than that of the single-shot system.

Author Contributions: Conceptualization, C.-T.H.; Data curation, X.-W.C. and W.-W.F.; Formal analysis, X.-W.C. and W.-W.F.; Funding acquisition, C.-T.H.; Investigation, C.-T.H., X.-W.C. and W.-W.F.; Methodology, C.-T.H., X.-W.C. and W.-W.F.; Software, X.-W.C. and W.-W.F.; Validation, W.-W.F.; Writing—original draft, C.-T.H.; Writing—review & editing, C.-T.H. All authors have read and agreed to the published version of the manuscript.

Funding: This research was funded by Ministry of Science and Technology of Taiwan (Project number: MOST 108-2221-E-032-013-MY2) and CoreTech System (Moldex3D) Co. Ltd.

Acknowledgments: The authors would like to thank Ministry of Science and Technology of Taiwan (Project number: MOST 108-2221-E-032-013-MY2) and CoreTech System (Moldex3D) Co. Ltd. for partial financial supports to this research. The authors also would like to thank Tzu Hung (Dusty) Lin and his team from MCL Multiscale X-ray CT laboratory, Industrial Technology Research Institute (Taiwan) for great technology supports in μ-CT scan measuring experiments.

Conflicts of Interest: The authors declare no conflict of interest.

References

1. Hammond, L.-A. BMW i3 Carbon Fiber Aluminum Body Frame a New Ara in Electromobility. Available online: https://www.drivingthenation.com/bmw-i3-carbon-fiber-aluminum-body-frame-a-new-era-in-electromobility/, (accessed on 10 December 2019).
2. Jacob, A. Carbon fibre and cars-2013 in review. In Reinforced Plastics. 2014. Available online: https://www.materialstoday.com/carbon-fiber/features/carbon-fibre-and-cars-2013-in-review/, (accessed on 10 December 2019).
3. Amaechi, C.-V.; Agbomerie, C.-O.; Orok, E.-O.; Job, S.; Ye, J. Economic aspects of fiber reinforced polymer composite recycling. In *Reference Module in Materials Science and Materials Engineering*; Elsevier: Amsterdam, The Netherlands, 2019. [CrossRef]
4. Wood, L. Global Composites Market Report 2019: $40.2 Billion Market Trends, Forecast and Competitive Analysis 2013–2018 & 2019–2024, Dublin. Available online: https://www.globenewswire.com/news-release/2019/04/12/1803326/0/en/Global-Composites-Market-Report-2019-40-2-Billion-Market-Trends-Forecast-and-Competitive-Analysis-2013-2018-2019-2024.html (accessed on 10 December 2019).
5. Witten, E.; Mathes, V.; Sauer, M.; Kuhnel, M. Composites Market Report 2018: Market Developments, Trends, Outlooks and Challenges. AVK & Carbon Composites. Available online: https://eucia.eu/userfiles/files/20181115_avk_ccev_market_report_2018_final.pdf (accessed on 8 December 2019).
6. Fenin, K.A.; Akinlabi, E.-T. Recycling of fiber reinforced composites: A review of current technologies. In Proceedings of the DII-2017-28, FDII-2017 Conference, Livingstone, Zambia,, 30 August–1 September 2017.
7. Barnes, F. Recycled Carbon Fiber: Its Time Has Come, Composite World. Available online: https://www.compositesworld.com/columns/recycled-carbon-fiber-its-time-has-come- (accessed on 29 June 2016).
8. Seldén, R. Co-injection molding: Effect of processing on material distribution and mechanical properties of a sandwich molded plate. *Polym. Eng. Sci.* **2000**, *40*, 1165. [CrossRef]

9. Messaoud, D.; Sanchagrin, B.; Derdouri, A. Study on Mechanical Properties and Material Distribution of Sandwich Plaques Molded by Co-injection. *Polym. Compos.* **2005**, *26*, 265–275. [CrossRef]
10. Gomes, M.; Martino, D.; Pontes, A.J.; Viana, J.C. Co-injection molding of immiscible polymers: skin-core structure and adhesion studies. *Polym. Eng. Sci.* **2011**, *51*, 2398–2407. [CrossRef]
11. Sun, S.-P.; Hsu, C.-C.; Huang, C.-T.; Huang, K.-C.; Tseng, S.-C. Sandwich injection molding: core breakthrough and flow imbalance studies. *SPE Tech. Pap.* **2013**, 1575837.
12. Yang, W.M.; Yokoi, H. Visual analysis of the flow behavior of core material in a fork portion of plastic sandwich injection moulding. *Polym. Test.* **2003**, *22*, 37–43. [CrossRef]
13. Huang, C.-T. Numerical visualization and optimization on the core penetration in multi-cavity co-injection molding with a bifurcation runner structure. *Int. J. Adv. Manuf. Technol.* **2017**, *92*, 2545–2557. [CrossRef]
14. Huang, C.-T.; Tzeng, W.-J.; Chen, C.-H.; Lin, G.-G.; Hsu, C.-C.; Chang, R.-Y.; Tseng, S.-C. Study on the main driving force to cause the variation of the core material penetration behavior in different materials arranged co-injection molding. In Proceedings of the 34th International Conference of the Polymer Processing Society (PPS-34), Taipei, Taiwan, 21–25 May 2018.
15. Thomason, J.L.; Vlug, M.A. Influence of fiber length and concentration on the properties of glass fiber-reinforced polypropylene: Part 1-Tensile and flexural modulus. *Composites* **1996**, *27A*, 477–484. [CrossRef]
16. Thomason, J.L. The influence of fibre length and concentration on the properties of glass fibre reinforced polypropylene: Interface strength and fibre strain in injection moulded long fibre PP at high fibre content. *Compos. Part A Appl. Sci. Manuf.* **2007**, *38*, 210–216. [CrossRef]
17. Fu, S.-Y.; Lauke, B. Effects of fiber length and fiber orientation distributions on the tensile strength of short-reinforced polymers. *Compos. Sci. Technol.* **1996**, *56*, 1179–1190. [CrossRef]
18. Wang, C.; Yang, S. Thermal, tensile and dynamic mechanical properties of short carbon fibre reinforced polypropylene composites. *Polym. Polym. Compos.* **2013**, *21*, 65–71. [CrossRef]
19. Lafranche, E.; Krawczak, P.; Ciolczyk, J.P.; Maugey, J. Injection moulding of long glass fibre reinforced poly-amide 6-6: Guidelines to improve flexural properties. *Express Polym. Lett.* **2007**, *1*, 456–466. [CrossRef]
20. Cilleruelo, L.; Lafranche, E.; Krawczak, P.; Pardo, P.; Lucas, P. Injection moulding of long glass fibre reinforced poly(ethylene terephtalate): Influence of carbon black and nucleating agents on impact properties. *Express Polym. Lett.* **2012**, *6*, 706–718. [CrossRef]
21. Folgar, F.; Tucker, C.L. Orientation behavior of fibers in concentrated suspensions. *J. Reinf. Plast. Compos.* **1984**, *3*, 98–119. [CrossRef]
22. Advani, S.G.; Tucker, C.L. The use of tensors to describe and predict fiber orientation in short fiber composites. *J. Rheol.* **1987**, *31*, 751–784. [CrossRef]
23. Advani, S.G. *Flow and Rheology in Polymer Composites Manufacturing*; Elsevier: New York, NY, USA, 1994.
24. Wang, J.-J.; O'Gara, J.-F.; Tucker, C.-L. An objective model for slow orientation kinetics in concentrated fiber suspensions: Theory and rheological evidence. *J. Rheol.* **2008**, *52*, 1179–1200. [CrossRef]
25. Tseng, H.-C.; Chang, R.-Y.; Hsu, C.-H. Phenomenological improvements to predictive models of fiber orientation in concentrated suspensions. *J. Rheol.* **2013**, *57*, 1597–1631. [CrossRef]
26. Tseng, H.-C.; Chng, R.-Y.; Hsu, C.-H. Method and Computer Readable Media for Determining Orientation of Fibers in a Fluid. U.S. Patent No. 8,571,828, 29 October 2013.
27. Tseng, H.-C.; Wang, T.-C.; Chang, Y.-J.; Hsu, C.-H.; Chang, R.-Y. Progress on fiber concentration for injection molding simulation of fiber reinforced thermoplastics. *SPE Tech. Pap.* **2014**, 1669–1673.
28. Bernasconi, A.; Cosmi, F.; Hine, P.J. Analysis of fibre orientation distribution in short fibre reinforced polymers: A comparison between optical and tomographic methods. *Compos. Sci. Technol.* **2012**, *72*, 2002–2008. [CrossRef]
29. Gandhi, U.; Sebastian, D.B.; Kunc, V.; Song, Y. Method to measure orientation of discontinuous fiber embedded in the polymer matrix from computerized tomography scan data. *J. Thermoplast. Mater.* **2016**, *29*, 1696–1709. [CrossRef]
30. Goris, S.; Gandhi, U.; Song, Y.-Y.; Osswald, T.A. Analysis of the process-induced microstructure in injection molding of long glass fiber-reinforced thermoplastics. *SPE Tech. Pap.* **2016**, 348–356.
31. Grellmann, W.; Seidler, S. *Polymer Testing*, 2nd ed.; Carl Hanser Verlag: Munich, Germany, 2013.

polymers

MDPI

Article

Utilization of Waste Bamboo Fibers in Thermoplastic Composites: Influence of the Chemical Composition and Thermal Decomposition Behavior

Chin-Hao Yeh and Teng-Chun Yang *

Department of Forestry, National Chung Hsing University, Taichung 402, Taiwan; harrison19960219@gmail.com
* Correspondence: tcyang.04@nchu.edu.tw

Received: 21 February 2020; Accepted: 9 March 2020; Published: 11 March 2020

Abstract: In this study, four types of waste bamboo fibers (BFs), Makino bamboo (*Phyllostachys makinoi*), Moso bamboo (*Phyllostachys pubescens*), Ma bamboo (*Dendrocalamus latiflorus*), and Thorny bamboo (*Bambusa stenostachya*), were used as reinforcements and incorporated into polypropylene (PP) to manufacture bamboo–PP composites (BPCs). To investigate the effects of the fibers from these bamboo species on the properties of the BPCs, their chemical compositions were evaluated, and their thermal decomposition kinetics were analyzed by the Flynn–Wall–Ozawa (FWO) method and the Criado method. Thermogravimetric results indicated that the Makino BF was the most thermally stable since it showed the highest activation energy at various conversion rates that were calculated by the FWO method. Furthermore, using the Criado method, the thermal decomposition mechanisms of the BFs were revealed by diffusion when the conversion rates (α) were below 0.5. When the α values were above 0.5, their decomposition mechanisms trended to the random nucleation mechanism. Additionally, the results showed that the BPC with Thorny BFs exhibited the highest moisture content and water absorption rate due to this BF having high hemicellulose content, while the BPC with Makino BFs had high crystallinity and high lignin content, which gave the resulting BPC better tensile properties.

Keywords: bamboo-plastic composites (BPCs); waste bamboo fibers; chemical composition; physico-mechanical properties; thermal decomposition kinetics

---

1. Introduction

Owing to the depletion of fossil fuels and the growth of environmental awareness, the effective utilization of forestry waste residues is a notable issue. In Taiwan, the waste residues that are produced from bamboo or woody processing are buried, incinerated, or burned in boilers [1]. According to the literature [2], this waste can be recycled and reused to contribute more economic and social benefits. Bamboo is a renewable material and grows quickly compared to other plants. Additionally, bamboo possesses approximately 60% cellulose with high lignin content and a longitudinal alignment of fibers, which includes a relatively small microfibrillar angle [3–5], resulting in highly specific mechanical properties. In Taiwan, Makino bamboo (*Phyllostachys makinoi*), Moso bamboo (*Phyllostachys pubescens*), Ma bamboo (*Dendrocalamus latiflorus*), and Thorny bamboo (*Bambusa stenostachya*) are common economical and popular bamboo species. Among these bamboo species, Moso bamboo is the most globally harvested bamboo. Therefore, several studies have investigated the chemical, anatomical, physical, and mechanical properties of Moso bamboo [4–7]. In Taiwan, production from Makino bamboo has accounted for more than 80% of gross bamboo production in the past decade [8]. Chung and Wang [6] reported that the flexural properties of Makino bamboo were greater than those of Moso bamboo since the chemical composition of Makino bamboo has higher holocellulose and α-cellulose contents. To effectively utilize waste bamboo residues, polymer composites composed of natural fibers are of significant interest and have been identified as emerging trends in composite science. Furthermore, the addition of natural fibers leads the composite to being an eco-friendly material and reduces the cost of the final composite products due to the numerous advantages of the natural fiber, including low density, high toughness, good specific strength properties, biodegradability, and renewability [9–13]. These composites are widely applied in residential markets and construction industries as window framing, decking, and fencing. Among several natural fibers, bamboo fiber (BF) reinforcement has significant potential for improving the properties of polymer composites due to its excellent characteristics [14–17]. Previous studies [18–20] indicated that the chemical composition and morphology of the fiber, fiber-matrix stress transfer efficiency, and microstructure and void content of the composite are factors that significantly affect the physical and mechanical properties of wood-plastic composites (WPCs). Similarly, the chemical components, including cellulose, hemicellulose, lignin, and extractives, of different bamboo fibers could result in distinct differences in the performance of bamboo–polypropylene composites (BPCs). Moreover, thermal degradation of natural fibers occurs during the manufacturing process of a composite [21,22]. Hence, the thermal decomposition mechanisms from kinetic analyses are crucial for providing information on the thermal degradation processes of fibers. Furthermore, the kinetic modeling of decomposition could help the design of composite processes and is useful for further understanding the thermal stability of the composite. The thermal decomposition kinetics can be evaluated by the isoconversional method, which includes model-free kinetics for determining the activation parameters [23,24]. Some studies have indicated that the thermal decomposition kinetics are influenced by the properties of the fibers such as chemical composition, moisture content, density, and crystallinity [25–27]. Criado et al. [28] proposed several kinetic equations to explain the thermal decomposition mechanisms of solid-state reactions, such as diffusion, nucleation and growth, random nucleation, and phase boundary control. To the best of our knowledge, there is little information available on the properties of BPCs with various waste BFs obtained from these four bamboo species in Taiwan. Accordingly, the aim of the present study was to focus on the effects of the chemical composition and thermal decomposition behavior of different BFs on the physical and mechanical properties of BPCs. Additionally, the thermal stability and kinetic mechanism of the BF were determined using thermogravimetric (TG) analysis by the isoconversional method.

2. Materials and Methods

2.1. Materials

Waste bamboo shavings from various 3-year-old Makino bamboo (*Phyllostachys makinoi*), Moso bamboo (*Phyllostachys pubescens*), Ma bamboo (*Dendrocalamus latiflorus*), and Thorny bamboo (*Bambusa stenostachya*) culms were provided by a local bamboo-processing factory (Nantou County, Taiwan). The BFs were prepared by hammer-milling and sieving between 6 and 16 mesh (ϕ1.00–3.35 mm). The polypropylene (PP) used in this study was purchased from Yung Chia Chemical Industries Co., Ltd. (Taipei, Taiwan). The density, melting temperature, and melt flow index of the PP were 915 kg/m^3, 145 °C, and 4–8 g/10 min, respectively. The PP pellets were ground in an attrition mill to reduce their size to between 20 and 80 mesh (ϕ180–850 μm). The solvents (methanol and toluene) and chemicals (glacial acetic acid, sodium chlorite, and sulfuric acid) were purchased from Sigma-Aldrich Chemical Co. (St. Louis, MO, USA).

2.2. Manufacturing Process of the Bamboo–Polypropylene Composites (BPCs)

The weight ratio of oven-dried BF (moisture content < 3%) to PP was 50/50 for manufacturing the BPCs through the flat-platen pressing process, designated BPC_{Makino}, BPC_{Moso}, BPC_{Ma}, and BPC_{Thorny}. All the BPCs were produced in a two-step pressing process as follows: (1) hot pressing (2.9 MPa) at 180 °C for 3 min and (2) finishing by cold pressing until the temperature decreased to 50 °C.

The expected density of the BPCs was 0.8 g/cm^3. The expected dimensions of the BPCs were 300 mm × 200 mm with a thickness of 3 mm.

2.3. Chemical Composition Analysis

According to ASTM D1107-96, ASTM D1104-56, and ASTM D1106-96, the contents of extractives, holocellulose, and Klason lignin were determined for the various BFs. The chemical composition contents were expressed as a percentage of the initial oven-dried weight.

2.4. X-ray Diffraction (XRD)

X-ray diffractograms (XRD) were collected with an MAC science MXP18 instrument (Tokyo, Japan) using Ni-filtered CuK$_{\alpha 1}$ radiation (λ = 0.1542 nm) at 40 kV and 30 mA. The intensities of the XRD patterns were recorded in the 2θ range of 4–40° with a scan rate of 2°/min. The crystallinity index (CrI) of the BF was calculated according to the following equation [29]:

$$\text{CrI (\%)} = 100 \times (I_{200} - I_{\text{am}})/I_{200} \quad (1)$$

where I_{am} is the intensity of diffraction of the amorphous material at 2θ = 18.3°, and I_{200} is the intensity of the 200 lattice reflection of the cellulose crystallographic form at 2θ = 22°.

2.5. Thermal Decomposition Kinetics Analysis

A Perkin Elmer Pyris 1 instrument (Shelton, CT, USA) was used to investigate the thermal properties of various BFs. A total of 3 mg of BF was heated in a nitrogen atmosphere (20 mL/min) from 50 to 600 °C at various constant heating rates of 5, 10, 20, 30, and 40 °C/min. The data obtained from the TG curves were used to calculate the kinetic parameters. The conversion rate (α) can be defined as:

$$\alpha = \frac{m_0 - m_t}{m_0 - m_f} \quad (2)$$

where m_0 is the initial weight of the sample, m_f is the final residual weight, and m_t is the weight of the pyrolyzed sample at time t. The fundamental equation for a dynamic TG analysis in a nonisothermal experiment can be generally written as follows:

$$\frac{d\alpha}{dT} = \frac{A}{\beta}\exp\left(-\frac{E_a}{RT}\right)f(\alpha) \quad (3)$$

where T is the absolute temperature (K), A is the pre-exponential factor (min^{-1}), β is the heating rate (= dT/dt), E_a is the activation energy (kJ/mol), R is the gas content (8.314 J/ K/mol), and $f(\alpha)$ is the reaction model. Additionally, the integrated form of Equation (3) with a constant heating rate can be expressed in Equation (4):

$$g(\alpha) = \int_0^{\alpha}\frac{d\alpha}{f(\alpha)} = \int_0^{T}\frac{A}{\beta}\exp\left(-\frac{E_a}{RT}\right)dT = \frac{E_a}{\beta R}\frac{\exp(-x)}{x}\pi(x) \quad (4)$$

where $x = E_a/RT$ and $\pi(x)$ is the rational approximation of Senum and Yang [30]. According to Equation (4), the isoconversional Flynn–Wall–Ozawa (FWO) method can be transformed to estimate the activation energy (E_a) value for the thermal decomposition process of the BF. This method is represented by the following equation [31,32]:

$$\log\beta = \log\left(\frac{AE_a}{g(\alpha)R}\right) - 2.315 - 0.4567\frac{E_a}{RT} \quad (5)$$

For various heating rates (β) and a given conversion rate (α), a linear relationship is observed by plotting log β versus 1/T, and the E_a value is calculated from the slope of the straight line [31–35]. The reaction mechanism of the decomposition process is determined by the Criado method [28,36], which assumes that the $Z_m(\alpha)$ master curves are a convolution of the functions $f(\alpha)$ and $g(\alpha)$ corresponding to the different models listed in Table 1 [37,38]:

$$Z_m(\alpha) = f(\alpha)g(\alpha) \tag{6}$$

Table 1. Algebraic expressions of the kinetic models for $f(\alpha)$ and $g(\alpha)$ for kinetic mechanisms of solid-state processes [28,36].

Kinetic Mechanism	Kinetic Model	Algebraic Expression	
		$f(\alpha)$	$g(\alpha)$
Nucleation and growth			
Avrami equation	A_2	$2(1-\alpha)[-\ln(1-\alpha)]^{1/2}$	$[-\ln(1-\alpha)]^{1/2}$
Avrami equation	A_3	$3(1-\alpha)[-\ln(1-\alpha)]^{2/3}$	$[-\ln(1-\alpha)]^{1/3}$
Avrami equation	A_4	$4(1-\alpha)[-\ln(1-\alpha)]^{3/4}$	$[-\ln(1-\alpha)]^{1/4}$
Geometrical: Phase boundary-controlled reaction			
Linear contraction	R_1	1	α
Contracting area	R_2	$2(1-\alpha)^{1/2}$	$1-(1-\alpha)^{1/2}$
Contracting volume	R_3	$3(1-\alpha)^{2/3}$	$1-(1-\alpha)^{1/3}$
Diffusion			
One-dimensional	D_1	$(1/2)\alpha$	α^2
Two-dimensional (Valensi equation)	D_2	$[-\ln(1-\alpha)]^{-1}$	$(1-\alpha)\ln(1-\alpha)+\alpha$
Three-dimensional (Jander equation)	D_3	$(3/2)(1-\alpha)^{2/3}[1-(1-\alpha)^{1/3}]^{-1}$	$[1-(1-\alpha)^{1/3}]^2$
Three-dimensional (Ginstling-Brounshtein equation)	D_4	$(3/2)[(1-\alpha)^{-1/3}-1]^{-1}$	$[1-(2/3)\alpha]-(1-\alpha)^{2/3}$
Reaction-order: Random nucleation on the individual particle			
1st order (One nucleus)	F_1	$(1-\alpha)$	$-\ln(1-\alpha)$
2nd order (Two nuclei)	F_2	$(1-\alpha)^2$	$[(1-\alpha)^{-1}]-1$
3rd order (Three nuclei)	F_3	$(1-\alpha)^3$	$(1/2)[(1-\alpha)^{-2}]-1$

On the other hand, the experimental $Z_e(\alpha(T))$ function can be obtained from Equation (6) by combining Equations (3) and (4):

$$Z_e(\alpha(T)) = \left[\frac{d\alpha}{dT}\exp\left(\frac{E_a}{RT}\right)\right]\left[\frac{E_a}{R}\frac{\exp(-x)}{x}\pi(x)\right] \tag{7}$$

In this study, the fourth degree rational expression of Senum and Yang [30] was used, in which the percentage deviation is less than $10^{-5}\%$ when $x > 20$. This $\pi(x)$ is expressed as follows:

$$\pi(x) = \frac{x^3 + 18x^2 + 86x + 96}{x^4 + 20x^3 + 120x^2 + 240x + 120} \tag{8}$$

2.6. Determination of BPC Properties

The density, moisture content (MS), and water absorption rate (WAR) were determined according to ASTM D2395-17 and ASTM D1037-12. The ASTM D 638-14 and ASTM D 790-17 methods were implemented using a universal testing machine (Shimadzu AG-10kNX, Tokyo, Japan) to determine the tensile properties and flexural properties of the BPCs, respectively. The tensile properties, including the tensile strength (TS) and tensile modulus (TM), were assessed with type I dumbbell-shaped samples at a loading speed of 5 mm/min and a gage length of 57 mm. The modulus of rupture (MOR) and modulus of elasticity (MOE) were obtained using a three-point static flexural test with dimensions of 80 mm × 10 mm × 3 mm at a loading speed of 1.28 mm/min and a span of 48 mm. All the samples were conditioned at 20 °C and 65% relative humidity (RH) for two weeks prior to testing.

2.7. Analysis of Variance

All of the results are expressed in terms of the mean ± the standard deviation (SD). The significance of the differences was calculated using Scheffe's test; $p < 0.05$ was considered to be significant.

3. Results and Discussion

3.1. Chemical Composition and Thermal Stability of Various BFs

The chemical compositions of various BFs are listed in Table 2. The extractive contents of the Makino, Moso, Ma, and Thorny BFs were 2.9, 3.8, 8.5, and 6.9%, respectively, which yields an order of Makino < Moso < Thorny < Ma. It is noted that their holocellulose contents were in the order of Makino > Ma > Moso > Thorny (the values were 62.5, 58.0, 57.1, and 56.0%, respectively). This result indicated that the Makino BF had the highest holoceullose content among all of the samples. Additionally, it is worth noting that the Moso BF revealed the lowest lignin content (24.5%), while there were no signifcant differences among the lignin contents of Makino, Ma, and Thorny BFs, which were in the range of 28.1% to 30.7%. As shown in Figure 1, the crystallinity indexes (CrIs) of various BFs were calculated by the XRD patterns. The major peaks of cellulose crystal diffraction were observed for all the samples at approximately 15.9° (101/101 lattice planes) and 22° (200 lattice plane), whereas the diffraction value at 18.3° represented the amorphous region [8]. According to Equation (1) described above, the CrI values were 53.2%, 41.8%, 40.9%, and 35.0% for Makino, Moso, Ma, and Thorny BFs, respectively. A parameter termed the CrI has been used to describe the relative amount of crystalline material in cellulose. Therefore, this result illustrated that the Makino BF had the highest ordered cellulose content among all the samples, while the lowest amount of ordered cellulose appeared for the Thorny BF.

Table 2. Chemical compositions of various bamboo fibers (BFs).

Bamboo Species	Chemical Composition		
	Holocellulose (%)	**Lignin (%)**	**Extractives (%)**
Makino	62.5 ± 0.8 [a]	30.7 ± 1.0 [a]	2.9 ± 0.5 [d]
Moso	57.1 ± 0.6 [b,c]	24.5 ± 1.2 [b]	3.8 ± 0.4 [c]
Ma	58.0 ± 0.7 [b]	30.3 ± 0.5 [a]	8.5 ± 0.5 [a]
Thorny	56.0 ± 0.6 [c]	28.1 ± 0.9 [a]	6.9 ± 0.3 [b]

Values are the mean ± SD ($n = 3$). Different letters (a, b, c, and d) indicate significant differences ($p < 0.05$).

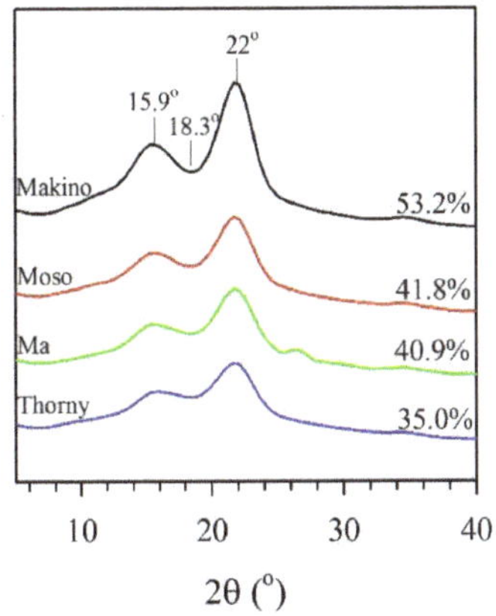

Figure 1. X-ray diffraction (XRD) patterns and crystallinity indexes (CrIs) of various BFs.

In this study, TG analysis was used at various thermal decomposition temperatures to interpret the preliminary evaluation of the thermal stabilities of the BFs and their BPCs during composite processing. Figure 2 shows the residual weight (RW) and differential RW curves of the different BFs from the results of the TG analysis. Órfão et al. [26] and Yang et al. [39] revealed that the thermal decomposition of wood is separated into three stages in the differential RW curve. The first stage corresponds to water evaporation, with a temperature range of 60–120 °C. The second stage simultaneously includes the total decomposition of hemicellulose and cellulose and the partial decomposition of lignin, at temperatures from 210–370 °C. During the third stage in the range of 370–480 °C, the remaining lignin decomposition and the combustion of the residues occur. As shown in detail in Figure 2a, the RW curves showed that various BFs started to exhibit significant weight loss at approximately 140–180 °C. Additionally, the temperatures at which the sample lost 3% ($T_{3\%}$) of its weight were 226, 219, 196, and 204 °C for the Makino, Moso, Ma, and Thorny BFs, respectively. The results demonstrated that the Ma BF exhibited the lowest $T_{3\%}$ value among all the samples. This result is related to the Ma BF having the highest extractive content, as shown in Table 2. It is well-known that extractive decomposition occurs at lower temperatures since extractives are compounds with lower molecular weights than the other chemical components. Previous studies [40–42] indicated that extractives could promote the ignitability of wood at lower temperatures and accelerate wood degradation due to their higher volatility. Furthermore, the Makino BF had the highest $T_{3\%}$ value (226 °C) and thus had higher thermal stability relative to the other BFs. This result is associated with the Makino BF having the lowest content of extractives. Figure 2b presents the differential RW curves of various BFs. These curves displayed a slight shoulder in the range of 250–300 °C, and remarkable peaks were observed in the range of 310–340 °C. According to the pyrolysis behaviors of the three main components (cellulose, hemicellulose, and lignin) in the differential RW curves, hemicellulose and cellulose were decomposed in active pyrolysis in the ranges of 220–315 °C and 315–400 °C, respectively [39]. The thermal decomposition of lignin occurs in active and passive pyrolysis over a wide temperature range from 160–900 °C without characteristic peaks [39]. As shown in Figure 2b, the hemicellulose decomposition in the Thorny BFs occurred at relatively low temperatures (215–285 °C), and the temperature of its characteristic peak (315 °C) was lower than those of the other BFs. This behavior may be related to the higher content of hemicellulose in Thorny BFs relative to those in the other BFs. Hence, this result indicates that a higher content of hemicellulose causes cellulose decomposition at a lower temperature and accelerates the thermal degradation of a BF. John and Thomas [43] explained that hemicellulose comprises a random amorphous structure and is more easily degraded than the other components at its thermal decomposition temperatures (220–315 °C).

3.2. Thermal Decomposition Kinetics of Various BFs

To further understand the thermal properties of different BFs, the activation energy (E_a) was determined by the FWO method. Figure 3 shows the plots of the application of the FWO method for conversion rates (α) from 10% to 70%. Highly linear fits are remarkably obtained from the plots of logβ vs. $1/T$ for various BFs, as validated by the square of the correlation coefficient (R^2), which was greater than 0.99 (Table 3). Thus, this method is suitable for determining the E_a at different α for the thermal decomposition processes of various BFs. At a given conversion rate, the E_a values were calculated from the slope of the linear portion and the intercept of the curve according to Equation (5). According to a previous study [35], the temperature used for composite processing does not exceed 10% conversion. Therefore, the E_a value at 10% conversion was used to evaluate the thermal stability of the BFs in this study. As shown in Table 3, when the conversion rate reached 10% (α = 10%), the temperature was approximately 260 to 290 °C (at heating rates of 5–40 °C/min), which occurred with the decomposition of hemicellulose and the amorphous area of cellulose and the partial decomposition of lignin. At this conversion rate, the E_a value was approximately 188 kJ/mol for the Makino BF, whereas lower E_a values (171–173 kJ/mol) were observed for the Moso BF and Ma BF. The higher E_a value for the Makino BF is related to that this BF contained the lowest extractive content (2.9%). In addition, the lower lignin content (24.5%) for Moso BF and the high quantities of extractives (8.5%) for Ma BF caused lower E_a values. Accordingly, this result indicated that the Moso BF and the Ma BF were the least thermally stable, while the Makino BF was the most thermally stable. Moreover, these results revealed that the lower lignin content or higher extractive content promoted the thermal decomposition of BFs at relatively low temperatures, reducing the thermal stability of the BF. Furthermore, when α = 30–40%, the temperature reached approximately 300–350 °C, corresponding to the decomposition of hemicellulose and cellulose. The Makino BF had an E_a value close to 200 kJ/mol at conversion rates of 30–40%. However, the E_a values for the Moso, Ma, and Thorny BFs were nearly 200 kJ/mol up to conversion rates of 60% to 70%. This result confirmed that the Makino BF had the highest thermal stability among all the samples. This result is attributed to the higher CrI value of Makino BF, indicating that the Makino BF contained a higher quantity of ordered cellulose.

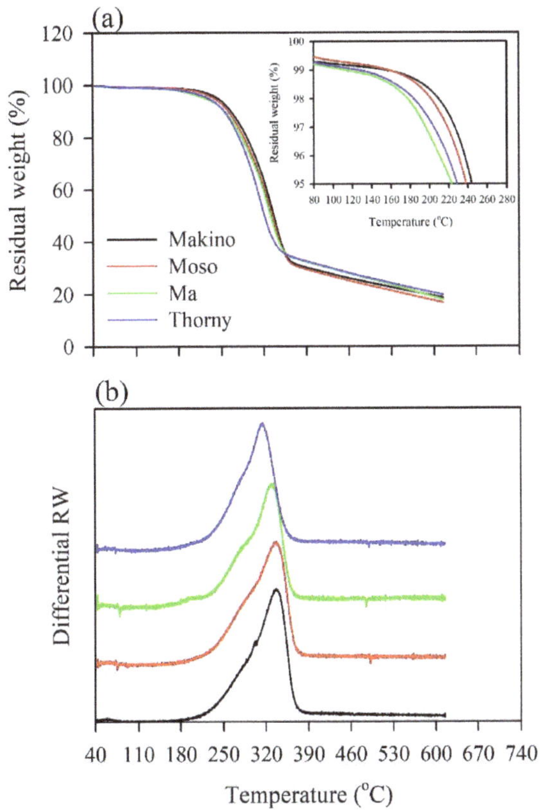

Figure 2. Thermogravimetric (TG) curves of various BFs at a heating rate of 5 °C/min. (**a**) Residual weight (RW) curves; (**b**) differential RW curves.

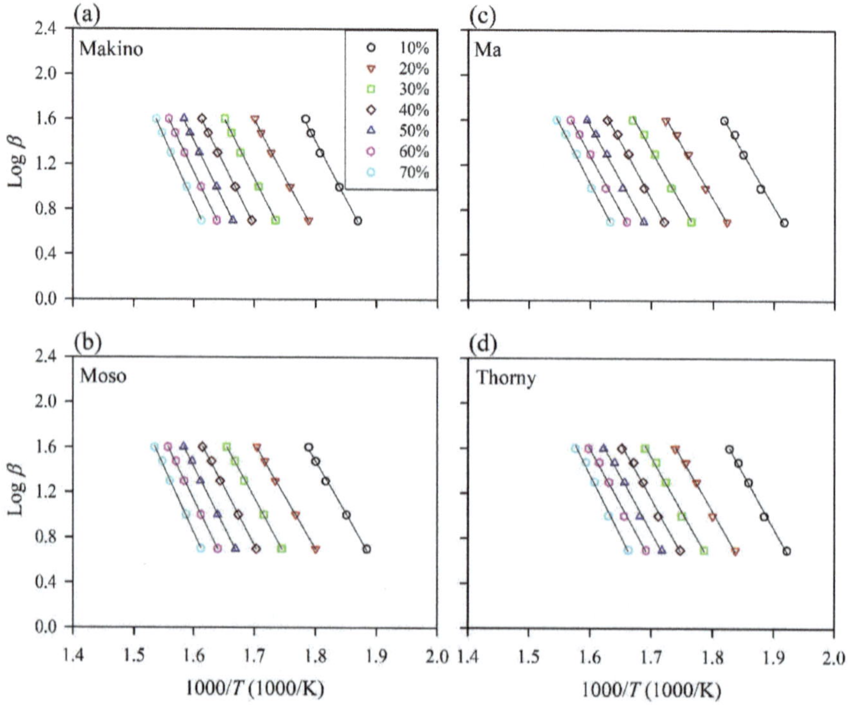

Figure 3. Typical isoconversional plots of various BFs using the Flynn–Wall–Ozawa (FWO) method. (**a**) Makino, (**b**) Moso, (**c**) Ma, and (**d**) Thorny.

Table 3. Apparent activation energies of various BFs calculated by the FWO method.

Bamboo Species	Items	Conversion Rate (α)						
		10%	20%	30%	40%	50%	60%	70%
Makino	E_a (kJ/mol)	188	186	193	198	202	206	217
	R^2	0.9960	0.9992	0.9992	0.9995	0.9994	0.9999	0.9997
Moso	E_a (kJ/mol)	171	172	180	188	196	201	214
	R^2	0.9979	0.9998	0.9990	0.9988	0.9988	0.9990	0.9986
Ma	E_a (kJ/mol)	173	170	176	181	181	183	191
	R^2	0.9941	0.9966	0.9939	0.9969	0.9971	0.9969	0.9969
Thorny	E_a (kJ/mol)	179	173	177	178	178	182	198
	R^2	0.9954	0.9948	0.9934	0.9927	0.9937	0.9928	0.9918

To analyze the thermal decomposition mechanisms of the BFs in depth, the Criado method was used in this study. In this method, the reference theoretical curves, which are called the mater curves ($Z_m(\alpha)$), were obtained from Equation (6). The $Z_m(\alpha)$ curves are derivatives of algebraic expressions ($f(\alpha)$ and $g(\alpha)$) that represent four groups (A_n, R_n, D_n, and F_n) of theoretical mechanisms in Table 1. The experimental data, $Z_e(\alpha(T))$, for various BFs were calculated by applying the E_a values obtained from the FWO method (Equation (7)), and these values were determined using a heating rate of 5–40 °C/min. The experimental data and the master curves are compared, and thereby, the mechanism type of the thermally degraded BF can be identified. Figure 4 presents the master curves of the kinetic models and the experimental data for various BFs at conversion rates from 0.1 to 0.7. Comparison of the similarity of these curves yields the kinetic mechanisms throughout the thermal decomposition processes of the different BFs. Furthermore, the experimental data for all of the samples showed the same tendency, regardless of the conversion rate (α) reached or the heating rate (β) used. At the beginning ($0.1 \leq \alpha < 0.3$), the plots of the experimental data matched the D_n (diffusion-controlled mechanism) curves, which refer to one-, two-, and three-dimensional diffusion. For $0.3 \leq \alpha < 0.5$, the $Z_e(\alpha(T))$ curves reflected the trend of the D_3 and D_4 curves, which corresponds to diffusion in three dimensions (Jander equation and Ginstling–Brounshtein equation). These results implied that the diffusion mechanism becomes prominent for a lower conversion rate ($\alpha < 0.5$). According to previous studies [38,44–46], diffusion models are ascribed to the diffusion of gaseous products from thermally degraded samples. Accordingly, the thermal decomposition rates of waste BFs depend on the heat diffusion from the heating source and the diffusion of the formed gases throughout the sample at a lower conversion rate. For higher conversion rates ($0.5 \leq \alpha \leq 0.7$), there was a gradual change to the F_3 mechanism for the experimental data of all the BFs. This mechanism is associated with random nucleation with three nuclei on an individual particle. For these conversion rates, the temperature was higher than 350 °C. Accordingly, the higher temperatures accelerated the cellulose polymer chain into shorter chains. These chains with lower molecular weights could act as sites for random nucleation and growth for degradation reactions. Similar results were reported in the work of Poletto et al. [38] and Singh et al. [46].

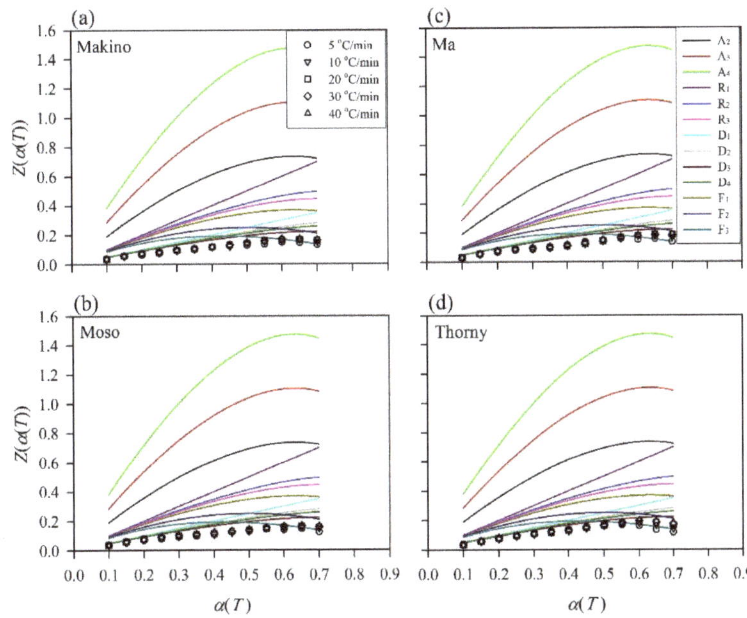

Figure 4. Master curves of kinetic models and experimental data obtained from the Criado method for various BFs. (**a**) Makino, (**b**) Moso, (**c**) Ma, and (**d**) Thorny.

3.3. Characteristic Properties of the BPCs

The density, moisture content (MC), and water absorption rate (WAR) of the BPCs with various BFs are listed in Table 4. Generally, the mechanical properties of materials can be directly influenced by the density. None of the densities of the composites were significantly different (approximately 0.77–0.80 g/cm). In addition, the BPC with Thorny BFs exhibited the highest MC (3.25%) and WAR (10.5%) values. This phenomenon may be affected by the high hemicellulose content in the Thorny BF, which was determined by the TG analysis results. It is well known that hemicellulose is very hydrophilic and has the highest capacity for water absorption, followed by cellulose and lignin [43]. Furthermore, the effects of the various BFs on the mechanical responses of the BPCs are presented in Figure 5 and Table 5. BPC_{Makino} showed the highest tensile strength (TS) and tensile modulus (TM) values at 11.7 MPa and 1130 MPa, respectively, whereas BPC_{Moso} had the lowest tensile properties (10.0 MPa for TS and 987 MPa for TM). For the flexural properties, the moduli of rupture (MORs) of all the samples were approximately 25.0–27.7 MPa; no significant differences were observed among these results. Moreover, the lowest modulus of elasticity (MOE) was found in the BPC with Moso BFs (1482 MPa), followed by BPC_{Makino} (1741 MPa), BPC_{Thorny} (1808 MPa), and BPC_{Ma} (1958 MPa). These results seem to reveal that the mechanical properties of the BPCs were influenced by the intrinsic properties of the BFs such as their structural stiffness and their chemical composition (cellulose and lignin). Jarvis [47] indicated that cellulose has a tight, high strength, and high stiffness crystalline edifice owing to a complex network of hydrogen bonds. The strength and stiffness of natural fibers depend on the crystallinity index, which can represent the cellulose content [20]. Lignin is a branched hydrophobic heteropolymer as a matrix that bonds cellulose fibers together. Additionally, lignin provides strength and stiffness to fiber walls and transfers stress between the cellulose fibers and the matrix [48]. In this study, the Makino BF had a high crystallinity index (53.2%) and lignin content (30.7%), resulting in BPC_{Makino} having the better tensile properties. In contrast, the BPC_{Moso} had the lowest TS, TM, and MOE due to having the lowest lignin content (24.5%) among all the samples. These results illustrated that the Makino BF is a desired reinforcing filler, providing the better tensile properties for a BPC.

Table 4. Physical properties of the BPCs with various BFs.

Code	Density (g/cm^3)	MC (%)	WAR After 24 Soaking (%)
BPC_{Makino}	0.78 ± 0.02 ^{a}	2.81 ± 0.27 ^{b}	6.7 ± 1.7 ^{b}
BPC_{Moso}	0.80 ± 0.02 ^{a}	2.87 ± 0.16 ^{b}	5.5 ± 0.3 ^{b}
BPC_{Ma}	0.77 ± 0.05 ^{a}	3.09 ± 0.17 ^{a,b}	7.3 ± 1.8 ^{b}
BPC_{Thorny}	0.78 ± 0.02 ^{a}	3.25 ± 0.10 ^{a}	10.5 ± 1.6 ^{a}

Values are the mean ± SD (n = 5). Different letters (a and b) indicate significant differences ($p < 0.05$).

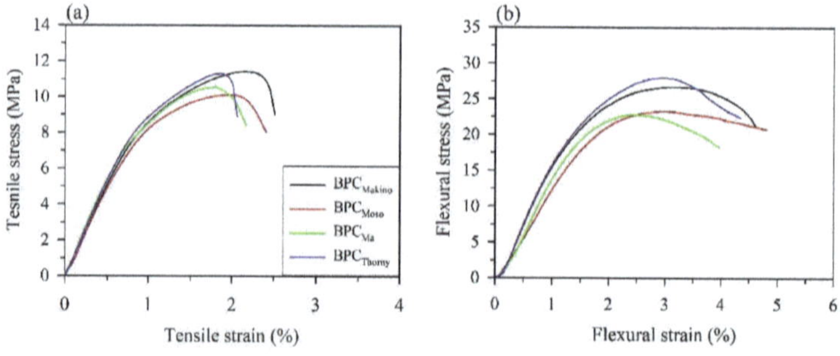

Figure 5. Stress-strain curves obtained from the tensile (**a**) and flexural (**b**) tests for various BPCs.

Table 5. Mechanical properties of the BPCs with various BFs.

Code	Tensile properties		Flexural properties	
	TS (MPa)	TM (MPa)	MOR (MPa)	MOE (MPa)
BPC_{Makino}	11.7 ± 1.0 ^{a}	1130 ± 106 ^{a}	25.0 ± 2.3 ^{a}	1741 ± 156 ^{a,b}
BPC_{Moso}	10.0 ± 0.4 ^{b}	987 ± 68 ^{b}	25.1 ± 1.3 ^{a}	1482 ± 210 ^{b}
BPC_{Ma}	10.9 ± 0.7 ^{a,b}	1110 ± 85 ^{a,b}	27.7 ± 4.5 ^{a}	1958 ± 291 ^{a}
BPC_{Thorny}	11.0 ± 1.0 ^{a,b}	1097 ± 65 ^{a,b}	26.0 ± 2.1 ^{a}	1808 ± 198 ^{a}

Values are the mean ± SD (n = 8). Different letters (a and b) indicate significant differences ($p < 0.05$).

4. Conclusions

The chemical composition analysis showed that the Makino BF contained the highest holocellulose content and the lowest extractive content, whereas the Moso BF contained the lowest amount of lignin. According to the XRD patterns, the Makino BF had the highest crystallinity index among all of the samples, which means that this BF had the highest cellulose content. Moreover, according to the results of TG analysis, higher hemicellulose and extractive contents promoted the thermal decomposition of the BF at lower temperatures. Moreover, the activation energy (E_a) and kinetic mechanisms for various BFs under the controlled heating of the TG analyses were determined by the FWO method and the Criado method. Among all the BFs used in this study, the Makino BF exhibited the highest E_a values at various conversion rates, indicating that this BF was the most thermally stable. Using the Criado method, it is noted that the diffusion-controlled mechanism (D_n mechanism) was dominant at lower conversion rates ($0.1 \leq \alpha < 0.5$). When the conversion rate was above 0.5, the degradation of the BFs was governed by the 3rd order random mechanism (a F_3 mechanism). Furthermore, the BPC with Thorny BFs exhibited the highest moisture content and water absorption rate due to the higher hemicellulose content of the Thorny BF. When a BPC was manufactured with Makino BF, which had high crystallinity and high lignin content, the tensile properties of the composite were high. These results indicated that the amounts of various chemical components within BFs affect the physical and mechanical properties of a BPC. Furthermore, the results of this study offer information for optimizing polymer composites, and the reinforcement of waste BFs needs to be precisely selected in the future.

Author Contributions: Conceptualization, T.-C.Y.; Formal analysis, C.-H.Y. and T.-C.Y.; Funding acquisition, T.-C.Y.; Investigation, C.-H.Y. and T.-C.Y.; Resources, T.-C.Y.; Supervision, T.-C.Y.; Validation, T.-C.Y.; Visualization, T.-C.Y.; Writing—original draft, C.-H.Y. and T.-C.Y.; Writing—review and editing, C.-H.Y. and T.-C.Y. All authors have read and agreed to the published version of the manuscript.

Funding: This research was funded by the Ministry of Science and Technology, Taiwan (MOST 108-2313-B-005-014-MY3).

Conflicts of Interest: The author declares no conflict of interest.

References

1. Chung, M.J.; Wang, S.Y. Physical and mechanical properties of composites made from bamboo and woody wastes in Taiwan. *J. Wood Sci.* **2019**, *65*, 57. [CrossRef]
2. Obata, Y.; Takeuchi, K.; Soma, N.; Kanayama, K. Recycling of wood waste as sustainable industrial resources—Design of energy saving wood-based board for floor heating systems. *Energy* **2006**, *31*, 2341–2349. [CrossRef]
3. Li, Y.; Yin, L.; Huang, C.; Meng, Y.; Fu, F.; Wang, S.; Wu, Q. Quasi-static and dynamic nanoindentation to determine the influence of thermal treatment on the mechanical properties of bamboo cell walls. *Holzforschung* **2015**, *69*, 909–914. [CrossRef]
4. Liu, H.; Jiang, Z.; Fei, B.; Hse, C.; Sun, Z. Tensile behaviour and fracture mechanism of moso bamboo (*Phyllostachys pubescens*). *Holzforschung* **2015**, *69*, 47–52. [CrossRef]
5. Liu, H.; Wang, X.; Zhang, X.; Sun, Z.; Jiang, Z. In situ detection of the fracture behaviour of moso bamboo (*Phyllostachys pubescens*) by scanning electron microscopy. *Holzforschung* **2016**, *70*, 1183–1190. [CrossRef]
6. Chung, M.J.; Wang, S.Y. Effects of peeling and steam-heating treatment on basic properties of two types of bamboo culms (*Phyllostachys makinoi* and *Phyllostachys pubescens*). *J. Wood Sci.* **2017**, *63*, 473–482. [CrossRef]
7. Obataya, E.; Kitin, P.; Yamauchi, H. Bending characteristics of bamboo (*Phyllostachys pubescens*) with respect to its fiber–foam composite structure. *Wood Sci. Technol.* **2007**, *41*, 385–400. [CrossRef]
8. Yang, T.-C.; Lee, T.-Y. Effects of density and heat treatment on the physico-mechanical properties of unidirectional round bamboo stick boards (UBSBs) made of Makino bamboo (*Phyllostachys makinoi*). *Constr. Build. Mater.* **2018**, *187*, 406–413. [CrossRef]
9. Alsaeed, T.; Yousif, B.F.; Ku, H. The potential of using date palm fibres as reinforcement for polymeric composites. *Mater. Des.* **2013**, *43*, 177–184. [CrossRef]
10. Bledzki, A.K.; Reihmane, S.; Gassan, J. Thermoplastics reinforced with wood fillers: A literature review. *Polym. Plast. Technol. Eng.* **1998**, *37*, 451–468. [CrossRef]

11. Dittenber, D.B.; Ganga Rao, H.V.S. Critical review of recent publications on use of natural composites in infrastructure. *Compos. Part A* **2012**, *43*, 1419–1429. [CrossRef]
12. Kumar, V.; Tyagi, L.; Sinha, S. Wood flour—reinforced plastic composites: A review. *Rev. Chem. Eng.* **2011**, *27*, 253–264. [CrossRef]
13. Saba, N.; Paridah, M.T.; Jawaid, M. Mechanical properties of kenaf fibre reinforced polymer composite: A review. *Constr. Build. Mater.* **2015**, *76*, 87–96. [CrossRef]
14. Hsu, C.-Y.; Yang, T.-C.; Wu, T.-L.; Hung, K.-C.; Wu, J.-H. Effects of a layered structure on the physicomechanical properties and extended creep behavior of bamboo-polypropylene composites (BPCs) determined by the stepped isostress method. *Holzforschung* **2018**, *72*, 589–597. [CrossRef]
15. Yang, G.; Zhang, Y.; Shao, H.; Hu, X. A comparative study of bamboo Lyocell fiber and other regenerated cellulose fibers. *Holzforschung* **2009**, *63*, 18–22. [CrossRef]
16. Yang, T.-C.; Wu, T.-L.; Hung, K.-C.; Chen, Y.-L.; Wu, J.-H. Mechanical properties and extended creep behavior of bamboo fiber reinforced recycled poly(lactic acid) composites using the time–temperature superposition principle. *Constr. Build. Mater.* **2015**, *93*, 558–563. [CrossRef]
17. Yu, Y.; Tian, G.; Wang, H.; Fei, B.; Wang, G. Mechanical characterization of single bamboo fibers with nanoindentation and microtensile technique. *Holzforschung* **2011**, *65*, 113–119. [CrossRef]
18. Ashori, A.; Nourbakhsh, A. Reinforced polypropylene composites: Effects of chemical compositions and particle size. *Bioresour. Technol.* **2010**, *101*, 2515–2519. [CrossRef]
19. Bouafif, H.; Koubaa, A.; Perré, P.; Cloutier, A. Effects of fiber characteristics on the physical and mechanical properties of wood plastic composites. *Compos. Part A* **2009**, *40*, 1975–1981. [CrossRef]
20. Ou, R.; Xie, Y.; Wolcott, M.P.; Sui, S.; Wang, Q. Morphology, mechanical properties, and dimensional stability of wood particle/high density polyethylene composites: Effect of removal of wood cell wall composition. *Mater. Des.* **2014**, *58*, 339–345. [CrossRef]
21. Bledzki, A.K.; Sperber, V.E.; Faruk, O. *Natural and Wood Fiber Reinforcement in Polymers*; Rapra Technology LTD.: Birmingham, UK, 2002.
22. Nabi Saheb, D.; Jog, J.P. Natural fiber polymer composites: A review. *Adv. Polym. Technol.* **1999**, *18*, 351–363. [CrossRef]
23. Vyazovkin, S. Advanced isoconversional method. *J. Therm. Anal.* **1997**, *49*, 1493–1499. [CrossRef]
24. Vyazovkin, S.; Dollimore, D. Linear and nonlinear procedures in iso-conversional computations of the activation energy of nonisothermal reactions in solids. *J. Chem. Inf. Comput. Sci.* **1996**, *36*, 42–45. [CrossRef]
25. Di Blasi, C. Modeling chemical and physical process and biomass pyrolysis. *Prog. Energy Combust. Sci.* **2008**, *34*, 47–90. [CrossRef]
26. Órfão, J.J.M.; Antunes, F.J.A.; Figueiredo, J.L. Pyrolysis kinetics of lignocellulosic materials–three independent reactions model. *Fuel* **1999**, *78*, 349–358. [CrossRef]
27. Popescu, M.C.; Popescu, C.M.; Lisa, G.; Sakata, Y. Evaluation of morphological and chemical aspects of different wood species by spectroscopy and thermal methods. *J. Mol. Struct.* **2011**, *988*, 65–72. [CrossRef]
28. Criado, J.M.; Málek, J.; Ortega, A. Applicability of the master plots in kinetic analysis of non-isothermal data. *Themochim. Acta* **1989**, *147*, 377–385. [CrossRef]
29. Hung, K.-C.; Wu, J.-H. Mechanical and interfacial properties of plastic composite panels made from esterified bamboo particles. *J. Wood Sci.* **2010**, *56*, 216–221. [CrossRef]
30. Pérez-Maqueda, L.A.; Criado, J.M. The accuracy of Senum and Yang's approximation to the Arrhenius integral. *J. Therm. Anal. Calorim.* **2000**, *60*, 909–915. [CrossRef]
31. Flynn, J.H.; Wall, L.A. General treatment for the thermogravimetry of polymers. *J. Res. Nat. Bureau Stand.* **1966**, *70A*, 487–523. [CrossRef]
32. Ozawa, T. A new method of analyzing thermogravimetric data. *Bull. Chem. Soc. Jap.* **1965**, *38*, 1881–1886. [CrossRef]
33. Hung, K.-C.; Wu, J.-H. Characteristics and thermal decomposition kinetics of wood-SiO_2 composites derived by the sol-gel process. *Holzforschung* **2017**, *71*, 233–240. [CrossRef]
34. Hung, K.-C.; Wu, J.-H. Comparison of physical and thermal properties of various wood-inorganic composites (WICs) derived by the sol-gel process. *Holzforschung* **2018**, *72*, 379–386. [CrossRef]
35. Hung, K.-C.; Yeh, H.; Yang, T.-C.; Wu, T.-L.; Xu, J.-W.; Wu, J.-H. Characterization of wood-plastic composites made with different lignocellulosic materials that vary in their morphology, chemical composition and thermal stability. *Polymers* **2017**, *9*, 726. [CrossRef] [PubMed]

36. Núñez, L.; Fraga, F.; Núñez, M.R.; Villanueva, M. Thermogravimetric study of the decomposition process of the system BADGE ($n = 0$)/1,2 DCH. *Polymer* **2000**, *41*, 4635–4641. [CrossRef]
37. Bianchi, O.; Martins, J. DeN.; Fiorio, R.; Oliveira, R.V.B.; Canto, L.B. Changes in activation energy and kinetic mechanism during EVA crosslinking. *Polym. Test.* **2011**, *30*, 616–624. [CrossRef]
38. Poletto, M.; Zattera, A.J.; Santana, R.M.C. Thermal decomposition of wood: Kinetics and degradation mechanisms. *Bioresour. Technol.* **2012**, *126*, 7–12. [CrossRef]
39. Yang, H.; Yan, R.; Chen, H.; Lee, D.H.; Zheng, C. Characteristics of hemicellulose, cellulose and lignin pyrolysis. *Fuel* **2007**, *86*, 1781–1788. [CrossRef]
40. Grønli, M.G.; Várhegyi, G.; Di Blasi, C. Thermogravimetric analysis and devolatilization kinetics of wood. *Ind. Eng. Chem. Res.* **2002**, *41*, 4201–4208. [CrossRef]
41. Poletto, M.; Zattera, A.J.; Forte, M.M.C.; Santana, R.M.C. Thermal decomposition of wood: Influence of wood components and cellulose crystallite size. *Bioresour. Technol.* **2012**, *109*, 148–153. [CrossRef]
42. Shebani, A.N.; van Reenen, A.J.; Meincken, M. The effect of wood extractives on the thermal stability of different wood species. *Thermochim. Acta* **2008**, *471*, 43–50. [CrossRef]
43. John, M.J.; Thomas, S. Biofibres and biocomposites. *Carbohydr. Polym.* **2008**, *71*, 343–364. [CrossRef]
44. Doddapaneni, T.R.K.C.; Konttinen, J.; Hukka, T.I.; Moilanen, A. Influence of torrefaction pretreatment on the pyrolysis of Eucalyptus clone: A study on kinetics, reaction mechanism and heat flow. *Ind. Crop. Prod.* **2016**, *92*, 244–254. [CrossRef]
45. Mishra, G.; Kumar, J.; Bhaskar, T. Kinetic studies on the pyrolysis of pinewood. *Bioresour. Technol.* **2015**, *182*, 282–288. [CrossRef]
46. Singh, S.; Chakraborty, J.P.; Mondal, M.K. Intrinsic kinetics, thermodynamic parameters and reaction mechanism of non-isothermal degradation of torrefied *Acacia nilotica* using isoconversional methods. *Fuel* **2020**, *259*, 116263. [CrossRef]
47. Jarvis, M. Chemistry: Cellulose stacks up. *Nature* **2003**, *426*, 611–612. [CrossRef]
48. Ali, A.; Shaker, K.; Nawab, Y.; Jabbar, M.; Hussain, T.; Militky, J.; Baheti, V. Hydrophobic treatment of natural fibers and their composites-A review. *J. Ind. Text.* **2018**, *47*, 2153–2183. [CrossRef]

polymers

MDPI

Article

Effect of Various Surface Treatments on the Performance of Jute Fibers Filled Natural Rubber (NR) Composites

Kumarjyoti Roy [1], Subhas Chandra Debnath [2], Lazaros Tzounis [3], Aphiwat Pongwisuthiruchte [1,4] and Pranut Potiyaraj [1,4,*]

1 Department of Materials Science, Faculty of Science, Chulalongkorn University, Bangkok 10330, Thailand; kukumarjyotiroy@gmail.com (K.R.); apw.pongwisuthiruchte@gmail.com (A.P.)
2 Department of Chemistry, University of Kalyani, Kalyani, Nadia 741235, India; scd@klyuniv.ac.in
3 Department of Materials Science & Engineering, University of Ioannina, 45110 Ioannina, Greece; latzounis@gmail.com
4 Center of Excellence on Petrochemical and Materials Technology, Chulalongkorn University, Bangkok 10330, Thailand
* Correspondence: pranut.p@chula.ac.th

Received: 4 December 2019; Accepted: 17 January 2020; Published: 7 February 2020

Abstract: In the present study, the suitability of various chemical treatments to improve the performance of jute fibers (JFs) filled natural rubber (NR) composites was explored. The surface of JFs was modified by three different surface treatments, namely, alkali treatment, combined alkali/stearic acid treatment and combined alkali/silane treatment. Surface modified JFs were characterized by X-ray diffraction (XRD) pattern, Fourier transform infrared (FTIR) spectroscopy and field emission scanning electron microscopy (FESEM). The reinforcing effect of untreated and surface treated JFs in NR composites was comparatively evaluated in terms of cure, mechanical, morphological and thermal properties. Combined alkali/silane treated JFs filled NR composite showed considerably higher torque difference, tensile modulus, hardness and tensile strength as compared to either untreated or other surface treated JFs filled NR systems. A crosslink density measurement suggested effective rubber-fibers interaction in combined alkali/silane treated JFs filled NR composite. Morphological analysis confirmed the improvement in the interfacial bonding between NR matrix and JFs due to combined alkali/silane treatment allowing an efficient "stress-transfer" mechanism. As a whole, combined alkali/silane treatment was found to be most efficient surface treatment method to develop strong interfacial adhesion between NR matrix and JFs.

Keywords: rubber; short jute fibers; surface treatments; mechanical properties; scanning electron microscopy

1. Introduction

Short fiber reinforced rubber composites are commonly used for the manufacturing of several industrial products such as hoses, seals, tire treads, V-belts, complex-shaped goods, etc. [1,2]. The overall performances of short fiber reinforced rubber composite are closely related to some factors like strong adhesion between rubber and fibers, aspect ratio of fibers, degree of dispersion of fibers within the rubber matrix and control of fibers orientation [2,3]. Natural fibers are bio-based renewable materials with some interesting features like high specific strength, low self-weight, unrestricted formability and resistance to corrosion [4,5]. Among the various non-petroleum based materials, natural occurring jute fibers (JFs) are one of the most promising alternatives to traditional petroleum based fillers for the development of environmentally friendly rubber composites.

In the last two decades, due to upward environmental awareness, there is an increasing demand for the development of natural fiber based green and sustainable rubber composites. Many researchers reported the designing of advanced rubber composites based on different types of natural fiber such as short jute [2,6–9], bamboo [10,11], short coir [12–14], sisal [15,16], oil palm [15–17], kenaf [18], grass [19,20], hemp [21,22], pineapple leaf [23,24], etc. Compared to petroleum based materials, natural fibers have some additional advantages, i.e., availability, biodegradability, light-weight, low-cost, renewability and non-toxic nature [9,25]. However, the proper dispersion of short natural fibers in natural rubber (NR) composites is a challenging task due to the poor compatibility between the hydrophilic natural fibers and the hydrophobic rubber matrix.

Among the different natural fibers, jute is a commercially cheap lignocellulosic fiber with large cellulose content [7,25]. Initially, Murty et al. introduced short JFs as new reinforcing material in rubber composites [6]. In recent years, very few research articles have been published regarding the use of short JFs as a green reinforcing filler to improve the performance of NR compounds [2,8,9]. Surface treatment of JFs is the key strategy to enhance the interfacial adhesion between hydrophilic jute and hydrophobic NR matrix [8]. Tzounis et al. [9] investigated the effect of carbon nanotube modified JFs (JF-CNT) as "hierarchical" multiscale reinforcements on the mechanical and thermal properties of NR composites. According to the authors, JF-CNT filler had better hydrophobic character than unmodified JFs as filler, while the nanostructured surface of the JFs due to the roughness endowed by the CNTs facilitated a mechanical interlocking mechanism between the fiber–matrix components. As a result, JF-CNT filled NR composites exhibited considerably higher tensile properties and thermal stability as compared to those of unmodified JFs filled NR samples at same filler loading level. Recently, Roy et al. [2] reported on the role of stearic acid modified nanoclay (SANC) on the cure, mechanical and thermal properties of alkali treated JFs filled NR composites. Actually, SANC was able to increase the hydrophobic character of JFs due to the hydrogen bonding interaction between the carboxyl group of SANC and surface hydroxyl groups of JFs. Thus, the tensile strength and storage modulus of JFs filled NR composites were greatly improved in presence of SANC. Nevertheless, the research work concerning the use of JFs as green and natural occurring filler for the development of both environmental and industrial friendly rubber composites is in the preliminary stage.

The main aim of the present study was to achieve a novel concept for the development of high performance and low-cost natural JFs reinforced NR composites. The performance of JFs filled NR composites was closely connected to the "engineered" and by design adhesive interface between rubber and fibers in this study. More precisely, the JFs surface modification endowed enhanced interfacial compatibility between NR and JFs with an enhanced interfacial strength as supported by the chemical interaction mechanism that have been illustrated as well as the SEM fracture surfaces. Undoubtedly, the present study demonstrates a facile, versatile, scalable and unique protocol to develop natural fibers based green rubber composites making use of surface science and technology.

2. Materials and Methods

2.1. Materials

NR (RMA-1X), zinc oxide (surface area 5–6 m^2/g, Merck, Kenilworth, NJ, USA), stearic acid (LobaChemie, Mumbai, India), sulfur (LobaChemie, Mumbai, India), sodium hydroxide (Merck) and toluene (Merck) were used as received. Tetra methyl thiuram disulfide (TMTD) was procured from Thailand Rubber Research Institute (Bangkok, Thailand). JFs (TD 4 grade) were obtained from Gloster Jute Mills, Howrah, India. Bis[3-(triethoxysilyl)propyl]tetrasulfide (TESPT) (Sigma-Aldrich, Dorset, UK) were used as received.

2.2. Surface Modification of JFs

2.2.1. Alkali Treatment

At first, short JFs were cut and ground to powder form using a mixer grinder. Then, the JFs were ultrasonicated in 1 wt % aqueous solution of sodium hydroxide (NaOH) for 1 h to eliminate lignin and hemicelluloses. Next, the alkali treated JFs were washed with distilled water followed by neutralization with diluted acetic acid until the fibers surface becomes completely free from the unreacted alkali [9]. Then, the JFs were again washed with distilled water. Finally, the alkali treated JFs were dried in a hot air oven at 70 °C for 24 h. The untreated JFs were designated hereafter as JFun. The alkali treated JFs were also designated as A-JF. The mechanism of alkali surface treatment of JFs is shown in Figure 1a.

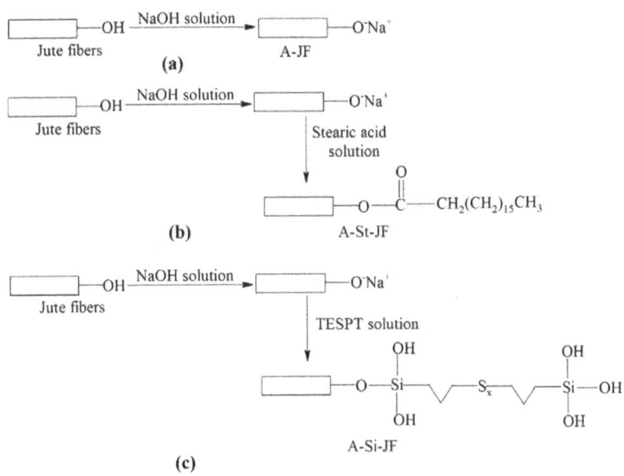

Figure 1. Mechanism of surface modification of jute fibers (JFs) by different chemical treatments, (**a**) alkali treatment, (**b**) combined alkali/stearic acid treatment, (**c**) combined alkali/silane treatment.

2.2.2. Combined Alkali/Stearic Acid Treatment

Initially, 1 wt % stearic acid was added to water and heated to prepare a homogeneous solution. After that, the alkali modified JFs (A-JF) were treated with 1 wt % stearic acid solution in an ultrasonication bath for 1 h. Then, the surface treated JFs were washed with toluene followed by methanol to remove excess unreacted stearic acid. Finally, combined alkali/stearic acid treated JFs were dried in a hot air oven at 70 °C for 24 h. The combined alkali/stearic acid treated JFs were designated as A-St-JF. The mechanism of surface treatment of JFs by combined alkali/stearic acid is shown in Figure 1b.

2.2.3. Combined Alkali/Silane Treatment

At first, 1 wt % solution of silane coupling agent, i.e., TESPT was prepared in water:ethanol mixture (water:ethanol = 40:60). Subsequently, the alkali treated jute fibers (A-JF) were treated with 1 wt % solution of TESPT in an ultrasonication bath for 1 h. Finally, the surface treated JFs were washed with distilled water to remove the unreacted TESPT followed by drying in a hot air oven at 70 °C for 24 h. The combined alkali/silane treated JFs were designated as A-Si-JF. The mechanism of surface treatment of JFs by combined alkali/silane is shown in Figure 1c.

2.3. Preparation of NR Composites

Various NR composites were prepared in a two-roll mixing mill according to the formulation shown in Table 1. The compounding of various NR composites was carried out at room temperature. During mixing process, the speed of one roll was kept at 20 rpm and that of the other roll was maintained at 24 rpm to attain the friction ratio of 1:1.2.

Table 1. The formulation of studied vulcanizates in parts per hundred parts of rubber (phr).

Ingredients	Compound Designation				
	Unfilled NR	NR/JF_{un}	NR/A-JF	NR/A-St-JF	NR/A-Si-JF
NR	100	100	100	100	100
ZnO	5	5	5	5	5
Stearic acid	2	2	2	2	2
TMTD	2.16	2.16	2.16	2.16	2.16
Sulfur	0.5	0.5	0.5	0.5	0.5
JF_{un}	-	10	-	-	-
A-JF	-	-	10	-	-
A-St-JF	-	-	-	10	-
A-Si-JF	-	-	-	-	10

2.4. Characterization Techniques

X-ray diffraction (XRD) patterns of untreated and surface treated JFs were recorded on Xpertpro-Panalytical X-ray diffractometer (Malvern, UK). Scanning electron microscopy (SEM) images of JFs and tensile fracture surfaces of NR composites were obtained using field emission scanning electron microscopy (FESEM, JEOL, JSM-7610 F, Tokyo, Japan). Fourier transform infrared (FTIR) investigations of untreated and surface treated JFs were performed using Perkin-Elmer L 120-000A spectrometer (Waltham, MA, USA) (ν_{max} in cm^{-1}) on KBr disks. The cure characteristics like minimum torque (ML), maximum torque (MH), scorch time (t_2) and optimum cure time (t_{90}) were measured on a moving die rheometer (rheoTech MD+, Model no. A022S, Ajpha Technologies, Akron, OH, USA) at 160 °C. Different compounded NR samples were cured according to their optimum cure time at 160 °C in a hot press. For various NR sheets, the mechanical properties like modulus at 100% (M100) elongation, tensile strength (T.S.) and elongation at break (E.B. in %) were measured using Amsler (Göteborg, Sweden) tensile tester according to ASTM D 412-51 T. Hardness (shore A) of different NR samples was calculated by a Hiroshima Hardness Tester (Aishwarya, Telangana, India) according to ASTM D 2240. Crosslink density values and solvent uptake properties of unfilled and filled NR samples were measured from swelling experiment according to the method given by Roy et al. [26]. Thermo-gravimetric analysis (TGA) was carried out using a TGA instrument (Mettler Toledo, Columbus, OH, USA), TGA/DSC 3+ under nitrogen flow from 50 to 600 °C with a heating rate of 10 °C/min.

3. Results and Discussion

3.1. Confirmation of Surface Modification of JFs

The XRD patterns of untreated and surface treated JFs are represented in Figure 2. In both untreated and surface treated JFs, the common peaks at about 17 and 22.5° were attributed to the (101) and (002) planes of cellulose [27]. Thus, there was no noticeable change in the macromolecular chain structure of JFs due to the surface treatments [27]. However, diffraction peak intensities of surface treated JFs showed clear increment as compared to untreated JFs. This result was due to the increase in the ratio of crystalline cellulose resulting from the removal of amorphous waxy substances after surface treatments [27].

The surface morphology of various JFs was examined using FESEM analysis. The FESEM images of untreated and surface treated JFs are presented in Figure 3. As shown in Figure 3a, JFun had smooth surface owing to the presence of waxy substances like lignin and hemicelluloses [27,28]. On the other hand, several grooves were formed along the structure of chemically treated JFs due to the elimination of waxy substances after surface treatments. As a result, all surface modified JFs, namely, A-JF, A-St-JF and A-Si-JF showed higher surface roughness than JFun (Figure 3b–d). The rough surface morphology was the key factor for the enhancement of interfacial strength due to mechanical interlocking between surface treated JFs and rubber matrix [27,28].

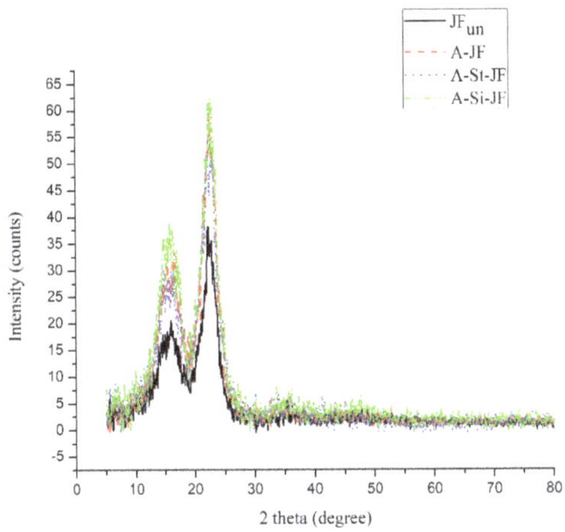

Figure 2. XRD pattern of unmodified and surface modified JFs.

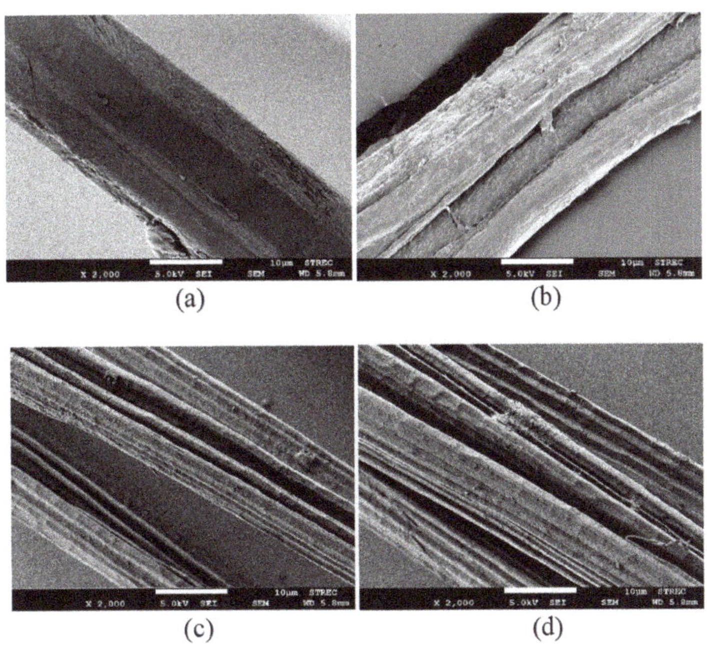

Figure 3. FESEM images of (**a**) JF_{un}, (**b**) A-JF, (**c**) A-St-JF and (**d**) A-Si-JF.

FTIR spectra of untreated and chemically treated JFs are shown in Figure 4. Both, untreated and surface treated JFs showed common peak at about 900 cm^{-1} due to the C–H bending mode of cellulose [27]. As shown in Figure 4a, there was a significant peak around 1737 cm^{-1} in the FTIR spectrum of JFun due to the C=O stretching vibration resulting from carboxyl and acetyl groups in hemicelluloses. This peak of hemicelluloses disappeared in the FTIR spectra of surface treated JFs. This result indicated the proper removal of waxy substances from JFs due to the surface treatments. As shown in Figure 4c, the FTIR spectrum of A-St-JF showed some characteristics band of stearic acid at about 2923 and 2856 cm^{-1} due to C–H stretching vibration of methylene groups [28]. Some interesting peaks were observed in the FTIR spectrum of A-Si-JF. As shown in Figure 4d, a clear peak at about 2924 cm^{-1} was attributed to the existence of C–H stretching vibration of CH_3 and CH_2 groups of TESPT [29]. In addition, the presence of a peak around 1100 cm^{-1} was associated with the formation of silicon-oxygen bond in the combined alkali/silane treated JFs [29,30].

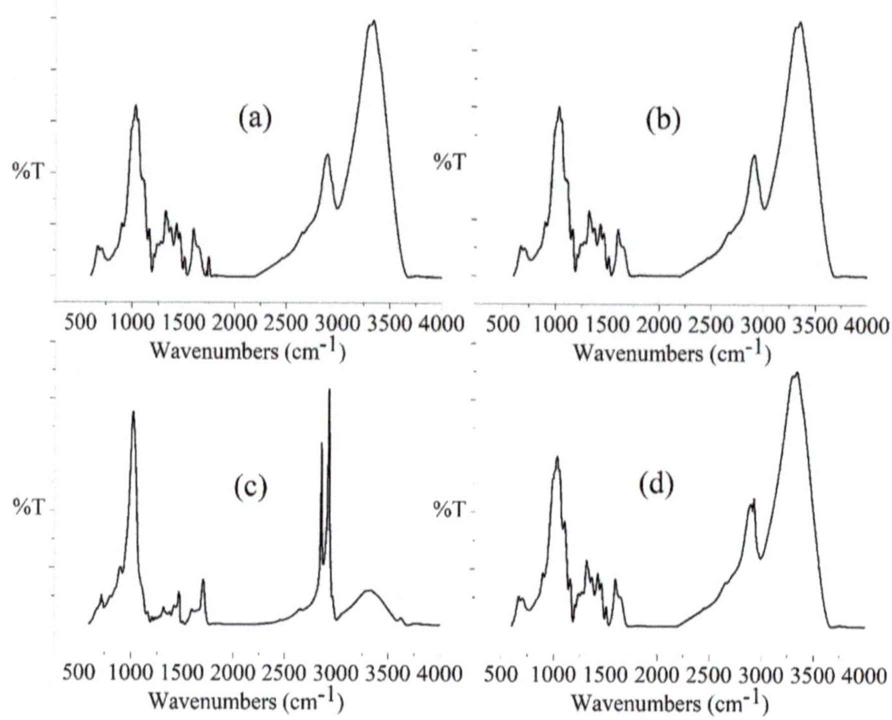

Figure 4. FTIR spectra of (**a**) JF_{un}, (**b**) A-JF, (**c**) A-St-JF and (**d**) A-Si-JF.

3.2. *Cure Characteristics*

The cure curves of unfilled and JFs filled NR composites are represented in Figure 5. Cure characteristics of various NR composites are also displayed separately in Table 2. The value of maximum torque (MH) of NR composites showed considerable increment due to the addition of both untreated and surface treated JFs as filler, which implies the restriction of the mobility of the NR chains in presence of filler [31]. In other words, the increase in the MH value of filled NR compounds was closely related to the increase in the stiffness of NR composites due to the incorporation of both untreated and surface treated JFs. Again, the value of torque difference i.e., the difference between maximum torque (MH) and minimum torque (ML) is the indirect measure of crosslink density for rubber composites [31,32]. As shown in Table 2, both untreated and surface treated JFs were able to increase the torque difference value of NR composites. Thus, there was a successful enrichment in the crosslink density of NR composites in presence of both untreated and surface treated JFs. Among the various filled NR composites, the torque difference value was found to be maximum for NR composite filled with A-Si-JF. This result might be explained by considering the exceptional improvement in the interfacial interaction between NR matrix and JFs after combined alkali/silane treatment [26].

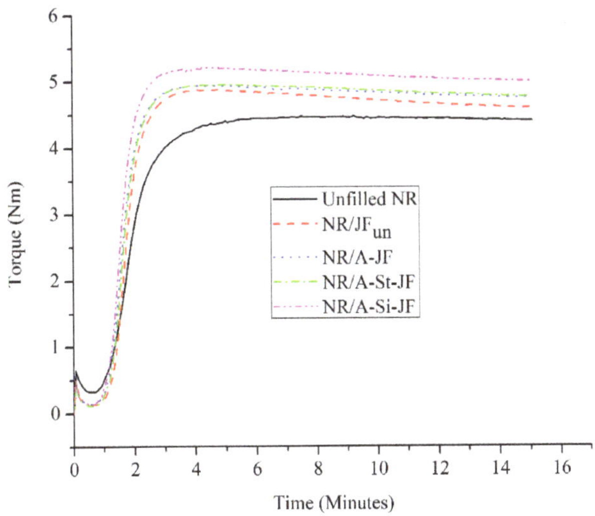

Figure 5. Cure curves of unfilled and JFs filled natural rubber (NR) composites.

Table 2. Cure properties of NR composites.

Formulation	Maximum Torque (Nm)	Torque Difference (Nm)	Scorch Time, t_2 (min)	Optimum Cure Time, t_{90} (min)
Unfilled NR	4.48	4.07	1.85	3.17
NR/JF$_{un}$	4.88	4.73	1.66	2.48
NR/A-JF	4.94	4.77	1.55	2.34
NR/A-St-JF	4.96	4.78	1.6	2.4
NR/A-Si-JF	5.23	4.98	1.5	2.26

The values of scorch time (t_2) and optimum cure time (t_{90}) decreased markedly due to the addition of both untreated and chemically treated JFs into the NR matrix. Actually, a rubber sample remains for a greater time on the mixing mill during preparation of filled rubber composites. Thus, for filled NR composites, the decreasing trend of t_2 and t_{90} was closely related to the generation of greater amount of heat due to additional friction [10]. NR/A-Si-JF had shorter t_2 and t_{90} than other JFs filled NR composites, which is attributed to the optimum dispersion of A-Si-JF within the NR matrix [10].

3.3. Mechanical Properties

The mechanical properties of NR composites in the presence of untreated and surface treated JFs are summarized in Table 3. The value of modulus at 100% elongation (M100) of NR composites showed clear increment due to the addition of both untreated and surface treated JFs. However, surface modified JFs were more effective to increase the M100 value of NR composites as compared to unmodified raw JFs. NR/A-Si-JF exhibited highest M100 value among the various JFs filled NR composites. The value of M100 was found to increase by 113% for NR/A-Si-JF system in comparison to an unfilled NR system. This result confirms the outstanding enhancement in the crosslink density of

NR/A-Si-JF system due to the excellent interfacial interaction between the NR matrix and A-Si-JF [30]. On the other hand, both untreated and surface treated JFs had a similar effect on the hardness of the NR composites. The hardness of the NR composites increased rapidly in the presence of JFs as filler, which indicates the formation of stiffer NR composites due to the addition of filler materials into the rubber matrix [32]. The variation of hardness followed the same trend as that of the M100 of filled NR composites and NR/A-Si-JF had higher hardness value as compared to other JFs filled NR composites. This might be due to the better crosslink density of NR/A-Si-JF composite than all other filled NR composites.

Table 3. Mechanical properties of NR composites.

Formulation	M_{100} (MPa)	Hardness (Shore A)	Tensile Strength (MPa)	Elongation at Break (%)	Crosslink Density $\times 10^5$ (mol cm^{-3})
Unfilled NR	0.83 ± 0.04	47 ± 2	12.07 ± 0.58	835 ± 15	7.21
NR/JF$_{un}$	1.20 ± 0.09	55 ± 2	10.52 ± 0.69	765 ± 15	8.54
NR/A-JF	1.49 ± 0.07	58 ± 1	14.21 ± 0.89	750 ± 20	8.76
NR/A-St-JF	1.63 ± 0.12	59 ± 1	15.91 ± 0.44	750 ± 20	9.01
NR/A-Si-JF	1.77 ± 0.11	62 ± 2	17.04 ± 0.52	770 ± 20	10.88

The tensile strength values of NR composites in presence of various types of JFs are also comparatively depicted in Table 3. The value of tensile strength showed a clear reduction due to incorporation of JFun into the NR matrix, which is attributed to the poor dispersion of untreated hydrophilic JFs within the hydrophobic NR matrix. However, surface modified JFs had slightly positive effect on the tensile strength of NR composites. The tensile strength value suggested the understandable reinforcing effect of various surface treated JFs in NR compounds. This result might be due to the improved dispersion of surface modified JFs within the NR matrix. Further, the tensile strength of NR/A-Si-JF was notably higher as compared to either NR/A-St-JF or NR/A-JF. Thus, combination of alkali and silane treatment was the most effective technique to improve the dispersion level of JFs within NR matrix. The value of elongation at break was lower for various JFs filled NR composites as compared to unfilled NR composite, which is due to the increase of stiffness and brittleness of NR composites in presence of filler [33].

3.4. Crosslink Density

Crosslink density is a unique parameter, which is closely linked with the cure and mechanical performances of filled rubber composites. The crosslink density values of unfilled and filled NR composites are displayed in Table 3. It was found that surface modified JFs based NR composites exhibited noticeably higher crosslink density as compared to unmodified JFs based NR composite, indicating excellent interfacial interaction between NR matrix and modified JFs. Among the three surface modified JFs, A-Si-JF facilitated greater crosslink density for NR composite as compared to either A-St-JF or A-JF. Therefore, TESPT had a clear role on the crosslink density of A-Si-JF filled NR composite. The plausible mechanism of cross-linking between A-Si-JF and NR chains is schematically illustrated in Figure 6. The variation of crosslink density was found to be in good agreement with the torque difference, hardness and tensile modulus of NR composites.

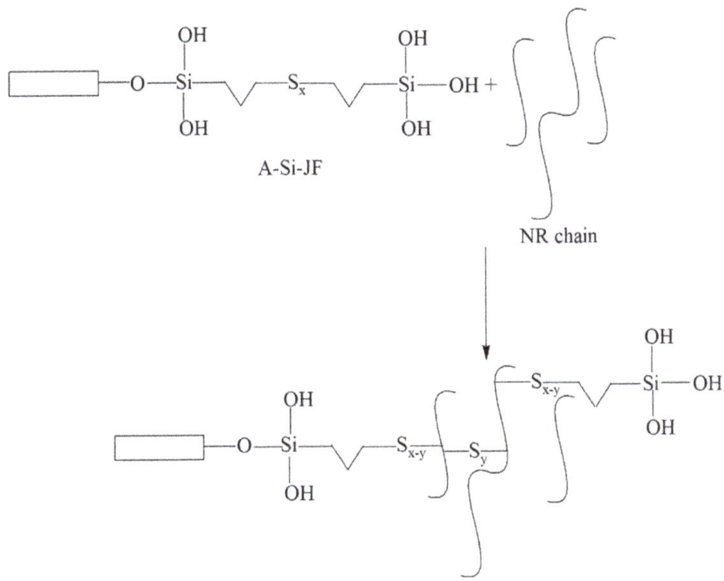

S_x= Polysulfide site, x = 2-8
S_y= Free sulfur, y = 1-7

Figure 6. The probable mechanism of cross-linking between A-Si-JF and NR chains.

3.5. Morphology of Composite Fracture Surfaces

Filler dispersion is an important parameter regarding the mechanical performances of filled rubber composites [34]. FESEM analysis was performed to compare the degree of dispersion of untreated and surface treated JFs in NR composites. Figure 7 shows the morphological characteristics of untreated and surface treated JFs filled NR composites fractured surfaces. As shown in Figure 7a, many holes were generated due to fibers pulling out from NR matrix in NR/JFun composite. This was due to the poor interfacial adhesion between hydrophobic NR matrix and hydrophilic JFs. As a result, the tensile strength of NR/JFun composite was lower than the unfilled NR composite. On the other hand, the fracture surfaces of NR/A-JF and NR/A-St-JF showed better wetting and dispersion of JFs within NR matrix (Figure 7b,c). Thus, moderate improvement in the mechanical properties was observed in the cases of NR/A-JF and NR/A-St-JF composites. As shown in Figure 7d, NR/A-Si-JF composite had very smooth and continuous surface, which indicates excellent interfacial adhesion between NR matrix and A-Si-JF. The remarkable enhancement in the tensile strength, modulus and crosslink density was reflected to the efficacious interfacial adhesion between NR matrix and A-Si-JF in NR/A-Si-JF composite.

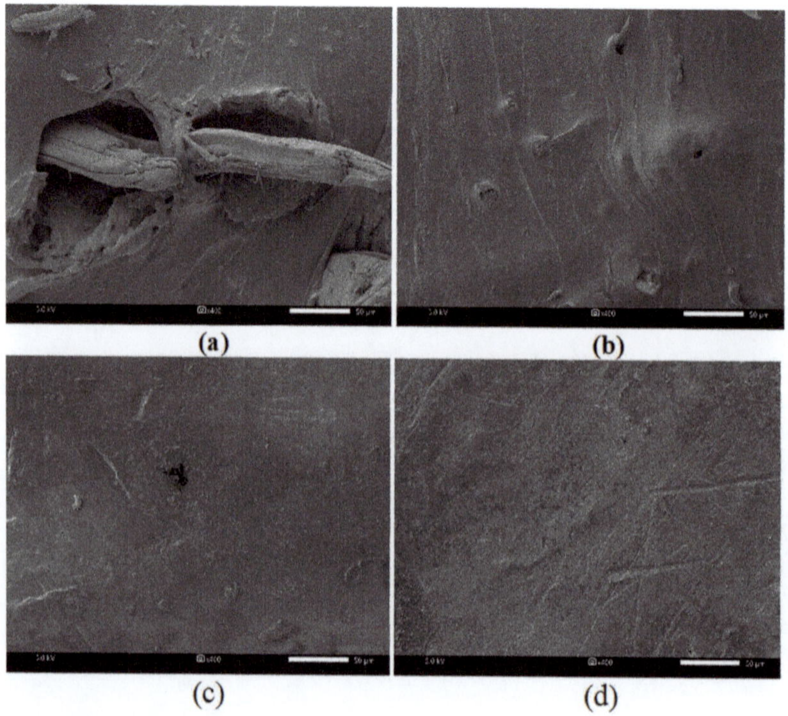

Figure 7. FESEM micrographs of (**a**) NR/JF_{un}, (**b**) NR/A-JF, (**c**) NR/A-St-JF and (**d**) NR/A-Si-JF.

3.6. Solvent Uptake Behaviour of JFs Filled NR Composites

Solvent uptake behavior provides indirect information about the interaction between cellulose based JFs and NR matrix [26]. The lowering of solvent uptake is related to the good interaction between JFs and NR matrix [26]. The plots of solvent uptake (weight percent) vs. (time) 1/2 of different JFs filled NR composites are illustrated in Figure 8. All the solvent uptake plots showed a similar pattern with speedy solvent uptake at smaller time region. The values of equilibrium solvent uptake (weight percent) of unfilled and JFs filled NR composites are also presented in Figure 9. The values of equilibrium solvent uptake of surface treated JFs filled NR composites were considerably lower as compared to either untreated JFs filled NR composite or unfilled NR composite. Moreover, the lowest value of equilibrium solvent uptake was found for an NR composite filled with A-Si-JF. Hence, the rubber-JFs interaction was improved obviously due to the surface modification of fibers by combined alkali/silane treatment.

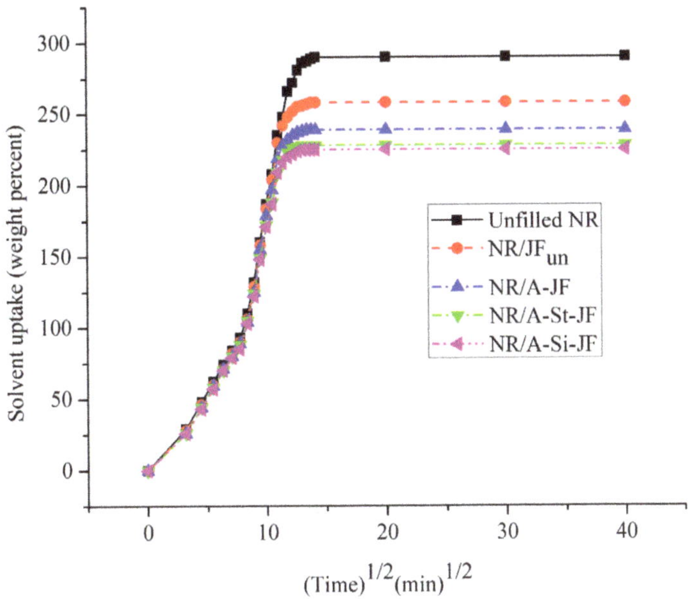

Figure 8. Solvent uptake behavior of unfilled and JFs filled NR composites.

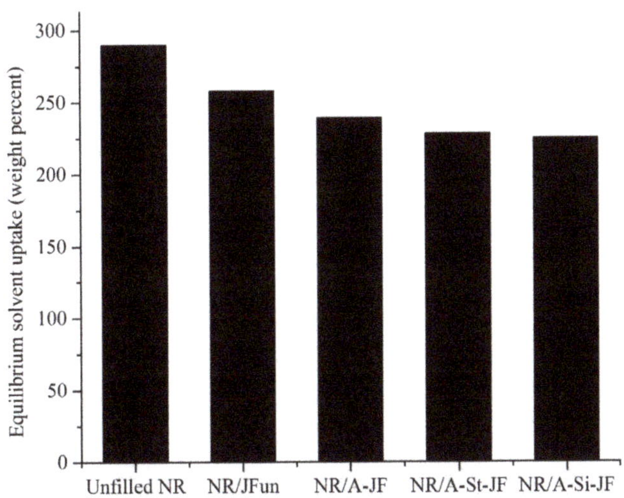

Figure 9. The values of equilibrium solvent uptake of different NR composites.

3.7. Thermal Properties of JFs Filled NR Composites

TGA study was utilized to compare the thermal stability of untreated and surface treated JFs filled NR composites. Different TGA curves of unfilled and filled NR samples are shown in Figure 10. The TGA results of various NR composites are summarized in Table 4. The thermal stabilities of various NR composites were examined in terms of temperature corresponds to 10% weight loss ($T_{10\%}$), temperature corresponds to 20% weight loss ($T_{20\%}$), temperature corresponds to 50% weight loss ($T_{50\%}$) and temperature corresponds to 80% weight loss ($T_{80\%}$). There was no improvement in the values of $T_{10\%}$ and $T_{20\%}$ of NR compounds in the presence of both untreated and surface treated JFs. The

NR/A-Si-JF composite showed a little increment in the values of $T_{50\%}$ and $T_{80\%}$ as compared to either unfilled NR or NR/JFun composites, which indicates a slight improvement in the thermal stability of NR composite in presence of A-Si-JF. Actually, the mobility of NR chains was restricted in the vicinity of A-Si-JF due to the presence of strong rubber-fibers interaction [35]. Thus, the diffusion of the degradation products from the NR/A-Si-JF system was little bit more difficult as compared to both unfilled NR and NR/JFun systems.

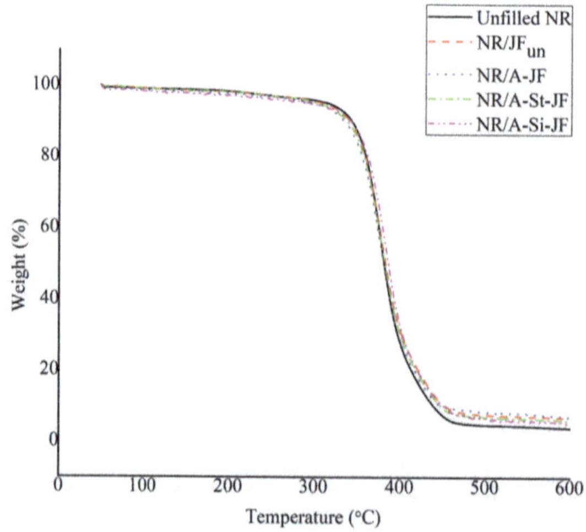

Figure 10. TGA curves of unfilled and JFs filled NR composites.

Table 4. Temperatures at different stages of degradation of NR composite.

Formulation	Temperature (°C)			
	$T_{10\%}$	$T_{20\%}$	$T_{50\%}$	$T_{80\%}$
Unfilled NR	343	359	381	413
NR/JF$_{un}$	341	359	382	420
NR/A-JF	335	355	381	417
NR/A-St-JF	338	357	382	419
NR/A-Si-JF	342	361	387	421

4. Conclusions

The main goal of the present study was to elucidate the potential of different surface treated JFs as a non-petroleum based filler for the advancement of commercially serviceable rubber technology. For this purpose, the surface of JFs was modified by using three different chemical approaches, i.e., alkali treatment, combined alkali/stearic acid treatment and combined alkali/silane treatment. The cure characteristics, mechanical, morphological, solvent uptake and thermal properties were comparatively assessed for NR composites filled with untreated and surface treated JFs. The value of the torque difference increased clearly due to the incorporation of both untreated and chemically treated JFs into NR matrix. Among the various JFs filled NR composites, NR/A-Si-JF provided highest value of torque difference. On the other hand, surface modified JFs based NR composites showed better mechanical properties as compared to either unmodified JFs based NR composite or unfilled NR composite. More importantly, the increment of mechanical properties was more predominant in case of NR/A-Si-JF composite as compared to either NR/A-St-JF or NR/A-JF. In addition, this was corroborated by the corresponding morphological analysis demonstrating an outstanding interfacial adhesion between NR matrix and A-Si-JF. The morphological observation was found thus to be in good agreement with the variation of mechanical properties in JFs filled NR composites. Moreover, A-Si-JF offered slightly better thermal stability for NR compounds than untreated JFs. For the first time, A-Si-JF was utilized as a novel green filler in NR based rubber composites. Finally, it could be envisaged that A-Si-JF may find important place as a suitable filler for the progress of commercially viable and environmentally friendly green rubber technology.

Author Contributions: Conceptualization, K.R., S.C.D. and P.P.; methodology, K.R. and A.P.; investigation, K.R., S.C.D. and A.P.; writing—original draft preparation, K.R. and L.T.; writing—review and editing, K.R., S.C.D. and P.P. All authors have read agreed to the published version of the manuscript.

Funding: This research received no external funding.

Acknowledgments: One of the authors, Kumarjyoti Roy would like to thank senior postdoctoral fellowship supported by Ratchadaphiseksomphot Endowment Fund, Chulalongkorn University for fellowship assistance.

Conflicts of Interest: The authors declare no conflict of interest.

References

1. Andideh, M.; Naderi, G.; Ghoreishy, M.H.R.; Soltani, S. Effects of Nanoclay and Short Nylon Fiber on Morphology and Mechanical Properties of Nanocomposites Based on NR/SBR. *Fibers Polym.* **2014**, *15*, 814–822. [CrossRef]
2. Roy, K.; Debnath, S.C.; Das, A.; Heinrich, G.; Potiyaraj, P. Exploring the synergistic effect of short jute fiber and nanoclay on the mechanical, dynamic mechanical and thermal properties of natural rubber composites. *Polym. Test.* **2018**, *67*, 487–493. [CrossRef]
3. Soltani, S.; Naderi, G.; Mohseniyan, S. Mechanical, Morphological and Rheological Properties of Short Nylon Fiber Reinforced Acrylonitrile-Butadiene Rubber Composites. *Fibers Polym.* **2014**, *15*, 2360–2369. [CrossRef]
4. Rana, A.K.; Mandal, A.; Bandyopadhyay, S. Short jute fiber reinforced polypropylene composites: Effect of compatibiliser, impact modifier and fiber loading. *Compos. Sci. Technol.* **2003**, *63*, 801–806. [CrossRef]
5. Gurunathan, T.; Mohanty, S.; Nayak, S.K. A review of the recent developments in biocomposites based on natural fibres and their application perspectives. *Compos. Part A Appl. Sci. Manuf.* **2015**, *77*, 1–25. [CrossRef]
6. Murty, V.M.; De, S.K. Short jute fiber reinforced rubber composites. *Rubber Chem. Technol.* **1982**, *55*, 287–308. [CrossRef]
7. Zaman, H.U.; Khan, R.A.; Haque, M.E.; Khan, M.A.; Khan, A.; Huq, T.; Noor, N.; Rahman, M.; Rahman, K.M.; Huq, D.; et al. Preparation and mechanical characterization of jute reinforced polypropylene/natural rubber composite. *J. Reinforc. Plast. Compos.* **2010**, *29*, 3064–3065. [CrossRef]
8. Pantamanatsopa, P.; Ariyawiriyanan, W.; Meekeaw, T.; Suthamyong, R.; Arrub, K.; Hamada, H. Effect of modified jute fiber on mechanical properties of Green rubber composite. *Energy Procedia* **2014**, *56*, 641–647. [CrossRef]
9. Tzounis, L.; Debnath, S.; Rooj, S.; Fischer, D.; Mäder, E.; Das, A.; Stamm, M.; Heinrich, G. High performance natural rubber composites with a hierarchical reinforcement structure of carbon nanotube modified natural fibers. *Mater. Des.* **2014**, *58*, 1–11. [CrossRef]
10. Ismail, H.; Shuhelmy, S.; Edyham, M.R. The effects of a silane coupling agent on curing characteristics and mechanical properties of bamboo fibre filled natural rubber composites. *Eur. Polym. J.* **2002**, *38*, 39–47. [CrossRef]
11. Ismail, H. The effects of filler loading and a silane coupling agent on the dynamic properties and swelling behaviour of bamboo filled natural rubber compounds. *J. Elastom. Plast.* **2003**, *35*, 149–159. [CrossRef]
12. Geethamma, V.G.; Mathew, K.T.; Lakshminarayanan, R.; Thomas, S. Composite of short coir fibres and natural rubber: Effect of chemical modification, loading and orientation of fibre. *Polymer* **1998**, *39*, 1483–1491. [CrossRef]
13. Geethamma, V.G.; Kalaprasad, G.; Groeninckx, G.; Thomas, S. Dynamic mechanical behavior of short coir fiber reinforced natural rubber composites. *Compos. Part A Appl. Sci. Manuf.* **2005**, *36*, 1499–1506. [CrossRef]
14. Xu, Z.H.; Kong, Z.N. Mechanical and thermal properties of short-coirfiber-reinforced natural rubber/polyethylene composites. *Mech. Compos. Mater.* **2014**, *50*, 353–358. [CrossRef]
15. Jacob, M.; Thomas, S.; Varughese, K.T. Mechanical properties of sisal/oil palm hybrid fiber reinforced natural rubber composites. *Compos. Sci. Technol.* **2004**, *64*, 955–965. [CrossRef]

16. John, M.J.; Varughese, K.T.; Thomas, S. Green composites from natural fibers and natural rubber: Effect of fiber ratio on mechanical and swelling characteristics. *J. Nat. Fibers* **2008**, *5*, 47–60. [CrossRef]
17. Joseph, S.; Joseph, K.; Thomas, S. Green composites from natural rubber and oil palm fiber: Physical and mechanical properties. *Int. J. Polym. Mater.* **2006**, *55*, 925–945. [CrossRef]
18. Azammi, A.M.N.; Sapuan, S.M.; Ishak, M.R.; Sultan, M.T.H. Mechanical and Thermal Properties of Kenaf Reinforced Thermoplastic Polyurethane (TPU)-Natural Rubber (NR) Composites. Fiber. *Polymer* **2018**, *19*, 446–451.
19. De, D.; De, D.; Adhikari, B. The effect of grass fiber filler on curing characteristics and mechanical properties of natural rubber. *Polym. Adv. Technol.* **2004**, *15*, 708–715. [CrossRef]
20. De, D.; De, D.; Adhikari, B. Curing characteristics and mechanical properties of alkali-treated grass-fiber-filled natural rubber composites and effects of bonding agent. *J. Appl. Polym. Sci.* **2006**, *101*, 3151–3160. [CrossRef]
21. Manaila, E.; Stelescu, M.D.; Doroftei, F. Polymeric composites based on natural rubber and hemp fibers. *Iran. Polym. J.* **2015**, *24*, 135–148. [CrossRef]
22. Moonart, U.; Utara, S. Effect of surface treatments and filler loading on the properties of hemp fiber/natural rubber composites. *Cellulose* **2019**, *26*, 7271–7295. [CrossRef]
23. Hariwongsanupab, N.; Thanawan, S.; Amornsakchai, T.; Vallat, M.F.; Mougin, K. Improving the mechanical properties of short pineapple leaf fiber reinforced natural rubber by blending with acrylonitrile butadiene rubber. *Polym. Test.* **2017**, *57*, 94–100. [CrossRef]
24. Yantaboot, K.; Amornsakchai, T. Effect of preparation methods and carbon black distribution on mechanical properties of short pineapple leaf fiber-carbon black reinforced natural rubber hybrid composites. *Polym. Test.* **2017**, *61*, 223–228. [CrossRef]
25. Rahman, M.R.; Huque, M.M.; Islam, M.N.; Hasan, M. Improvement of physico-mechanical properties of jute fiber reinforced polypropylene composites by post-treatment. *Compos. Part A Appl. Sci. Manuf.* **2008**, *39*, 1739–1747. [CrossRef]
26. Roy, K.; Potiyaraj, P. Development of high performance microcrystalline cellulose based natural rubber composites using maleated natural rubber as compatibilizer. *Cellulose* **2018**, *25*, 1077–1087. [CrossRef]
27. Liu, X.; Hao, S.J.; Cui, Y.H.; Chen, H. Improvement on the interfacial compatibility of jute fiber-reinforced polypropylene composites by different surface treatments. *J. Ind. Text.* **2018**. [CrossRef]
28. Dilfi, K.F.A.; Balan, A.; Bin, H.; Xian, G.; Thomas, S. Effect of Surface Modification of Jute Fiber on the Mechanical Properties and Durability of Jute Fiber-Reinforced Epoxy Composites. *Polym. Compos.* **2018**, *39*, E2519–E2528. [CrossRef]
29. Roy, K.; Alam, M.N.; Mandal, S.K.; Debnath, S.C. Surface modification of sol–gel derived nano zinc oxide (ZnO) and the study of its effect on the properties of styrene–butadiene rubber (SBR) nanocomposites. *J. Nanostruct. Chem.* **2014**, *4*, 133–142. [CrossRef]
30. Roy, K.; Potiyaraj, P. Exploring the comparative effect of silane coupling agents with different functional groups on the cure, mechanical and thermal properties of nano-alumina (Al_2O_3)-based natural rubber (NR) compounds. *Polym. Bull.* **2019**, *76*, 883–902. [CrossRef]
31. Ismail, H.; Shaari, S.M. Curing characteristics, tensile properties and morphology of palm ash/halloysite nanotubes/ethylene-propylene-diene monomer (EPDM) hybrid composites. *Polym. Test.* **2010**, *29*, 872–878. [CrossRef]
32. Intiya, W.; Thepsuwan, U.; Sirisinha, C.; Sae-Oui, P. Possible use of sludge ash as filler in natural rubber. *J. Mater. Cycles Waste Manag.* **2017**, *19*, 774–781. [CrossRef]
33. Ismail, H.; Rusli, A.; Rashid, A.A. Maleated natural rubber as a coupling agent for paper sludge filled natural rubber composites. *Polym. Test.* **2005**, *24*, 856–862. [CrossRef]
34. Roy, K.; Debnath, S.C.; Potiyaraj, P. A critical review on the utilization of various reinforcement modifiers in filled rubber composites. *J. Elastom. Plast.* **2019**. [CrossRef]
35. Visakh, P.M.; Thomas, S.; Oksman, K.; Mathew, A.P. Crosslinked natural rubber nanocomposites reinforced with cellulose whiskers isolated from bamboo waste: Processing and mechanical/thermal properties. *Compos. Part A Appl. Sci. Manuf.* **2012**, *43*, 735–741. [CrossRef]

polymers MDPI

Article

Experimental Evaluation of Low Velocity Impact Properties and Damage Progression on Bamboo/Glass Hybrid Composites Subjected to Different Impact Energy Levels

Ain Umaira Md Shah [1,2,*], Mohamed Thariq Hameed Sultan [1,2,3,*] and Syafiqah Nur Azrie Safri [1]

1 Laboratory of Biocomposite Technology, Institute of Tropical Forestry and Forest Products (INTROP), UPM Serdang 43400, Selangor Darul Ehsan, Malaysia; snasafri@gmail.com
2 Department of Aerospace Engineering, Faculty of Engineering, Universiti Putra Malaysia, UPM Serdang 43400, Selangor Darul Ehsan, Malaysia
3 Aerospace Malaysia Innovation Centre (944751-A), Prime Minister's Department, MIGHT Partnership Hub, Jalan Impact, Cyberjaya 63000, Selangor Darul Ehsan, Malaysia
* Correspondence: ainumaira91@gmail.com (A.U.M.S.); thariq@upm.edu.my (M.T.H.S.)

Received: 20 March 2020; Accepted: 18 April 2020; Published: 4 June 2020

Abstract: Six impact energy values, ranging from 2.5 J to 10 J, were applied to study the impact properties of neat epoxy and bamboo composites, while six impact energy values, ranging from 10 J to 35 J, were applied on bamboo/glass hybrid composites. Woven glass fibre was embedded at the outermost top and bottom layer of bamboo powder-filled epoxy composites, producing sandwich structured hybrid composites through lay-up and molding techniques. A drop weight impact test was performed to study the impact properties. A peak force analysis showed that neat epoxy has the stiffest projectile for targeting interaction, while inconsistent peak force data was collected for the non-hybrid composites. The non-hybrid composites could withstand up to 10 J, while the hybrid composites showed a total failure at 35 J. It can be concluded that increasing the filler loading lessened the severity of damages in non-hybrid composites, while introducing the woven glass fibre could slow down the penetration of the impactor, thus lowering the chances of a total failure of the composites.

Keywords: low velocity impact; hybrid composites; damage progression; glass fibre; bamboo

---

1. Introduction

Generally, mechanical properties are the most important information that needs to be measured in materials [1]. However, in real life situations, impact is one of the very common phenomena experienced by all materials and structures. An instantaneous load applied on a surface during an impact event can cause unpredictable damages, which can sometimes lead to total structural failure. It is even worse if the low velocity impact event caused non-visible damages; the accumulated damages after several repeated events can lead to serious failures [2,3]. Experimentally, low velocity impact can be simulated using the drop test rig instrument. The Izod and Charpy impact testers are more descriptive of fracture toughness, which can be considered as the mechanical properties of materials [4].

It was cited from one source that low velocity impact is an impact event below 10 m/s, while intermediate, high and hypervelocity impacts correspond to the range of 10 m/s–50 m/s, 50 m/s–1000 m/s and 2 km/s–5 km/s respectively. Different ranges of impact velocities are very important to analyse, rather than saying that only high velocity needs more attention, as each structure has its own surroundings and working environment [5]. Different mechanisms of damage initiations can be observed being subjected to impact loading, and these observations illustrate the dissipation energy from the impact force to the materials [6].

Compared to metal materials, worse damages were experienced by fibre-reinforced composites as damages that could occur in a wider form, and which are matrix cracking, lamina splitting, fibre-matrix interfacial disbanding, and the last failing mechanism of fibre breakage. These situations lead to a complex analysis of composites compared to metals [7,8]. A study was specifically conducted to examine the effects of thickness, stacking sequence and scaling technique on the barely visible impact damage of composite laminates. Interlaminar damages in terms of delamination are one of the observations analysed using the ultrasonic phased array inspection technique in this study [9]. Further analyses on the impacted samples were conducted to determine the residual strength through compression after impact testing. It was in a good agreement, as samples impacted with a higher impact energy possessed a lower residual compressive strength due to the higher degree of damages that occurred [10]. An overall study on the impact properties of sugar palm/glass reinforced epoxy composites covered the impact testing, damage analysis using c-scan technique and post-impact properties through compression after impact testing. The damage area of sugar palm/glass hybrid composites increased proportionally with an increase in the impact energy levels [11]. Different parameters for stitched and unstitched flax fibre-reinforced epoxy composites were studied for their impact damages. The overall findings showed that delamination was not the main damage mode in both types of composites. Besides that, the propagation of in-plane cracks within the composites was not enhanced by stitching fibres compared to unstitched fibres [12].

The use of bamboo in structural applications as well as daily life utensils had been explored since a hundred years back. The light-weight advantage of natural bamboo culms, with a comparable strength to mild steel for some species, suits the current demand for producing environmentally friendly materials with acceptable properties comparable to conventional ones. The distinctive fast-growing features of bamboo plant make it among the most promising sources of fibre supplies in the continuous production of composites [13]. Similarly to any other natural fibres, different methods of extraction give different types of bamboo fibres with strength variations and, by some measures, lower strength properties compared to the bamboo culms [14].

Bamboo-Polyvinyl Chloride (PVC) composites were developed in an attempt to replace the wood-PVC composites. The inclusion of bamboo particles in PVC had significantly improved the flexural modulus of elasticity compared to neat PVC, which indirectly enhanced the potential of bamboo-PVC composites as a replacement to wood-PVC composites [15]. In a different study, increasing the concentration of tetraethyl orthosilicate (TEOS) in ethanol solution used to modify the bamboo flour slightly increased the shear modulus of bamboo/polyamide composites. However, the effects of different concentration of TEOS was not observed on the percentage of crystallinity of bamboo/polyamide composites through the differential scanning calorimetry analysis conducted in the similar study [16]. Adding 20% bamboo fibre in polypropylene (PP) increased all the tensile, flexural and impact strengths of bamboo/PP composites compared to neat PP. Further research with the inclusion of hollow glass microspheres in the bamboo/PP composites suggested that the hybrid composites can be used in light-weight and high-strength components in engineering applications [17].

Many studies have been conducted by researchers around the world to widen the use of bamboo fibres in composites, yet most of the studies have focused on the mechanical properties with only a small area covering the impact properties. The current study is the continuation of a previous study on the mechanical properties of bamboo/glass hybrid composites [18]. In the present study, the damage progressions on short bamboo fibre composites and bamboo/glass hybrid composites were analysed. The initiative of using bamboo powder as fillers in some measure filled the gap of utilising all forms of bamboo fibres in the development of bamboo composites.

2. Experimental Section

2.1. Materials

Epoxy matrix typed Epoxamite 100 with 103 slow hardeners were used in this study. Bamboo from the species of Bambusa vulgaris was collected from Raub, Pahang in Peninsular Malaysia, and the extraction process was carried out in several laboratories within Universiti Putra Malaysia. The preparation of bamboo powder was mentioned in the previous study [19]. Woven glass fibre-typed E 600 was chosen to be hybrid with bamboo powder, as it possesses a higher impact strength when compared to other types of glass fibre [20]. Table 1 lists the general properties of the epoxy matrix and E-glass fibres obtained from the suppliers.

Table 1. Different impact energies applied to the EP, EP-BF composites and EP/G-BF composites.

Materials	Epoxy Matrix	E-Glass Fibre
Density (g/cm^3)	1.10	2.58
Tensile Strength (MPa)	54	3445
Tensile Modulus (GPa)	3.2	72.3

2.2. Composite Fabrication

The neat epoxy, short bamboo fibre composites and bamboo/glass hybrid composites were fabricated as reported in the previous study [18]. The fabricated composites were listed as neat epoxy (EP), non-hybrid bamboo composites with 10% loading (EP-BF10) and 30% loading (EP-BF30), and bamboo /glass hybrid composites with 10% loading (EP/G-BF10) and 30% loading (EP/G-BF30).

2.3. Characterisation of Impact Properties

The low velocity impact was simulated using an instrumented drop weight impact test machine, model IMATEK IM10, at the Faculty of Engineering, Universiti Putra Malaysia. The impact testing was conducted with five repeatability samples. The instrument was equipped with IMATEK Impact Analysis software, to record and process the impact results data. A hemispherical tip striker with a radius of 5 mm attached to a variable weight resulting in a total weight of 5.101 kg was dropped from several desired heights onto the clamped sample. Different heights were used to represent different magnitudes of the impact energy applied onto the samples, based on the following equation:

$$E_I = mgh \tag{1}$$

where m is the total mass of the impactor, 5.101 kg; g is the gravitational acceleration, 9.81 m/s^2; and h is the height of the impactor.

The raw data of the force, time, displacement, velocity and energy absorbed by the samples were recorded and calculated by the installed software. Table 2 shows the impact energy applied on the EP, EP-BF composites and EP/G-BF composites.

Table 2. Different impact energies applied on the EP, EP-BF composites and EP/G-BF composites.

Composites	EP and EP-BF Composites					EP/G-BF Composites					
Impact Energy (J)	2.50	3.75	4.40	5.0	10.0	10.0	15.0	20.0	25.0	30.0	35.0

Different ranges of impact energy were applied to the non-hybrid and hybrid composites so as to achieve the maximum impact energy withstood by each type of composite.

2.4. Characterisation of Impact Damages

The dye penetrant method was applied to observe the damages on the impacted samples. The length of the matrix cracking was measured to characterise the damage propagation as the response to different impact energies on the EP-BF composites, while the area of damage was measured for the EP/G-BF composites [21].

3. Results and Discussion

3.1. Force Displacement Analysis

Important information regarding the damage progression within the sample during an impact event can be obtained from the force-displacement graph. The movement of the impactor and the deformation of the impacted surface of the sample during contact with the impactor are marked as displacement values in the graph [22]. Figure 1 shows the force-displacement graphs of EP at the lowest impact energy level of 2.5 J.

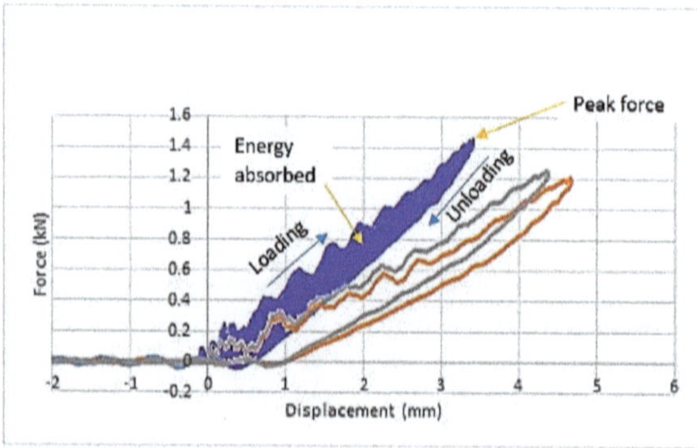

Figure 1. Force-displacement graphs of EP at the 2.5 J impact energy level showing closed curves.

The overlapping of the three graphs in the figure shows the repeatability of the three samples for EP under the same magnitude of impact energy. Testing conducted on all samples, for each impact energy level, was also repeated three times to confirm the repeatability of the results. The closed curve from the force-displacement graph indicates the non-full penetration damage of the sample tested, which indirectly explains that the full penetration of the impactor into the sample will produce an open curve in the force-displacement graph [23].

The ascending and descending parts of the closed curve explain the loading and unloading conditions, respectively. The ascending part also provides information about the impact bending stiffness of the samples. The greater the peak force, the stiffer the projectile-to-target interaction, thus shortening the contact period of the impactor onto the surface [24]. The relationship between a greater peak force and a stiffer target is a representation of a situation in which greater force is needed to initiate damage in stiffer materials. Besides the target stiffness, the peak force also depends on the magnitude of the impact energy from the impactor.

The peak deflection or peak deformation occurring in the sample is the value of the peak displacement from the force-displacement graph. In most graphs, the point of the peak deformation value is almost the same as the point for the peak force. However, it is clearly understood that the peak deformation can be identified as the turning point at which the force curve returns to zero after the loading condition or ascending curve, while the peak force is the maximum value in the vertical direction of the graph. It is clear that both the peak force and the peak deformation values were obtained from two different points [25].

Another important value extracted from the force-displacement graph is the energy absorbed by the samples, which can be determined from the area under this graph. The energy absorbed is the kinetic energy transferred from the impactor to the samples during impact [24]. Both the energy absorbed throughout the impact event and the energy absorbed up to the peak deformation can be obtained from the force-displacement graph as shown in Figure 1. An example of a force-displacement graph for fully penetrated or perforated samples during impact can be seen in Figure 2.

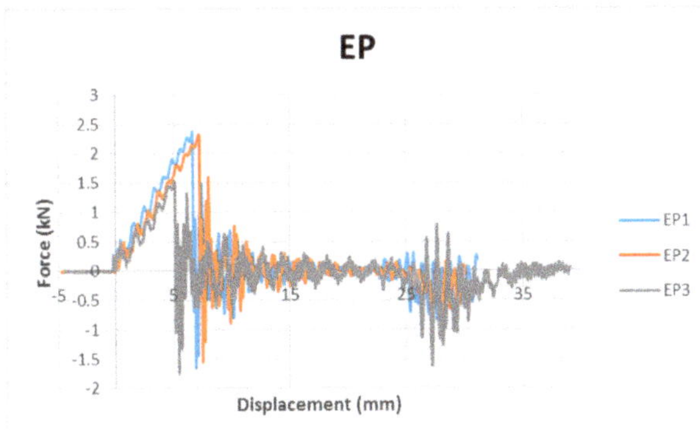

Figure 2. Force displacement graphs of EP at a 15 J impact energy level showing open curves indicating full penetration.

The open curves show that the displacement increased monotonically with a decreasing force. The penetration of the samples demonstrates a situation where the force applied exceeded the maximum allowable force for the samples [24]. Throughout the low velocity impact analysis, the full penetration events will not be discussed. Therefore, the force displacement graphs of all the samples were first analysed to exclude the data for the penetrated samples. This explained the different maximum impact energy levels for each type of sample discussed in this study. The perfect overlapping graphs in Figure 2 confirmed the repeatability of the tests and the consistency of the samples. The excluded data for the broken samples was confirmed after these three repeatability tests.

It can be seen that different samples can withstand different maximum impact energy levels. Obviously, the second range of impact energy, as listed in Table 2, was higher when compared to the first range, as the inclusion of woven glass fibre was assumed to slow down the penetration of the impactor, thus increasing the maximum allowable force impacted on the surface of the composites. For all composites, the data analysed from the force-displacement graphs is presented in Tables 3 and 4, which list the first and second ranges of impact energy applied on the samples, respectively. The relationship between the tabulated values will be further discussed in the following sections.

3.2. Peak Force Variation with Impact Energy

Figure 3 shows the variation of the peak force for the EP and EP-BF composites at the first range of impact energy levels.

In Figure 3, it can be seen that the EP has the highest peak force when compared to EP-BF10 and EP-BF30 for all impact energy levels. This leads to the first conclusion that epoxy has the stiffest projectile-to-target interaction [24]. Comparing EP-BF10 and EP-BF30 gives a different trend at different impact energy levels. At an impact energy of 2.5 J and 4.4 J, the EP-BF10 shows a higher peak force compared to EP-BF30. However, at an impact energy of 3.75 J, the EP-BF10 has a lower peak force compared to EP-BF30 with a significant value. These inconsistent values were caused by the random orientation of the bamboo powder in the epoxy and the agglomeration of powder that might be occurring, which results from the poor distribution of the bamboo powder during fabrication [18].

Based on the first conclusion for epoxy, a comparison of the different loadings of bamboo composites made at impact energies of 2.5 J and 4.4 J is more consistent when compared to the comparison made at an impact energy of 3.75 J; these comparisons show that lower bamboo filler loading gives a stiffer projectile-to-target interaction [24]. However, at an impact energy of 5 J, the EP-BF30 could still withstand the force without breaking, while the EP-BF10 experienced total damage. This situation shows that EP-BF30 has good strength but lower stiffness, while the EP-BF10 shows the opposite trend. In the second range of impact energy levels, the hybrid composites respond consistently to the force applied, as illustrated in Figure 4.

Table 3. Data analysed from the force displacement graphs of EP, EP-BF10 and EP-BF30 at the first range of the impact energy level.

Sample	Impact Energy (J)	Peak Force (kN)	Energy Absorbed (J)	Peak Deformation (mm)
EP	2.50	1.31 (0.03)	0.85 (0.01)	4.16 (0.05)
	3.75	1.46 (0.02)	1.35 (0.02)	5.19 (0.05)
	4.40	1.57 (0.02)	1.61 (0.02)	5.51 (0.05)
	5.00	1.78 (0.03)	1.78 (0.03)	5.45 (0.03)
	10.00	2.57 (0.02)	4.72 (0.04)	7.23 (0.05)
EP-BF10	2.50	1.14 (0.01)	2.12 (0.04)	4.88 (0.03)
	3.75	0.65 (0.02)	3.32 (0.03)	9.22 (0.04)
	4.40	1.02 (0.04)	2.70 (0.03)	6.94 (0.04)
	5.00	-	-	-
	10.00	-	-	-
EP-BF30	2.50	0.91 (0.02)	1.76 (0.03)	3.99 (0.02)
	3.75	1.42 (0.03)	2.39 (0.05)	3.99 (0.03)
	4.40	0.98 (0.02)	2.95 (0.02)	6.43 (0.02)
	5.00	1.11 (0.02)	3.21 (0.02)	5.93 (0.05)
	10.00	-	-	-

Table 4. Data analysed from the force displacement graphs of EP/G-BF10 and EP/G-BF30 at the second range of the impact energy level.

Sample	Impact Energy (J)	Peak Force (kN)	Energy Absorbed (J)	Peak Deformation (mm)
EP/G-BF10	10	3.54 (0.04)	5.01 (0.03)	4.91 (0.04)
	15	3.78 (0.05)	9.29 (0.04)	6.41 (0.5)
	20	4.28 (0.05)	12.80 (0.07)	7.63 (0.05)
	25	4.87 (0.06)	17.05 (0.07)	8.53 (0.04)
	30	4.97 (0.05)	21.42 (0.07)	9.87 (0.05)
	35	5.24 (0.04)	27.92 (0.05)	11.32 (0.07)
EP/G-BF30	10	3.99 (0.03)	5.02 (0.05)	4.01 (0.05)
	15	4.55 (0.06)	8.10 (0.05)	4.89 (0.05)
	20	5.04 (0.05)	11.03 (0.04)	5.97 (0.04)
	25	5.55 (0.04)	14.85 (0.07)	6.63 (0.04)
	30	5.73 (0.05)	20.26 (0.05)	7.83 (0.06)
	35	6.10 (0.05)	25.87 (0.08)	8.96 (0.08)

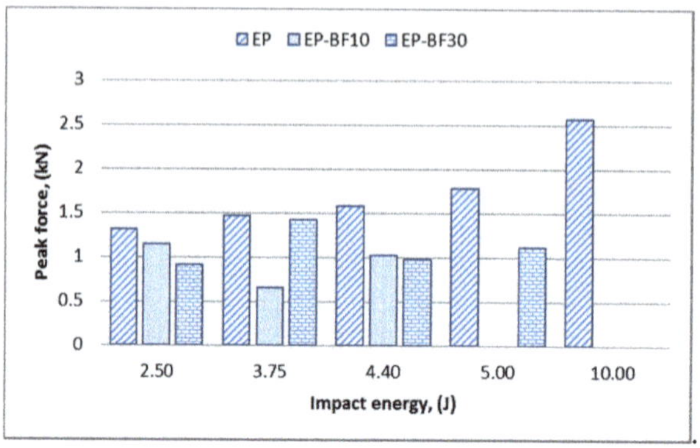

Figure 3. Variation of the peak force for the EP and EP-BF composites at the first range of impact energy levels.

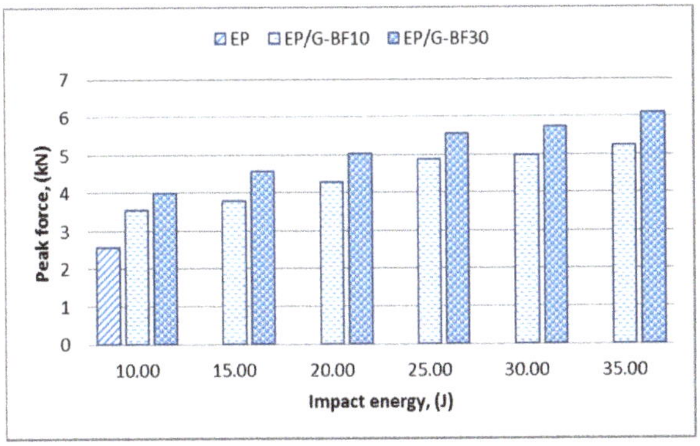

Figure 4. Variation of the peak force for the EP and EP/G-BF composites in the second range of impact energy levels.

The peak force increased as the impact energy increased for both loadings of hybrid composites. The inclusion of woven-type glass fibre on the outermost surface of the composites helps with a better force distribution when compared to the random orientation of fibre, resulting in a more consistent data trend [22]. The highest impact energy marked from the first range of impact energy levels in Figure 3, which was 10 J, was the lowest value in the second range of impact energy levels in Figure 4. At this impact energy, only the EP was comparable to the EP/G-BFC, while the EP-BFC samples failed and experienced total damage. From the values shown in both Figures 3 and 4, the hybrid composites tend to have a higher peak force to initiate damages as expected. Moreover, woven-type glass fibres are good in impact resistant, and it is expected that a woven-type fibre of any material will have better resistance towards impact when compared to other types of fibre such as random and unidirectional fibres [26].

3.3. Energy Absorbed Variation with Impact Energy

The impact energy supplied during an impact event is converted into two fractions, which are the loss elastic energy and the energy absorbed by the sample. The absorbed energy is presented by the damage mechanisms on the sample or structure, where more severe damage can be an indication of more energy being absorbed [24,26]. The severity of the damage is subject to the mechanical properties of the reinforcement and the matrix, the shape of the impactor tip, the fibre orientation, the sample's geometry and the impact energy levels [21].

Each sample absorbed a different amount of energy at each impact energy level, and thus a direct relationship of impact energy with the amount of absorbed energy was not advisable for explaining the severity of the damage on the samples. It is understood that a higher impact energy will cause a higher amount of impact energy to be absorbed, as depicted in Tables 1 and 2 [25]. Therefore, the percentage of energy absorbed by each sample at the respective impact energy level presents a better relationship of the impact energy with the absorbed energy and explains the severity of damage on the samples.

Figures 5 and 6 present the percentage of energy absorbed by the EP, EP-BF and EP/G-BF composites, respectively.

It was seen that an inconsistent trend was represented by the EP-BF composites compared to the EP in Figure 5. This is due to the random orientation of the short bamboo fibres, since the force being applied cannot be distributed evenly. At the same time, the agglomeration of bamboo powder that might be located within the structure tends to either slow down the damage propagation or worsen the damage by integrating a larger surface damage [8,13]. The inconsistent response of randomly oriented bamboo composites towards impact causes an inconsistent energy to be absorbed by the samples. This concept is illustrated in Figure 7.

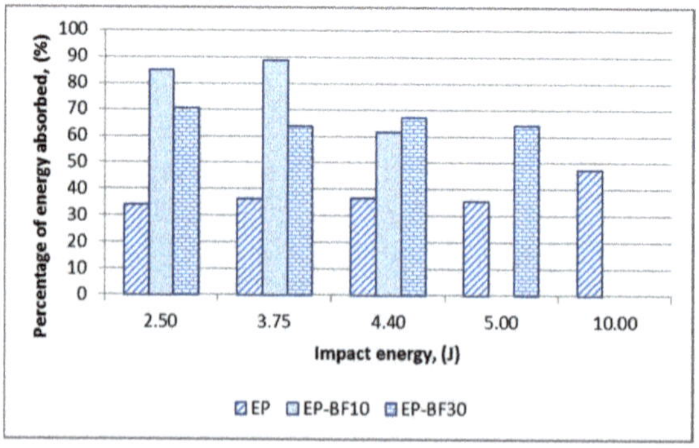

Figure 5. Percentage of energy absorbed by the EP and EP-BF composites at the first range of impact energy levels.

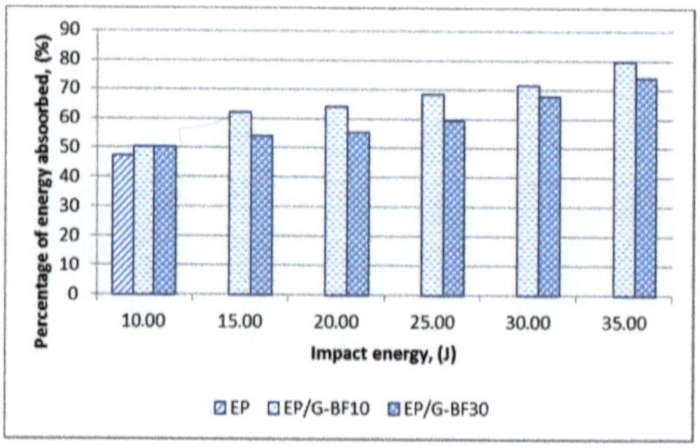

Figure 6. Percentage of energy absorbed by the EP and EP/G-BF composites at the second range of impact energy levels.

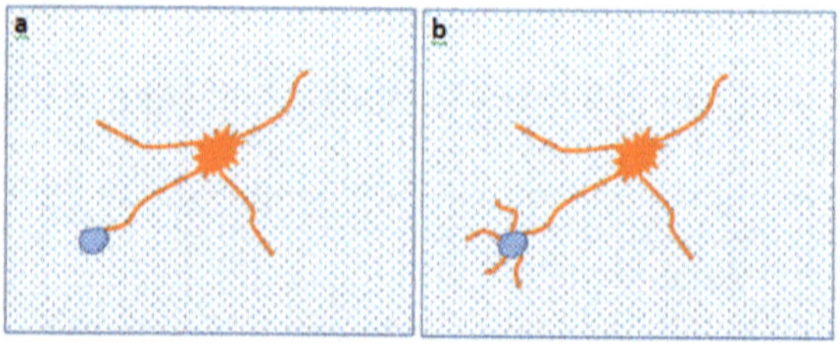

Figure 7. Agglomeration in bamboo composites can either (**a**) stop the damage progression or (**b**) cause more severe damage in the sample.

Compared to non-hybrid composites, a clearer relationship between the percentage of energy absorbed and the different impact energy values of hybrid composites can be seen in Figure 6. As the impact energy increased, the percentage of energy absorbed increased. Besides this, at all impact energy levels, the percentage of energy absorbed for the EP/G-BF30 is lower than for the EP/G-BF10. This is in good agreement with the damage found on the samples, where EP/G-BF10 experienced more severe damage when compared to EP/G-BF30.

3.4. Damage Analysis on The Impacted Samples

The damage analysis will be separated based on the types of samples and their response towards different impact energy levels. Figure 8 shows the damages on the EP samples.

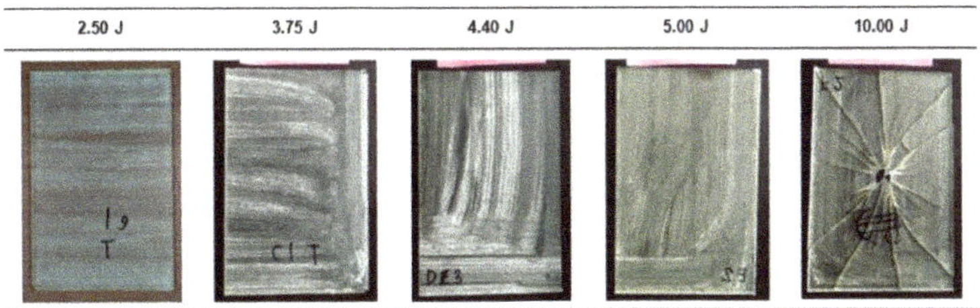

Figure 8. The EP samples after low velocity impacts at different impact energy levels.

From the lowest energy values of 2.50 J to 5.00 J, the samples were free from cracks, and no dented surface was visible. However, at an impact energy level of 10.00 J, the EP sample was totally broken. This observation leads to the conclusion that neat polymer plates will experience total failure at a certain impact energy level without experiencing minor cracks [21]. This situation therefore lowers the dependency and safety of a product in real life applications. It might withstand a higher load, but an unexpected total failure might happen at any limit without any preliminary sign of damage.

The damage on the non-hybrid EP-BF composites after the low velocity impact was observed with the aid of the dye penetrant and is shown in Figures 9 and 10.

As depicted in Figure 9, matrix cracking was detected on the EP-BF10 samples at 2.50 J, 3.75 J and 4.40 J impact energy levels. The matrix cracking propagates from the top surface to the bottom, and from the centre (where the impactor was dropped) to the sides of the rectangular plates. The EP-BF10 samples can withstand the first three impact energy levels; the propagation of matrix cracking stopped before reaching the sides of the samples. However, 10.00 J of impact energy enabled the matrix cracking to propagate until the end sides of the samples, thus breaking the rectangular plates into pieces. Compared to the visual look of the broken EP samples in Figure 9, the broken sample of EP-BF10 indicates a less severe damage as it broke into four large pieces that can be laid out, instead of numerous smaller pieces. Micro-sized fibres with a random orientation limit the analysis of the damage, as no trend can be suggested concerning the relationship of the impact energy levels with the severity of the damage caused [27].

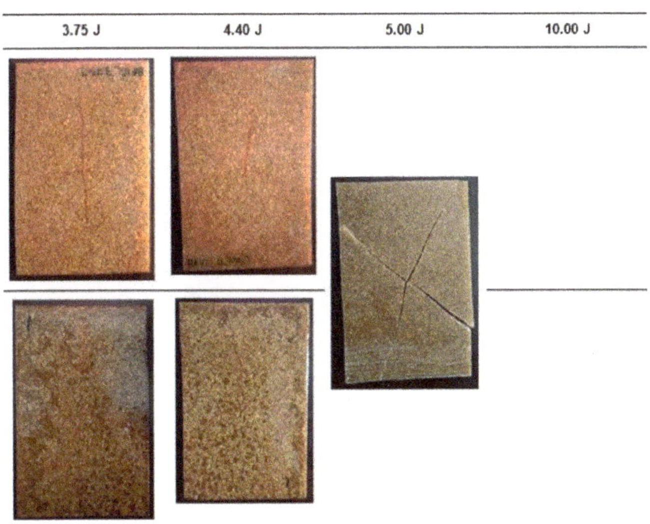

Figure 9. The damage on the impacted samples of EP-BF10 was observed with the help of the dye penetrant.

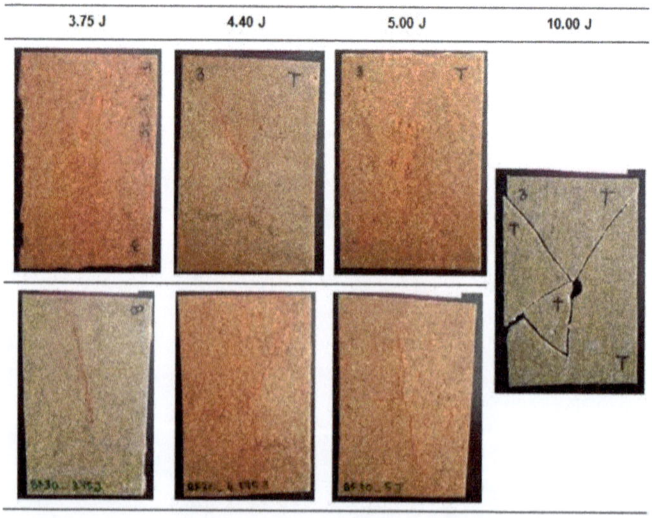

Figure 10. The damage on the impacted samples of EP-BF30 was observed with the help of dye penetrant.

Although no trend can be suggested regarding the propagation of matrix cracking, the analysis was presented in terms of the distance of the matrix cracking propagation from the centre to the side of the rectangular samples. It was found that the distance increased as the impact energy increased from 2.50 J to 4.40 J. These distances were measured on the bottom surface of the EP-BF10 samples, and the longest distance from the centre is recorded in Table 5.

Table 5. Distance travelled by the matrix cracking on the bottom surface of the EP-BF10 sample at different impact energy levels.

Energy Levels (J)	2.50	3.75	4.40	5.00
Distance (mm)	53	61	70	break

Similar damage behaviour was observed on the EP-BF30 samples compared to the previous EP-BF10. Dye penetrant was used, and the observation of the damage is shown in Figure 10.

Matrix cracking was detected as propagating from the top to the bottom and from the centre to the sides of the impacted samples. However, small differences can be seen between the damage on the EP-BF30 and that on the EP-BF10, i.e., the number of lines of matrix cracking is lower than the number found on the EP-BF10. For the EP-BF30, only two obvious lines of matrix cracking were observed from the centre of impact, while in the EP-BF10 the damage propagated into four observable lines from the centre of impact. Figure 10 shows that the EP-BF30 withstood the 5.00 J impact energy but failed at 10.00 J, which is one level higher than for the EP-BF10. The damage analysis for the EP-BF30 is presented in Table 6.

Table 6. Distance travelled by the matrix cracking on the bottom surface of EP-BF30 at different impact energy levels.

Energy Levels (J)	2.50	3.75	4.40	5.00	10.00
Distance (mm)	27	40	59	63	*break*

The shorter distance of the damage propagation for the EP-BF30 suggested that increasing the bamboo fibre loading in the epoxy matrix improved the impact resistance of the composites. The impact resistance of the hybrid EP/G-BFC was expected to be higher compared to the non-hybrid EP-BF composites. The inclusion of woven glass fibres at the top and bottom outermost layers of the composites was believed to help in slowing down the impact absorption into the plates, thus reducing the damage [13].

Figures 11 and 12 show the damage propagation on the hybrid EP/G-BF10 and on the EP/G-BF30, respectively. Dye penetrant was not used as the damage is clearly visible and the area can directly be calculated from the surface of the impacted samples.

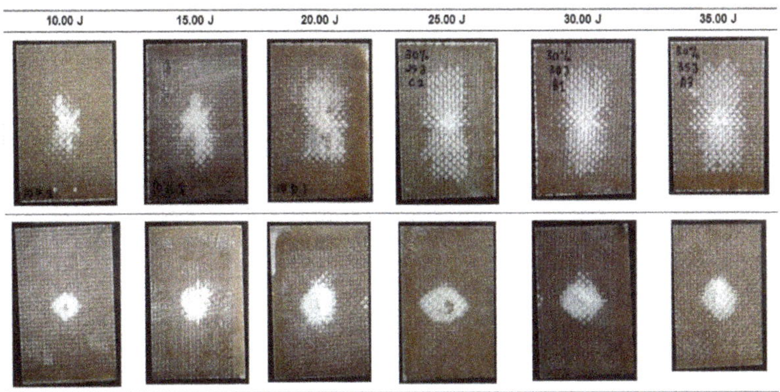

Figure 11. Visible damage on the impacted samples of EP/G-BF10.

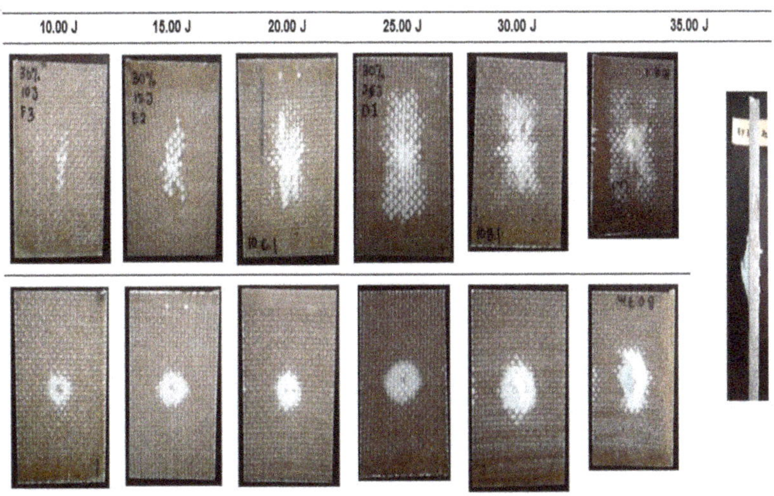

Figure 12. Visible damage on the impacted samples of the EP/G-BF30.

Generally, the damaged areas on both surfaces, top and bottom, were seen to increase as the impact energy level increased. A significant difference in the area was clearly observed on the top surfaces, while on the bottom perforation was detected [24,26]. Compared to the non-hybrid EP-BFC, the EP/G-BFC samples did not break into pieces during the full penetration event. The inclusion of woven glass fibres improved the properties of the composites in terms of impact damage resistance. This improvement is very beneficial for real life applications as the severity of failure is lower when compared to the total failure in the non-hybrid composites. The woven glass fibres can hold the structure in one piece during the highest impact incident. The damage area of the hybrid EP/G-BFC is presented in Table 7.

Table 7. Damage area for EP/G-BF10 and EP/G-BF30 at different energy levels.

Energy Levels (J)		10	15	20	25	30	35
Area (mm^2)	EP/G-BF10	730	1700	3800	5500	5500	5600
	EP/G-BF30	300	700	1760	4000	3300	1000

For both types of samples, EP/G-BF10 and EP/G-BF30, the damage area increased significantly from 10 J to 20 J, and a smaller difference was calculated as the energy level increased further. From 20 J to 35 J, the damage was seen to propagate more towards the bottom surface compared to the propagation from the centre to the sides of the samples.

4. Conclusions

The greater the peak force, the stiffer the projectile-to-target interaction, thus shortening the contact period of the impactor on the surface of the composites. The non-hybrid EP-BF10 composites have a stiffer projectile-to-target interaction compared to the EP-BF30 composites. However, neat epoxy samples showed the stiffest projectile-to-target interaction among the three samples, EP, EP-BF10, EP-BF30, at all impact energy levels. The non-hybrid EP-BF10 composites exhibited good stiffness but lower strength, while the EP-BF30 composites showed the opposite relation. Damage initiation and propagation in the EP-BF30 composites was slower and less severe when compared to the EP-BF10 composites. The non-visible damage in the bamboo composites, which occurred after the low velocity impact, can be analysed through the force undulations in the force-time graphs and can be observed using the dye penetrant method.

A significant improvement was observed with the inclusion of woven glass fibres in the composites. The non-hybrid composites broke into pieces during the highest impact energy that is applied, while the hybrid composites experienced only perforation and the structure did not totally break. The distance of the matrix cracking was shorter for the EP-BF30 when compared to the EP-BF10 at the same impact energy level, suggesting that increasing the bamboo fibre loading can improve the impact resistance, although in short fibre-reinforced composites.

Author Contributions: Formal analysis, A.U.M.S.; Funding acquisition, M.T.H.S.; Methodology, A.U.M.S; Writing—original draft, A.U.M.S.; Writing—review & editing, S.N.A.S. All authors have read and agreed to the published version of the manuscript.

Funding: This research was funded by Ministry of Education Malaysia, grant number FRGS 5540320.

Acknowledgments: The authors would like to thank Ministry of Education Malaysia for the financial support through the Fundamental Research Grant Scheme FGRS/1/2019/STG07/UPM/02/2 (5540320). The authors would like to thank the Department of Aerospace Engineering, Faculty of Engineering, Universiti Putra Malaysia and Laboratory of Biocomposite Technology, Institute of Tropical Forestry and Forest Product (INTROP), Universiti Putra Malaysia (HICOE) for the close collaboration in this research.

Conflicts of Interest: The authors declare no conflict of interest.

References

1. Salman, S.D.; Leman, Z.; Sultan, M.; Ishak, M.R.; Cardona, F. The Effects of Orientation on the Mechanical and Morphological Properties of Woven Kenaf-reinforced Poly Vinyl Butyral Film. *Bioresources* **2015**, *11*, 1176–1188. [CrossRef]
2. Mao, C.; Zhang, C. Numerical analysis of influence factors on low-velocity impact damage of stitched composite laminates. *Mech. Adv. Mater. Struct.* **2018**, 1–10. [CrossRef]
3. Salman, S.D.; Leman, Z.; Sultan, M.T.H.; Ishak, M.R.; Cardona, F. Effect of kenaf fibers on trauma penetration depth and ballistic impact resistance for laminated composites. *Text. Res. J.* **2016**, *87*, 1–15. [CrossRef]

4. Srivastava, V.K. Impact Behaviour of Sandwich GFRP-Foam-GFRP Composites. *Int. J. Compos. Mater.* **2012**, *2*, 63–66. [CrossRef]
5. Safri, S.N.A.; Sultan, M.T.H.; Yidris, N.; Mustapha, F. Low velocity and high velocity impact test on composite materials—A review. *Int. J. Eng. Sci.* **2014**, *3*, 50–60.
6. Nor, A.F.M.; Sultan, M.T.H.; Jawaid, M.; Azmi, A.M.R.; Shah, A.U.M. Analysing impact properties of CNT filled bamboo/glass hybrid nanocomposites through drop-weight impact testing, UWPI and compression-after-impact behavior. *Compos. Part B* **2019**, *168*, 166–174. [CrossRef]
7. Spronk, S.W.F.; Kersemans, M.; De Baerdemaeker, J.C.A.; Gilabert, F.A.; Sevenois, R.D.B.; Garoz, D.; Kassapoglou, C.; Paepegem, W.V. Comparing damage from low-velocity impact and quasi-static indentation in automotive carbon/epoxy and glass/polyamide-6 laminates. *Polym. Test.* **2018**, *65*, 231–241. [CrossRef]
8. Mostafa, N.H.; Ismarrubie, Z.N.; Sapuan, S.M.; Sultan, M.T.H. Fibre prestressed composites: Theoretical and numerical modelling of unidirectional and plain-weave fibre reinforcement forms. *Compos. Struct.* **2017**, *159*, 410–423. [CrossRef]
9. Caminero, M.A.; Garcia-Moreno, I.; Rodriguez, G.P. Damage resistance of carbon fibre reinforced epoxy laminates subjected to low velocity impact: Effects of laminate thickness and ply-stacking sequence. *Polym. Test.* **2017**, *63*, 530–541. [CrossRef]
10. Caminero, M.A.; Garcia-Moreno, I.; Rodriguez, G.P. Experimental study of the influence of thickness and ply-stacking sequence on the compression after impact strength of carbon fibre reinforced epoxy laminates. *Polym. Test.* **2018**, *66*, 360–370. [CrossRef]
11. Safri, S.N.A.; Sultan, M.T.H.; Jawaid, M.; Majid, M.S. Analysis of dynamic mechanical, low velocity impact and compression after impact behavior of benzoyl treated sugar palm/glass/epoxy composites. *Compos. Struct.* **2019**, *226*, 111308. [CrossRef]
12. Ravandi, M.; Teo, W.S.; Tran, L.Q.N.; Yong, M.S.; Tay, T.E. Low velocity impact performance of stitched flax/epoxy composite laminates. *Compos. Part B* **2017**, *117*, 89–100. [CrossRef]
13. Zakikhani, P.; Zahari, R.; Sultan, M.T.H.; Majid, D.L.A.A. Thermal degradation of four bamboo species. *Bioresources* **2016**, *11*, 414–425. [CrossRef]
14. Zakikhani, P.; Zahari, R.; Sultan, M.T.H.; Majid, D.L. Extraction and preparation of bamboo fibre-reinforced composites. *Mater. Des.* **2014**, *63*, 820–828. [CrossRef]
15. Bahari, S.A.; Krause, A. Utilizing Malaysian bamboo for use in thermoplastic composites. *J. Clean. Prod.* **2016**, *110*, 16–24. [CrossRef]
16. Haddou, G.; Dandurand, J.; Dantras, E.; Maiduc, H.; Thai, H.; Giang, N.V.; Trung, T.H.; Ponteins, P.; Lacabanne, C. Physical structure and mechanical properties of polyamide/bamboo composites. *J. Therm. Anal. Calorim.* **2017**, *129*, 1463–1469. [CrossRef]
17. Kumar, N.; Mireja, S.; Khandelwal, V.; Arun, B.; Manik, G. Light-weight high-strength hollow glass microsphere and bamboo fiber based hybrid polypropylene composite: A strength analysis and morphological study. *Compos. Part B* **2017**, *109*, 277–285. [CrossRef]
18. Shah, A.U.M.; Sultan, M.T.H.; Jawaid, M. Sandwich-structured bamboo powder/glass fibre-reinforced epoxy hybrid composites—Mechanical performance in static and dynamic evaluations. *J. Sandw. Struct. Mater.* **2019**, 1–18. [CrossRef]
19. Shah, A.U.M.; Sultan, M.T.H.; Cardona, F.; Jawaid, M.; Talib, A.R.A.; Yidris, N. Thermal analysis of bamboo fibre and its composites. *Bioresources* **2017**, *12*, 2394–2406.
20. Safri, S.N.A.; Sultan, M.T.H.; Aminanda, Y. Impact characterization of glass fibre reinforced polymer (GFRP) type C-600 and E-800 using a drop weight machine. *Appl. Mech. Mater.* **2017**, *629*, 461–466. [CrossRef]
21. Naebe, M.; Abolhasani, M.M.; Khayyam, H.; Amini, A.; Fox, B. Crack damage in polymers and composites: A review. *Polym. Rev.* **2016**, *56*, 31–69. [CrossRef]
22. Simeoli, G.; Acierno, D.; Meola, C.; Sorrentino, L.; Iannace, S.; Russo, P. The role of interface strength on the low velocity impact behavior of PP/glass fibre laminates. *Compos. Part B* **2014**, *62*, 88–96. [CrossRef]
23. Ismail, K.I.; Sultan, M.T.H.; Shah, A.U.M.; Jawaid, M.; Safri, S.N.A. Low velocity impact and compression after impact properties of hybrid bio-composites modified with multi-walled carbon nanotubes. *Compos. Part B* **2019**, *163*, 455–463. [CrossRef]
24. Ahmad, F.; Hong, J.W.; Choi, H.S.; Park, S.J.; Park, M.K. The effects of stacking sequence on the penetration-resistant behaviours of T800 carbon fiber composite plates under low velocity impact loading. *Carbon Lett.* **2015**, *16*, 107–115. [CrossRef]

25. Safri, S.N.A.; Chan, T.Y.; Sultan, M.T.H. An experimental study of low velocity impact (LVI) on fibre glass reinforced polymer (FGRP). *Int. J. Eng. Sci.* **2014**, *3*, 1–10.
26. Hung, P.; Lau, K.; Cheng, L.; Leng, J.; Hui, D. Impact response of hybrid carbon/glass fibre reinforced polymer composites designed for engineering applications. *Compos. Part B* **2018**, *133*, 86–90. [CrossRef]
27. Zivkovic, I.; Fragassa, C.; Pavlovic, A.; Brugo, T. Influence of moisture absorption on the impact properties of flax, basalt and hybrid flax/basalt fiber reinforced green composites. *Compos. Part B* **2017**, *111*, 148–164. [CrossRef]

polymers

MDPI

Article

Pinned Hybrid Glass-Flax Composite Laminates Aged in Salt-Fog Environment: Mechanical Durability

Luigi Calabrese [1], Vincenzo Fiore [2,*], Paolo Bruzzaniti [1], Tommaso Scalici [3] and Antonino Valenza [2]

1. Department of Engineering, University of Messina, Contrada Di Dio (Sant'Agata), 98166 Messina, Italy; lcalabrese@unime.it (L.C.); pgbruzzaniti@unime.it (P.B.)
2. Department of Engineering, University of Palermo, Viale delle Scienze, Edificio 6, 90128 Palermo, Italy; antonino.valenza@unipa.it
3. School of Mechanical and Aerospace Engineering, Queen's University Belfast, Ashby Building, Stranmillis Road, BT9 5AH Belfast; t.scalici@qub.ac.uk

* Correspondence: vincenzo.fiore@unipa.it; Tel.: +39-091-23863721

Received: 12 December 2019; Accepted: 22 December 2019; Published: 26 December 2019

Abstract: The aim of the present paper is to study the mechanical performance evolution of pinned hybrid glass-flax composite laminates under environment aging conditions. Hybrid glass-flax fibers/epoxy pinned laminates were exposed to salt-spray fog environmental conditions up to 60 days. With the purpose of assessing the relationship between mechanical performances and failure mechanisms at increasing aging time, single lap joints at varying joint geometry (i.e., hole diameter D and hole distance E from free edge) were characterized after 0 days (i.e., unaged samples), 30 days, and 60 days of salt-fog exposition. Based on this approach, the property–structure relationship of the composite laminates was assessed on these critical environmental conditions. In particular, a reduction of failure strength for long-aging-time-aged samples was observed in the range 20–30% compared to unaged one. Due to the natural fiber degradation in a salt-fog environment, premature catastrophic fractures mode due to shear-out and net-tension were found, related to reduced joint fracture strength. This behavior identifies that this type of joint requires a careful design in order to guarantee an effective mechanical stability of the composite hybrid joint under long-term operating conditions in an aggressive environment.

Keywords: bearing; salt fog aging; glass-flax hybrid coposites; pinned joints; failure modes

1. Introduction

In the last years, hybrid composite materials were addressed as effective approach to optimize structural design in several industrial fields, such as construction, aerospace or automotive [1,2]. The engineering choice of these materials has shown significant advantages compared to conventional non-hybrid composites [3]. In such a context, the high mechanical properties of synthetic fibers and the environmental compatibility of natural fibers represent an effective and reliable combination to develop hybrid composite materials with marked performances of sustainability and mechanical/structural effectiveness. Several research activates highlighted that the engineering design of natural fiber based composite materials is a potentially suitable key factor to guarantee an effective synergy between mechanical performance and green sustainable material [2,4].

Mechanical joint is one of the industrially applied methods to assemble composite structures. Bolted joining is the most common approach to join similar and dissimilar materials thanks to their low cost and easiness in assembling and disassembling. However, structural joints, although often required at the design level, are high stress concentration points that make this region sensitive to damage. Several factors can influence the composite joint strength, such as the preload moment [5], temperature [6], environmental conditions [7], and geometric parameters [8]. The pinned joints with different geometrical parameters highlight different failure modes, and the damage activation phenomena in the composite materials can be foreseen using failure criteria [9].

Several research activities highlighted that the hybridization of lignocellulosic fibers with synthetic counterparts (i.e., glass or carbon fibers) enhances the mechanical properties of composite laminates, favoring the improvement of their tensile, bending, and impact strength [10–13].

Furthermore, the hybridization of natural fibers allows to exalt the durability in critical conditions, preventing the limited durability of the natural materials in moist or wet environments [14–17]. In fact, it is well known that the performances reduction of GFRP (Glass Fiber Reinforced Plastics) composites in wet condition stabilizes after moisture saturation level [18]. Calabrese et al. [19] showed that the hybridization of lignocellulosic flax fibers with glass ones represents an effective and suitable compromise in terms of environmental impact, mechanical properties, aging resistance, and cost between flax and glass composites, for its use in marine applications. Analogously, Saidane et al. [20] evidenced that the water uptake and diffusion are limited by the addition of glass fiber layers to flax laminates. In particular, the flax-glass hybridization offers a positive contribute in the Young's modulus and the tensile strength. However, an aspect that needs attention requiring an improvement of knowledge is the evaluation of the mechanical stability in aggressive environments of composite joints based on hybrid natural fibers laminates. Natural fibers exhibit in usual working conditions an acceptable mechanical reliability, configuring these materials as suitable reinforcing material in several applied composites [21,22]. However, due to their relevant hydrophilic nature, they are very sensitive to damp or wet environments, implying a significant performance depletion of the resulting composites, thus addressing premature fracture modes. This makes the mechanical stability and durability design of hybrid joints in aggressive environments very complex. The evaluation of strength variation must be synergistically integrated with the occurred fracture mechanism in the joint [6,23]. The former should be excluded or limited in time, and the latter should preferably be preserved. Instead, there is often the triggering of premature fracture mechanisms, induced by aging, which enhance the limit strength the hybrid joint.

Esendemir and Cabioglu [24] showed that different environmental aging conditions do not modify the failure modes (i.e., net tension, shear-out, bearing, and mixed) for woven glass epoxy pinned composites. Nevertheless, they evidenced that the bearing strength of the joints is noticeable affected by the applied environment conditions.

Karakuzu et al. [25] investigated pinned glass fiber-reinforced composites due to aging up to 12 months of immersion in seawater. Their study evidenced a detrimental effect of this environmental aging condition on failure strength and fracture mechanism of the joint at varying joint geometry.

In a previous paper, the authors assessed the effect of salt-fog environment on the bearing behavior of pinned flax/epoxy composites [26], showing that salt-fog exposition severely influences the mechanical performances of pinned composite. In particular, a progressive modification of damage mechanism occurred, favoring premature and catastrophic shear out and net tension failure mechanisms despite the preferred bearing one and thus, as a consequence, limiting the effective mechanical durability of the mechanical joint. In such a context, the joint geometrical configuration (i.e., hole diameter, edge distance, etc.) has been a relevant role in the damage activation and propagation.

At the same time, some promising results were obtained evaluating the durability of glass composite laminates in salt-fog environments [27]. In this case, salt-fog exposition did not modify noticeably the mechanical performances of the pinned flax/epoxy composite. Thanks to the quite good stability in wet environments of glass fibers, only a slight reduction in the performance of the pinned joint was found with no relevant change in fracture mechanisms at increasing aging exposition time. These results encouraged assessment of the hybridization of pinned flax/epoxy composites with glass fibers as an effective approach to enhance the mechanical stability and durability of the joint. Moreover, this would allow improving the knowledge in the hybrid joint design, introducing in this context useful information for a prediction of the component service life based on their real environmental conditions.

By considering the wide use of composites laminates for marine applications, this issue is extremely interesting to evaluate how the exposition aging time can influence the predominant failure mechanism and, consequently, the structural performance of the hybrid joint.

In the present paper, pinned glass-flax hybrid reinforced laminates were exposed to salt-fog spray environment up to 60 days with the aim of studying the mechanical performances and failure mechanism modification induced by aging conditions. Different joint geometrical configurations at varying E/D (i.e., hole center to laminate free edge distance over hole diameter) and W/D (i.e., sample width over hole diameter) ratios were also investigated, thus identifying critical values of these parameters that altered the mechanical stability of the joint in a marine environment.

2. Materials and Methods

A composite panel (350 × 350 mm^2) was produced by using vacuum-assisted resin infusion technique. The was cured at 25 °C for 24 h and post-cured at 50 °C for 8 h. A DEGBA epoxy resin, SX8 EVO (supplied by Mates Italiana, Segrate, Italy) was used as matrix, and 6 layers of 2 × 2 twill weave woven flax fabric with nominal areal weight of 318 g/m^2 (Lineo, Saint Martin du Tilleul, France) were used as natural fiber reinforcement. In order to improve strength, stiffness, and environmental durability of the composite laminate, 3 layers of plain weave woven glass fabrics with nominal areal weight of 200 g/m^2 (Mike Compositi, Milano, Italy) were used as external fiber reinforced skins. A full number of 12 reinforced layers (6 internal flax layers and 6 external glass layers) were applied. The stacking sequence of the hybrid composite laminate is $(G_3/F_3)_S$.

Prismatic samples with length 150 mm (preferred width 15 mm) were obtained by cutting the panel by using a band saw. Afterward, a single hole was made in each sample by using at first undersized drilling bits and then a mill tool to obtain the hole diameter without edge defects. In order to assess the effect of the pinned joint geometry on mechanical performance and failure mechanisms, hole diameter (D) and its free edge distance (E) were varied in order to obtain samples with large W/D and E/D geometrical ratios.

A double-lap pinned joint was considered for experimental bearing tests, according to ASTM D5961/D standard (procedure A), using a universal testing machine Z250 (Zwick-Roell, Ulm, Germany), equipped with a 250 kN load cell, and setting a displacement rate of 0.5 mm/min. A scheme of the sample geometry and bearing test setup was drawn in Figure 1.

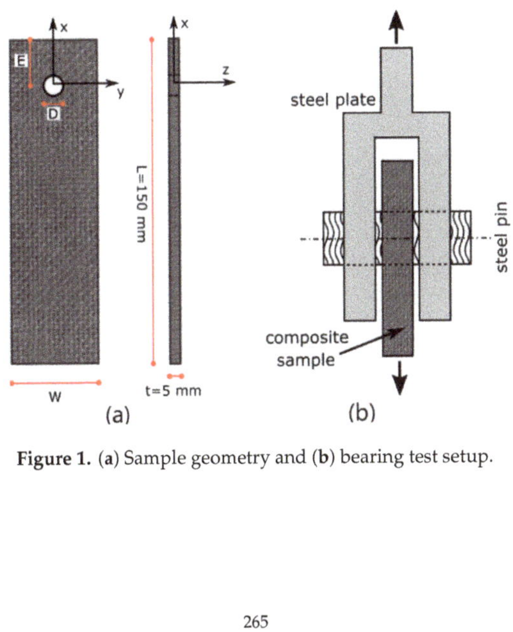

Figure 1. (**a**) Sample geometry and (**b**) bearing test setup.

The average fibre and void contents of the laminate, its nominal thickness, the samples geometry, and the bearing test set-up are detailed in our previous paper [28], and it will not reported here for the sake of brevity.

Aging exposition was carried out by using a DCTC 600 climatic chamber (Angelantoni, Massa Martana (PG), Italy), according to the ASTM B 117 standard. In order to assess the effect of salt-fog exposition on mechanical performances stability of hybrid composite laminates, 30 and 60 aging days were applied on the selected batches.

All samples are codified as "GFA", "GFB", and "GFC" depending on if the hybrid laminates were aged for 0, 30, or 60 days, respectively. Furthermore, this code was coupled with a lot of numbers xx-yy-zz related to the hole diameter (*D*), the edge distance (*E*), and sample width (*W*), respectively. For example, GFB-6-12-15 indicates hybrid laminates exposed to salt-fog for 30 days, having 6 mm hole diameter, 12 mm edge distance, and 15 mm sample width.

In the present paper, the results and discussion comparison among the three batches was performed by using the bearing stress determined according to following the expression:

$$\sigma = P/(D * s) \quad (1)$$

where P, D, and s are the applied load, hole diameter, and sample thickness, respectively. Detailed images of failure mechanism were carried out on fractured samples by means of a 3D digital microscope (Hirox KH-8700).

3. Results and Discussion

3.1. Pinned Joints with 4-mm Hole Diameter

Figure 2 shows the bearing stress versus displacement trend evolution for hybrid glass-flax laminates having diameter D = 4 mm, width W = 15 mm, and edge distance E = 12 mm (i.e., W/D = 3.75 and E/D = 2.75), as a function of time exposition to salt-fog environment. Three stages can be identified:

- In the beginning phase, all curves are characterized by a linear relationship between the bearing stress and displacement. The slope of this trend could be related to the joint stiffness;
- Afterwards, at increasing displacement, a progressive deviation from the linear trend can be highlighted depending on aging time. This behavior is due to compression collapse of the matrix in correspondence of the composite laminate just behind the small hole/pin contact area;
- Eventually, all specimens tend to reach a plateau zone where the bearing stress is roughly constant, beyond which its failure occurs. All the specimens showed a bearing failure mechanism.

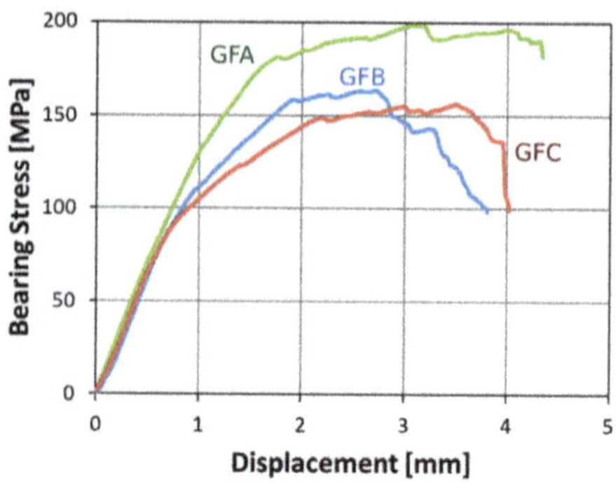

Figure 2. Stress-displacement curves at increasing aging time for pin-loaded laminates (D = 4 mm; E = 12 mm; W = 15 mm).

It is worth noting that a gradual modification of the stress-displacement trend can be highlighted at increasing aging time.

The unaged glass-flax laminate (i.e., GFA sample) evidences effective mechanical performance as confirmed by the high stress level reached before failure. Furthermore, this batch highlights a quite large linearity region for bearing stress values up to about 125 MPa. Then, bearing stress progressively increases at increasing displacement until a stabilization is reached at about 180–190 MPa. A significant stress fluctuation in the bearing stress plateau can be identified due to the compression collapse of the epoxy matrix just behind the pin/hole contact area. A maximum stress at about 200 MPa was reached beyond which the sample failure occurred.

On the contrary, GFB and GFC samples evidenced a reduction of the maximum bearing stress (i.e., about 18% and 21% lower than unaged sample, respectively) due to salt-fog aging conditions. Moreover, an evident change in plateau region can be observed, i.e., the unaged sample is characterized by a wide plateau region extending in the 1.8–4.2 mm displacement range whereas the stress-displacement curves are characterized in a less extended stabilization region of the stress due to aging condition. This phenomenon is evident already after 30 days of aging exposition (i.e., batch GFB). Flax fibres undergo a gradual absorption of water in salt-fog environment, which leads to a progressive reduction of their mechanical properties. The hydrophilic nature of these fibres also speeds up the water permeation at the fibre-matrix interphase, thus weakening this area and, as a consequence, implying a reduction of both strength and stiffness of the composites due to the limited transfer of stresses [29]. Hence, the residual resistance of the composite laminate decreases, thus promoting the premature compressive collapse of the sample in the pin-hole contact region due to lower stress/deformation limits than unaged hybrid composite laminate. This relevant change of the damage evolution also favours a relevant modification on the occurring failure mechanism. In particular, a transition from full bearing mode to mixed bearing/net tension failure mechanism takes place at increasing exposition time to the salt-fog environment, i.e., 60-day aged samples experienced a clearly mixed net tension and bearing failure mode.

Further interesting information can be extrapolated by analysing the fracture images of the samples at varying aging time, as reported in Figure 3.

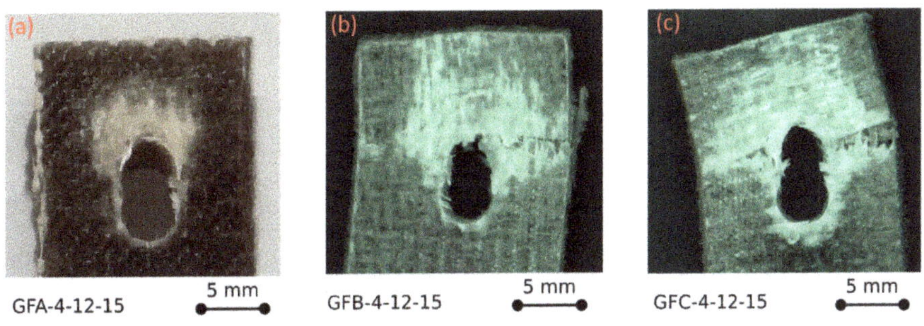

Figure 3. Fracture images of (**a**) GFA-4-12-15, (**b**) GFB-4-12-15, and (**c**) GFC-4-12-15 glass-flax samples.

The fracture image of GFA-4-12-15 unaged sample (Figure 3a), fails through bearing mode, as can be evidenced by the large compression collapse area on the pin/hole contact surface area. Furthermore, by analysing the detail of Figure 3a, kink bands that evolve radially from the hole edge can be observed, thus highlighting a progressive extension of the damaged area, defined as failure process zone [28]. This damage phenomenon favours at the same time the triggering of synergistic damage mode, such as delamination, that promotes the mechanical collapse of the pinned laminate joint [30]. In fact, kink bands develop due to plastic micro-buckling phenomena in the compression area and favour large local deformations at the fibre-matrix interface, inducing the triggering of premature secondary failure mechanisms [31]. Therefore, a progressive process of damage accumulation is activated by coupling different damage modes, such as fibre kink bands and shear cracks at the layers interface that evolve to large-scale delamination phenomena, resulting in a reduction in the mechanical stability of the joint [32].

As already stated, the salt-fog exposition induces a significant modification of the failure mechanism. For GFB and GFC samples, a mixed bearing/net-tension fracture was observed (Figure 3b,c). The failure process zone due to bearing is still evident, with a large delaminated area. In these cases, large kink bands toward the free edges of the sample are not evident, indicating that other premature failure mechanisms compete in the damage propagation phenomenon. This can be related to the matrix softening due to environmental aging that limits the local brittle fracture mode such as kink bands and stimulating interfacial dominated fracture mechanisms [33]. The reduced displacement at failure found for GFB and GFC specimens is a consequence of these factors. In particular, the limited plateau region for the specimen GFB (characterized by intermediate exposure times in a salt spray chamber) can be attributed to two competing mechanisms. On the one hand, the onset of a catastrophic fracture mechanism, as net-tension, stimulates the reduction of displacement at failure. On the other hand, the exposure of the sample in a wet environment favors softening phenomena in the composite that reduce the joint stiffness. However, this contribution is mainly relevant for the GFC batch, exposed for 60 days in a salt spray chamber, for which a much larger bearing stress plateau region is detected.

Summarizing, a reduction of about 18% of the bearing stress was evidenced already at 30 days of aging time. Although glass fibres in hydrothermal environments show a quite good mechanical stability, flax fibres, on the other hand, have a limited durability following the reduction of the resistance. These play a key role in triggering damage phenomena that subsequently evolve along the interface surface between dissimilar flax-glass laminae. By observing Figure 3b,c, the appearance of fibre breaking mainly ascribed to the detrimental reduction of the flax tensile strength caused by salt-fog exposition can be evidenced.

Further considerations can be argued by evaluating the maximum bearing stress evolution at increasing edge distance (Figure 4) for samples having constant *D* and *W* (equal to 4 mm and 15 mm, respectively) at varying aging time. The observed experimental trend of all batches (i.e., 0, 20, and 60 aging days) highlights a progressive increase of the maximum bearing stress (σb) at increasing E distance until a critical threshold value, above which its stabilization has been reached, as showed previously [34,35].

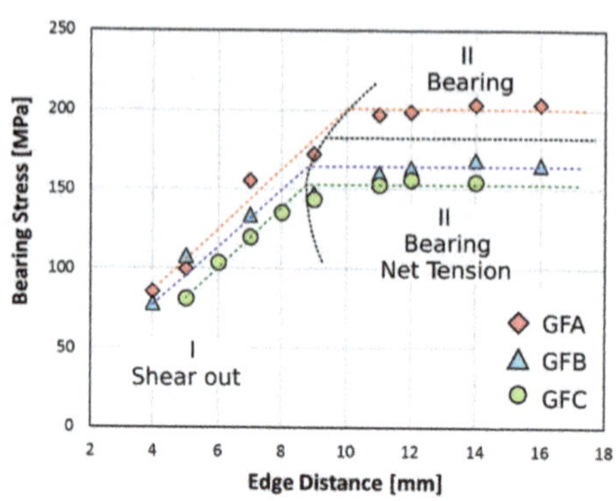

Figure 4. Bearing stress evolution at increasing edge distance E for pin-loaded laminates with hole diameter *D* = 4 mm and width *W* = 15 mm (*W*/*D* = 3.75) for all batches.

For all batches, two linear segments were identified. In particular, F, the pinned composite failure, occurs at very low stress values for small E distance since, the very limited area just behind the pin that suffers the applied stress favours the activation and propagation of premature failure damage by shear out mode (region I). However, as evidenced by the high slope of the first fitting linear segment, the stress is very sensitive to E, i.e., a slight increase in edge distance implies a significant increase in the maximum stress. As already discussed, a threshold E value at about 8–10 mm can be identified.

For edge distances higher than 10 mm (i.e., E/D > 2.5), a plateau is reached and the bearing stress becomes constant (region II). In this region, the pinned composite joint fails mainly through bearing failure mode [32]. However, after 1 month of aging exposition, glass-flax laminates (i.e., batches B and C) highlighted a combined net-tension and bearing fracture mode with the former dominating on the latter one. Due to salt-fog exposition, hybrid laminates undergo both physical and chemical degradation phenomena that significantly worsen their mechanical behaviour. It is widely known that hydrophilic fibres such as flax ones greatly suffer humid environmental conditions in evidencing gradual reductions in their mechanical properties up to 40% [36]. This sensitivity to damp or wet environments is due to the intrinsic microstructure of the flax fibre, which can be classified as a hierarchical structure reinforced by cellulose micro-fibrils grouped in bundles to form meso-fibrils [37,38]. Thanks to their high elastic modulus (in the range 134 to 160 GPa [39]), these act as reinforcements of the fibers structure. Due to water sorption experienced by the amorphous fraction of cellulose and other polysaccharides such as hemicellulose, a relevant decrease of the mechanical properties of natural fibers occurs [40]. In particular, the cellulose structure is destroyed by the water molecules that lead to a reduction in stiffness, i.e., water acts as a plasticizer increasing the fiber flexibility [41]. Furthermore, the presence of Na+ and Cl- ions in the solution helps to speed up the degradation of epoxy matrix, flax fibers, and their interface by improving the osmotic diffusion of water at the fiber/matrix interface [15].

The hydrophilic behavior of the flax fibers enhances the diffusion of water in natural fiber-reinforced laminae. Consequently, this implies an acceleration of the degradation phenomena at the interface between flax and glass reinforced laminae. Furthermore, the water could reach glass fibers, thus partially reducing their mechanical performances due to local dissolution of fiber surface [42]. At the same time. this stimulates a wide activation of softening phenomena of the thermoset matrix [33]. The flax fibers induce the activation of preferential diffusive pathways that enhance the activation and propagation of the conventional degradation phenomena of composites in wet environments. The combination of these phenomena leads to a decrease of the tensile strength due to the water absorption, inducing a significant decrease of net tension failure stress of the aged hybrid composite laminates as early as 1 month of salt-fog exposition. This is confirmed by the reduction of plateau bearing stress for GFB and GFC batches (18% and 24%, respectively) compared to an unaged GFA one. This abrupt stress decrease at high E/D values can be related to the triggering of a premature fracture by net tension that synergistically contributes coupled with bearing on the failure mechanism for this specific joint geometry configuration.

3.2. Pinned Joints with 8-mm Hole Diameter

The mechanical behavior of glass-flax hybrid laminates can be deeper analyzed by evaluating the stress-displacement trends and fracture mechanisms of pinned composite joints having hole diameter larger than 4 mm. The different geometrical configuration of the joint represents a relevant aspect to be taken into consideration in order to identify the possible causes of triggering and propagation of premature damage on the joint. In particular, Figure 5 shows the evolution of the stress versus displacement curves for pinned joint at increasing aging cycles characterized by hole diameter (D) and edge distance (E) equal to 8 mm and 9 mm, respectively.

It is worth noting that, for this geometrical configuration, no noticeable decrease of the maximum bearing stress can be observed comparing the three curves although a gradual increase of the displacement at failure can be highlighted at increasing aging time exposition (i.e., 146% and 151% higher than that of unaged one, for batches B and C, respectively). This large deflection for aged samples can be ascribed to two factors. The first is related to a joint stiffness decrease, as evidenced by the reduction of the stress/displacement slope. Moreover, a progressive and not catastrophic damage mechanism occurs at larger deflection values. This favors the stress stabilization, and large deflection is required to induce the final fracture of the aged samples. This behavior, similar to that found for the specimens with a 4-mm hole size, can be attributed to a progressive softening of the matrix and/or matrix-fiber interface. As already stated, this phenomenon is strongly conveyed by the significant sensitivity of flax fibers to water. This is also experimentally confirmed by the reduction of the stress/displacement slope (which may be indirectly related to the joint stiffness) with increasing aging time. After 60 days under salt-fog spray condition, sample GFC-8-9-15 shows a displacement at failure more than double than GFA-8-9-15 unaged sample.

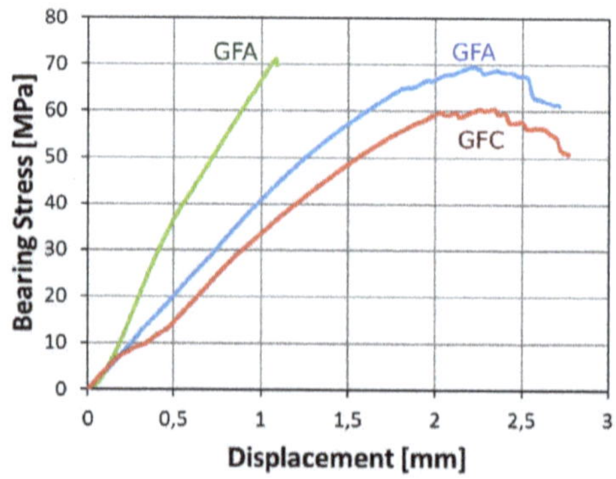

Figure 5. Stress-displacement curves at increasing aging time for pin-loaded laminates (D = 8 mm; E = 9 mm; and W = 15 mm).

For a GFA-8-9-15 sample, a catastrophic fracture type was obtained, identifiable by the abrupt and sudden drop of the bearing stress when the stress-displacement trend still has a quite linear relationship. This catastrophic and brittle fracture mechanism can be considered typical for brittle thermoset matrix-based composites [43]. On the other hand, the joint failure takes place prematurely at low stress level for aged samples. Furthermore, the stress-displacement curve does not evidence an abrupt reduction of the stress but a slight and progressive decrease in stress when its maximum value is reached, i.e., this different behavior becomes evident already after 30 days of exposition in the salt-fog environment. As already stated, it can be related to the epoxy matrix softening coupled to the chemical and physical degradation phenomena that can be triggered due to the presence of hydrophilic natural fibers in the hybrid composite laminate (i.e., water diffusion into the composite toward preferential flow pathways, fiber swelling, etc.)

Figure 6 summarizes the fracture surface for GFA-8-9-15, GFB-8-9-15, and GFC-8-9-15 glass-flax samples).

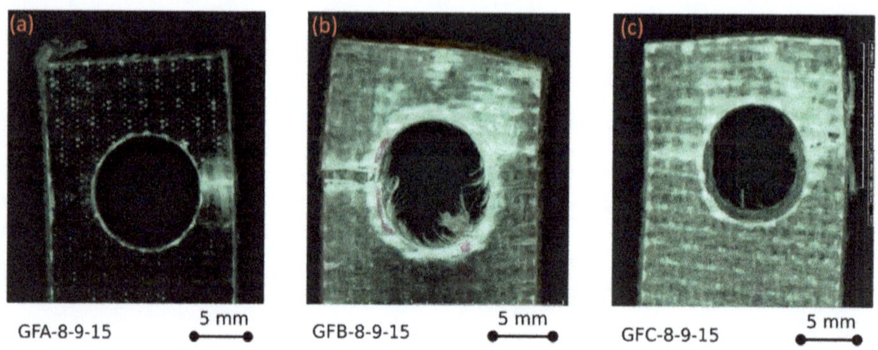

Figure 6. Fracture images of (**a**) GFA-8-9-15, (**b**) GFB-8-9-15, and (**c**) GFC-8-9-15 glass-flax samples.

It is found that, for all the compared samples, a net tension fracture mode mainly occurred. In particular, Figure 6a evidences that an unaged sample fails through a neat fracture surface, typical of tensile failures of fiber reinforced thermoset polymers. No further competing failure mechanisms can be identified. On the other hand, by analyzing Figure 6b,c related to GFB-8-9-15 and GFC-8-9-15 aged samples, respectively, a more complex fracture mechanism can be shown. The net tension fracture area is characterized by local fiber detachments from the matrix coupled to the net-tension crack, local layer detachments, and weakening of the fibers, probably due to the worsening of the fiber-matrix adhesion; to the matrix softening; and to the reduced tensile properties of the natural fibers. At the same time, fracture cracks rivers along the load direction in the area of the composite laminate immediately behind the pinned hole emerge and become more relevant as the exposition time in salt-spray chamber increases. These local damages can be attributed to the onset of joint damaging phenomena due to shear-out mode, which acts synergistically to net-tension one, thus reducing the mechanical strength of the joint. Consequently, this type of fracture although dominated by the fracture for net-tension can be ascribed to an incipient cleavage fracture mode.

Figure 7 shows the maximum bearing stress evolution at increasing edge distance E for pin-loaded laminates with hole diameter D = 8 mm and width W = 15 mm, for all batches.

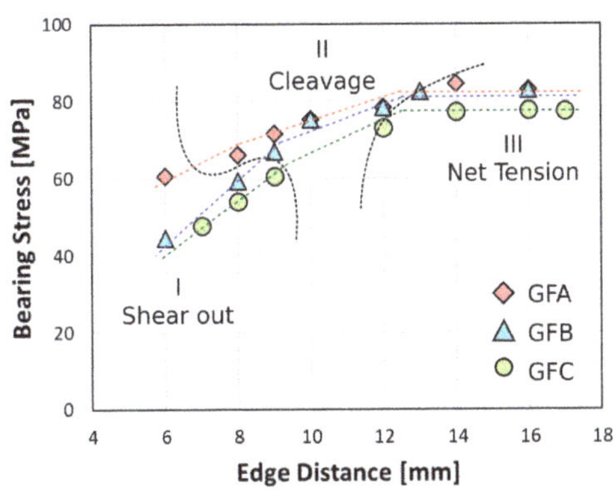

Figure 7. Bearing stress evolution at increasing edge distance E for pin-loaded laminates with hole diameter D = 8 mm and width W = 15 mm for all batches.

For low edge distances (i.e., E < 9 mm), the pinned joints evidenced premature fracture at low stress levels. In particular, for this joint geometrical configuration, shear-out mode is the failure mechanism (region I) due to short free edge distance. The maximum stress increases quite linearly at increasing edge distance up to reaching a threshold value (for E = 9). Afterwards, a slight modification of slope trend can be observed, indicating that competing failure phenomena are emerging when pin hole distance from the free edge increases. For all hybrid batches, a transition from shear out to cleavage failure mode occurred (region II). This trend for the GFA batch was maintained up to E = 12 mm and then a new failure region due to net tension mode was defined (region III). The curve trend reaches a plateau, indicating that the maximum stress becomes independent from E. It is worth noting that aged samples (i.e., GFB and GFC) evidenced a different failure mechanism at large edge distance: i.e., a mixed mode between net-tension and cleavage. This behavior can be ascribed to a progressive degradation of the composite laminate due to environmental aging that has more markedly affected the laminate shear strength than the tensile strength limit. This is confirmed by the large cleavage failure area evidenced for GFB and GFC samples in Figure 6b,c. Further confirmation of this evidence can be argued by evaluating the relevant slope change in the maximum stress trend in region I (i.e., shear-out region) already at 30 days of exposure in a salt spray chamber (Figure 7). The reduction of the stress value in this region identifies a significant reduction in the shear-out strength of hybrid composite laminates.

Instead, only a slight modification of maximum stress was shown at large E distance, highlighting a still suitable mechanical stability for this joint configuration, even after long time under salt-fog environment.

The presence of six internal laminae reinforced with hydrophilic flax fibers in the stacking sequence can be considered the main cause of the high water absorption evidenced by hybrid laminates (~6% [19]) during the salt-fog exposition. Furthermore, the water diffusion into the laminate is also influenced by matrix voids and micro cracks due to aging environmental exposure and/or the manufacturing process that speed up the phenomenon [44].

Due to this, a progressive degradation of the hydrophilic flax fibers as well as a weakening of the interfacial adhesion with the surrounding matrix occur during the salt-fog exposition. This justifies the decrement of the specimens' shear resistance, thus making shear out mode predominant for a wider range of geometrical conditions of aged samples (i.e., GFB and GFC), so remaining competitive with the net tension even at high at large edge distance.

3.3. *Evolution of Failure Mechanisms at Varying Joint Geometry*

In order to better discriminate the damage mechanisms that compete at varying joint geometry, in Figure 8, different topological failure plots obtained evaluating the maximum stress at varying the geometric parameters E (edge distance) and D (hole diameter) were schemed.

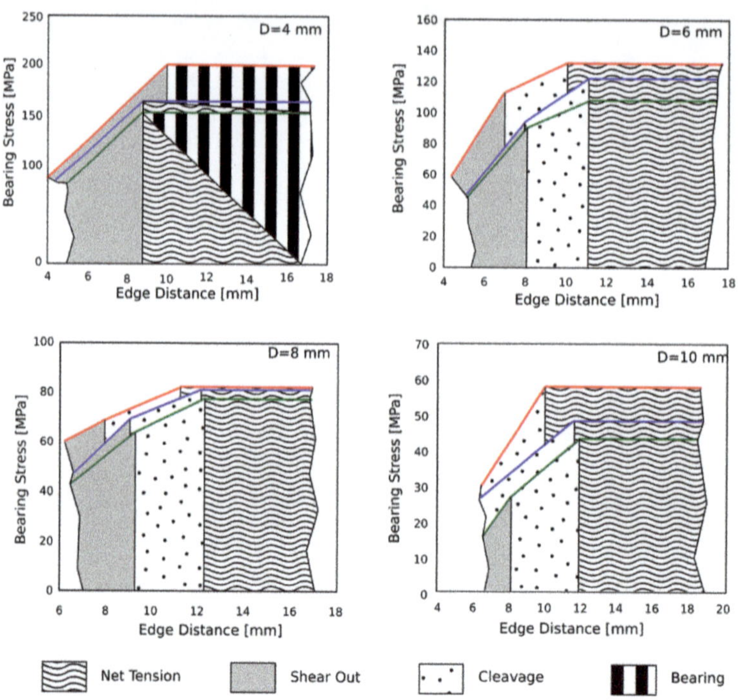

Figure 8. Scheme of failure mechanisms evolution in bearing stress vs. edge distance E plot for all laminates.

Some consideration can be addressed based on this schematization:

- D = 4 mm: For a small hole diameter, the fracture mechanism evolves from shear-out (i.e., low E values) to bearing mode (i.e., large E values). Nevertheless, a premature fracture caused by net-tension fracture mode was highlighted for large edge distance value due to environmental aging. For the aged samples, a mixed bearing/net tension fracture was identified. The maximum stress decreases at increasing aging time. This behavior is more sensitive for joint geometry having large edge distance and, therefore, large E/D ratio.
- D = 6–8 mm: For intermediate hole diameters, a dual failure damage mechanism occurs. A transition from shear-out to net-tension failure mechanism takes place at increasing edge distance. For this sample geometry, the cleavage mode is relevant, especially for aged specimens. This behavior can be related to the reduced tensile and shear strengths of the aged flax fibers that influences the triggering of combined failure mechanisms. GFB and GFC batches, compared to the unaged GFA one, are characterized by a clearly lower stress in the shear-out region (mainly for samples with hole diameter equal to 6 mm). This significant stress reduction is related to the specific joint geometry since the difference between hole diameter and sample width is still relevant, thus limiting net-tension fracture mechanism only at high E values. The low stress evidenced in the shear-out failure region could be related to a relevant reduction of the interfacial adhesion at the fiber/matrix interface that favors shear cracks activation and propagation in the pin/hole contact area. This stimulates a premature fracture by shear-out, thus preventing a mixed failure mechanism for cleavage.
- D = 10 mm: For large hole diameters, net-tension is the main failure mechanism for E > 10–12 mm for aged and unaged samples. This experimental evidence indicates that all samples, regardless of the applied aging time, exhibit quite similar fracture transitions. Nevertheless, a relevant discrepancy on the mechanical strength was observed. In particular, GFB and GFC samples showed a reduction of maximum stress of about 20% and 30% compared to unaged one for E > 12 mm, respectively. This trend can be attributed to the reduced tensile strength in the hydrophilic aged flax fibers. This degradation phenomenon is less relevant at low E/D ratio, where the shear-out mode is the main failure mechanisms.

These results show that the hybridization procedure of flax fibers using glass fibers allows to obtain composite materials with an acceptable aging stability up to 60 days of exposure to salt spray environment. In particular, a reduction of the maximum stress of about 20–30% can be identified due to aging. The presence of not hydrophilic glass fibers, placed externally in the stacking sequence, made it possible to limit the phenomenon of water absorption [15]. Hybridization on flax composites improves the durability of the composite. This allowed to obtain an acceptable mechanical stability [26]. These results are in agreement with Reference [19], where the hybridization of lignocellulosic fibers (i.e., flax) with synthetic ones (i.e., glass) was possibly considered if accurately designed and suitably applied for marine applications based on its compromise in terms of environmental impact, mechanical properties, aging resistance, and cost between flax and glass composites.

However, the triggering of premature fracture mechanisms at low stress levels, such as shear-out and net-tension, are a symptom of the partial interaction of the hybrid composite with wet environments. Thus, these environmental conditions are at potentially critical state at long exposure times.

Summarizing, due to the natural fiber degradation in salt-fog environment, at varying D and E dimensions, premature catastrophic fractures mode due to shear-out and net-tension were found, related to reduced joint fracture strength. This behavior identifies that this type of joint requires a careful mechanical durability design in an aggressive environment in order to guarantee an effective mechanical stability of the composite hybrid joint under long-term operating conditions.

4. Conclusions

In the present paper, the effect of salt-fog environment on the mechanical behavior of pinned hybrid glass-flax/epoxy composites was evaluated. Samples at varying geometrical joint configuration were exposed to a salt-fog spray test up to 60 aging days, according to ASTM B 117 standard.

The experimental results highlighted that salt-fog environmental conditions significantly modify the mechanical performances and failure mechanism of pinned hybrid composites. For low hole diameter (i.e., $D = 4$ mm), a reduction of the bearing strength of about 18–21% was observed due to environmental aging, i.e., from 198.7 to 156.1 MPa for GFA-4-12-15 and GFC-4-12-15 samples, respectively. Furthermore, it was evidenced a noticeable modification of damage mode at increasing the aging exposition time. In particular, a reduction of the bearing phenomenon thus favoring premature and catastrophic mechanisms such as shear out (i.e., for low E values) and net tension (i.e., for high E values) was highlighted and, as a consequence, limited the effective mechanical durability of the mechanical joint.

This greater sensitivity to net-tension and shear-out fracture evolution is exalted also for joints having larger hole dimensions. In particular, for hole size D equal to 10 mm, it was observed a bearing strength reduction of about 30% when the fracture dominant mechanism was both shear out (i.e., for E values lower than 8 mm) and net-tension (i.e., for E values higher than 12 mm), respectively.

In conclusion, the main goal of the present paper consists in the evaluation of the effect of glass-flax hybridization on the durability of pinned composite laminates, i.e., a deeper knowledge of the stability of hybrid composite components in salt-fog environments was achieved. In particular, the experimental approach may help the design phase with the aim of optimizing the performance of the composite structures in terms of mechanical stability and durability. Future studies will be addressed to improve this knowledge by investigating the mechanical stability of these composite materials in other critical environmental aging conditions complementary to salt spray setup (e.g., wet/dry cycles).

Author Contributions: Data curation, L.C. and V.F.; formal analysis, L.C., A.V. and T.S.; investigation, L.C. and P.B.; methodology, A.V., L.C. and V.F.; validation, T.S.; writing—Original draft, L.C. and V.F., writing—Review and editing, L.C. and V.F.; visualization, L.C. and V.F.; supervision, A.V. All authors have read and agreed to the published version of the manuscript.

Funding: No founding acknowledgement is required.

Conflicts of Interest: The authors declare no conflict of interest.

References

1. Pervaiz, M.; Panthapulakkal, S.; KC, B.; Sain, M.; Tjong, J. Emerging Trends in Automotive Lightweighting through Novel Composite Materials. *Mater. Sci. Appl.* **2016**, *7*, 26–38. [CrossRef]
2. Fragassa, C. Marine applications of natural fibre-reinforced composites: A manufacturing case study. In *Advances in Applications of Industrial Biomaterials*; Springer International Publishing: Berlin/Heidelberg, Germany, 2017; pp. 21–47. ISBN 9783319627670.
3. Pošvář, Z.; Růžička, M.; Kulíšek, V.; Mareš, T.; Doubrava, K.; Uher, O. Design of Composite Hybrid Structures "on Request". *Mater. Today Proc.* **2016**, *3*, 1129–1134.
4. Mastura, M.T.; Sapuan, S.M.; Mansor, M.R.; Nuraini, A.A. Environmentally conscious hybrid bio-composite material selection for automotive anti-roll bar. *Int. J. Adv. Manuf. Technol.* **2017**, *89*, 2203–2219. [CrossRef]
5. Zou, P.; Cheng, H. Effect of interference-fit percentage and preload on the mechanical behaviour of single-shear lap composite joint. In Proceedings of the IOP Conference Series: Materials Science and Engineering; Institute of Physics Publishing, Suzhou, China, 29–31 March 2019; Volume 531.
6. Calabrese, L.; Fiore, V.; Scalici, T.; Bruzzaniti, P.; Valenza, A. Failure maps to assess bearing performances of glass composite laminates. *Polym. Compos.* **2019**, *40*, 1087–1096. [CrossRef]
7. Fiore, V.; Calabrese, L.; Proverbio, E.; Passari, R.; Valenza, A. Salt spray fog ageing of hybrid composite/metal rivet joints for automotive applications. *Compos. Part B Eng.* **2017**, *108*, 65–74. [CrossRef]
8. Lv, J.; Xiao, Y.; Xue, Y. Time–temperature-dependent response and analysis of preload relaxation in bolted composite joints. *J. Reinf. Plast. Compos.* **2018**, *37*, 460–474. [CrossRef]
9. Valenza, A.; Fiore, V.; Calabrese, L. Three-Point Flexural Behaviour of GFRP Sandwich Composites: A Failure Map. *Adv. Compos. Mater.* **2010**, *19*, 79–90. [CrossRef]
10. Zhang, Y.; Li, Y.; Ma, H.; Yu, T. Tensile and interfacial properties of unidirectional flax/glass fiber reinforced hybrid composites. *Compos. Sci. Technol.* **2013**, *88*, 172–177. [CrossRef]

11. Mandal, S.; Alam, S. Dynamic mechanical analysis and morphological studies of glass/bamboo fiber reinforced unsaturated polyester resin-based hybrid composites. *J. Appl. Polym. Sci.* **2012**, *125*, E382–E387. [CrossRef]
12. Fiore, V.; Valenza, A.; Di Bella, G. Mechanical behavior of carbon/flax hybrid composites for structural applications. *J. Compos. Mater.* **2012**, *46*, 2089–2096. [CrossRef]
13. Ramesh, M.; Palanikumar, K.; Reddy, K.H. Influence of fiber orientation and fiber content on properties of sisal-jute-glass fiber-reinforced polyester composites. *J. Appl. Polym. Sci.* **2016**, *133*. [CrossRef]
14. Pan, Y.; Zhong, Z. The effect of hybridization on moisture absorption and mechanical degradation of natural fiber composites: An analytical approach. *Compos. Sci. Technol.* **2015**, *110*, 132–137. [CrossRef]
15. Fiore, V.; Calabrese, L.; Di Bella, G.; Scalici, T.; Galtieri, G.; Valenza, A.; Proverbio, E. Effects of aging in salt spray conditions on flax and flax/basalt reinforced composites: Wettability and dynamic mechanical properties. *Compos. Part B Eng.* **2016**, *93*, 35–42. [CrossRef]
16. Retegi, A.; Arbelaiz, A.; Alvarez, P.; Llano-Ponte, R.; Labidi, J.; Mondragon, I. Effects of hygrothermal ageing on mechanical properties of flax pulps and their polypropylene matrix composites. *J. Appl. Polym. Sci.* **2006**, *102*, 3438–3445. [CrossRef]
17. Fiore, V.; Scalici, T.; Sarasini, F.; Tirilló, J.; Calabrese, L. Salt-fog spray aging of jute-basalt reinforced hybrid structures: Flexural and low velocity impact response. *Compos. Part B Eng.* **2017**, *116*, 99–112. [CrossRef]
18. Garcia-Espinel, J.D.; Castro-Fresno, D.; Parbole Gayo, P.; Ballester-Muñoz, F. Effects of sea water environment on glass fiber reinforced plastic materials used for marine civil engineering constructions. *Mater. Des.* **2015**, *66*, 46–50. [CrossRef]
19. Calabrese, L.; Fiore, V.; Scalici, T.; Valenza, A. Experimental assessment of the improved properties during aging of flax/glass hybrid composite laminates for marine applications. *J. Appl. Polym. Sci.* **2019**, *136*, 47203. [CrossRef]
20. Saidane, E.H.; Scida, D.; Assarar, M.; Sabhi, H.; Ayad, R. Hybridisation effect on diffusion kinetic and tensile mechanical behaviour of epoxy based flax-glass composites. *Compos. Part A Appl. Sci. Manuf.* **2016**, *87*, 153–160. [CrossRef]
21. Peças, P.; Carvalho, H.; Salman, H.; Leite, M. Natural Fibre Composites and Their Applications: A Review. *J. Compos. Sci.* **2018**, *2*, 66. [CrossRef]
22. Li, X.; Tabil, L.G.; Panigrahi, S. Chemical treatments of natural fiber for use in natural fiber-reinforced composites: A review. *J. Polym. Environ.* **2007**, *15*, 25–33. [CrossRef]
23. Machado, J.J.M.; Gamarra, P.M.R.; Marques, E.A.S.; da Silva, L.F.M. Improvement in impact strength of composite joints for the automotive industry. *Compos. Part B Eng.* **2018**, *138*, 243–255. [CrossRef]
24. Esendemir, Ü.; Ayşe, A.M. Investigating bearing strength of pin-loaded composite plates in different environmental conditions. *J. Reinf. Plast. Compos.* **2013**, *32*, 1685–1697. [CrossRef]
25. Karakuzu, R.; Kanlioğlu, H.; Deniz, M.E. Effect of seawater on pin-loaded laminated composites. *Mater. Test.* **2018**, *60*, 85–92. [CrossRef]
26. Fiore, V.; Calabrese, L.; Scalici, T.; Valenza, A. Evolution of the bearing failure map of pinned flax composite laminates aged in marine environment. *Compos. Part B Eng.* **2020**. submitted.
27. Calabrese, L.; Fiore, V.; Bruzzaniti, P.G.; Scalici, T.; Valenza, A. An Aging Evaluation of the Bearing Performances of Glass Fiber Composite Laminate in Salt Spray Fog Environment. *Fibers* **2019**, *7*, 96. [CrossRef]
28. Fiore, V.; Calabrese, L.; Scalici, T.; Bruzzaniti, P.; Valenza, A. Bearing strength and failure behavior of pinned hybrid glass-flax composite laminates. *Polym. Test.* **2018**, *69*, 310–319. [CrossRef]
29. Fiore, V.; Scalici, T.; Calabrese, L.; Valenza, A.; Proverbio, E. Effect of external basalt layers on durability behaviour of flax reinforced composites. *Compos. Part B Eng.* **2016**, *84*, 258–265. [CrossRef]
30. Sola, C.; Castanié, B.; Michel, L.; Lachaud, F.; Delabie, A.; Mermoz, E. On the role of kinking in the bearing failure of composite laminates. *Compos. Struct.* **2016**, *141*, 184–193. [CrossRef]
31. Opelt, C.V.; Cândido, G.M.; Rezende, M.C. Compressive failure of fiber reinforced polymer composites–A fractographic study of the compression failure modes. *Mater. Today Commun.* **2018**, *15*, 218–227. [CrossRef]
32. Fiore, V.; Calabrese, L.; Scalici, T.; Bruzzaniti, P.; Valenza, A. Experimental design of the bearing performances of flax fiber reinforced epoxy composites by a failure map. *Compos. Part B Eng.* **2018**, *148*, 40–48. [CrossRef]
33. Malmstein, M.; Chambers, A.R.; Blake, J.I.R. Hygrothermal ageing of plant oil based marine composites. *Compos. Struct.* **2013**, *101*, 138–143. [CrossRef]

34. Abd-El-Naby, S.F.M.; Hollaway, L. The experimental behaviour of bolted joints in pultruded glass/ polyester material. Part 1: Single-bolt joints. *Composites* **1993**, *24*, 531–538.
35. Turvey, G. Failure of single-lap single-bolt tension joints in pultruded glass fibre reinforced plate. In Proceedings of the 6th International Conference on Composites in Construction Engineering (CICE), Rome, Italy, 13–15 June 2012.
36. Assarar, M.; Scida, D.; El Mahi, A.; Poilâne, C.; Ayad, R. Influence of water ageing on mechanical properties and damage events of two reinforced composite materials: Flax-fibres and glass-fibres. *Mater. Des.* **2011**, *32*, 788–795. [CrossRef]
37. Baley, C.; Le Duigou, A.; Bourmaud, A.; Davies, P. Influence of drying on the mechanical behaviour of flax fibres and their unidirectional composites. *Compos. Part A Appl. Sci. Manuf.* **2012**, *43*, 1226–1233. [CrossRef]
38. Bos, H.L.; Van Den Oever, M.J.A.; Peters, O.C.J.J. Tensile and compressive properties of flax fibres for natural fibre reinforced composites. *J. Mater. Sci.* **2002**, *37*, 1683–1692. [CrossRef]
39. Mohanty, A.K.; Misra, M.; Hinrichsen, G. Biofibres, biodegradable polymers and biocomposites: An overview. *Macromol. Mater. Eng.* **2000**, *276–277*, 1–24. [CrossRef]
40. Akil, H.M.; Cheng, L.W.; Mohd Ishak, Z.A.; Abu Bakar, A.; Abd Rahman, M.A. Water absorption study on pultruded jute fibre reinforced unsaturated polyester composites. *Compos. Sci. Technol.* **2009**, *69*, 1942–1948. [CrossRef]
41. Dhakal, H.N.; Zhang, Z.Y.; Richardson, M.O.W. Effect of water absorption on the mechanical properties of hemp fibre reinforced unsaturated polyester composites. *Compos. Sci. Technol.* **2007**, *67*, 1674–1683. [CrossRef]
42. Wei, B.; Cao, H.; Song, S. Degradation of basalt fibre and glass fibre/epoxy resin composites in seawater. *Corros. Sci.* **2011**, *53*, 426–431. [CrossRef]
43. Ouarhim, W.; Zari, N.; Bouhfid, R.; Qaiss, A. el kacem Mechanical performance of natural fibers–based thermosetting composites. In *Mechanical and Physical Testing of Biocomposites, Fibre-Reinforced Composites and Hybrid Composites*; Elsevier: Amsterdam, The Netherlands, 2019; pp. 43–60. ISBN 978-0-08-102292-4.
44. Fiore, V.; Scalici, T.; Badagliacco, D.; Enea, D.; Alaimo, G.; Valenza, A. Aging resistance of bio-epoxy jute-basalt hybrid composites as novel multilayer structures for cladding. *Compos. Struct.* **2017**, *160*, 1319–1328.

polymers

MDPI

Article

Evaluation of the Compatibility of Organosolv Lignin-Graphene Nanoplatelets with Photo-Curable Polyurethane in Stereolithography 3D Printing

Fathirrahman Ibrahim [1,2], Denesh Mohan [1,2], Mohd Shaiful Sajab [1,2,*], Saiful Bahari Bakarudin [3] and Hatika Kaco [4]

1 Research Center for Sustainable Process Technology (CESPRO), Faculty of Engineering and Built Environment, Universiti Kebangsaan Malaysia, Bangi 43600, Selangor, Malaysia; fatir96@gmail.com (F.I.); denesh.mohan@gmail.com (D.M.)

2 Chemical Engineering Programme, Faculty of Engineering and Built Environment, Universiti Kebangsaan Malaysia, Bangi 43600, Selangor, Malaysia

3 Institute of Microengineering and Nanoelectronics (IMEN), Universiti Kebangsaan Malaysia, Bangi 43600, Selangor, Malaysia; saifulbahari@ukm.edu.my

4 Kolej GENIUS Insan, Universiti Sains Islam Malaysia, Bandar Baru Nilai, Nilai 71800, Negeri Sembilan, Malaysia; hatikakaco@usim.edu.my

* Correspondence: mohdshaiful@ukm.edu.my; Tel.: +60-3-8921-6425

Received: 27 August 2019; Accepted: 21 September 2019; Published: 23 September 2019

Abstract: In this study, lignin has been extracted from oil palm empty fruit bunch (EFB) fibers via an organosolv process. The organosolv lignin obtained was defined by the presence of hydroxyl-containing molecules, such as guaiacyl and syringyl, and by the presence of phenolic molecules in lignin. Subsequently, the extracted organosolv lignin and graphene nanoplatelets (GNP) were utilized as filler and reinforcement in photo-curable polyurethane (PU), which is used in stereolithography 3D printing. The compatibility as well as the characteristic and structural changes of the composite were identified through the mechanical properties of the 3D-printed composites. Furthermore, the tensile strength of the composited lignin and graphene shows significant improvement as high as 27%. The hardness of the photo-curable PU composites measured by nanoindentation exhibited an enormous improvement for 0.6% of lignin-graphene at 92.49 MPa with 238% increment when compared with unmodified PU.

Keywords: 3D printing; composites; DLP; lignocellulose; nanoindentation

1. Introduction

Apart from being fast and precise, 3D printing enables a product to be easily modified and customized. This technology is suitable for applications in a field with profound individual differences, such as biomedicine [1,2]. The rise of 3D printing is also expected to increase with the use of the polymer. Resins used in stereolithography are thermosetting plastics, engineered to cure when receiving energy typically from a laser beam or ultraviolet rays [3]. The thermoset nature of stereolithography fabricated parts, with a high crosslink density, results in brittle fracture with low elongation properties [4].

Polyurethane (PU) is usually used in 3D printing because its mechanical properties can be tuned by changing its chemical structure [2]. Microbes are unable to decompose UV-curable adhesives because of their unique chemical composition and characteristics obtained after UV curing, which often lead to long-term retention issues [5]. One solution is to develop a polymer with good biodegradability via a composite approach. Composites can produce new materials with improved performance and biodegradability when a natural material is used as a filler [6].

In 2017, Malaysia accounted for 29% of the global palm oil production and 55.5% of the national export sector [7]. Despite the success of the oil palm industry, major drawbacks include the generation and utilization of empty fruit bunches (EFB). In fact, EFB has become a major threat to the industry and needs to be rectified quickly. Nevertheless, this biomass waste can be a renewable alternative to petroleum-based sources, fulfilling the energy demand and reducing pollution. Lignocellulosic materials predominantly contain a mixture of carbohydrate polymers such as cellulose, hemicelluloses, and lignin [8]. Lignin, which provides mechanical support to the cell walls of plants, also can be used to support in polymer composites. Although lignin is the second most abundant natural polymer on earth, it is under-utilized.

Lignin is categorized according to type of lignocellulose and extraction method; it can be of sulfonate or non-sulfonate type. Sulfonate lignin is produced via commercial methods such as the sulfur process and kraft process. Non-sulfur lignin is produced via the organosolv, soda, and hydrolysis processes. Extraction of lignin using sulfuric acid (H_2SO_4) via the kraft process is widely employed by the pulp and paper industry [9]. Organosolv is commonly utilized in laboratories owing to the easy recovery of the used acid [10]. Organosolv lignin has been utilized because of its significantly better solubility relative to those of other lignins, such as kraft and sulfur lignins [11]. Additionally, lignin produced by organosolv processes has also been reported to be high-quality, technical lignin [12]. Owing to its highly reactive polyphenolic structures, lignin can reduce the cost of producing other, more expensive, filler materials [13]; however, the sole incorporation of lignin will not boost the mechanical property of a polymer nanocomposite. The integration of filler-reinforcement of graphene also has been introduced in polymeric composites; graphene exhibited electrical, thermal, and mechanical properties as nanofiller to the reinforcement of the polymeric materials [14–16].

In this study, the isolation of lignin from oil palm EFB fibers was carried out using formic acid (FA) at different concentrations. The best condition in lignin isolation was further characterized by FTIR, FT-NMR, and GC-MS analyses. Subsequently, the isolated organosolv lignin was used and modified with the graphene nanoplatelets as filler-reinforcement for photo-curable PU in stereolithography 3D printing. In addition to the stress-strain behavior of the polymers, the hardness of the 3D-printed specimen was measured using nanoindentation technique for in-depth analysis of the effect of lignin-graphene compatibility in photo-curable PU resin.

2. Materials and Methods

2.1. Materials

Oil palm EFB fibers were procured from Szetech Engineering Sdn. Bhd. (Selangor, Malaysia). These were milled and sieved to obtain fibers of diameters 106–500 μm. Fractionation of lignin was performed using 90% formic acid (Merck, Darmstadt, Germany) and the lignin content was determined using 98% sulfuric acid (Merck). Graphene nanoplatelets (Sigma Aldrich, Darmstadt, Germany) was used as reinforced for the printed materials. In stereolithography 3D printing, photo-curable resin with the major composition of 45–47 wt% polyurethane acrylate, 34–36 wt% morpholine, and 15–17 wt% tripropylene glycol diacrylate was provided by Wanhao Precision Casting Co. Ltd. (Jinhua, China) and isopropyl alcohol (Merck) was used to remove the excess resin on the printed object.

2.2. Lignin Extraction

Lignin was extracted from oil palm EFB fibers via an organosolv extraction method, whereby 300 mL of 40–90% FA was added to 10 g of EFB, maintaining an EFB-solution ratio of 1:30. Using a magnetic stirrer, the solution was stirred at 95 °C for 2 h. Next, the solution was filtered using a vacuum filter to remove the pulp from the organosolv lignin. Subsequently, the isolation was carried out using rotary evaporator (RE 600 with VR 300 vacuum controller, Yamato Scientific Co. Ltd., Tokyo, Japan). The resulting black liquor was evaporated to separate the organosolv solution from the extracted lignin. Briefly, a round-bottomed flask was half-filled with lignin organosolv and then placed in a

96 °C water bath. The flask was connected to the evaporator using a clip, and the evaporator was left to operate until the solvent was completely removed. The isolated lignin was then repeatedly washed with water and centrifuged multiple times to remove the excess formic acid. Subsequently, the lignin was recovered via oven-drying.

2.3. Preparation of the Photo-Curable Resin Composites

In anticipation of the filler-reinforcement in photo-curable PU, lignin-graphene was prepared with the 10 wt% of graphene nanoplatelets (G) mixed continuously with the extracted organosolv lignin using a homogenizer for 30 min (IKA T 25 Ultra-Turrax Digital High-Speed Homogenizer). Afterwards, photo-curable PU resin was mixed with lignin and lignin-graphene at a different weight ratio (0.2, 0.4, 0.6, 0.8, 1.0, and 3.0 wt%). As a reference, PU was homogenized with 0.02 wt% of graphene, equivalent to the graphene comprised in PU-0.2Lignin/G. The resin compounds were homogeneously mixed using a homogenizer for 10 min or until the particles were well dispersed. The prepared resin compounds were kept in the dark container until further use.

2.4. Stereolithography 3D Printing

The UV curable polyurethane with the lignin-graphene mixture was then added into a Duplicator 7 V1.5 by Wanhao 3D printer—a digital light-processing (DLP) system with a 405 nm UV lamp as the curing agent. The STL file model was followed by standard tensile specimen according to the ASTM D638 type IV with the slight modification on the thickness of the tensile specimen, which was set at 0.5 mm, width 7.5 mm, and neck length 24 mm. The density was measured by the dimension and weight of printed samples. Prior to curing under UV light of wavelength 405 nm, the final product was washed in an isopropyl alcohol solution to remove the excess polyurethane resin.

2.5. Characterization

The extracted acid-soluble and insoluble lignin was measured by the compositional analysis of TAPPI T222 os-06 (2006) standard. The morphological structure of the sample before and after processes was observed by a variable pressure scanning electron microscope, VPSEM (Merlin Compact, Zeiss Pvt Ltd., Oberkochen, Germany). To determine the types of functional groups present in the lignin as well as the effectiveness of the recovery process, the oil palm EFB fibers and lignin were characterized via Fourier transform infrared spectroscopy (FTIR, Bruker, Billerica, MA, USA) at a resolution of 1 cm^{-1} in the range of 4000 to 650 cm^{-1}. Meanwhile, Fourier transform nuclear magnetic resonance (FT-NMR) was performed to analyze the molecular structure of the lignin. In this process, the lignin was dissolved in dimethyl sulphoxide-d6 (DMSO). The components of the extracted lignin were identified by gas chromatography/mass spectrometry (GC-MS, Agilent 7890 GC/5975 MSD), with ethyl acetate as the solvent and DB-5 the column. The viscosity of the photo-curable resin composites was examine using Brookefield Ametek D1 viscometer with spindle type DV1HA at 100 rpm. The tensile test analysis was performed in the Instron® Electromechanical Universal Testing Systems 3300 Series at 500 mm/min with a load cell of 1 kN, following which the tensile strength and elongation were recorded. Nanoindentation behavior of the materials was performed by Nano TestTM (Micro Materials, Wrexham, UK) at the maximum load of 50 mN. Tensile test and nanoindentation analyses were repeated five times for each parameter.

3. Results and Discussion

3.1. Optimization of Organosolv Lignin Extraction

In the preliminary study, the control parameters of FA concentrations on lignocellulosic fractionation were conducted early to maximize lignin that can be extracted. The initial lignin content for the untreated oil palm EFB fibers was recorded at 21.7 wt%. As seen in Figure 1, the amount of acid-soluble lignin extracted from oil palm EFB fibers was increased with the increment of

FA concentration. The lignin isolation shows the highest activity at 90% of FA, which extracted ~52.6% from the total lignin in oil palm EFB fibers. Although delignification is the method of breaking down the chemical structure of lignin to make it soluble in a liquid, formic acid has a higher rate of results of dissolution when compared to acetic acid or others with longer carbon chains attributable to the decrease in lignin solubility with increasing alkyl chain length [17,18]. Thus, at higher FA concentration, it provided an effective lignin dissolution through cleavage of ether bonds in lignin macromolecules using acid [19]. The extracted lignin was further isolated by rotary evaporator, repeatedly washed with water to remove the excess formic acid and further modification with graphene nanoplatelets as filler-reinforcement in photo-curable polyurethane resin in the next stages, as seen in Figure 1a.

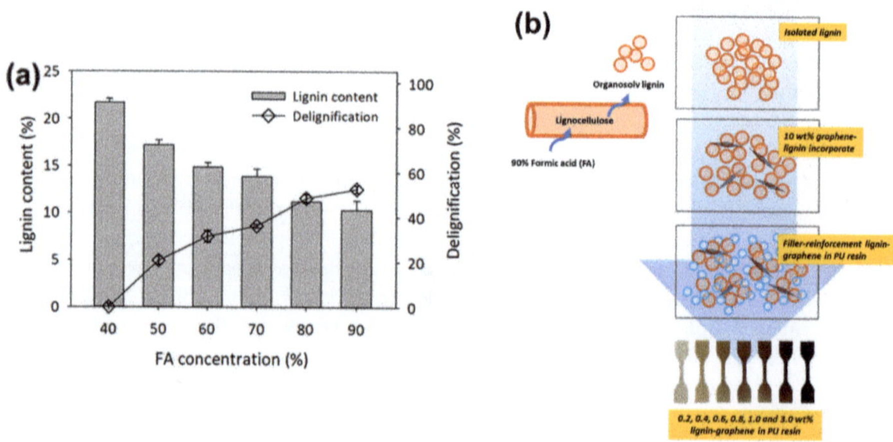

Figure 1. (**a**) The effect of different formic acid (FA) concentration on delignification of empty fruit bunch (EFB) (EFB to aqueous ratio: 1:30, Operating temperature: 90 °C, FA concentration: 40–90%) and (**b**) the further techniques involved in the utilization of organosolv lignin-graphene nanoplatelets with photo-curable polyurethane (PU).

The oil palm EFB fibers show significantly morphological changes after the organosolv lignin extraction, as seen in Figure 2a. The untreated oil palm EFB fibers indicate silica bodies covering the surface structure of the fibers. The treated fibers after organosolv lignin extraction exposed the rough surface structure, an exposed lumen, and noticeable micro cavities due to the removal of lignin and dislodgement of the silica bodies [20]. Additionally, as the advantage of the organosolv extraction, the separation of FA-lignin through rotary evaporator provided 89.9% of the acid recovery, which can be used for the next cycle of the organosolv extraction.

As seen in Figure 2b, the FTIR spectrum at wavenumber of 3400 cm^{-1} exhibited the greater abundance of O–H bonds in oil palm EFB fibers relative to the glucose monomers of cellulose which was significantly changed in organosolv pulp of oil palm EFB fibers and extracted organosolv lignin [20]. Additionally, the peak at 1371 cm^{-1} was present in oil palm EFB fibers but not in lignin is because of alterations in the C–H bond following the removal of cellulose and hemicellulose. Meanwhile, the intensity at 2900 cm^{-1} denoted the C–H of the methyl (CH_2) and methylene (CH_3) groups in the monomer of each compound [21]. This finding was also consistent with the fact that the molecules that made up lignocellulose were aromatic and, accordingly, contained a small amount of alkane groups. Typically, the double bond of aromatic carbonyl (C=O) group, C=C aromatic group and ether linkages were observed at 1700 cm^{-1}, 1500–1600 cm^{-1}, and 1038 cm^{-1} were abundant in lignin, cellulose, and hemicellulose [22,23]. In comparison, syringyl and guaiacyl units of lignin can be identified at 1117 cm^{-1} and 1271 cm^{-1} [24,25].

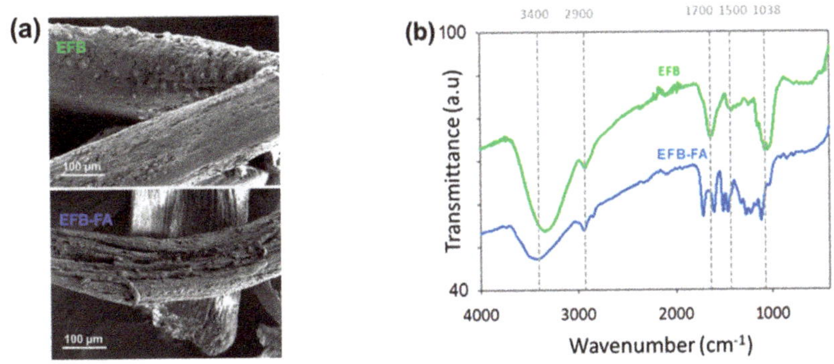

Figure 2. The (**a**) micrograph of VPSEM and (**b**) FTIR spectrums of oil palm EFB fibers before and after the organosolv lignin extraction.

3.2. Characterization of Organosolv Lignin

The ^{1}H NMR spectrum of extracted organosolv lignin is shown in Figure 3a. The 3.14 ppm peak shows the presence of hydrogen C-β in β-1 and β-β was stronger than that of the ether linkage because the ether linkages were the easiest to hydrolyze during the formic acid organosolv process [26]. Peaks at 3.758–3.763 ppm revealed the presence of methoxyl proton (–OCH_3). The peak at 8.133 ppm revealed the presence of aromatic protons in guaiacyl (G) and syringyl (S) units and 6.524–6.826 ppm aromatic protons in the syringyl-propane units verify the presence of the main monomers of lignin in the sample. Furthermore, the stronger signal at 6.8 ppm relative to that at 7.0 ppm indicates that the lignin contained more syringyl than guaiacyl monomers [27,28]. Other peaks were detected at (1) 7.792–7.868 ppm ortho-hydrogen within the carbonyl groups; (2) peak at 4.898 ppm β-H within the β-O-4 linkages; (3) 2.18–2.216 ppm phenolic protons in lignin; (4) 1.272 ppm and 0.861–0.895 ppm methylene and methyl groups respectively within the lignin side chains; as well as (5) 2.500–2.216 ppm protons in the solvent (DMSO-d6) [22,26,29,30].

The ^{1}C NMR provided in Figure 3b shows significant signals at 56.56 ppm, indicating the presence of –OCH_3 in the syringyl units, 39.72–40.97 DMSO-d6 (i.e., the solvent), as well as 29.34 β-methylene in the n-propyl side chains. On the contrary, the absence of peaks at 57–103 ppm implied that the lignin did not contain a noticeable amount of polysaccharide. It may be concluded that the ester and ether bonds in lignin and hemicellulose have been successfully cleaved during organosolv extraction [28].

As seen in Figure 3c, the GC-MS spectrum revealed the presence of a phenolic molecule and 1,1-bifenil-2,3-diol, both of which gave rise to a phenylpropane unit, the building block of lignin [31]. This was attributable to the chemical modification of the β-O-4 linkage during extraction [32]. The abundant presence of these two components also showed that this sample was rich in phenolic molecules. Apart from phenolic monomers, acyclic hydrocarbons like nonadecane, 1-dodecene, 6-tetradecene, hexadecene, (Z), 2-bromononane, tetratetracontane, 2,6-dimethyldecane, and methylnonadecane were also present. Other acyclic components included oxygenated components, such as 1-heptacosanol, oxalic acid, n-hexadecanoic acid, 9,12-octadecadienoic acid (Z,Z)-, octadecanoic acid, and danbromoacetic acid, which were also present in the lignin [33]. The reason behind the presence of n-hexadecanoic acid or palmitic acid in the lignin sample was that the acid is a typical component of the oil within palm fruits [34].

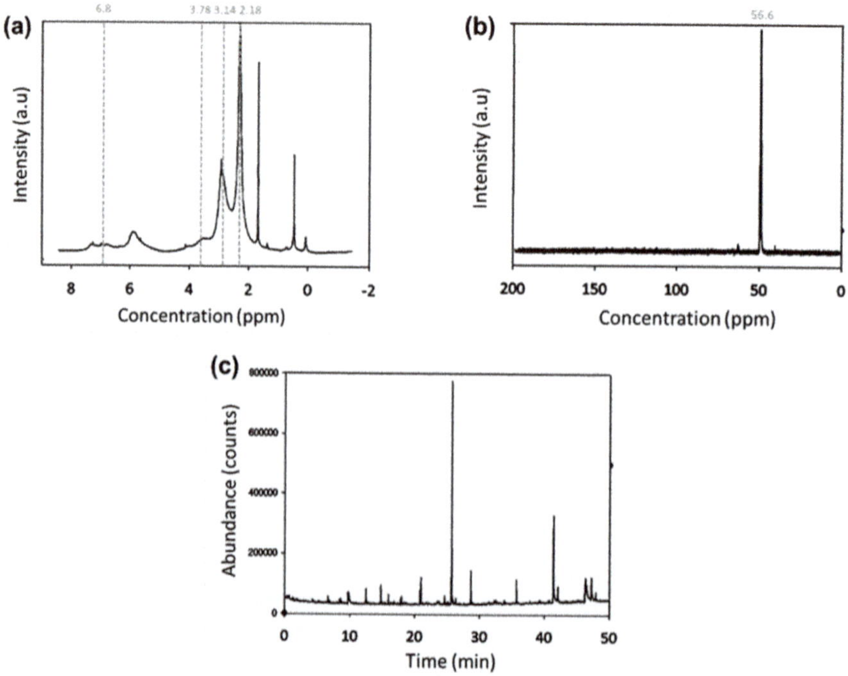

Figure 3. Chemical characterization of organosolv lignin extracted through (**a**) ^{1}H NMR, (**b**) ^{1}C NMR, and (**c**) GC-MS spectrums.

3.3. *Tensile Strength of 3D-Printed Composites*

As seen in the graph in Figure 4a, the composites which contained 1% and 3% of lignin had lower tensile strength than the resin. Meanwhile, the composite whose lignin concentration was 0.6% had a tensile strength close to that of resin. When lignin was added to the PU resin, the proportion of resin decreased; however, the presence of hydroxyl- and phenolic-rich lignin gave rise to a 3-dimensional structure with polyurethane and retained the tensile strength of the [35]. Likewise, the tensile strength of the PU-0.6% Lignin/G composite was much greater than that of the PU resin and lignin incorporated, showing that the mechanical property of the composite was enhanced when the concentration of lignin-graphene increased until 0.8% addition. The increase in lignin-graphene content increased the hardness of the composite and reduced the flexibility of the polymer. The 3D printed composited shown in Figure 4 displays the darkening of the samples with an increment of the lignin and lignin/G. The maximum load the filler-reinforcement was able to attain was 3%; higher loading interrupted the curing behavior of the photo-curable PU resin and defected the orientation of the printed samples.

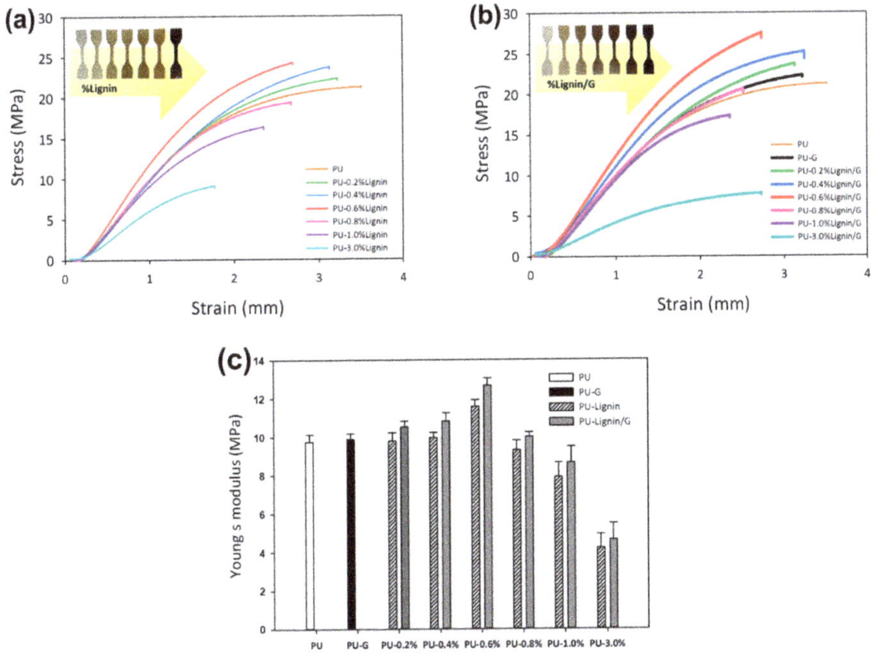

Figure 4. Mechanical properties of the photo-curable PU at stress-strain curves of (**a**) composited with lignin, (**b**) lignin/graphene, and (**c**) Young's modulus behaviours of the materials.

Figure 4c shows the Young's modulus of the sample with the increment of lignin and lignin-graphene as well as the data summarized in Table 1. The low loading lignin-graphene shows exceptional performance: as tensile strength increased, the ductility of the materials was compromised [16]. In contrast, the addition of lignin-graphene improved the Young's modulus of the composite. The high filler-reinforcement load often resulted in particle agglomeration; however, the viscosity measurement tabulated in Table 1 shows slight changes (~0.55% viscosity increase) even at the highest concentration of the additional filler-reinforcement of lignin-graphene [15].

Table 1. The effect of resin viscosity towards the tensile Stress-Strain and Young's modulus of photo-curable PU composited with lignin and lignin/graphene.

Sample	Viscosity (cP)	Stress (MPa)	Strain (mm)	Young's Modulus (MPa)
PU	50.63 ± 0.05	21.15 ± 0.21	3.51 ± 0.02	9.77 ± 0.15
PU-0.2%Lignin	50.72 ± 0.07	22.19 ± 0.11	3.21 ± 0.05	9.81 ± 0.13
PU-0.6%Lignin	50.76 ± 0.09	24.50 ± 0.32	2.69 ± 0.03	11.59 ± 0.14
PU-3.0%Lignin	50.91 ± 0.04	9.19 ± 0.22	1.77 ± 0.06	4.21 ± 0.09
PU-G	50.64 ± 0.03	22.56 ± 0.29	3.22 ± 0.09	9.89 ± 0.13
PU-0.2%Lignin/G	50.73 ± 0.04	23.87 ± 0.24	3.11 ± 0.02	10.53 ± 0.16
PU-0.6%Lignin/G	50.76 ± 0.04	27.35 ± 0.30	2.73 ± 0.05	12.68 ± 0.17
PU-3.0%Lignin/G	50.93 ± 0.06	8.39 ± 0.16	7.80 ± 0.03	4.63 ± 0.13

3.4. Nanoindentation Behavior of 3D-Printed Composites

Nanoindentation technique on the photo-curable PU composites has been carried out at the maximum load of 50 mN. The loading-unloading curves of the sample can be seen in Figure 5 and the calculated data is summarized in Table 2. The contribution of lignin as a filler in PU resin provided a minimum reformation of the sample hardness.

Table 2. The loading-unloading behavior of photo-curable PU composited with lignin and lignin/graphene.

Sample	Max Depth (nm)	Plastic Depth (nm)	Hardness (MPa)
PU	10,573 ± 15	8708 ± 12	27.33 ± 0.3
PU-0.2%Lignin	9665 ± 3	8181 ± 7	38.21 ± 0.3
PU-0.6%Lignin	8830 ± 11	5079 ± 8	38.96 ± 0.2
PU-3.0%Lignin	9655 ± 7	8011 ± 6	38.12 ± 0.1
PU-G	8573 ± 5	6154 ± 4	54.51 ± 0.4
PU-0.2%Lignin/G	10,799 ± 8	8715 ± 5	27.29 ± 0.3
PU-0.6%Lignin/G	5609 ± 12	4714 ± 8	92.49 ± 0.4
PU-3.0%Lignin/G	14,801 ± 8	11,951 ± 11	14.54 ± 0.3

In the additional of lignin-graphene, typical loading-unloading behavior in comparison with stress-strain curve was observed, which demonstrated the depth indentation decreased and the curves shifted to the left due to the increment of the hardness [36]. The filler-reinforcement of lignin-graphene showed the highest improvement at 0.6% Lignin/G as the sample provided tremendous enhancement of the hardness at 92.49 MPa (238% increment from unmodified photo-curable PU resin). The integration of graphene nanoplatelets was shown to be the main element for the improvement of the hardness of the PU. The resistance of the graphene towards deformation of the material provided better elasticity [36]. In contrast, the chemical interaction between the isocyanate group and oxygenated groups of PU-graphene provided the compatibility of the lignin-graphene in photo-curable PU resin [14]. However, the increment of the graphene up to 3% showed a similar trend as a stress-strain curve, which exhibited the low hardness profile of the materials.

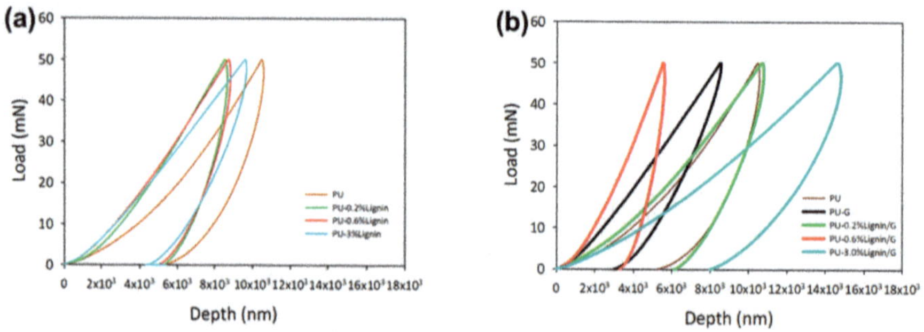

Figure 5. Nanoindentation of the photo-curable PU composited with (**a**) lignin and (**b**) lignin/graphene.

3.5. The Compatibility of PU-Lignin with Graphene

As seen in Figure 6, the chemical compatibility of photo-curable PU with lignin and graphene was analyzed through FTIR spectrums. In cured PUs, the infrared region indicates fewer isocyanates, discovered at 2312 cm^{-1}, suggesting that there is no surplus moisture and isocyanates in pre-polymer reactants [37]. Thus, curing of the polymer is optimal as no major peak was observed at 2312 cm^{-1}. The broad peak was observed at 3200–3600 cm^{-1}, indicating the presence of alcoholic and phenolic –OH absorptions. Peaks observed at 1400–1600 cm^{-1} indicated the presence of aromatic structures present in the PU [38]. After the addition of the lignin-graphene, a strong interfacial interaction between functional groups of lignin-graphene and polyurethane occurred resulting in better compatibility, suggesting a strong hydrogen bond between polyurethane and lignin-graphene [39].

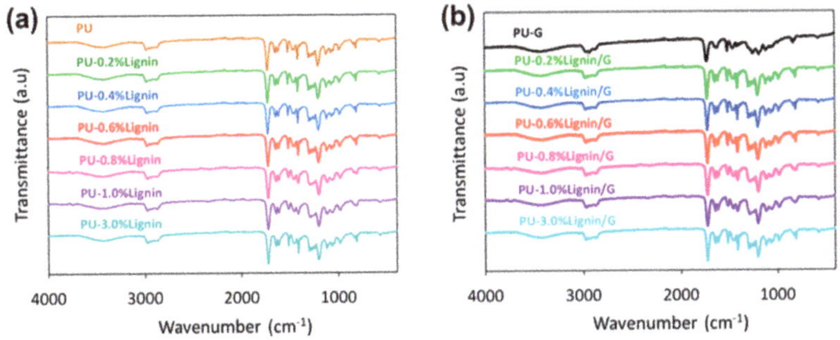

Figure 6. FTIR spectrums of the photo-curable PU composited with (**a**) lignin and (**b**) lignin/graphene.

The micrograph of the cross-section 3D-printed composites after tensile testing shows the effect of lignin-graphene in favor of the better compatibility of filler-reinforcement in photo-curable PU, as seen in Figure 7. PU samples indicated a wide line of micro cracks and homogeneously flat surface area. In Figure 7b, graphene nanoplatelets were distinctly spotted on the PU-G surface with non-uniform blending between photo-curable PU. In the addition of lignin-graphene, the blending between PU-lignin-graphene was well distributed in the polymeric matrix [40].

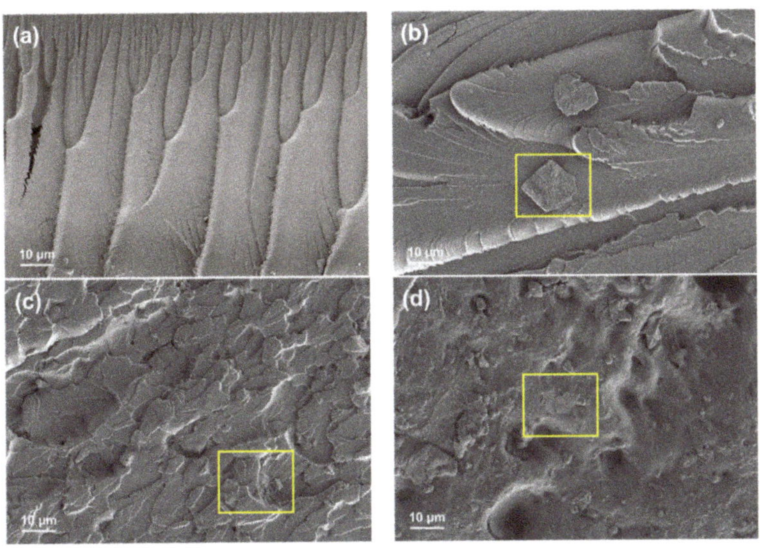

Figure 7. Micrograph images of the fracture surface after tensile testing (**a**) PU, (**b**) PU-G, (**c**) PU-0.6% Lignin/G, and (**d**) top surface of PU-0.6% Lignin/G.

A small poor dispersion of graphene nanoplatelets will provide more concentrated stress locally, affecting the mechanical properties of the material [41]. The clear morphological structure from the top of the PU-0.6% Lignin/G which has been exposed to the UV light, as seen in Figure 7d, exhibited the homogeneity of graphene nanoplatelets with the assistance of lignin as filler reinforcement in photo-curable PU. The synergistic interaction seen in lignin-graphene, lignin-polyurethane, graphene-polyurethane, and lignin-graphene in polyurethane resin plays a major role in the formation of compatible photo-curable resin composites. In comparison with lignin-polyurethane blends, the molecular structure of lignin tends to form a lignin-lignin interaction; the viscosity of the resin composite gradually increases as the composition of the lignin increases, as seen in Figure 1a. Unlike graphene oxide, which has a nanolayer formation, the graphene nanoplatelets particles require proper mechanical homogenization method to distribute in the polyurethane system [36,39,42]. Accordingly, the composition of lignin-graphene in polyurethane system facilitated greater distribution, significantly increasing the mechanical properties of the 3D-printed composites.

4. Conclusions

In this study, the extracted lignin at the best condition of organosolv extraction reveals the presence of syringyl, guaiacyl, and hydroxyl molecules, which is highly compatible with the photo-curable PU. The ability of the organosolv lignin as a compatibilizer for graphene nanoplatelets was successfully demonstrated in the mechanical properties of the PU-Lignin/G. Substantial improvement by filler-reinforcement of lignin-graphene in photo-curable PU, identified at 0.6% Lignin/G as the stress-strain and loading-unloading behavior, shows higher tensile strength and the resistance against the deformation of the material.

Author Contributions: Conceptualization, M.S.S.; formal analysis, F.I., D.M. and S.B.B.; writing—original draft preparation, M.S.S and D.M.; writing—review and editing, M.S.S. and H.K.; funding acquisition, M.S.S.

Funding: This research was funded by Universiti Kebangsaan Malaysia through the grant provided, DIP-2017-017 and GUP-2018-104.

Conflicts of Interest: The authors declare no conflict of interest.

References

1. Liu, J.; Cheng, F.; Grenman, H.; Spoljaric, S.; Seppala, J.; Eriksson, J.E.; Willfor, S.; Xu, C. Development of nanocellulose scaffolds with tunable structures to support 3D cell culture. *Carbohydr. Polym.* **2016**, *148*, 259–271. [CrossRef] [PubMed]
2. Hung, K.C.; Tseng, C.S.; Hsu, S.H. *3D Printing of Polyurethane Biomaterials*; Elsevier Ltd.: Duxford, UK, 2016; ISBN 9780081006221.
3. Rinaldi, M.; Esposti, A.; Mottola, A.; Ganz, S. *Chapter 3—Computer-Assisted Implant Surgery*; Elsevier Inc.: Atlanta, GA, USA, 2016.
4. Eng, H.; Wei, J.; Choong, Y.Y.C.; Yu, S.; Tan, C.L.C.; Su, P.C.; Maleksaeedi, S.; Wiria, F.E. 3D Stereolithography of Polymer Composites Reinforced with Orientated Nanoclay. *Procedia Eng.* **2018**, *216*, 1–7. [CrossRef]
5. Huang, J.; Sun, J.; Zhang, R.; Zou, R.; Liu, X.; Yang, Z.; Yuan, T. Improvement of biodegradability of UV-curable adhesives modified by a novel polyurethane acrylate. *Prog. Org. Coat.* **2016**, *95*, 20–25. [CrossRef]
6. Yin, Q.; Yang, W.; Sun, C.; Di, M. Preparation and properties of lignin-epoxy resin composite. *BioResources* **2012**, *7*, 5737–5748. [CrossRef]
7. Varga, S. Essential Palm Oil Statistics Palm Oil Analytics. *Palm Oil Anal.* **2017**, *1*, 4–26.
8. Binod, P.; Pandey, A. Introduction. *Pretreat. Biomass Process. Technol.* **2014**, *1*, 3–6.
9. Olivares, M.; Guzmán, J.A.; Natho, A.; Saavedra, A. Kraft lignin utilization in adhesives. *Wood Sci. Technol.* **1988**, *22*, 157–165. [CrossRef]
10. Lange, H.; Decina, S.; Crestini, C. Oxidative upgrade of lignin - Recent routes reviewed. *Eur. Polym. J.* **2013**, *49*, 1151–1173. [CrossRef]
11. Park, Y.; Doherty, W.O.S.; Halley, P.J. Developing lignin-based resin coatings and composites. *Ind. Crops Prod.* **2008**, *27*, 163–167. [CrossRef]
12. Stephen, J.D.; Mabee, W.E.; Saddler, J.N. Will second-generation ethanol be able to compete with first-generation ethanol? Opportunities for cost reduction. *Biofuels Bioprod. Biorefining* **2012**, *6*, 159–176.
13. Pouteau, C.; Baumberger, S.; Cathala, B.; Dole, P. Lignin-polymer blends: Evaluation of compatibility by image analysis. *Comptes Rendus Biol.* **2004**, *327*, 935–943. [CrossRef]
14. Cai, D.; Yusoh, K.; Song, M. The mechanical properties and morphology of a graphite oxide nanoplatelet/polyurethane composite. *Nanotechnology* **2009**, *20*, 1–9. [CrossRef] [PubMed]
15. Young, R.J.; Liu, M.; Kinloch, I.A.; Li, S.; Zhao, X.; Vallés, C.; Papageorgiou, D.G. The mechanics of reinforcement of polymers by graphene nanoplatelets. *Compos. Sci. Technol.* **2018**, *154*, 110–116. [CrossRef]
16. Gao, Y.; Picot, O.T.; Bilotti, E.; Peijs, T. Influence of filler size on the properties of poly(lactic acid) (PLA)/graphene nanoplatelet (GNP) nanocomposites. *Eur. Polym. J.* **2017**, *86*, 117–131. [CrossRef]

17. Rashid, T.; Kait, C.F.; Regupathi, I.; Murugesan, T. Dissolution of kraft lignin using Protic Ionic Liquids and characterization. *Ind. Crops Prod.* **2016**, *84*, 284–293. [CrossRef]
18. Bajpai, P. Biermann's Handbook of Pulp and Paper: Raw Material and Pulp Making. *Elsevier* **2018**, *12*, 295–351.
19. Mohaiyiddin, M.S.; Lin, O.H.; Owi, W.T.; Chan, C.H.; Chia, C.H.; Zakaria, S.; Villagracia, A.R.; Akil, H.M. Characterization of nanocellulose recovery from Elaeis guineensis frond for sustainable development. *Clean Technol. Environ. Policy* **2016**, *18*, 2503–2512. [CrossRef]
20. Santanaraj, J.; Sajab, M.S.; Mohammad, A.W.; Harun, S.; Chia, C.H.; Zakari, S.; Kaco, H. Enhanced delignification of oil palm empty fruit bunch fibers with in situ fenton-oxidation. *BioResources* **2017**, *12*, 5223–5235. [CrossRef]
21. Hashim, S.N.A.S.; Zakaria, S.; Chia, C.H.; Pua, F.L.; Jaafar, S.N.S. Chemical and thermal properties of purified kenaf core and oil palm empty fruit bunch lignin. *Sains Malaysiana* **2016**, *45*, 1649–1653.
22. Whetten, R.; Sederoff, R. Lignin Biosynthesis. *Plant Cell* **2007**, *7*, 1001. [CrossRef]
23. Ramezani, N.; Sain, M. Thermal and Physiochemical Characterization of Lignin Extracted from Wheat Straw by Organosolv Process. *J. Polym. Environ.* **2018**, *26*, 3109–3116. [CrossRef]
24. Mohamad Ibrahim, M.N.; Zakaria, N.; Sipaut, C.S.; Sulaiman, O.; Hashim, R. Chemical and thermal properties of lignins from oil palm biomass as a substitute for phenol in a phenol formaldehyde resin production. *Carbohydr. Polym.* **2011**, *86*, 112–119. [CrossRef]
25. García, A.; González Alriols, M.; Spigno, G.; Labidi, J. Lignin as natural radical scavenger. Effect of the obtaining and purification processes on the antioxidant behaviour of lignin. *Biochem. Eng. J.* **2012**, *67*, 173–185. [CrossRef]
26. Klein, A.P.; Beach, E.S.; Emerson, J.W.; Zimmerman, J.B. Accelerated solvent extraction of lignin from aleurites moluccana (candlenut) nutshells. *J. Agric. Food Chem.* **2010**, *58*, 10045–10048. [CrossRef]
27. Rashid, T.; Gnanasundaram, N.; Appusamy, A.; Kait, C.F.; Thanabalan, M. Enhanced lignin extraction from different species of oil palm biomass: Kinetics and optimization of extraction conditions. *Ind. Crops Prod.* **2018**, *116*, 122–136. [CrossRef]
28. She, D.; Xu, F.; Geng, Z.C.; Sun, R.C.; Jones, G.L.; Baird, M.S. Physicochemical characterization of extracted lignin from sweet sorghum stem. *Ind. Crops Prod.* **2010**, *32*, 21–28. [CrossRef]
29. Esteves Costa, C.A.; Coleman, W.; Dube, M.; Rodrigues, A.E.; Rodrigues Pinto, P.C. Assessment of key features of lignin from lignocellulosic crops: Stalks and roots of corn, cotton, sugarcane, and tobacco. *Ind. Crops Prod.* **2016**, *92*, 136–148. [CrossRef]
30. Coral Medina, J.D.; Woiciechowski, A.L.; Zandona Filho, A.; Bissoqui, L.; Noseda, M.D.; de Souza Vandenberghe, L.P.; Zawadzki, S.F.; Soccol, C.R. Biological activities and thermal behavior of lignin from oil palm empty fruit bunches as potential source of chemicals of added value. *Ind. Crops Prod.* **2016**, *94*, 630–637. [CrossRef]
31. Takada, D.; Ehara, K.; Saka, S. Gas chromatographic and mass spectrometric (GC-MS) analysis of lignin-derived products from Cryptomeria japonica treated in supercritical water. *J. Wood Sci.* **2004**, *50*, 253–259. [CrossRef]
32. Villaverde, J.J.; Li, J.; Ek, M.; Ligero, P.; De Vega, A. Native lignin structure of Miscanthus x giganteus and its changes during acetic and formic acid fractionation. *J. Agric. Food Chem.* **2009**, *57*, 6262–6270. [CrossRef]
33. Singh, S.K.; Ekhe, J.D. Towards effective lignin conversion: HZSM-5 catalyzed one-pot solvolytic depolymerization/hydrodeoxygenation of lignin into value added compounds. *RSC Adv.* **2014**, *4*, 27971–27978. [CrossRef]
34. Harsono, H.; Putra, A.S.; Maryana, R.; Rizaluddin, A.T.; H'ng, Y.Y.; Nakagawa-izumi, A.; Ohi, H. Preparation of dissolving pulp from oil palm empty fruit bunch by prehydrolysis soda-anthraquinone cooking method. *J. Wood Sci.* **2016**, *62*, 65–73. [CrossRef]
35. Wang, Z.; Yang, X.; Zhou, Y.; Liu, C. Mechanical and thermal properties of polyurethane films from peroxy-acid wheat straw lignin. *BioResources* **2013**, *8*, 3833–3843. [CrossRef]
36. Shokrieh, M.M.; Hosseinkhani, M.R.; Naimi-Jamal, M.R.; Tourani, H. Nanoindentation and nanoscratch investigations on graphene-based nanocomposites. *Polym. Test.* **2013**, *32*, 45–51. [CrossRef]
37. Yang, Z.; Peng, H.; Wang, W.; Liu, T. Crystallization behavior of poly(ε-caprolactone)/layered double hydroxide nanocomposites. *J. Appl. Polym. Sci.* **2010**, *116*, 2658–2667. [CrossRef]

38. Derkacheva, O.; Sukhov, D. Investigation of lignins by FTIR spectroscopy. *Macromol. Symp.* **2008**, *265*, 61–68. [CrossRef]
39. Aqlil, M.; Moussemba Nzenguet, A.; Essamlali, Y.; Snik, A.; Larzek, M.; Zahouily, M. Graphene oxide filled lignin/starch polymer bionanocomposite: structural, physical, and mechanical studies. *J. Agric. Food Chem.* **2017**, *65*, 10571–10581. [CrossRef]
40. Bafana, A.P.; Yan, X.; Wei, X.; Patel, M.; Guo, Z.; Wei, S.; Wujcik, E.K. Polypropylene nanocomposites reinforced with low weight percent graphene nanoplatelets. *Compos. Part B Eng.* **2017**, *109*, 101–107. [CrossRef]
41. Wei, J.; Saharudin, M.S.; Vo, T.; Inam, F. Dichlorobenzene: An effective solvent for epoxy/graphene nanocomposites preparation. *R. Soc. Open Sci.* **2017**, *4*, 1–9. [CrossRef]
42. Al-Shahrani, D.; Love, S.; Salas-de la Cruz, D. The role of reduced graphene oxide toward the self-assembly of lignin-based biocomposites fabricated from ionic liquids. *Int. J. Mol. Sci.* **2018**, *19*, 3518. [CrossRef]

polymers MDPI

Article

Synergistic Effect of Maleated Natural Rubber and Modified Palm Stearin as Dual Compatibilizers in Composites Based on Natural Rubber and Halloysite Nanotubes

Nabil Hayeemasae [1,*], Zareedan Sensem [1], Indra Surya [2], Kannika Sahakaro [1] and Hanafi Ismail [3]

1 Department of Rubber Technology and Polymer Science, Faculty of Science and Technology, Prince of Songkla University, Pattani Campus, Pattani 94000, Thailand; zareedan@gmail.com (Z.S.); kannika.sah@psu.ac.th (K.S.)
2 Department of Chemical Engineering, Faculty of Engineering, Universitas Sumatera Utara, Medan 20155, Sumatera Utara, Indonesia; isurya@usu.ac.id
3 School of Materials and Mineral Resources Engineering, Engineering Campus, Universiti Sains Malaysia, Nibong Tebal 14300, Penang, Malaysia; ihanafi@usm.my
* Correspondence: nabil.h@psu.ac.th

Received: 28 February 2020; Accepted: 25 March 2020; Published: 1 April 2020

Abstract: The performance of rubber composite relies on the compatibility between rubber and filler. This is specifically of concern when preparing composites with very different polarities of the rubber matrix and the filler. However, a suitable compatibilizer can mediate the interactions. In this study, composites of natural rubber (NR) with halloysite nanotubes (HNT) were prepared with maleated natural rubber (MNR) and modified palm stearin (MPS) as dual compatibilizers. The MPS dose ranged within 0.5–1.5 phr, while the MNR dose was fixed at 10 phr in all formulations. It was found that the mixed MNR/MPS significantly enhanced modulus, tensile strength, and tear strength of the composites. The improvements were mainly due to improved rubber-HNT interactions arising from hydrogen bonds formed in the presence of these two compatibilizers. This was clearly verified by observing the Payne effect. Apart from that, the MPS also acted as a plasticizer to provide improved dispersion of HNT. It was clearly demonstrated that MNR and MPS as dual compatibilizers improved rubber-HNT interactions and reduced filler-filler interactions, which then improved tensile and tear strengths, as well as dynamical properties. Therefore, the mix of MNR and MPS had a great potential to compatibilize non-polar rubber with HNT filler.

Keywords: natural rubber; maleated natural rubber; palm stearin; halloysite nanotubes

1. Introduction

Enhanced properties of rubber can be obtained by adding a small amount of nanofillers. This technique has drawn considerable attention during the last decades [1–3]. The improvements of physical and other related properties of rubber depend on several factors, such as the filler aspect ratio, filler's compatibility, the degree of dispersion, and the alignment of the particulates. Halloysite nanotubes (HNT) are a type of nanofiller that has been recently tested in many types of matrix [4–7]. This is because of the very special characteristics of this material formed by surface weathering of aluminosilicate minerals and composed of aluminum, silicon, hydrogen, and oxygen. Due to the unique surface chemistry of HNT, it is not compatible with non-polar rubbers, such as natural rubber (NR). Scientists have been trying to address this drawback by several approaches to improve their compatibility. These include using silane coupling agents [8], adjusting the preparation methods [9], and using compatibilizers [10].

The silane coupling agent has been widely used and reported for the last decades. The use of silane is known but considered to be expensive and requires a high mixing temperature to obtain effective silanization. The chemistry behind the enhancement of the composites has also been very well studied. In the meantime, adjusting the processing methods has also been focused on the preparation of rubber composites. As, for example, Varghese and Karger-Kocsis [9] prepared natural rubber/layered silicates through latex compounding method. They found out that the latex route was promising, but it might not be of great practical relevance in comparison to the melt-compounding route. Besides, the acting shear forces strongly favor the dispersion of filler and is a strong argument for melt compounding with rubbers. Thus, searching for an alternative and effective compatibilizer for rubber/HNT composite is of great interest. A compatibilizer tends to affect the overall structure of a composite. However, most common compatibilizers are synthetic chemicals, and there has been less focus on the utilization of natural-based compatibilizers. In this study, two types of compatibilizer were used to modify the compatibility of NR with HNT: one being a modified natural rubber, and another being modified palm stearin.

As for the modified natural rubber, the compatibility of NR and HNT can also be improved by some functional groups. Pasbakhsh et al. [11] prepared maleic anhydride (MA) grafted ethylene propylene diene rubber (EPDM) or EPDM-g-MA, to increase the compatibility of EPDM and HNT. It was obvious that the use of EPDM-g-MA reduced HNT agglomeration and hence improved the HNT dispersion. This was attributed to interactions between the hydroxyl groups on HNT surfaces and succinic anhydride groups of the EPDM-g-MA. Similar approaches can be found in the literature [12,13]. They have also proposed possible interactions between hydroxyl groups of the paper sludge and succinic anhydride groups.

Despite adding modified natural rubber as a compatibilizer, further improvements in material properties can be obtained by introducing another compatibilizer. Palm stearin is an interesting material that has been useful in natural rubber compounds. Palm stearin is derived from extensive processing in the palm oil factory. It is fractionated from the refinery processing of crude palm oil [12], and its value is lower compared to the main product (palm olein). Due to its waxy character, this material can act as a plasticizer and improve the processability of rubber. Apart from the physical appearance of palm stearin, the chemical substances available in palm stearin are also interesting. It consists of highly saturated fats and triglycerides [13]. These components react with amines to produce unique chemical substances called fatty acid amides [14,15]. From the structural point of view, the interactions between NR and HNT could be improved by modified palm stearin. Recently, Surya et al. [14] reported on the use of modified palm stearin in a NR/silica composite, and it was found that the modified palm stearin gave shorter scorch and cure times, whereby the torque difference, tensile modulus, tensile strength, hardness, and crosslink density increased up to 5 phr doses of modified palm stearin. Similar observations have been reported for carbon black filled NR composites, showing that the mechanical properties have been improved by modified palm stearin [16].

The aim of this study was to use modified palm stearin (MPS) as a mixed compatibilizer with maleated natural rubber (MNR) for NR/HNT composites. Based on the chemical structures of both MNR and MPS, they are anticipated to provide better compatibility, especially at the outer layers of HNT (silanol and/or siloxane groups). To date, no reports have been published of detailed investigations concerning the use of dual compatibilizers from MNR and MPS to improve the mechanical and morphological evolution of NR/HNT composites. This study would bring a scientific perspective on the role of MNR and MPS as dual compatibilizers for NR/HNT composites and provide detailed information for manufacturing rubber products with HNT filler.

2. Materials and Methods

2.1. Materials

The main NR matrix used in this experiment was STR 5L, which was manufactured by Chalong Latex Industry Co., Ltd. (Songkhla, Thailand). The HNT was manufactured by Imerys Ceramics Limited, (Kerikeri, New Zealand). HNT consists of the following components: SiO_2 (49.5 wt %), Al_2O_3 (35.5 wt %), Fe_2O_3 (0.29 wt %), TiO_2 (0.09 wt %), as well as CaO, MgO, K_2O, and Na_2O as traces. The palm stearin was fractionated and supplied by Chumporn Palm Oil Industry PCL. (Chumporn, Thailand). The curing activators ZnO and stearic acid were purchased from Global Chemical Co., Ltd. (Samut Prakan, Thailand) and Imperial Chemical Co., Ltd. (Bangkok, Thailand), respectively. N-cyclohexyl-2-benzothiazole sulfenamide (CBS), used as an accelerator, was supplied by Flexsys America L.P. (Creve Coeur, MO, USA), and sulfur, used as vulcanizing agent, was bought from Siam Chemical Co., Ltd. (Samut Prakan, Thailand). Other chemicals involved in the preparation of MNR and MPS, such as maleic anhydride, sodium methoxide, diethanolamine, diethyl ether, and saturated sodium chloride, were purchased from Sigma-Aldrich (Thailand) Co., Ltd. (Bangkok, Thailand).

2.2. Synthesis of MPS

The synthesis of MPS followed the procedure described by Surya et al. [14]. This was done in a reaction kettle fitted with a stirrer at atmospheric pressure. The methanol was initially mixed together with sodium methoxide while stirring. The mixture of palm stearin and diethanolamine was then added to the mixture while stirring. The reaction was carried out at 70 °C for 5 h. The mixture was later extracted and washed with diethyl ether and saturated sodium chloride solution. Finally, the crude MPS was purified with anhydrous sodium sulfate and concentrated in a rotary evaporator prior to use. The MPS was stored in a desiccator prior to characterization with Fourier transform infrared spectroscopy (FTIR) to assess changes in functionalities. The reaction to make MPS is shown in Figure 1. The MPS appeared as a cream-colored wax and was used as a compatibilizer to improve the compatibility in NR/HNT composites.

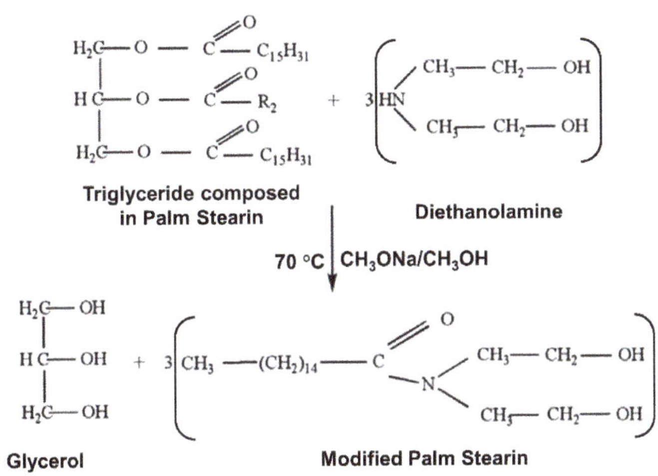

Figure 1. The chemical reaction between triglyceride-based palm stearin and diethanolamine (modified from Surya et al. [14]).

2.3. Synthesis of MPS

Grafting of MA onto NR was done by mixing the NR with 4 phr of MA in a Brabender Plasticorder at 145 °C at a rotor speed of 60 rpm under a normal atmosphere. The mixing lasted for

10 min. The resulting rubber was purified by reprecipitation. This was done just for the purpose of characterization by FTIR. The resulting MNR was then purified to confirm the grafting of MA onto NR. This was carried out by dissolving the rubber sample in toluene at room temperature for 24 h and then at 60 °C for 2 h. The soluble part was collected and precipitated in acetone. The sample was dried in a vacuum oven at 40 °C for 24 h. The purified MNR was finally characterized by the FTIR spectrum.

2.4. Preparation of Composites based on NR and HNT

Table 1 depicts the main ingredients to prepare the rubber composites, and MNR was used in all the modified compounds. Here, reference compound is denoted for the composite without compatibilizer. MPS 0 phr is the composite with only MNR as compatibilizer, MPS 0.5–1.5 phr are the composites with MNR/MPS as dual compatibilizers at MPS contents from 0.5–1.5 phr respectively. The entire amounts of additives were mixed in a Brabender (Plastograph®EC Plus, Mixer W50EHT 3Z, Brabender ®GmbH & Co. KG, Duisburg, Germany), and, just after dumping, the compounds were passed through a two-roll mill to avoid overheating prior to determining curing characteristics. The compounds were then compressed into certain shapes using a hydraulic hot press, with the vulcanizing times obtained by using a moving-die rheometer (MDR) as described later.

Table 1. Compounding ingredients used to prepare composites.

Ingredient	Compounding Code and Amounts in phr				
	Reference	MPS 0 phr	MPS 0.5 phr	MPS 1.0 phr	MPS 1.5 phr
NR	100	90	90	90	90
MNR *	-	10	10	10	10
ZnO	5	5	5	5	5
Stearic acid	1	1	1	1	1
CBS	2	2	2	2	2
Sulfur	2	2	2	2	2
HNT	10	10	10	10	10
MPS	-	-	0.5	1	1.5

Remark: * MNR prepared at 4 phr of maleic anhydride. MPS, modified palm stearin; NR, natural rubber; MNR, maleated natural rubber; HNT, halloysite nanotubes; CBS, N-cyclohexyl-2-benzothiazole sulfenamide.

2.5. Attenuated Total Reflection-Fourier Transform Infrared Spectroscopy (ATR-FTIR)

The FTIR spectra of MNR were analyzed using a Bruker FTIR spectrometer (Tensor 27, Bruker Optik GmbH, Baden-Württemberg, Germany) with a smart durable single bounce diamond in the ATR cell. Each spectrum was recorded in transmission mode after 32 scans per spectrum, with 4 cm^{-1} resolution from 4000 to 400 cm^{-1}.

2.6. Determination of Curing Characteristics

A moving die rheometer or MDR (Rheoline, Mini MDR Lite, Prescott instruments Ltd., Gloucestershire, UK) was utilized to determine the curing characteristics of the composites. The tests were carried out according to ASTM D5289 at 150 °C. The data recorded were torques, scorch time (t_{s1}), and curing time (t_{c90}).

2.7. Measurement of Mechanical Properties

The samples were cut into a dumbbell shape, according to ASTM D412. The tensile tests were conducted using a universal tensile machine (Tinius Olsen, H10KS, Tinius Olsen TMC, Horsham, PA, USA) at a cross-head speed of 500 mm/min. This was done to determine 100% modulus, 300% modulus, tensile strength, and elongation at break. Further, tear strengths of the composites were also tested using the same machine by following ASTM D624 with a cross-head speed of 500 mm/min. The tear strength recorded was the average of five repeated tests for each compound.

2.8. Dynamic Properties

The dynamic properties of the NR/HNT composites in the presence of dual compatibilizers were studied using a Rubber Process Analyzer model D-RPA 3000 (MonTech Werkstoffprüfmaschinen GmbH, Buchen, Germany). The composite sample was cured at 150 °C for the curing time obtained from Rheoline Mini MDR Lite (Prescott Instruments Ltd., Gloucestershire, UK). Then, the sample was cooled down to 60 °C and deformed at 10 Hz frequency, varying the strain in the range from 0.5 to 100%. The raw outputs storage modulus (G') and damping characteristics (tan δ) were recorded, and the rubber-filler interactions in the composites were assessed by the Payne effect. The Payne effect was calculated as follows.

$$\text{Payne effect} = G'_i - G'_f \quad (1)$$

where G'_i is G' at 0.5% strain, and G'_f is the G' at 100% strain. A larger Payne effect indicates lesser rubber-filler interactions.

2.9. Scanning Electron Microscopy

Fractured samples from tensile testing were used to assess the microdefects. Imaging was carried out using a scanning electron microscope (FEI Quanta 400 ESEM, Thermo Fisher Scientific, Waltham, MA, USA) to obtain information on the dispersion of the HNT filler throughout the NR matrix, in both the absence and presence of MNR/MPS as compatibilizers. The fractured pieces were sputter-coated with gold–palladium to eliminate electrostatic charge buildup during imaging.

3. Results and Discussion

3.1. Functionalities of Maleated Natural Rubber

The typical infrared spectra of unmodified and modified palm stearin are shown in Figure 2. The wavenumbers and their respective assignments are summarized in Table 2. Similar bands of C–H stretch were detected at wavenumbers 2922 and 2852 cm^{-1}, and CH_2 rocking bands appeared at 719 and 721cm^{-1} associated with long alkyl chains in the fatty acids; these were observed in both unmodified and modified palm stearin. Further evidence of a methyl group (CH_3) attached to a carbon atom was shown by the umbrella mode at 1352 cm^{-1} [17]. The distinct bands observed in the modified palm stearin distinguishing it from unmodified palm stearin are also shown in Figure 2. These included the strong band for O–H stretch at 3410 cm^{-1}, the C=O stretch at 1629 and 1556 cm^{-1}, and the amide C–N stretch at 1064 cm^{-1}, respectively. The spectrum clearly indicated the functional groups present in the proposed MPS chemical structure seen in Figure 1.

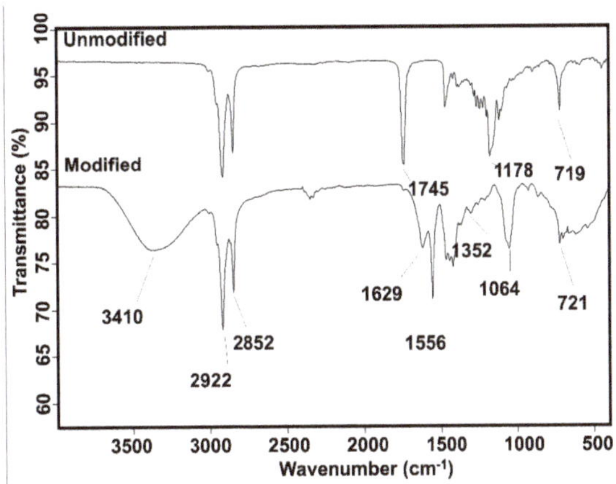

Figure 2. FTIR spectra of unmodified and modified palm stearin.

Table 2. The peaks and their assignments in spectra for unmodified and modified palm stearin.

Wavenumber (cm^{-1})	Assignment
3410	O–H stretch
2922, 2852	C–H stretch
1745	C=O stretch in ester
1629, 1556	C=O stretch in a modified structure
1352	CH_3 umbrella mode
1064	C–N stretch
719/721	CH_2 rocking

3.2. Functionalities of Maleated Natural Rubber

FTIR spectra of MNR at various MA contents are shown in Figure 3, while the peak assignments are listed in Table 3. A broad and intense band at 1787 cm^{-1} and a weak absorption band at 1875 cm^{-1} were observed. These bands could be assigned to the successfully grafted anhydride and were due to symmetric (strong) and asymmetric (weak) C=O stretching vibrations of succinic anhydride rings, respectively. The observed bands were clearly indicating succinic anhydride groups grafted onto NR molecules. Moreover, there was an important peak captured at wavenumber 1723 cm^{-1} due to the formation of carbonyl groups of opened ring structure succinic anhydride. The peaks seen in this study were quite similar to previous results in the literature [18,19].

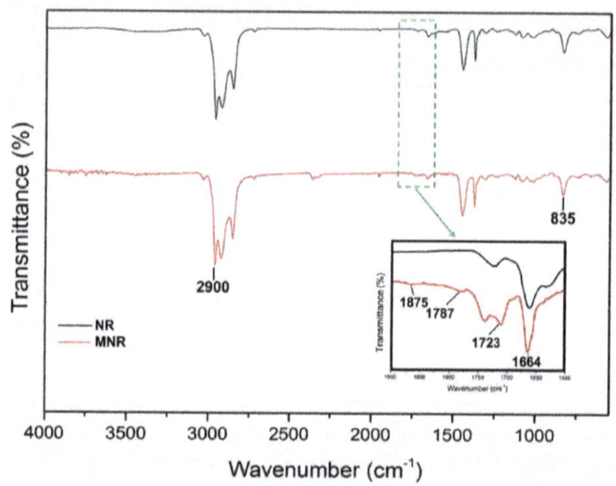

Figure 3. FTIR spectra of MNR prepared at various MA contents. NR, natural rubber; MNR, maleated natural rubber; MA, maleic anhydride.

Table 3. The peaks and their assignments in spectra for NR and MNR.

Wavenumber cm^{-1}	Assignment
2900	C–H stretch of NR
1875	C=O stretch of succinic anhydride (weak)
1787	C=O stretch of polymeric anhydride (weak)
1723	C=O stretch of a carbonyl group
1664	C=C stretch of NR
835	C–H out of plane bend of NR

3.3. Cure Characteristics

The curing curves of the NR/HNT composites with and without MNR/MPS as dual compatibilizers are illustrated in Figure 4, and the results are also summarized in Table 4. The maximum torque (M_H) decreased on adding MPS but then increased again with further increases in the MPS dose, showing that the MPS played an important role in improving the compatibility of the NR/HNT composites. A similar trend was observed for the torque difference ($M_H - M_L$), which is the difference between maximum torque (M_H) and minimum torque (M_L). This value is known to indicate the degree of cross-linking and/or interactions within the composite system [20], so this could indicate that the compatibility of NR with HNT was significantly enhanced when MPS was added to the composite. The amide groups in MPS also shortened the scorch (t_{s1}) and cure times (t_{c90}) of the composites. As mentioned before, MPS was synthesized from palm stearin and diethanolamine, which made MPS an alkaline substance. This increased the pH of the rubber compounds, which tended to increase the cure rate. Any chemical substance that makes the rubber compound more alkaline will increase the cure rate, while acidity tends to retard the reactivity of accelerators [21]. It was, therefore, expected that the amine in MPS could accelerate the cure rate of the composites.

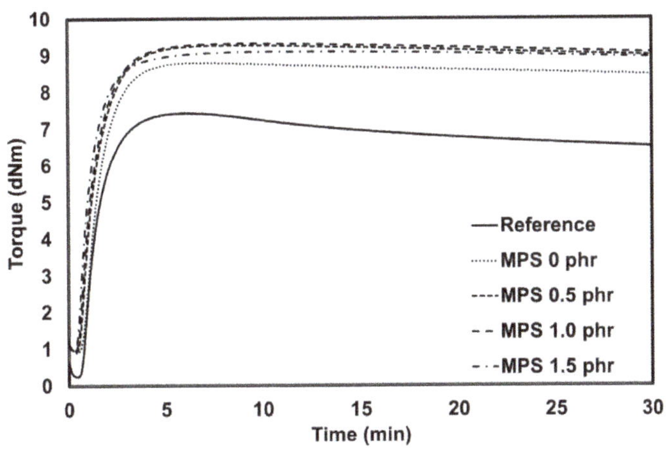

Figure 4. Curing curves of NR/HNT composites in the presence of MNR/MPS as a dual compatibilizer.MPS, modified palm stearin.

Table 4. Curing characteristics of NR/HNT composites in the presence of MNR/MPS as a dual compatibilizer.

Compound Code	M_H (dN.m)	$M_H - M_L$ (dN.m)	t_{S1} (min)	t_{c90} (min)
Reference	7.43	7.18	0.84	2.81
MPS 0 phr	7.33	6.38	1.25	3.16
MPS 0.5 phr	9.27	8.35	0.7	2.84
MPS 1.0 phr	9.32	8.37	0.64	2.76
MPS 1.5 phr	9.1	8.2	0.56	2.43

3.4. Mechanical Properties

Tensile strength and elongation at break of NR/HNT composites with and without MNR/MPS as a dual compatibilizer are shown in Figure 5. The tensile strength increased upon the incorporation of MPS. Improved tensile strength was attributed to the MPS itself, which has great potential to mediate interactions between NR and HNT. The mixed compatibilizers obviously influenced the tensile strength of this composite, as could be seen from the tensile strengths of the composites without (reference) and with dual compatibilizers (MPS 0–1.5 phr). Here, the interaction between NR and HNT was discussed for each compatibilizer used. Scheme 1 shows the proposed interaction obtained by the action of MNR, with two possible interactions in the composite either through the opened ring and/or cyclic structures. Grafting of the succinic anhydride groups onto NR molecules of the MNR increased the polarity of rubber and made it compatible with HNT. Pasbakshs et al. [11] also proposed the same interactions between hydroxyl groups of HNT and succinic anhydride groups of EPDM-g-MA. As for the MPS, the interactions were between amide groups of MPS and siloxane groups at the outer surfaces of HNT (see Scheme 2), and these reactions increased the reinforcing efficiency in the NR/HNT composites.

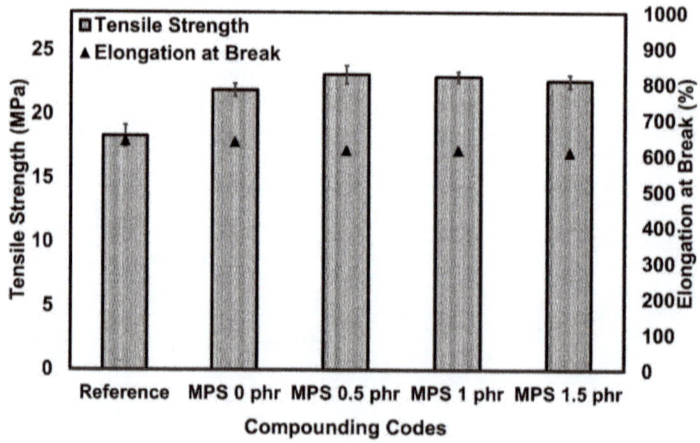

Figure 5. Tensile strength and elongation at break of NR/HNT composites in the presence of MNR/MPS as dual compatibilizers.

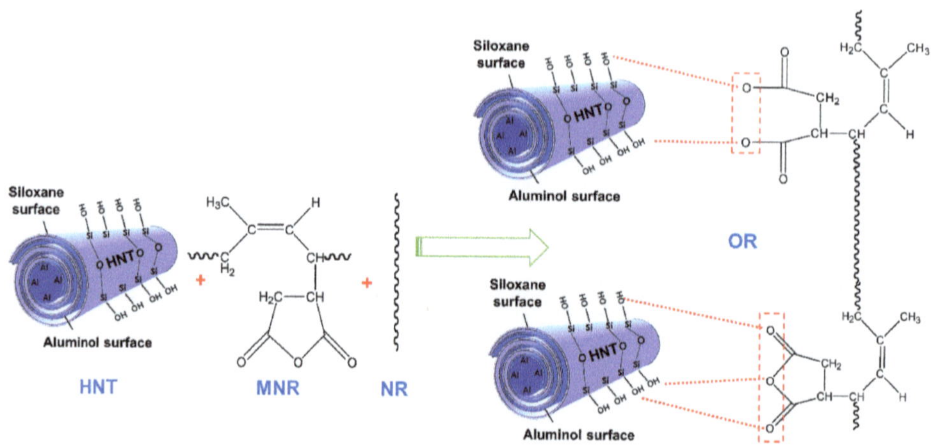

Scheme 1. Possible interactions between NR and HNT in the presence of MNR as compatibilizer.

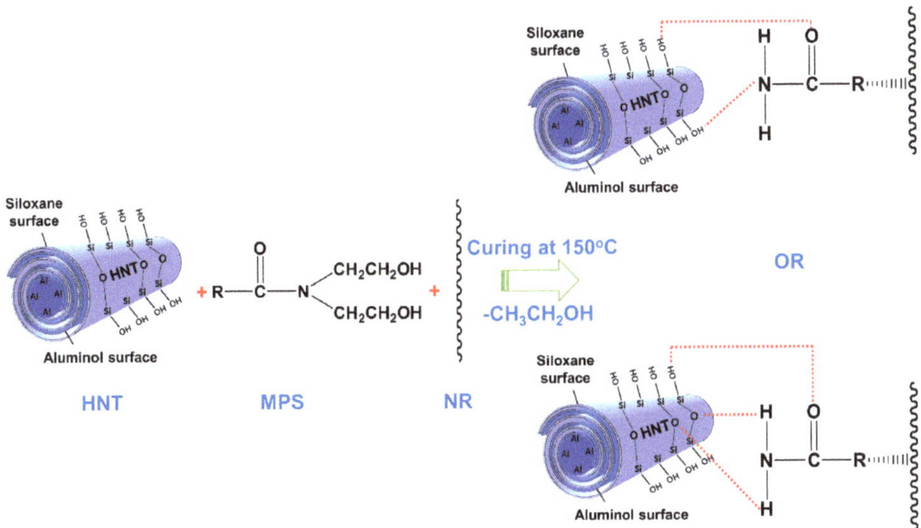

Scheme 2. Possible interactions between NR and HNT in the presence of MPS as compatibilizer.

The strong interactions of NR and HNT could be confirmed from the stresses at 100% and 300% strain (see Figure 6). It could be seen that the stresses at 100% and 300% elongations (M100 and M300) increased with MPS dose. As more MPS was added to the rubber, more interactions took place, resulting in stiffer and harder composites. This was in good agreement with the M_H and $M_H - M_L$ reported in the preceding section. In addition to this, the tear strength also showed remarkable improvement upon the inclusion of MPS, as could be seen in Figure 7. The improved tear strength was simply due to the strong interactions between NR and HNT, as well as the ability of MPS to improve the dispersion of HNT filler in the NR matrix.

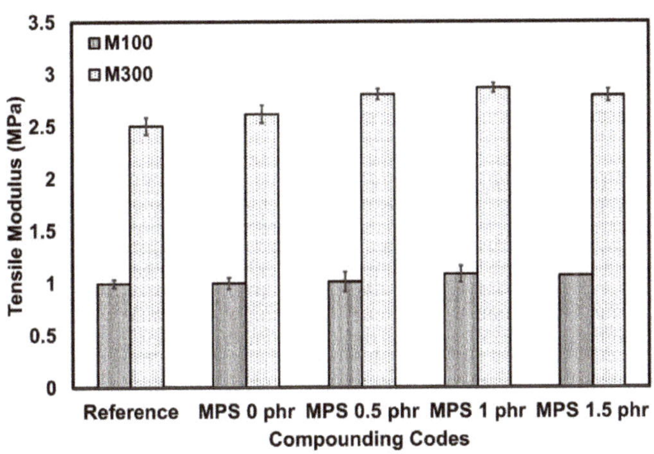

Figure 6. Tensile modulus of NR/HNT composites in the presence of MNR/MPS as a dual compatibilizer.

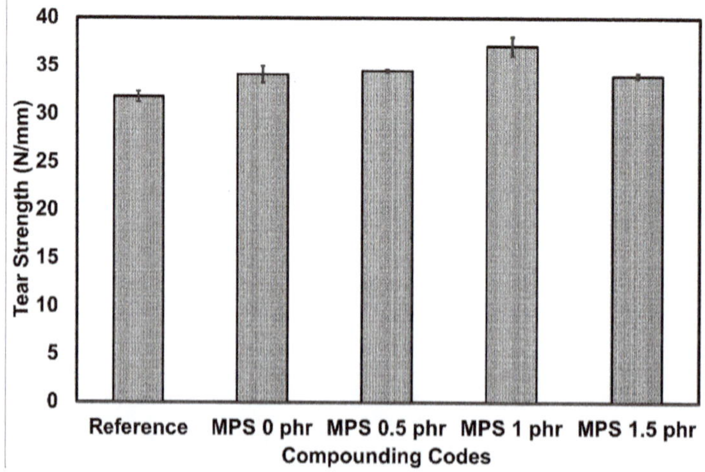

Figure 7. Tear strength of NR/HNT composites in the presence of MNR/MPS as dual compatibilizers.

3.5. *Dynamic Properties*

The dynamic properties of the composites were determined with a Rubber Process Analyzer to investigate the storage modulus and the Payne effect. Figures 8 and 9 show the storage modulus (G') and the Payne effect ($G'_i - G'_f$) of NR/HNT composites. It could be seen that the storage modulus of gum NR was constant in the low strain region but slightly decreased when the strain exceeded 50%. This is a common phenomenon for viscoelastic materials and is due to the molecular stability of rubber. In addition to that, the Payne effect was estimated as the difference between storage modulus at low and high strain amplitudes [22,23]. The level of the Payne effect for the composite without MNR/MPS as a dual compatibilizer was found to be 238.65 kPa, and this decreased to 172.40 kPa for the composite with solely MNR. This was a good indication that the interactions between NR and HNT were improved by MNR. Moreover, the Payne effect was reduced, on introducing MPS as a second compatibilizer, to 166.30, 153.61, and 141.02 kPa, respectively, for MPS loadings of 0.5, 1, and 1.5 phr. The lower Payne effect indicates lesser filler-filler interactions [24].

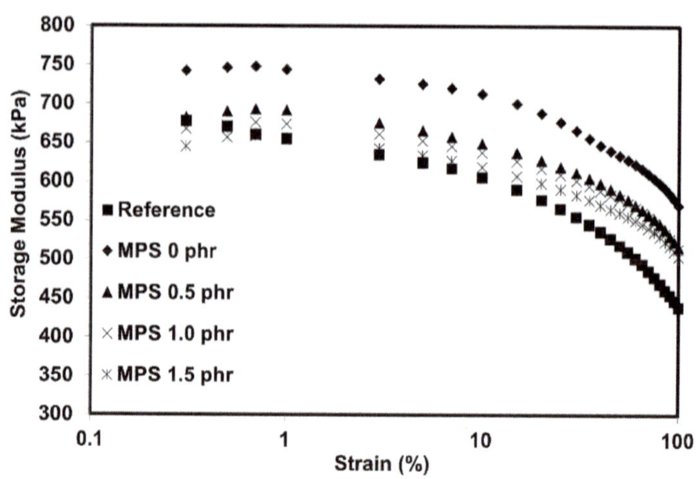

Figure 8. The storage modulus (G') of NR/HNT composites in the presence of MNR/MPS as a dual compatibilizer.

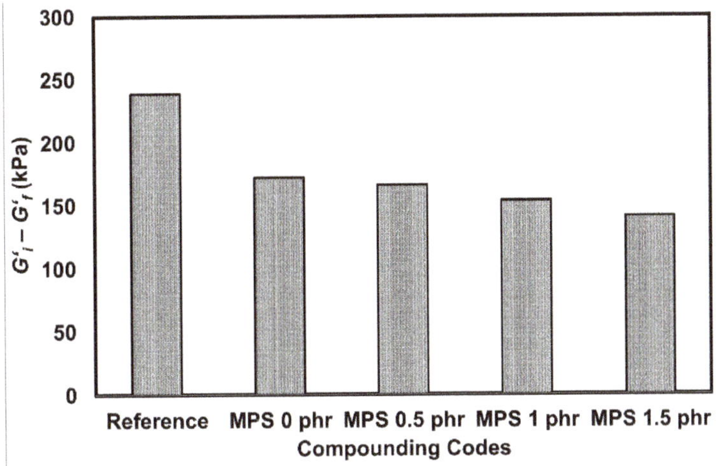

Figure 9. Payne effect ($G'_i - G'_f$) of NR/HNT composites in the presence of MNR/MPS as a dual compatibilizer.

Dependence of the damping characteristic (tan δ) on strain is shown in Figure 10. It is obvious that the composites exhibited low damping characteristics after the addition of MPS, indicating a considerable degree of molecular mobility. This was simply due to the better interactions between NR and HNT after adding MNR/MPS as a dual compatibilizer. The improved compatibility of rubber-HNT increased the interfacial adhesion and resulted in improved elastic properties of the vulcanizates.

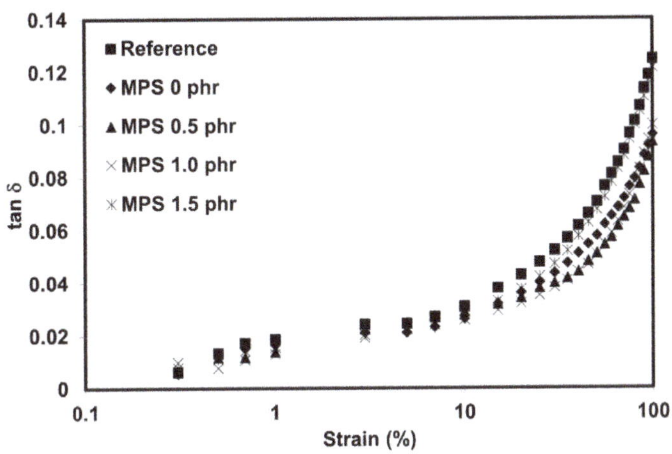

Figure 10. Damping characteristic (tan δ) of NR/HNT composites in the presence of MNR/MPS as a dual compatibilizer.

3.6. Scanning Electron Microscopy

Figure 11 shows SEM micrographs of the tensile fractured surfaces of the NR/HNT composites with and without MNR/MPS as a dual compatibilizer at 10k× magnification. From Figure 11A, agglomeration of HNT is seen at a fractured surface of the NR/HNT composite without MNR and MPS. Upon the inclusion of MNR (see Figure 11B), agglomeration was no longer seen due to the compatibilizing effect of MNR. Furthermore, the dispersion of HNT was enhanced by MPS, and the homogeneity of the composite was significantly improved, especially at 1.5 phr dose level (see Figure 11C,D). The improved dispersion of HNT was clearly responsible for the improved tensile strength and tear strength, with more energy needed to break the sample. Better dispersion of HNT throughout the matrix increased the stress at any given strain. Similar observations were previously reported for microfractured surfaces of filled NR composites in the presence of a compatibilizer [25,26].

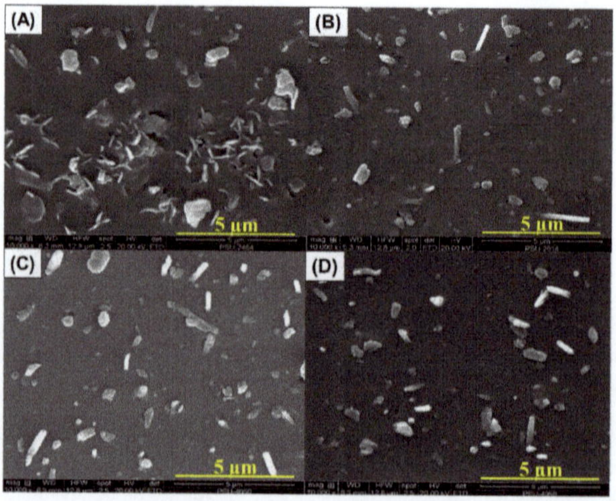

Figure 11. SEM images of tensile fractured surfaces of NR/HNT composites in the presence of MNR/MPS as a dual compatibilizer: Reference (**A**), MPS 0 phr (**B**), MPS 0.5 phr (**C**), and MPS 1.5 phr (**D**), all at 10,000× magnification.

4. Conclusions

The overall properties of composites based on NR and HNT were clearly improved by adding MNR/MPS as a dual compatibilizer. Both MNR and MPS had special functional groups that could form hydrogen bonds with the hydroxyl groups on HNT surfaces. MPS alone could also promote the dispersion of HNT filler in the NR matrix due to its waxy character. The filler-matrix interactions improved the mechanical properties, such as tensile strength, modulus, and tear strength of the composites, and this mechanism was corroborated by the decreased Payne effect observed in dynamic measurements. It was clearly demonstrated in this present work that the use of MNR and MPS improved the rubber-HNT interactions and reduced filler-filler interactions, providing great benefits to the mechanical and dynamical properties. This finding contributed to understanding the roles of MNR and MPS as dual compatibilizers for NR/HNT composites and could be a source of useful information for manufacturing such composites. To some extent, this could lead to modifying processing methods without requiring commercial compatibilizers that are costly and complicate the process.

Author Contributions: Conceptualization, N.H. and I.S.; methodology, N.H. and K.S.; software, Z.S.; validation, N.H., K.S., and H.I.; investigation, Z.S.; data curation, N.H.; writing—review and editing, N.H., I.S., K.S., and H.I.; supervision, N.H. All authors have read and agreed to the published version of the manuscript.

Funding: This research was funded by Prince of Songkla University through a Research Grant for Fiscal Year 2017. Grant No. SAT600101S.

Acknowledgments: The authors would like to thank the Research and Development Office (RDO), Prince of Songkla University and Seppo Karrila for assistance with manuscript preparation.

Conflicts of Interest: The authors declare no conflict of interest.

References

1. Arrighi, V.; McEwen, I.; Qian, H.; Prieto, M.S. The glass transition and interfacial layer in styrene-butadiene rubber containing silica nanofiller. *Polymers* **2003**, *44*, 6259–6266. [CrossRef]
2. Jovanović, V.; Samaržija-Jovanović, S.; Marković, G.; Marinović-Cincović, M.; Budinski-Simendić, J. Mechanical and Morphological Properties Rubber blends reinforced with nanofillers. *KGK Kautschuk Gummi Kunststoffe* **2011**, *9*, 52–56.
3. Chakravarty, S.; Chakravarty, A. Reinforcement of rubber compounds with nano-filler. *KGK Kautschuk Gummi Kunststoffe* **2007**, *60*, 619–622.
4. Hu, D.; Zhang, Z.; Liu, M.; Lin, J.; Chen, X.; Ma, W. Multifunctional UV-shielding nanocellulose films modified with halloysite nanotubes-zinc oxide nanohybrid. *Cellulose* **2020**, *27*, 401–413. [CrossRef]
5. Cavallaro, G.; Chiappisi, L.; Pasbakhsh, P.; Gradzielski, M.; Lazzara, G. A structural comparison of halloysite nanotubes of different origin by Small-Angle Neutron Scattering (SANS) and Electric Birefringence. *Appl. Clay Sci.* **2018**, *160*, 71–80. [CrossRef]
6. Feng, K.; Hung, G.-Y.; Liu, J.; Li, M.; Zhou, C.; Liu, M. Fabrication of high performance superhydrophobic coatings by spray-coating of polysiloxane modified halloysite nanotubes. *Chem. Eng. J.* **2018**, *331*, 744–754. [CrossRef]
7. Hu, D.; Zhong, B.; Jia, Z.; Lin, J.; Liu, M.; Luo, Y.; Jia, D. A novel hybrid filler of halloysite nanotubes/silica fabricated by electrostatic self-assembly. *Mater. Lett.* **2017**, *188*, 327–330. [CrossRef]
8. Rooj, S.; Das, A.; Thakur, V.; Mahaling, R.; Bhowmick, A.K.; Heinrich, G. Preparation and properties of natural nanocomposites based on natural rubber and naturally occurring halloysite nanotubes. *Mater. Des.* **2010**, *31*, 2151–2156. [CrossRef]
9. Varghese, S.; Karger-Kocsis, J. Natural rubber-based nanocomposites by latex compounding with layered silicates. *Polymers* **2003**, *44*, 4921–4927. [CrossRef]
10. Paran, S.; Naderi, G.; Ghoreishy, M. XNBR-grafted halloysite nanotube core-shell as a potential compatibilizer for immiscible polymer systems. *Appl. Surf. Sci.* **2016**, *382*, 63–72. [CrossRef]
11. Pasbakhsh, P.; Ismail, H.; Fauzi, M.A.; Bakar, A.A. Influence of maleic anhydride grafted ethylene propylene diene monomer (MAH-g-EPDM) on the properties of EPDM nanocomposites reinforced by halloysite nanotubes. *Polym. Test.* **2009**, *28*, 548–559. [CrossRef]
12. Norizzah, A.R.; Chong, C.; Cheow, C.; Zaliha, O. Effects of chemical interesterification on physicochemical properties of palm stearin and palm kernel olein blends. *Food Chem.* **2004**, *86*, 229–235. [CrossRef]
13. Che Man, Y.; Haryati, T.; Ghazali, H.; Asbi, B. Composition and thermal profile of crude palm oil and its products. *J. Am. Oil Chem. Soc.* **1999**, *76*, 237–242. [CrossRef]
14. Surya, I.; Ismail, H.; Azura, A. Alkanolamide as an accelerator, filler-dispersant and a plasticizer in silica-filled natural rubber compounds. *Polym. Test.* **2013**, *32*, 1313–1321. [CrossRef]
15. Kolancilar, H. Preparation of laurel oil alkanolamide from laurel oil. *J. Am. Oil Chem. Soc.* **2004**, *81*, 597–598. [CrossRef]
16. Surya, I.; Ismail, H. The effect of the addition of alkanolamide on properties of carbon black-filled natural rubber (SMR-L) compounds cured using various curing systems. *Polym. Test.* **2016**, *50*, 276–282. [CrossRef]
17. Bhargava, R.; Wang, S.-Q.; Koenig, J.L. *FTIR Microspectroscopy of Polymeric Systems, Liquid Chromatography/FTIR Microspectroscopy/Microwave Assisted Synthesis*; Springer: Berlin/Heidelberg, Germany, 2003; pp. 137–191.
18. Nakason, C.; Kaesaman, A.; Supasanthitikul, P. The grafting of maleic anhydride onto natural rubber. *Polym. Test.* **2004**, *23*, 35–41. [CrossRef]
19. Sahakaro, K.; Beraheng, S. Reinforcement of maleated natural rubber by precipitated silica. *J. Appl. Polym. Sci.* **2008**, *109*, 3839–3848. [CrossRef]
20. Rattanasom, N.; Saowapark, T.; Deeprasertkul, C. Reinforcement of natural rubber with silica/carbon black hybrid filler. *Polym. Test.* **2007**, *26*, 369–377. [CrossRef]
21. Coran, A. Chemistry of the vulcanization and protection of elastomers: A review of the achievements. *J. Appl. Polym. Sci.* **2003**, *87*, 24–30. [CrossRef]
22. Payne, A.; Whittaker, R. Low strain dynamic properties of filled rubbers. *Rubb. Chem. Technol.* **1971**, *44*, 440–478. [CrossRef]

23. Kaewsakul, W.; Sahakaro, K.; Dierkes, W.K.; Noordermeer, J.W. Cooperative effects of epoxide functional groups on natural rubber and silane coupling agents on reinforcing efficiency of silica. *Rubb. Chem. Technol.* **2014**, *87*, 291–310. [CrossRef]
24. Rooj, S.; Das, A.; Stöckelhuber, K.W.; Wang, D.-Y.; Galiatsatos, V.; Heinrich, G. Understanding the reinforcing behavior of expanded clay particles in natural rubber compounds. *Soft Matter* **2013**, *9*, 3798–3808. [CrossRef]
25. Nabil, H.; Ismail, H.; Azura, A. Recycled polyethylene terephthalate filled natural rubber compounds: Effects of filler loading and types of matrix. *J. Elast. Plast.* **2011**, *43*, 429–449. [CrossRef]
26. Waesateh, K.; Saiwari, S.; Ismail, H.; Othman, N.; Soontaranon, S.; Hayeemasae, N. Features of crystallization behavior of natural rubber/halloysite nanotubes composites using synchrotron wide-angle X-ray scattering. *Int. J. Polym. Anal. Charac.* **2018**, *23*, 260–270. [CrossRef]

polymers

MDPI

Communication

Role of Nanoparticle–Polymer Interactions on the Development of Double-Network Hydrogel Nanocomposites with High Mechanical Strength

Andrew Chang, Nasim Babhadiashar, Emma Barrett-Catton and Prashanth Asuri *

Department of Bioengineering, Santa Clara University, Santa Clara, CA 95053, USA; ahchang@scu.edu (A.C.); nbabhadiashar@scu.edu (N.B.); ebarrettcatton@scu.edu (E.B.-C.)

* Correspondence: asurip@scu.edu; Tel.: +1-408-551-3005

Received: 28 January 2020; Accepted: 16 February 2020; Published: 18 February 2020

Abstract: Extensive experimental and theoretical research over the past several decades has pursued strategies to develop hydrogels with high mechanical strength. Our study investigated the effect of combining two approaches, addition of nanoparticles and crosslinking two different polymers (to create double-network hydrogels), on the mechanical properties of hydrogels. Our experimental analyses revealed that these orthogonal approaches may be combined to synthesize hydrogel composites with enhanced mechanical properties. However, the enhancement in double network hydrogel elastic modulus due to incorporation of nanoparticles is limited by the ability of the nanoparticles to strongly interact with the polymers in the network. Moreover, double-network hydrogel nanocomposites prepared using lower monomer concentrations showed higher enhancements in elastic moduli compared to those prepared using high monomer concentrations, thus indicating that the concentration of hydrogel monomers used for the preparation of the nanocomposites had a significant effect on the extent of nanoparticle-mediated enhancements. Collectively, these results demonstrate that the hypotheses previously developed to understand the role of nanoparticles on the mechanical properties of hydrogel nanocomposites may be extended to double-network hydrogel systems and guide the development of next-generation hydrogels with extraordinary mechanical properties through a combination of different approaches.

Keywords: hydrogel mechanical properties; nanocomposites; double-network hydrogels; polymer–nanoparticle interactions

1. Introduction

Unique properties of hydrogels have enabled their use in a wide range of applications in biotechnology, bioengineering, and medicine [1,2]. Hydrogels have been used as drug delivery vehicles as they can encapsulate hydrophilic drugs and release them at a controlled rate within the body, through solute diffusion, or matrix swelling or degradation [3]. Researchers have also taken advantage of the ability of hydrogels to swell or shrink in response to external stimuli (e.g., pH, temperature) to develop biosensors for the detection of biomolecules [4,5]. Additionally, their highly porous and hydrated polymer structure mimics the extracellular cellular matrix and renders them highly suitable for in vitro cell culture, and several studies have demonstrated the successful use of hydrogels to encapsulate mammalian cells in a 3D physiological-like environment and develop in vitro models of cell proliferation, migration, and differentiation [6–10]. However, hydrogels have poor mechanical strength, which limits their broad applicability for tissue engineering [11]. For example, hydrogels cannot mimic stiffer tissues, which hinders their use for orthodontic or orthopedic applications [12,13]. Recent research has delved into several methods to improve mechanical properties of hydrogels. One approach that has gained significant interest over the past couple of decades is the incorporation of nanoparticles, which can enable the formation of additional crosslinks within the polymer network and contribute to enhancements in the mechanical strength of hydrogels. It has been shown that nanoparticle-mediated physical crosslinking can complement chemical crosslinking and that the combination of chemical and physical crosslinking can lead to enhancements in an elastic modulus that is greater than with either alone [14–16]. However, previous work suggests that there might be an upper limit to enhancements in mechanical strength achievable through the addition of nanoparticles, possibly due to the existence of a saturation point in the gains achievable through a combination of chemical and physical crosslinking [15,17,18]. Another method to improve the mechanical strength of hydrogels is to create hybrid hydrogels by combining two different polymers with contrasting properties [19]. The resulting double-network (DN) hydrogel has been shown to possess improved mechanical strength, due to the contrasting network structures and strong interpenetrating network entanglement [20,21]. Interestingly, researchers have also shown that these approaches may be combined and that the mechanical strength of DN hydrogels can be further improved with the addition of nanoparticles [22,23].

Our study sought to explore the role of polymer–nanoparticle interactions on the properties of DN hydrogel nanocomposites, and investigated if the hypotheses previously developed to understand the role of nanoparticles on the mechanical properties of (single-network) hydrogel nanocomposites may be applied to DN hydrogel systems. Specifically, rotational rheological measurements and swelling experiments were conducted to investigate if the observed saturation in nanoparticle-mediated enhancements in mechanical properties for single-network hydrogel systems also extend to double-network hydrogels. Double-network hydrogels composed of chemically crosslinked polyacrylamide (pAAm) as the primary network, and incorporating silica nanoparticles (SiNPs) were used as the model system. Previous studies have demonstrated that strong interactions between polyacrylamide chains and silica may facilitate the formation of SiNP-mediated pseudo crosslinks within the hydrogel network and mediate mechanical reinforcement of pAAm hydrogels [24,25]. Consistent with previous investigations of DN hydrogel nanocomposites, our data shows significant enhancements in the elastic modulus of DN hydrogels, as well as decreased swellability, upon the addition of nanoparticles. Our experiments also reveal that in order to observe the strongest impact of nanoparticles on the mechanical properties of DN hydrogels, both networks in the hydrogel must have strong physical adsorption to the nanoparticles. Finally, a saturation in the gains in the elastic modulus afforded by a combination of adding a second network and incorporation of nanoparticles was observed. In summary, these results indicated that nanoparticle-mediated enhancements in hydrogel mechanical properties may be a general phenomenon and similar theories may be used to describe the role of nanoparticles to improve the mechanical properties of both single- and double-network hydrogels.

2. Materials and Methods

2.1. Materials

Materials for preparation of the hydrogels, acrylamide (40% *w/v*), acrylic acid (sodium salt), N,N′-methylenebis(acrylamide) (2% *w/v*), alginic acid (sodium salt, low viscosity), ammonium persulfate (APS), N,N,N′,N′-tetramethylethylenediamine (TEMED), and calcium chloride ($CaCl_2$) were purchased from Sigma Aldrich (Saint Louis, MO, USA), and agarose (low melting) was purchased from Thermo Fisher Scientific (Waltham, MA, USA), and used as received. Tris–HCl buffer (pH 7.2) was obtained from Life Technologies (Carlsbad, CA, USA) and binzil silica nanoparticle colloid solution with mean particle size of 4 nm was obtained as a gift from AkzoNobel Pulp and Performance Chemicals Inc. (Marietta, GA, USA).

2.2. Polymerization Reaction

All hydrogel samples were prepared using an acrylic mold (1.6 mm thick and 6.5 mm in radius) at room temperature as previously described [17,24]. The polymerization reactions were performed between parallel plates of the mold to minimize exposure to air as oxygen inhibits the free radical polymerization reaction for pAAm and poly(sodium acrylate) (pNaAc), as well as to maintain consistency in sample size across experiments. For pAAm and pNaAc samples, the monomer (AAm or NaAc) and crosslinker (Bis) stocks were diluted to their desired concentrations in pH 7.2, 250 mM Tris–HCl buffer, followed by the addition of TEMED (0.1% of the final reaction volume) and 10% w/v APS solution (1% of the final reaction volume). Agarose hydrogels were prepared by first dissolving agarose in Tris–HCl buffer at 70 °C. The agarose stock solutions were then cooled down to and maintained at 37 °C before diluting to desired concentrations and adding to the acrylic molds to form the hydrogels. Alginate gels were prepared by first dissolving alginate in Tris–HCl buffer at room temperature. The alginate stock solutions were then diluted to the desired concentrations and added to the acrylic molds, followed by the addition of 100 mM $CaCl_2$ in Tris–HCl buffer to crosslink the alginate and form the hydrogels. For DN hydrogels composed of pAAm and pNaAc, NaAc monomer was added to the pAAm reaction mixture prior to the addition of APS and TEMED. Similarly, DN hydrogels composed of pAAm and alginate were prepared by adding alginate solutions to the pAAm reaction mixture prior to the addition of APS, TEMED, and $CaCl_2$. For DN hydrogels composed of pAAm and agarose, the pAAm reaction mixture was first warmed to 37 °C, before the addition of agarose solutions at 37 °C and subsequent addition of APS and TEMED. For nanocomposite hydrogels, various amounts of silica nanoparticles were added to the reaction mixture prior to the addition of APS and TEMED (and $CaCl_2$ for hydrogels made using alginate).

2.3. Rheological Measurements

Rheological measurements of neat hydrogels and hydrogel nanocomposites were carried out using the MCR302 rotational rheometer (AntorPaar, Austria), as described previously [17]. After waiting for 1 h to ensure complete gelation (gelation usually occurs within 20 min), the hydrogel discs were taken from the acrylic mold and transferred onto the lower plate of the rheometer. The samples were gently wiped with tissue paper to remove any excess water before lowering the top plate. The elastic modulus was then determined at 1 Hz and 1% strain and reported as an average of three independent measurements.

2.4. Measurement of the Swelling Properties

To study the swelling properties of neat hydrogels and hydrogel nanocomposites, the hydrogel disks prepared as described above were wiped with tissue paper to remove any excess water, weighed and then immersed in pH 7.2, 100 mM Tris–HCl buffer. Samples were withdrawn from the buffer at different time intervals (6, 12, and 24 h) and their weights were determined after first blotting excess buffer with tissue paper. The swelling ratios at different time intervals were calculated using the following equation:

$$Swelling\ ratio\ (\%) = \left[\frac{W_t - W_0}{W_0}\right] \times 100$$

where W_t is the weight of the hydrogel samples after a given time interval and W_0 is the initial weight (i.e., before immersing in buffer).

3. Results

3.1. Development and Characterization of the Model

To study the effect of SiNPs on the mechanical properties of DN hydrogels, a model system was first established using 5% *w*/*v* of pAAm as the primary network and 2% *w*/*v* of either agarose, alginate, or pNaAc as the second network. All these materials are well characterized, commercially available, and routinely used in a wide variety of industrial and scientific applications [26–29]. Moreover, they represent hydrogels that are formed through different crosslinking mechanisms: pAAm and pNaAc (free-radical copolymerization), agarose (thermosensitive sol-to-gel transition), and alginate (physical crosslinking). Characterization of the DN hydrogels using rotational rheometry revealed that the relative elastic modulus of pAAm hydrogels improved upon the addition of a second polymer network (Figure 1a). Moreover, consistent with previous investigations [20,21], DN hydrogels prepared by incorporating polymers with properties different to those of pAAm lead to better enhancements. However, it is important to note that pNaAc gels were significantly softer that alginate and agarose gels at similar concentrations (Figure S1 in Supplementary Materials), which may have an impact on the elastic modulus of DN hydrogels prepared by incorporating pNaAc as one of the network polymers.

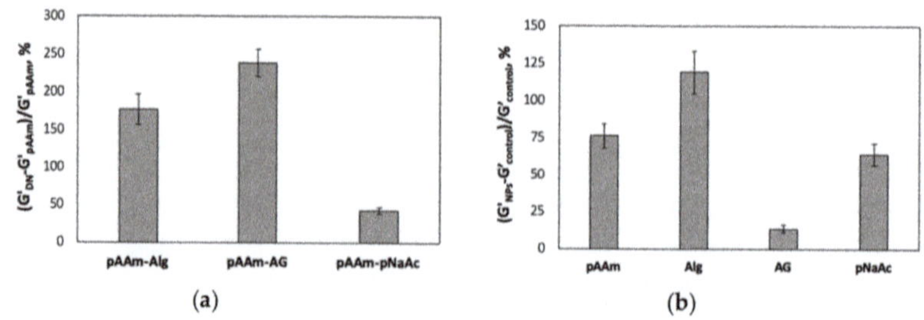

Figure 1. (**a**) Percent relative elastic moduli of pAAM-alginate (pAAM-Alg), pAAm-agarose (pAAM-AG), and pAAm-pNaAc DN hydrogels, prepared using 5% pAAM and 2% second polymer. Values for relative elastic modulus were calculated by normalizing the values for the DN hydrogels to pAAm hydrogels, and (**b**) percent relative elastic moduli of pAAM, alginate (Alg), agarose (AG), and pNaAc hydrogel nanocomposites prepared using 2% monomer and 2% 4 nm SiNPs. Values for relative elastic modulus were calculated by normalizing the values for the hydrogel nanocomposites to those for neat hydrogels not containing SiNPs. Data shown are the mean of triplicate measurements ± standard deviation and have been repeated at least three times with similar results.

The elastic moduli of the individual (or single-network) hydrogels (2% *w*/*v*) prepared with and without SiNPs were also measured using rotational rheology (Figure 1b). Unsurprisingly, the degree of nanoparticle-mediated enhancements was different for different hydrogels, with the highest benefit observed for alginate hydrogels. These differences in enhancements may be attributed to the differences in the extent to which SiNPs may interact with the polymer chains and contribute to the extent of crosslinking in the polymer network. For instance, while the remarkably low improvements in modulus for agarose due to the addition of SiNPs were unanticipated, this result may be explained by a no to low level of interactions between silica and agarose (to the best of our knowledge, there are no published studies that clearly demonstrate positive interactions between agarose and silica nanoparticles).

3.2. *Influence of Chemical Crosslinking on Nanoparticle Mediated Enhancements of Hydrogel Elastic Modulus*

The impact of incorporating nanoparticles on the mechanical properties of DN hydrogels was then assessed using rotational rheology, which revealed increases in elastic moduli for DN hydrogel nanocomposites relative to neat DN hydrogels not incorporating SiNPs (Figure 2a). Furthermore, it was interesting to note that the observed enhancements in elastic modulus were consistent with the observations made for single-network hydrogel nanocomposites. For instance, pAAM-agarose DN hydrogels did not significantly benefit from the addition of nanoparticles, at least not beyond the relative enhancements observed for pAAm-SiNP composites. This suggested that both hydrogels within the DN hydrogel should have strong interactions with the nanoparticles to realize the full extent of benefits afforded by the incorporation of nanoparticles.

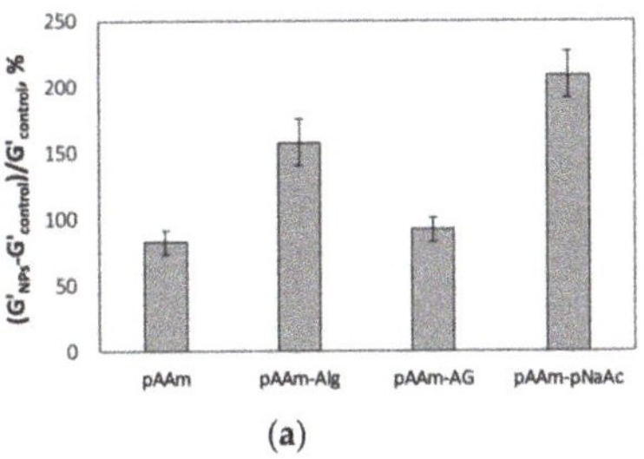

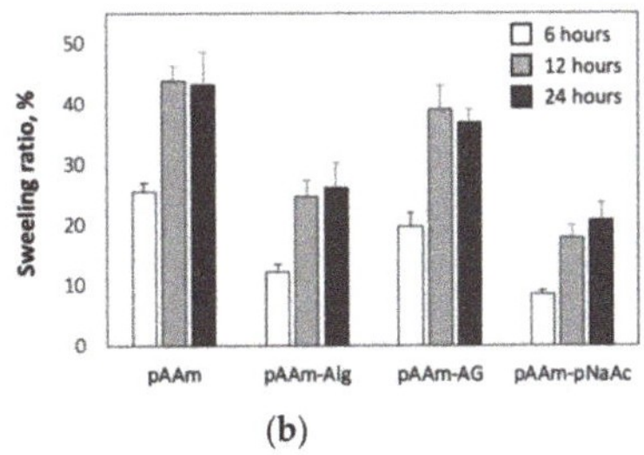

Figure 2. (**a**) Percent relative elastic moduli of pAAm, pAAM-alginate (pAAM-Alg), pAAm-agarose (pAAM-AG), and pAAm-pNaAc hydrogels, prepared using 5% pAAM, 2% second polymer (for DN hydrogel samples) and 2% 4 nm SiNPs. Values for relative elastic modulus were calculated by normalizing the values for the hydrogel nanocomposites to those for neat hydrogels not containing SiNPs, and (**b**) percent swelling ratios of pAAm, pAAM-alginate (pAAM-Alg), pAAm-agarose (pAAM-AG), and pAAm-pNaAc hydrogel nanocomposites, prepared using 5% pAAM, 2% second polymer (for DN hydrogel samples) and 2% 4 nm SiNPs at various time points—6 h (white bars), 12 h (light grey bars), and 24 h (dark grey bars). The values for swelling ratio were calculated by normalizing the values obtained for hydrogel samples at various time points to those obtained at time = 0 min. Data shown are the mean of triplicate measurements ± standard deviation and have been repeated at least three times with similar results.

Previous studies have indicated that nanoparticles facilitate enhancements in hydrogel elastic modulus by serving as pseudo crosslinkers and increasing the extent of crosslinking within the hydrogel network [17,25]. To check if the aforementioned differences in the enhancements in modulus for different DN hydrogel nanocomposites can be explained using a similar mechanism, swellability of DN hydrogel nanocomposites was measured. If hydrogel-nanoparticle interactions contribute to an increase in the DN hydrogel modulus through increases in the average crosslinking density, the swellability of DN hydrogel nanocomposites prepared using alginate or pNaAc should be lower than that of pAAm-agarose nanocomposites. As expected, these experiments showed decreased levels of relative swellability for pAAm-NaAc and pAAm-alginate nanocomposites compared to that for pAAm-agarose nanocomposites. Furthermore, consistent with the hypothesis and measurements of elastic modulus, the values for relative swellability of pAAm-agarose and pAAm nanocomposites were comparable. Together, these results indicated that the lack of enhancements in the elastic modulus of pAAm-agarose nanocomposites relative to pAAm nanocomposites can be explained by limited interactions between agarose and SiNPs, which dilutes the positive impact of nanoparticles on the modulus of DN hydrogels prepared using agarose.

3.3. Influence of Polymer–Nanoparticle Interactions on Enhancements in DN Hydrogel Modulus

Next, experiments were designed to further validate that the nanoparticles must exhibit strong physical interactions with both polymers in the DN hydrogel network to realize the strongest impact of nanoparticles on the DN hydrogel modulus. The enhancements afforded by the incorporation of SiNPs were compared for DN hydrogels prepared using high (2% *w*/*v*) and low (1% *w*/*v*) amounts of agarose or alginate (Figure 3). These experiments demonstrated a dependence of concentration of the second network on SiNP-mediated enhancements for DN hydrogels incorporating alginate, but not for DN hydrogels incorporating agarose. It is important to note that the elastic moduli of both pAAm-agarose and pAAM-alginate DN hydrogels increased with increasing concentration of the second polymer (Figure S2). These results offered additional evidence in support of previous observations that indicate the importance of strong polymer–nanoparticle interactions on the enhancements in DN hydrogel modulus. More importantly, diminishing impact of SiNPs on the DN hydrogel modulus at higher concentrations of alginate (Figure 3) suggested the existence of a saturation point in gains in hydrogel modulus achievable through a combination of incorporating a second polymer network and addition of nanoparticles.

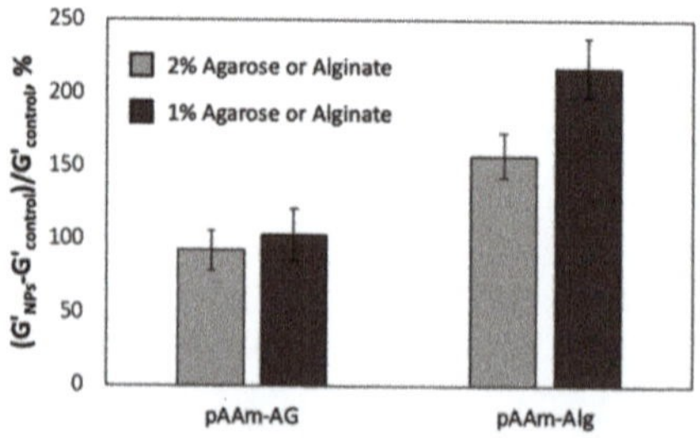

Figure 3. Relative elastic moduli of pAAm-agarose (pAAM-AG) and pAAM-alginate (pAAM-Alg) hydrogel nanocomposites, prepared using 5% pAAM, 2% (light grey bars) or 1% (dark grey bars) second polymer (agarose or alginate), and 2% 4 nm SiNPs. Values for relative elastic modulus were calculated by normalizing the values for the hydrogel nanocomposites to those for neat hydrogels not containing SiNPs. Data shown are the mean of triplicate measurements plus standard deviation and have been repeated at least three times with similar results.

To validate our observations that suggest a saturation point in the overall hydrogel modulus mediated by combining different polymers and incorporation of nanoparticles, the mechanical characterization studies were repeated for pAAm-alginate hydrogels and nanocomposites over a range of pAAm and alginate concentrations (Figure 4 and Figure S3). Not surprisingly, both pAAM and alginate concentrations played an important role on SiNP-mediated enhancements in the modulus of pAAm-alginate DN hydrogels. The decreasing role of nanoparticles with increasing polymer concentrations observed in this study, for pAAm-alginate DN hydrogels, was consistent with previous studies that demonstrate a decreased impact of nanoparticles on the hydrogel modulus at higher monomer concentrations [15].

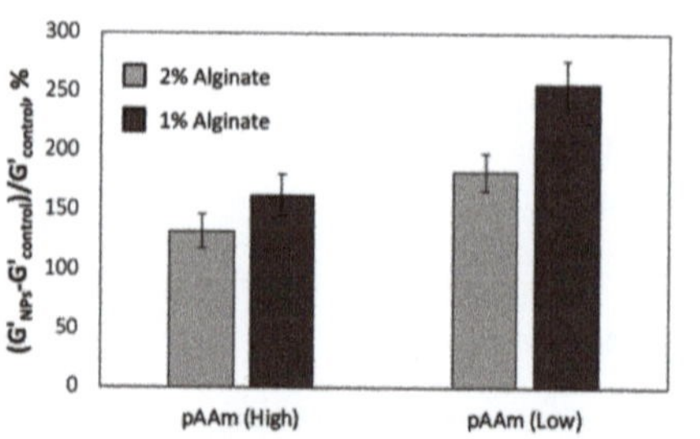

Figure 4. Relative elastic moduli of pAAM-alginate hydrogel nanocomposites, prepared using 5% (High) or 2.5% (Low) pAAm, 2% (light grey bars) or 1% (dark grey bars) alginate, and 2% 4 nm SiNPs. Values for relative elastic modulus were calculated by normalizing the values for the hydrogel nanocomposites to those for neat hydrogels not containing SiNPs. Data shown are the mean of triplicate measurements plus standard deviation and have been repeated at least three times with similar results.

4. Discussion

The aim of this study was to investigate the role of polymer–nanoparticle interactions on the mechanical properties of DN hydrogels using three different double-network hydrogel compositions. pAAm was used as the primary network and the second network was either alginate, agarose, or pNaAc, each one representing hydrogels with different mechanism of gelation and chemical properties. SiNPs were chosen as a model nanoparticle, given its ability to positively interact with pAAm and serve as a pseudo crosslinker to enhance the extent of crosslinking in the polymer network [25]. Results from the experimental analyses demonstrated the ability of nanoparticles to improve the mechanical properties of DN hydrogels, consistent with previous studies. However, significant enhancements in elastic modulus upon incorporation of SiNPs were only observed for pAAm-pNaAc and pAAM-alginate DN hydrogels, but not for pAAm-agarose. The rheological characterization studies were well supported by swelling studies that showed that the swellability of pAAm-pNaAc and pAAM-alginate hydrogel nanocomposites were significantly lower than that of pAAm-agarose hydrogel nanocomposites. These results suggested that the degree of improvement was not universal, and seemingly limited by the ability of nanoparticles to strongly interact with and contribute to additional crosslinking for both the polymers in the DN hydrogel network. Our experiments also revealed an inverse correlation between the polymer concentrations of DN hydrogels and the degree of nanoparticle-mediated enhancements in mechanical properties, i.e., DN hydrogels composed of higher concentrations of polymers benefited less due to the addition of nanoparticles. Further characterization of the DN hydrogel nanocomposites using morphology and microstructure analysis will be necessary to further confirm our observations, but are outside the scope of this investigation.

Overall, the results presented in this paper have two main outcomes. From an application standpoint, development of orthogonal strategies to enhance the elastic modulus is important to realize the broad applicability of hydrogels, especially in applications related to the development of stiff tissue mimics. Furthermore, the introduction of nanomaterials into the hydrogel network may allow the consideration of DN hydrogels for new applications; for example, researchers have already demonstrated the utility of DN hydrogel nanocomposites for applications in self-healing, shape memory, 3D printing, and dye removal [30–32]. Second, from a fundamental standpoint, this work provides a mechanistic insight into the role of nanoparticles in the mechanical properties of DN hydrogels, including elastic moduli and swellability. Our results indicated the existence of a 'global' saturation point for DN hydrogel nanocomposites, beyond which it becomes less plausible to enhance elastic modulus by simply increasing the concentration of the second network hydrogel or nanoparticles. A similar phenomenon previously reported for single-network hydrogels describes that gains in elastic modulus achievable through the addition of nanoparticles maybe limited by the saturations in the combined crosslinking density (i.e., the sum of chemical crosslinker- and nanoparticle-mediated crosslinking densities) [15]. Further experimental analyses may be necessary to fully elucidate the role of nanoparticles in DN hydrogels; however, our study indicates that the mechanistic understandings previously derived using single-network hydrogel systems may be applied to guide the design of DN hydrogel nanocomposites with high mechanical strengths.

Supplementary Materials: The following are available online at http://www.mdpi.com/2073-4360/12/2/470/s1, Figure S1: Elastic modulus for neat alginate, agarose, and pNaAc hydrogels prepared using 2% monomer, Figure S2: Elastic moduli of pAAM-alginate and pAAm-agarose hydrogels, prepared using 5% pAAM and 2% or 1% alginate or agarose, Figure S3: Elastic moduli of pAAM-alginate hydrogels, prepared using 5% or 2.5% pAAm and 2% or 1% alginate.

Author Contributions: P.A. conceived and designed the experiments; A.C. and N.B. performed the experiments; A.C., N.B., and P.A. analyzed the data; E.B.-C. and P.A. wrote the paper. All authors have read and agreed to the published version of the manuscript.

Funding: This research received no external funding.

Acknowledgments: This work was supported by the School of Engineering at Santa Clara University. Silica nanoparticles were a kind gift from AkzoNobel Pulp and Performance Chemicals Inc.

Conflicts of Interest: The authors declare no conflict of interest. The funders had no role in the design of the study; in the collection, analyses, or interpretation of data; in the writing of the manuscript, or in the decision to publish the results.

References

1. Caló, E.; Khutoryanskiy, V.V. Biomedical applications of hydrogels: A review of patents and commercial products. *Eur. Polym. J.* **2015**, *65*, 252–267. [CrossRef]

2. Annabi, N.; Tamayol, A.; Uquillas, J.A.; Akbari, M.; Bertassoni, L.E.; Cha, C.; Camci-Unal, G.; Dokmeci, M.R.; Peppas, N.A.; Khademhosseini, A. 25th anniversary article: Rational design and applications of hydrogels in regenerative medicine. *Adv. Mater.* **2014**, *26*, 85–124. [CrossRef] [PubMed]
3. Li, J.; Mooney, D.J. Designing hydrogels for controlled drug delivery. *Nat. Rev. Mater.* **2016**, *1*, 16071. [CrossRef] [PubMed]
4. Erfkamp, J.; Guenther, M.; Gerlach, G. Enzyme-functionalized piezoresistive hydrogel biosensors for the detection of urea. *Sensors* **2019**, *19*, 2858. [CrossRef]
5. Tavakoli, J.; Tang, Y. Hydrogel based sensors for biomedical applications: An updated review. *Polymers* **2017**, *9*, 364. [CrossRef]
6. Goldshmid, R.; Seliktar, D. Hydrogel modulus affects proliferation rate and pluripotency of human mesenchymal stem cells grown in three-dimensional culture. *ACS Biomater. Sci. Eng.* **2017**, *3*, 3433–3446. [CrossRef]
7. Banerjee, A.; Arha, M.; Choudhary, S.; Ashton, R.S.; Bhatia, S.R.; Schaffer, D.V.; Kane, R.S. The influence of hydrogel modulus on the proliferation and differentiation of encapsulated neural stem cells. *Biomaterials* **2009**, *30*, 4695–4699. [CrossRef]
8. Pebworth, M.-P.; Cismas, S.A.; Asuri, P. A novel 2.5D culture platform to investigate the role of stiffness gradients on adhesion-independent cell migration. *PLoS ONE* **2014**, *9*, e110453. [CrossRef]
9. DeLong, S.A.; Moon, J.J.; West, J.L. Covalently immobilized gradients of bFGF on hydrogel scaffolds for directed cell migration. *Biomaterials* **2005**, *26*, 3227–3234. [CrossRef]
10. Tsou, Y.-H.; Khoneisser, J.; Huang, P.-C.; Xu, X. Hydrogel as a bioactive material to regulate stem cell fate. *Bioact. Mater.* **2016**, *1*, 39–55. [CrossRef]
11. Calvert, P. Hydrogels for soft machines. *Adv. Mater.* **2009**, *21*, 743–756. [CrossRef]
12. Ahearne, M.; Liu, I.K.-K.; Yang, Y. Mechanical characterisation of hydrogels for tissue engineering applications. *Top. Tissue Eng.* **2008**, *4*, 1–16.
13. Willie, B.M.; Petersen, A.; Schmidt-Bleek, K.; Cipitria, A.; Mehta, M.; Strube, P.; Lienau, J.; Wildemann, B.; Fratzl, P.; Duda, G. Designing biomimetic scaffolds for bone regeneration: Why aim for a copy of mature tissue properties if nature uses a different approach? *Soft Matter* **2010**, *6*, 4976–4987. [CrossRef]
14. Dannert, C.; Stokke, B.T.; Dias, R.S. Nanoparticle-hydrogel composites: From molecular interactions to macroscopic behavior. *Polymers* **2019**, *11*, 275. [CrossRef]
15. Zaragoza, J.; Fukuoka, S.; Kraus, M.; Thomin, J.; Asuri, P. Exploring the role of nanoparticles in enhancing mechanical properties of hydrogel nanocomposites. *Nanomaterials* **2018**, *8*, 882. [CrossRef]
16. Bonhome-Espinosa, A.B.; Campos, F.; Rodriguez, I.A.; Carriel, V.; Marins, J.A.; Zubarev, A.; Duran, J.D.G.; Lopez-Lopez, M.T. Effect of particle concentration on the microstructural and macromechanical properties of biocompatible magnetic hydrogels. *Soft Matter* **2017**, *13*, 2928–2941. [CrossRef] [PubMed]
17. Zaragoza, J.; Babhadiashar, N.; O'Brien, V.; Chang, A.; Blanco, M.; Zabalegui, A.; Lee, H.; Asuri, P. Experimental investigation of mechanical and thermal properties of silica nanoparticle-reinforced poly(acrylamide) nanocomposite hydrogels. *PLoS ONE* **2015**, *10*, e0136293. [CrossRef]
18. Zaragoza, J.; Chang, A.; Asuri, P. Effect of crosslinker length on the elastic and compression modulus of poly(acrylamide) nanocomposite hydrogels. *J. Phys. Conf. Ser.* **2017**, *790*, 1–6. [CrossRef]
19. Gu, Z.; Huang, K.; Luo, Y.; Zhang, L.; Kuang, T.; Chen, Z.; Liao, G. Double network hydrogel for tissue engineering. *Wiley Interdiscip. Rev. Nanomed. Nanobiotechnol.* **2018**, *10*, e1520. [CrossRef] [PubMed]
20. Chen, Q.; Wei, D.; Chen, H.; Zhu, L.; Jiao, C.; Liu, G.; Huang, L.; Yang, J.; Wang, L.; Zheng, J. Simultaneous enhancement of stiffness and toughness in hybrid double-network hydrogels via the first, physically linked network. *Macromolecules* **2015**, *48*, 8003–8010. [CrossRef]
21. Gong, J.P. Why are double network hydrogels so tough? *Soft Matter* **2010**, *6*, 2583–2590. [CrossRef]
22. Gao, G.; Xiao, Y.; Wang, Q.; Fu, J. Synergistic toughening of nanocomposite double network hydrogels by physical adsorption and chemical bonding of polymer chains to inorganic nanospheres and nanorods: A comparative study. *RSC Adv.* **2016**, *6*, 37974–37981. [CrossRef]
23. Wang, Q.; Hou, R.; Cheng, Y.; Fu, J. Super-tough double-network hydrogels reinforced by covalently compositing with silica-nanoparticles. *Soft Matter* **2012**, *8*, 6048–6056. [CrossRef]
24. Lu, X.; Mi, Y. Characterization of the interfacial interaction between polyacrylamide and silicon substrate by fourier transform infrared spectroscopy. *Macromolecules* **2005**, *38*, 839–843. [CrossRef]

25. Wu, L.; Zeng, L.; Chen, H.; Zhang, C. Effects of silica sol content on the properties of poly(acrylamide)/silica composite hydrogel. *Polym. Bull.* **2012**, *68*, 309–316. [CrossRef]
26. Lee, K.Y.; Mooney, D.J. Alginate: Properties and biomedical applications. *Prog. Polym. Sci.* **2012**, *37*, 106–126. [CrossRef] [PubMed]
27. Martău, G.A.; Mihai, M.; Vodnar, D.C. The use of chitosan, alginate, and pectin in the biomedical and food sector—Biocompatibility, bioadhesiveness, and biodegradability. *Polymers* **2019**, *11*, 1837. [CrossRef]
28. Graham, S.; Marina, P.F.; Blencowe, A. Thermoresponsive polysaccharides and their thermoreversible physical hydrogel networks. *Carbohydr. Polym.* **2019**, *207*, 143–159. [CrossRef]
29. Xie, Y.; Wei, Y.; Huang, Y.; Lin, J.; Wu, J. Synthesis and characterization of poly(sodium acrylate)/ bentonite superabsorbent composite. *J. Phys. Conf. Ser.* **2012**, *339*, 012008. [CrossRef]
30. Xing, L.; Hu, C.; Zhang, Y.; Wang, X.; Shi, L.; Ran, R. A mechanically robust double-network hydrogel with high thermal responses via doping hydroxylated boron nitride nanosheets. *J. Mater. Sci.* **2019**, *54*, 3368–3382. [CrossRef]
31. Nurly, H.; Yan, Q.; Song, B.; Shi, Y. Effect of carbon nanotubes reinforcement on the polyvinyl alcohol—Polyethylene glycol double-network hydrogel composites: A general approach to shape memory and printability. *Eur. Polym. J.* **2019**, *110*, 114–122. [CrossRef]
32. Bahrami, Z.; Akbari, A.; Eftekhari-Sis, B. Double network hydrogel of sodium alginate/polyacrylamide cross-linked with POSS: Swelling, dye removal and mechanical properties. *Int. J. Biol. Macromol.* **2019**, *129*, 187–197. [CrossRef] [PubMed]

polymers

MDPI

Article

Effect of Na_2CO_3 on the Microstructure and Macroscopic Properties and Mechanism Analysis of PVA/CMC Composite Film

Jufang Zhu [1], Qiuying Li [1,*], Yanchao Che [2], Xingchen Liu [1], Chengcheng Dong [1], Xinyu Chen [1] and Chao Wang [1]

1 School of Material Science and Engineering, East China University of Science and Technology, 130 Meilong Rd, Shanghai 200000, China; zhujufang1995@163.com (J.Z.); 18721503568@163.com (X.L.); 18621039859@163.com (C.D.); nagewo1314@163.com (X.C.); chaowang826@163.com (C.W.)

2 Shanxi Novofluo New Material Science and Technology Co., LTD, Jinzhong 030600, China; liqy@ecust.edu.com

* Correspondence: liqiuying75713@163.com

Received: 18 December 2019; Accepted: 12 February 2020; Published: 14 February 2020

Abstract: Polyvinyl alcohol (PVA)/carboxyl methyl cellulose sodium (CMC)/Na_2CO_3 composite films with different contents of Na_2CO_3 were prepared by blending and solution-casting. The effect of Na_2CO_3 on the microstructure of PVA/CMC composite film was analyzed by Fourier-transform infrared (FTIR) spectroscopy, X-ray diffraction (XRD), differential scanning calorimetry (DSC), and atomic force microscopy (AFM). Its macroscopic properties were analyzed by water sorption, solubility, and dielectric constant tests. The results show that the microstructure of PVA/CMC/Na_2CO_3 composite films was different from that of PVA and PVA/CMC composite films. In addition, compared to PVA and PVA/CMC composite films, the water sorption of PVA/CMC/Na_2CO_3 composite films relatively increased, the solubility in water significantly decreased, and the dielectric properties significantly improved. All these results indicate that the hydrogen bonding interaction between PVA and CMC increased and the crystallinity of PVA decreased after the addition of Na_2CO_3. This was also a direct factor leading to increased water sorption, decreased solubility, and enhanced dielectric properties. The reaction mechanism of PVA, CMC, and Na_2CO_3 is proposed to further evaluate the effect of Na_2CO_3 on the microstructure and macroscopic properties of PVA/CMC/Na_2CO_3 composite films.

Keywords: PVA; CMC; Na_2CO_3; film

1. Introduction

With increasing difficulties in terms of waste disposal and global warming, severe environmental problems are raising concerns all over the world. In addition, petroleum shortage is also a serious threat throughout the world. Considering these two problems, all countries are trying to develop environmentally friendly materials from nonpetroleum resources. Biopolymers are considered as an ideal alternative to traditional plastic materials owing to their advantages such as biodegradability, nontoxicity, and potential widespread applications; at present, a lot of research is focused on biodegradable materials. Among the biopolymers, cellulose and its water-soluble derivatives are mostly preferred because of the abundance of cellulose on earth and relatively low cost of cellulose derivatives compared to synthetic polymers [1].

Carboxyl methyl cellulose sodium (CMC) is an anionic, linear, and water-soluble polymer; CMC is generally formed by the carboxymethylation of hydroxyl groups in cellulose [2]. CMC has strong water solubility, hydrophilicity, and chemical reactivity because of the presence of hydroxyl and polar carboxylate groups. Moreover, it also has many other characteristic properties such as biocompatibility, renewability, nontoxicity, and biodegradability [3–5]. In addition, CMC has many excellent features such as film formation, emulsification, suspension, water retention, and bonding properties. Therefore, it is the most preferred polymer for food, medicine, detergent, paint, paper, flocculant, sizing agent, textile, and other industries [6–11]. On the other hand, polyvinyl alcohol (PVA) is a synthetic water-soluble polymer with many applications including textile pulps, adhesives, coatings, dispersants, and a series of medical materials with medical functions owing to its excellent adhesion, flexibility, and film-forming properties [12,13]. PVA is very suitable for blending with natural polymers such as CMC because of its biocompatibility and water solubility [14]. Owing to the presence of hydroxyl groups in both PVA and CMC, stronger intermolecular hydrogen bonds can be formed. Films made from blends of biopolymers generally exhibit more excellent properties compared to films made from biopolymers alone. Therefore, studies on PVA and CMC blend films were a hot topic in recent years. For example, El-Sayed et al. conducted differential scanning calorimetry (DSC), thermogravimetric analysis (TGA), and dielectric spectroscopy of blends of PVA and CMC with different compositions [15]. Zhang et al. controlled the sorption and permeability of PVA/CMC composite films through cross-linking, a potential coating for controlled release of fertilizers [16].

Extensive studies showed that the water sorption and solubility of PVA/CMC composite films have great significance for their applications. At present, some studies showed that different pH environments have a certain effect on the water sorption and solubility of composite films. Ibrahima et al. immersed PVA/CMC and PVA/cellulose composite films in solutions of different pH to evaluate the effect of pH on the solubility of films; the results showed that the solubility of composite films was the lowest at pH 10 [17]. Bajpai et al. studied the swelling ratio of CMC/starch composite films in different pH solutions; the swelling ratio constantly increased with increasing pH [18]. However, these studies only evaluated the water sorption of composite films in different pH solutions, whereas they did not systematically study the microstructure and other macroscopic properties of composite films in different pH solutions.

In this study, PVA and CMC were used as a biopolymer matrix, and Na_2CO_3 was used as a filler to evaluate its effect on the microscopic and macroscopic properties of PVA/CMC composite films. A reaction mechanism of Na_2CO_3 with PVA and CMC is proposed, providing a theoretical basis for improving the water sorption, solubility, and dielectric properties of PVA/CMC composite films and promoting their application in biodegradable superabsorbent resins and polymer electrolytes.

2. Materials and Methods

2.1. Materials

PVA (>99%, 87–89% hydrolyzed) and Na_2CO_3 (99.5%) were purchased from Shanghai Aladdin Biochemical Technology Co., Ltd., China. CMC (≥99.5%, Degree of substitution: 0.7) was purchased from Shanghai Yuanye Biotechnology Co., Ltd., China. All chemicals were used as received without any further purification. Deionized water was used, with a conductivity of 0.0556 μs/cm at 25 °C. The structural formula of PVA is as follows:

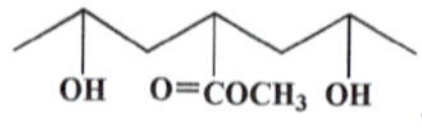

2.2. Preparation of Samples

Preparation of PVA/CMC blends: Firstly, a PVA solution was obtained by dissolving 1 g of PVA powder in 19 g of water under stirring at 90 °C for 3 h. Then, 0.4 g of CMC powder was dissolved in 19.6 g of water under stirring at 60 °C for 3 h. Finally, a homogeneous solution of PVA/CMC (2.5/1 *w/w*) was obtained by adding the CMC solution to the PVA solution and stirring at 85 °C for 3 h.

Preparation of PVA/CMC/Na_2CO_3 blends: Firstly, a PVA/Na_2CO_3 blend was obtained by adding an Na_2CO_3 solution to a PVA solution under stirring at 85 °C for 1 h. The mass of Na_2CO_3 was varied as 8 wt.%, 12 wt.%, and 16 wt.% of PVA. Then, a PVA/CMC/Na_2CO_3 solution was obtained by adding the CMC solution to the PVA/Na_2CO_3 solution under stirring at 85 °C for 3 h.

Preparation of composite films: Firstly, the obtained PVA, PVA/CMC, PVA/Na_2CO_3, and PVA/CMC/Na_2CO_3 film-forming solutions were ultrasonicated for 30 min. Next, films were prepared from these solutions on petri dishes using a solution-casting technique. Finally, the cast films were dried at 60 °C for 24 h. The formulations and abbreviations of the samples are shown in Table 1.

Table 1. The formulations and abbreviations of the samples. PVA—polyvinyl alcohol; CMC—carboxyl methyl cellulose sodium.

Sample	Sample Notation	PVA Solution		CMC Solution		Na_2CO_3 Solution	
		PVA (g)	Water (g)	CMC (g)	Water (g)	Na_2CO_3 (g)	Water (g)
PVA	PVA	1	19				
CMC	CMC			0.4	19.6		
PVA/CMC	P/C	1	19	0.4	19.6		
	P/C/N1	1	19	0.4	19.6	0.08	5
PVA/CMC/Na_2CO_3	P/C/N2	1	19	0.4	19.6	0.12	5
	P/C/N3	1	19	0.4	19.6	0.16	5
PVA/Na_2CO_3	P/N3	1	19			0.16	5

2.3. *Fourier-Transform Infrared (FTIR) Spectroscopy*

FTIR spectra were measured using a NICOLET 6700 spectrometer (Thermo Scientific Co., Waltham, MA, USA) from 4000 cm^{-1} to 400 cm^{-1} to study the interaction between PVA and CMC. Samples were ground into powder after drying.

2.4. *X-ray Diffraction (XRD)*

XRD analysis was carried out using a PANalytical X'Pert-Pro MPD (Almelo, The Netherlands) diffractometer with Cu-Kα radiation (λ = 1.5406 Å) to characterize the crystal form and crystallinity of samples. The diffraction angle ranged from 5° to 80°. Meanwhile, XRD deconvolution analysis was performed using Origin 9.0 software in order to deconvolute specific regions (crystals or amorphous peaks).

2.5. *Differential Scanning Calorimetry (DSC)*

Thermal properties were measured by DSC using a Netzsch PC200 instrument (Netzsch Inc., Selb, Germany). Specimens weighing 6–9 mg were heated at a rate of 10 °C/min from 25 °C to 250 °C under N_2 gas purging.

2.6. *Atomic Force Microscopy (AFM) Studies*

The morphology of composite films was observed using an atomic force microscope (Veeco, DI) operated in contact mode in air. The test specimen had dimensions of approximately 1.0 mm × 5.0 mm × 5.0 mm (thickness × width × length).

2.7. *Dielectric Properties*

The dielectric properties were analyzed using a broad-peak dielectric spectrometer (Concept 40, Novocontrol, Montabaur, Germany) in the frequency range from 10^{-1} to 10^{7} Hz at room temperature. Specimens in the form of a disc of size 30 mm × 2 mm (diameter × thickness) were cut from the dried films.

2.8. Water Sorption

Firstly, the films were weighed (W_1) and placed in 90% humidity at room temperature. Then, the films were periodically removed and weighed (W_2). The experiment was repeated twice, and the average weight was determined. The water sorption was calculated as follows [19]:

$$Water\ sorption = [(W_2 - W_1) \times 100\%]/W_1. \quad (1)$$

2.9. Solubility

Film solubility was determined using the following method: pieces of each film were firstly weighed (W_1) and then immersed in deionized water at room temperature for 24 h. Next, the films were taken out and dried in an oven at 60 °C to constant weight (W_2). The experiment was repeated twice, and the average weight was determined. The solubility was calculated using Equation (2) [20].

$$Solubility = [(W_1 - W_2) \times 100\%]/W_1. \quad (2)$$

3. Results and Discussion

3.1. FTIR Spectroscopy

Figure 1 shows the FTIR spectra of PVA, CMC, P/C, P/N, P/C/N, and Na_2CO_3. For pure PVA, the peaks at 3588 cm^{-1} and 1330 cm^{-1} were assigned to OH stretching and bending vibrations for PVA film [21,22]. The peak corresponding to methylene group (CH_2) asymmetric stretching vibration appeared at −2932 cm^{-1} [23]. The peak at −1727 cm^{-1} corresponded to the C=O symmetrical stretching vibration of the unhydrolyzed ester functional group present on the PVA backbone, and the peak at −1249 cm^{-1} could be attributed to the C–O–C asymmetric stretching vibration of the ester group [24,25]. For CMC, a strong broad peak was observed at 3541 cm^{-1}, which arose from OH stretching vibration [26]. The vibrational peak observed at 2915 cm^{-1} could be ascribed to CH asymmetric stretching [24]. The peak at 1607 cm^{-1} could be attributed to the asymmetric modes of stretching vibration of ester groups (COO^-) [26]. Two absorption peaks that appeared in the region of 1417 cm^{-1} and 1325 cm^{-1} corresponded to scissoring CH_2 and bending OH, respectively [27]. For Na_2CO_3, the peaks at 1426 cm^{-1}, 878 cm^{-1}, and 700 cm^{-1} were carbonate bands [28]. For P/C/N and P/N composite films, the absorption peaks at 878 cm^{-1} and 700 cm^{-1} indicated the presence of Na_2CO_3. The Na_2CO_3 absorption peak at 1426 cm^{-1} was not obvious, because it overlapped with the PVA absorption peak.

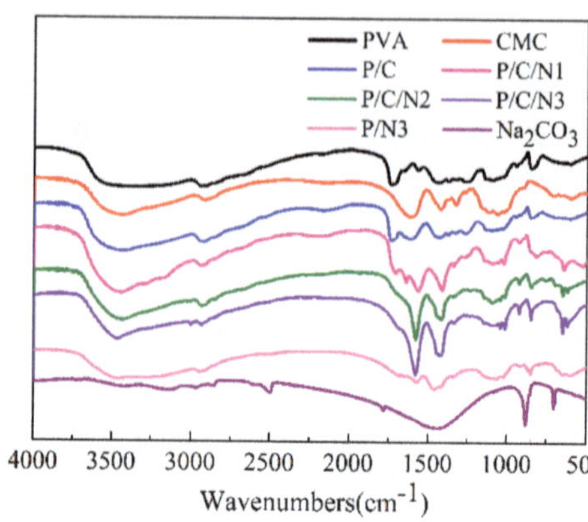

Figure 1. Fourier-transform infrared (FTIR) spectra of Na_2CO_3 and the obtained composite films.

The FTIR spectra were analyzed based on two potential regions in Figure 2 in the following ranges of wavenumbers: (a) 3750–2750 cm^{-1} and (b) 2000–1000 cm^{-1}. The spectrum of the P/C composite film exhibited, in general, the same peaks as those observed in the PVA and CMC films, indicating that no reaction occurred between PVA and CMC [29]. For P/C/N composite films, the absorption peak of OH of PVA moved further toward lower energies than those of the P/C film, exhibiting a shift from 3588 cm^{-1} to 3464 cm^{-1}. The COO^- peak of CMC had a similar trend, shifting from 1607 cm^{-1} to1578 cm^{-1}. At the same time, Figure 2b shows that the OH bending vibration peak of PVA and CMC was gradually weakened with the addition of Na_2CO_3. The changes in OH stretching vibration peak, OH bending vibration peak, and COO^- asymmetric stretching vibration indicated an enhancement of hydrogen bonding between PVA and CMC after the addition of Na_2CO_3, and the intensity of the absorption peak of C=O (1727 cm^{-1}) of PVA gradually decreased with the increase in Na_2CO_3 content, which could be attributed to the hydrolysis of vinyl acetate group of PVA in an alkaline environment. The acetate group was replaced with a hydroxyl group and, thus, the absorption of C=O of the acetate group in PVA was reduced, as shown in Equation (3). The intensity of the absorption peak of the methylene group (CH_2) for the P/C/N3 composite film was significantly reduced. Because OH is an electron-donating group, the electron cloud density around the C atom connecting the hydroxyl group was increased, and the electron cloud density around the C atom adjacent to the methylene group was affected, thus decreasing the absorption peak intensity of methylene group. For the P/N3 composite film, the absorption peak of OH moved to 3475 cm^{-1}, which was related to the formation of intermolecular hydrogen bonds in PVA. The disappearance of the C=O absorption peak in the spectrum of the P/N3 composite film confirmed the reaction shown in Equation (3).

(3)

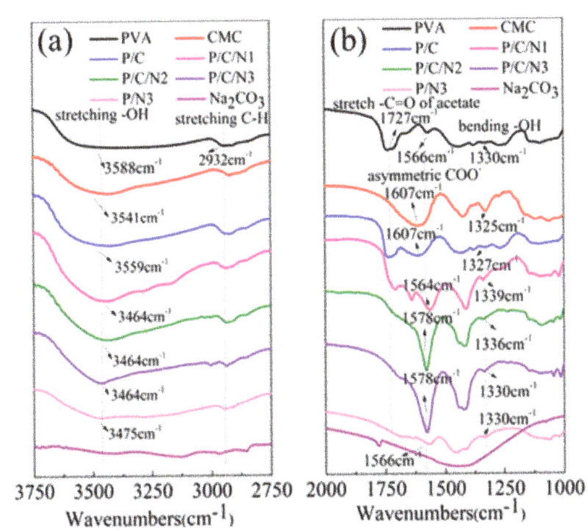

Figure 2. FTIR spectra of Na_2CO_3 and the obtained composite films: (**a**) region from 3750 cm^{-1} to 2750 cm^{-1}; (**b**) region from 2000 cm^{-1} to 1000 cm^{-1}.

3.2. XRD Analysis

The XRD patterns of PVA, CMC, P/C, P/N, P/C/N, and Na_2CO_3 are shown in Figure 3. The crystallinity (X_c) of the obtained films was determined from the equation $X_c = A_c/(A_c + A_a) \times 100$ (where A_c and A_a are the peak intensities of crystalline and amorphous cellulose, respectively) [30]. The results are shown in Table 2. Figure 3 shows that pure PVA had two sharp diffraction peaks at 19.7° and 40.8° corresponding to an orthorhombic lattice (110) reflection [31]. On the other hand, three diffraction peaks for pure CMC appeared at 20.7°, 31.8°, and 45.7°. Compared to pure PVA and CMC films, the position of the diffraction peak of the P/C composite film had no obvious movement. This indicated that there was no new crystal form, but the crystallinity of the P/C composite film compared to that of the PVA film was reduced, probably because the CMC macromolecular chains inhibited the movement of the PVA molecular chains [32]. For the P/N3 film, the crystallinity was significantly weakened compared to PVA, indicating that the addition of Na_2CO_3 destroyed the crystal structure of PVA. After adding CMC, the crystallinity of the P/C/N composite films continued to decrease. This was related to the hindrance of the CMC macromolecular chain. However, Figure 3 and Table 2 show no significant change in the peak position and crystallinity of P/C/N composite films compared to those of the P/C film. Considering that this was related to the use of Na_2CO_3 as a nucleating agent, the crystallinity of P/C/N composite films was improved. Figure 3 also shows that the XRD peak corresponding to crystalline Na_2CO_3 was not obvious in composite films; however, it can be speculated that the crystallinity of Na_2CO_3 in the composite films was greatly reduced [33,34].

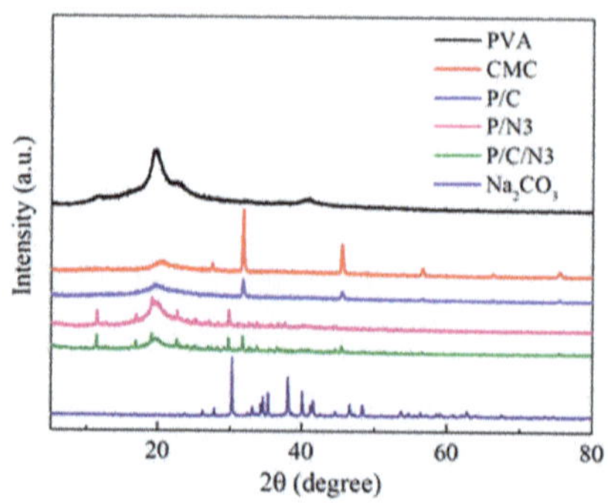

Figure 3. X-ray diffraction (XRD) patterns of Na_2CO_3 and the obtained composite films.

Table 2. Percentage of crystallinity of the obtained composite films.

Sample	A_c	A_a	X_c (%)
PVA	2270	4223	34.96
CMC	373	1834	17.65
P/C	124	1536	7.47
P/C/N3	214	953	18.34
P/N3	501	1979	20.20

3.3. DSC Analysis

Figure 4 shows the characteristic features of the DSC curves for the studied composite films. The crystallinity of composite films could be calculated using the formula $X_c = (\Delta H/w\Delta H_c) \times 100\%)$, where $\Delta H_c = 161$ J/g, and w is the mass fraction of PVA in the composite films [35]. The results are shown in Table 3. The crystallinity obtained from DSC was different from that obtained from XRD, but the trend of crystallinity changes was completely consistent. For the PVA film, DSC showed an endotherm melting transition at 192 °C, representing the melting temperature of PVA. The DSC thermogram of P/C showed an endotherm melting transition at 178 °C, lower than that of PVA. The melting temperature decreased due to a decrease in the crystallinity degree, consistent with the XRD and DSC results.

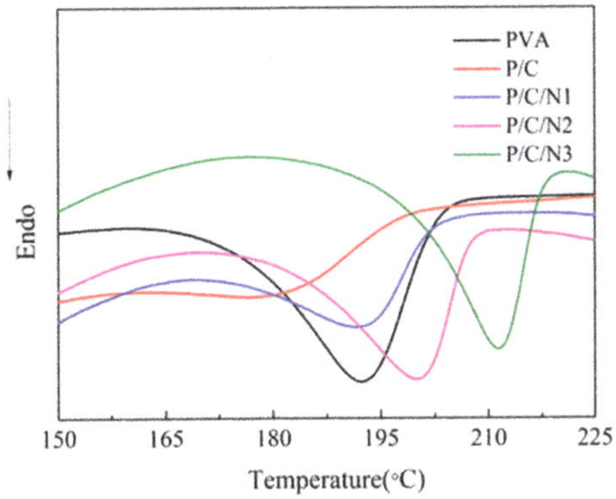

Figure 4. Differential scanning calorimetry (DSC) curves of the obtained composite films.

Table 3. DSC data for the obtained composite films.

Sample	PVA	P/C	P/C/N1	P/C/N2	P/C/N3
Tm (°C)	192	178	194	200	212
Melting enthalpy (J/g)	33.83	8.94	16.18	24.87	25.50
Crystallinity (%)	21.01	7.77	14.87	23.48	24.71

The P/C/N1 composite film exhibited an endotherm melting transition at 194 °C, which was higher than the melting temperature of PVA, and the crystallinity degree of the film was lower than that of PVA. This was because of an increase in the intermolecular hydrogen bonds of PVA and CMC, as indicated by the FTIR results. The hydrogen bonds between PVA and CMC inhibited the movement of PVA molecular chains, increasing the melting temperature and decreasing the crystallinity degree of P/C/N composite films. This also showed that the melting temperature of the P/C/N composite films increased with the increasing content of Na_2CO_3. This was because the crystal structure of PVA was gradually destroyed as the content of Na_2CO_3 increased. The PVA molecular chains became disordered, which strengthened the contact of PVA and CMC. At the same time, as Na_2CO_3 content increased, more ester groups were hydrolyzed, and the amount of hydroxyl groups increased. This generated more intermolecular hydrogen bonds and, thus, increased the melting temperature. Moreover, the crystallinity of the P/C/N composite films was higher than that of P/C composite film, consistent with the XRD results. At the same time, the crystallinity of the P/C/N composite films gradually increased with the increase in sodium carbonate content, which was also related to the role of sodium carbonate as a nucleating agent.

3.4. Dielectric Analysis

Figure 5 shows the patterns of dielectric constant as a function of frequency at room temperature for the resulting composite films. In general, the dielectric constant of pure PVA was low, and the dielectric constant of the P/C composite film was slightly increased. However, the dielectric constant of P/C/N composite films was significantly improved with the addition of Na_2CO_3, and it also gradually increased with increasing Na_2CO_3 content. Specially, the dielectric constant of the P/C/N3 composite film was 32,000 at 10^2 Hz, 25 times higher than that of the P/C composite film (the dielectric constant of the P/C composite film was 1210 at 10^2 Hz). This could be ascribed to the addition of Na_2CO_3, which promoted the generation of hydrogen bonds between PVA and CMC molecules. This led to a multimolecular dipole, significantly increasing the dielectric constant of P/C/N composite films.

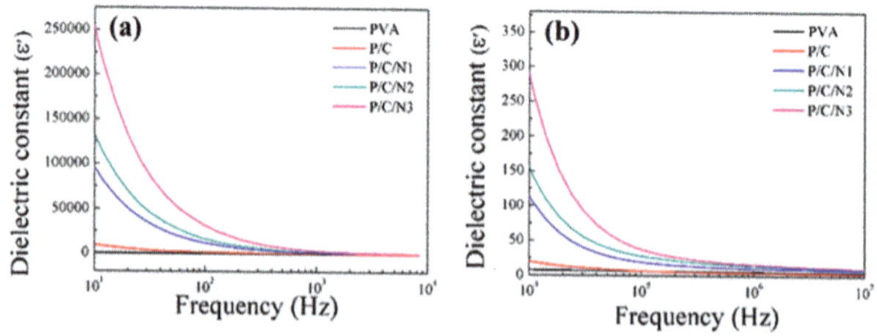

Figure 5. The dielectric constants of the obtained composite films at (**a**) low frequency and (**b**) high frequency.

At the same time, the dielectric constant of pure PVA slightly varied with frequency and remained constant. The change in dielectric constant for the P/C composite film was also very small. However, the dielectric constants of P/C/N composite films rapidly decreased with increasing frequency in the low-frequency range, while they decreased very slowly when the frequency was above 10^5 Hz. This was because, with the addition of Na_2CO_3, the composite films had a certain dielectric relaxation phenomenon in the low-frequency range, and this phenomenon was gradually attenuated in the high-frequency range [36].

3.5. Water Sorption Analysis

Figure 6 shows the water sorption of PVA, P/C, and P/C/N composite films, which were significantly different. The pure PVA film exhibited the lowest water sorption. For the P/C composite film, the water sorption ability was higher than that of pure PVA. This was because the actual mixing of two different types of molecules led to the deformation of the structure, making the network more hydrophilic [29]. In addition, the poor water sorption ability of the pure PVA film could be attributed to higher intermolecular attraction and strong intramolecular or intermolecular hydrogen bonding.

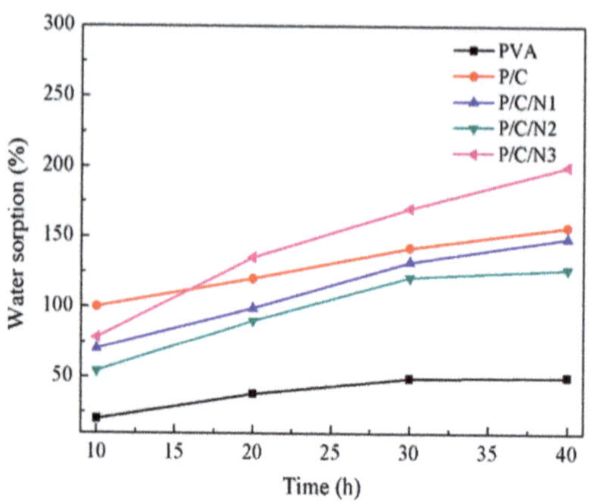

Figure 6. Water sorption of PVA, P/C, P/C/N1, P/C/N2, and P/C/N3 composite films.

The water sorption of P/C/N composite films was significantly higher than that of the pure PVA film, but lower than that of the P/C composite film. This was because the addition of Na_2CO_3 disrupted the crystal structure of PVA. The molecular chain of PVA changed from regular to disordered. This enhanced the contact with water molecules, making the water sorption of P/C/N composite films significantly higher than that of PVA film. Similar results were reported in the literature [18]. However, it also promoted the formation of hydrogen bonds between PVA and CMC, and the hydrogen bonds acted as a physical crosslinking point, promoting the formation of a much denser molecular structure. Therefore, its water sorption was significantly lower than that of the P/C composite film. The water sorption of the P/C/N2 composite film was lower than that of the P/C/N1 composite film because the hydrogen bonding between PVA and CMC was strengthened with the increase in Na_2CO_3 content, and a denser molecular structure was formed. The water sorption of the P/C/N3 composite film was slightly increased. This was related to excess Na_2CO_3, i.e., the PVA molecular chain was in sufficient contact with water molecules. At the same time, excess sodium carbonate also caused the ionization of the carboxyl group of CMC, and the COO/COOH ratio of CMC also increased because of the increasing ionization of carboxylic groups. This resulted in a greater mutual repulsion among the COO^--bearing CMC chains, consequently causing the CMC chains to undergo faster relaxation. This widened the mesh sizes of free volumes, resulting in a larger water sorption ratio [37].

3.6. Solubility Analysis

Figure 7 shows the solubility of PVA, P/C, P/C/N1, P/C/N2, and P/C/N3 composite films. Compared to the PVA film and P/C composite film, the solubility of the P/C/N composite films was significantly reduced with the addition of Na_2CO_3, further confirming the formation of intermolecular hydrogen bonding. The P/C/N2 composite film showed the lowest solubility of 24%, slightly lower than the solubility of the P/C/N1 composite film. This indicated that the hydrogen bonding network of the P/C/N2 composite film was denser. This result is similar to that of Ibrahima et al., i.e., the composite film showed the lowest solubility in a certain alkaline environment [17]. The solubility of the P/C/N3 composite film was slightly increased to 25%, which was related to the increased contact between the disordered molecular chains and water molecules. Secondly, the ionization of the CMC carboxyl group increased the mesh size of its free volume, which was also an important factor. A significant decrease in the solubility of the P/C/N composite films indicated that hydrogen bonding was the most important factor affecting the solubility of composite films, and slight changes in the solubility of the P/C/N3 composite film showed that the change in molecular structure also had a certain effect on the solubility of composite films.

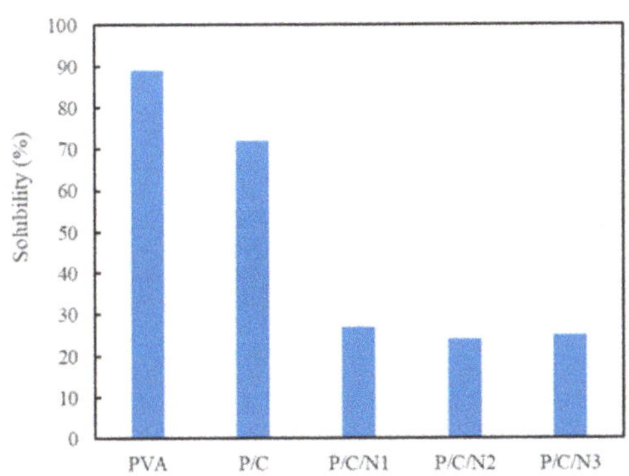

Figure 7. Solubility of the obtained composite films.

3.7. Surface Topography Analysis

The representative images obtained from AFM studies for the studied composite films are shown in Figure 8. Clearly, film surfaces with varying levels of surface roughness were present. The surfaces of the PVA film and P/C composite film were smooth and flat, and the surface of the P/C/N1 composite film was also relatively smooth. As the content of Na_2CO_3 increased, the surfaces of the P/C/N2 and P/C/N3 composite films became rough and uneven; Table 4 shows the corresponding roughness. According to the results of water sorption, although the roughness of P/C/N2 increased significantly, its water sorption still decreased, indicating that the roughness of the surface of composite films had a small effect on water sorption. As the main factors of water sorption, surface roughness, pores, and molecular structure are always mentioned, obviously, no pores were found in the composite films [38]. Thus, the interaction between PVA and CMC under the action of Na_2CO_3 was proven to be the main factor affecting the water sorption of composite films.

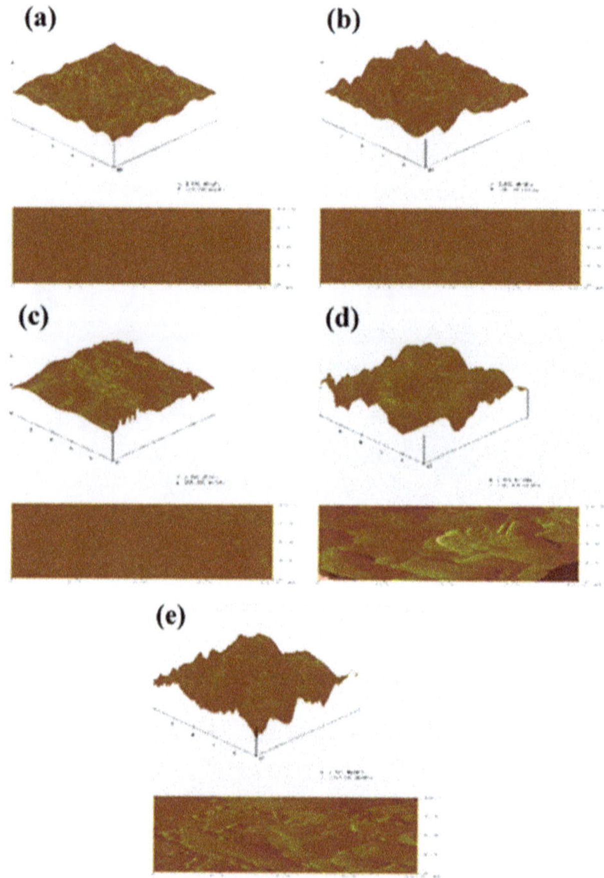

Figure 8. Representative atomic force microcopy (AFM) images of (**a**) PVA, (**b**) P/C, (**c**) P/C/N1, (**d**) P/C/N2, and (**e**) P/C/N3 composite films.

Table 4. Roughness of the obtained composite films.

	PVA	P/C	P/C/N1	P/C/N2	P/C/N3
R_q	10.205	19.963	16.160	295.52	264.09
R_a	7.756	15.115	11.991	233.14	210.62

3.8. Mechanism Analysis

FTIR, DSC, XRD, and the dielectric constant test results of P/C/N composite films were significantly different from those of the PVA film and P/C composite film. In addition, the solubility of the P/C/N composite films was much lower than that of the PVA film and P/C composite film, indicating the formation of a crosslinking structure in P/C/N composite films. The water sorption of the P/C/N composite films was higher than that of the PVA film, but lower than that of the P/C composite film, indicating that Na_2CO_3 promoted the relaxation of the PVA and CMC structure. Based on the above information, a reaction mechanism is proposed for Na_2CO_3, PVA, and CMC.

Na_2CO_3 is alkaline in water and ionizes OH groups. At a certain temperature (80 °C), the alkali can hydrolyze the vinyl acetate group in the partially alcoholic PVA; the acetate group is replaced with a hydroxyl group. The amount of hydroxyl groups in PVA increases, consistent with the weakening C=O absorption peak of P/C/N and P/N composite films in the FTIR spectrum. At the same time, the crystal structure of PVA is significantly destroyed in an alkaline environment, consistent with the XRD results of P/N film. The destruction of the crystal structure of PVA further stretches the molecular chain of PVA and strengthens the contact with the molecular chain of CMC, thus promoting the formation of intermolecular hydrogen bonds between the OH and COOH groups of CMC and the OH group of PVA. The shift in C=O and OH characteristic absorption peaks, the decrease in crystallinity, the increase in the melting point, and the increase in dielectric constant indicate the formation of intermolecular hydrogen bonds between PVA and CMC, as shown in Figure 9. In addition, the stretching of the PVA molecular chain enhances its contact with water molecules; the ionization of the CMC carboxyl group leads to an increase in the mesh size of its free volume, significantly increasing the water sorption of P/C composite films doped with Na_2CO_3. However, the increase in hydrogen bonding between PVA and CMC makes the molecular structure denser, which is the main reason for the decrease in water sorption and solubility.

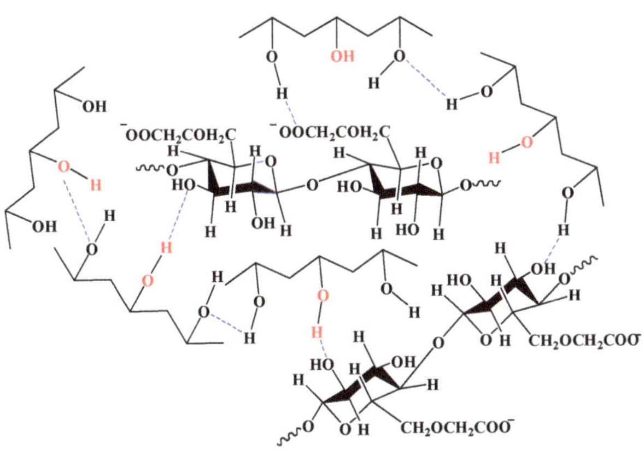

Figure 9. The reaction mechanism for PVA, CMC, and Na_2CO_3.

4. Conclusions

This study shows that Na_2CO_3 significantly affected the microstructure and macroscopic properties of P/C/N composite films. FTIR, XRD, AFM, DSC, and dielectric properties confirmed that Na_2CO_3 promoted the formation of intermolecular hydrogen bonds between PVA and CMC. The water sorption and solubility tests showed that the composite films had reasonable water sorption capacity and water-resisting properties, resulting from the synergistic interaction between the relaxation of the molecular chain and intermolecular hydrogen bonding between PVA and CMC. This study showed how the salt affects the microstructure and macroscopic properties of composite films. This has long-term significance in promoting the application of biodegradable superabsorbent resins and polymer electrolytes.

Author Contributions: Data curation, J.Z.; formal analysis, J.Z.; investigation, J.Z., X.L., C.D., X.C. and C.W.; methodology, J.Z.; project administration, Q.L. and Y.C.; writing—original draft, J.Z.; writing—review and editing, Q.L. All authors have read and agreed to the published version of the manuscript.

Funding: This research received no external funding.

Conflicts of Interest: The authors declare no conflict of interest.

References

1. Vlaia, L.; Coneac, G. Cellulose-derivatives-based hydrogels as vehicles for dermal and transdermal drug delivery. *Emerg. Concepts Anal. Appl. Hydrogels* **2016**. [CrossRef]
2. Kumar:, B.; Negi, Y.S. Water absorption and viscosity behaviour of thermally stable novel graft copolymer of carboxymethyl cellulose and poly(sodium 1-hydroxy acrylate). *Carbohydr. Polym.* **2018**, *181*, 862–870. [CrossRef] [PubMed]
3. Entcheva, E.; Bien, H.; Yin, L.; Chung, C.Y.; Farrell, M.; Kostov, Y. Functional cardiac cell constructs on cellulose-based scaffolding. *Biomaterials* **2004**, *25*, 5753–5762. [CrossRef] [PubMed]
4. Liu, H.; Huang, S.; Li, X.; Zhang, L.; Tan, Y.; Wei, C.; Lv, J. Facile fabrication of novel polyhedral oligomeric silsesquioxane/carboxymethyl cellulose hybrid hydrogel based on supermolecular interactions. *Mater. Lett.* **2013**, *90*, 142–144. [CrossRef]
5. Saha, N.; Shah, R.; Gupta, P.; Mandal, B.B.; Alexandrova, R.; Sikiric, M.D.; Saha, P. PVP-CMC hydrogel: An excellent bioinspired and biocompatible scaffold for osseointegration. *Mater. Sci. Eng.* **2019**, *95*, 440–449. [CrossRef]
6. Agbovi, H.K.; Wilson, L.D. Design of amphoteric chitosan flocculants for phosphate and turbidity removal in wastewater. *Carbohydr. Polym.* **2018**, *189*, 360–370. [CrossRef]
7. Holder, M.K.; Peters, N.V.; Whylings, J.; Fields, C.T.; Gewirtz, A.T.; Chassaing, B.; de Vries, G.J. Dietary emulsifiers consumption alters anxiety-like and social-related behaviors in mice in a sex-dependent manner. *Sci. Rep.* **2019**, *9*, 172. [CrossRef] [PubMed]
8. Mishra, S.; Usha Rani, G.; Sen, G. Microwave initiated synthesis and application of polyacrylic acid grafted carboxymethyl cellulose. *Carbohydr. Polym.* **2012**, *87*, 2255–2262. [CrossRef]
9. Toğrul, H.; Arslan, N. Carboxymethyl cellulose from sugar beet pulp cellulose as a hydrophilic polymer in coating of mandarin. *J. Food Eng.* **2004**, *62*, 271–279. [CrossRef]
10. Yang, Y.; Wang, R.; Feng, W.; Zhou, X.; Chen, Z.; Wang, T. Carboxymethylcellulose/pectin inhibiting structural folding of rice proteins via trinary structural interplays. *Int. J. Biol. Macromol.* **2019**, *133*, 93–100. [CrossRef]
11. Zhang, T.; Wang, Y.; Luo, G.; Bai, S. Preparation of highly dispersed precipitated nanosilica in a membrane dispersion microreactor. *Chem. Eng. J.* **2014**, *258*, 327–333. [CrossRef]
12. Harazono, K.; Nakamura, K. Decolorization of mixtures of different reactive textile dyes by the white-rot basidiomycete Phanerochaete sordida and inhibitory effect of polyvinyl alcohol. *Chemosphere* **2005**, *59*, 63–68. [CrossRef] [PubMed]
13. Zhu, Y.; Yu, X.; Zhang, T.; Wang, X. Constructing zwitterionic coatings on thin-film nanofibrous composite membrane substrate for multifunctionality. *Appl. Surf. Sci.* **2019**, *483*, 979–990. [CrossRef]
14. Follain, N.; Joly, C.; Dole, P.; Bliard, C. Properties of starch based blends. Part 2. Influence of poly vinyl alcohol addition and photocrosslinking on starch based materials mechanical properties. *Carbohydr. Polym.* **2005**, *60*, 185–192. [CrossRef]
15. El-Sayed, S.; Mahmoud, K.H.; Fatah, A.A.; Hassen, A. DSC, TGA and dielectric properties of carboxymethyl cellulose/polyvinyl alcohol blends. *Phys. B Condens. Matter* **2011**, *406*, 4068–4076. [CrossRef]
16. Zhang, L.; Zhang, G.; Lu, J.; Liang, H. Preparation and characterization of carboxymethyl cellulose/polyvinyl alcohol blend film as a potential coating material. *Polym.-Plast. Technol. Eng.* **2013**, *52*, 163–167. [CrossRef]
17. Ibrahim, M.M.; Koschella, A.; Kadry, G.; Heinze, T. Evaluation of cellulose and carboxymethyl cellulose/poly(vinyl alcohol) membranes. *Carbohydr. Polym.* **2013**, *95*, 414–420. [CrossRef]
18. Bajpai, A.K.; Shrivastava, J. In vitro enzymatic degradation kinetics of polymeric blends of crosslinked starch and carboxymethyl cellulose. *Polym. Int.* **2005**, *54*, 1524–1536. [CrossRef]
19. Paralikar, S.; Simonsen, J.; Lombardi, J. Poly(vinyl alcohol)/cellulose nanocrystal barrier membranes. *J. Membr. Sci.* **2008**, *320*, 248–258. [CrossRef]

20. Sivakumar, M.; Malaisamy, R.; Sajitha, C.J.; Mohan, D. Preparation and performance of cellulose acetate–polyurethane blend membranes and their applications—II. *J. Membr. Sci.* **2000**, *169*, 215–228. [CrossRef]
21. El-Sawy, N.M.; El-Arnaouty, M.B.; Ghaffar, A.M.A. γ-Irradiation effect on the non-cross-linked and cross-linked polyvinyl alcohol films. *Polym.-Plast. Technol. Eng.* **2010**, *49*, 169–177. [CrossRef]
22. Dai, H.; Huang, Y.; Huang, H. Eco-friendly polyvinyl alcohol/carboxymethyl cellulose hydrogels reinforced with graphene oxide and bentonite for enhanced adsorption of methylene blue. *Carbohydr. Polym.* **2017**, *185*, 1–11. [CrossRef] [PubMed]
23. Houssein, A.; Claude, D. Chemical modification of poly(vinyl alcohol) in water. *Appl. Sci.* **2015**, *5*, 840–850.
24. Kuanova, A.; Nurpeissova, Z.A.; Mangazbayeva, R.; Park, K. The obtaining of composite materials based on carboxymethylcellulose and polyvinyl alcohol. *Int. J. Biol. Chem.* **2017**, *10*, 62–68. [CrossRef]
25. Mansur, H.S.; Sadahira, C.M.; Souza, A.N.; Mansur, A.A. FTIR spectroscopy characterization of poly (vinyl alcohol) hydrogel with different hydrolysis degree and chemically crosslinked with glutaraldehyde. *Mater. Sci. Eng. C* **2008**, *28*, 539–548. [CrossRef]
26. Zhu, Y.; Xiao, S.; Li, M.; Chang, Z. Natural macromolecule based carboxymethyl cellulose as a gel polymer electrolyte with adjustable porosity for lithium ion batteries. *Power Sources* **2015**, *288*, 368–375. [CrossRef]
27. Biswal, D.; Singh, R. Characterisation of carboxymethyl cellulose and polyacrylamide graft copolymer. *Carbonhydrate Polym.* **2004**, *57*, 379–387. [CrossRef]
28. Mei, Y.; Wang, Z.; Fang, H. Na-containing mineral transformation behaviors during Na_2CO_3-catalyzed CO_2 gasification of high-alumina coal. *Energy Fuels* **2017**, *31*, 1235–1242. [CrossRef]
29. Lagarón, J.M.; López-Rubio, A.; José Fabra, M. Bio-based packaging. *J. Appl. Polym. Sci.* **2016**, 133. [CrossRef]
30. Zainuddin, N.K.; Samsudin, A.S. Investigation on the effect of NH_4Br at transport properties in k–carrageenan based biopolymer electrolytes via structural and electrical analysis. *Mater. Today Commun.* **2018**, *14*, 199–209. [CrossRef]
31. Alakanandana, A.; Subrahmanyam, A.R.; Siva Kumar, J. Structural and electrical conductivity studies of pure PVA and PVA doped with succinic acid polymer electrolyte system. *Mater. Today Proc.* **2016**, *3*, 3680–3688. [CrossRef]
32. El-Gamal, S.; El Sayed, A.M.; Abdel-Hady, E.E. Effect of cobalt oxide nanoparticles on the nano-scale free volume and optical properties of biodegradable CMC/PVA films. *J. Polym. Environ.* **2017**, *26*, 2536–2545. [CrossRef]
33. Uthayakumar, A.; Pandian, A.; Mathiyalagan, S. Interfacial effect of the oxygen-ion distribution on the conduction mechanism in strontium-added Ce0.8Sm0.2O2-δ/Na_2CO_3 nanocomposite. *Phys. Chem.* **2016**, *120*, 25068–25077. [CrossRef]
34. Raza, R.; Wang, X.; Ma, Y.; Liu, X.; Zhu, B. Improved ceria-carbonate composite electrolytes. *Hydrog. Energy* **2010**, *35*, 2684–2688. [CrossRef]
35. Blaine, R.L. Thermal applications note. TN 48. In *Polymer Heats of Fusion*; TA Instrument: New Castle, DE, USA, 2002.
36. El-Zawawy, W.K. Blended graft copolymer of carboxymethyl cellulose and poly(vinyl alcohol) with banana fiber. *J. Appl. Polym. Sci.* **2006**, *100*, 1842–1848. [CrossRef]
37. Peppas, N.A.; Wright, S.L. Solution diffusion in poly[(vinyl alcohol)/poly(acrylic acid)] interpenetrating networks. *Macromolecules* **1996**, *29*, 8798–8804. [CrossRef]
38. Baig, M.I.; Ingole, P.G.; Jeon, J.-D.; Hong, S.U.; Choi, W.K.; Jang, B.; Lee, H.K. Water vapor selective thin film nanocomposite membranes prepared by functionalized silicon nanoparticles. *Desalination* **2019**, *451*, 59–71. [CrossRef]

polymers

MDPI

Article

Preparation and Properties of Rubber Blends for High-Damping-Isolation Bearings

Tuo Lei [1], Yong-Wang Zhang [2,*], Dong-Liang Kuang [3] and Yong-Rui Yang [1]

1 School of Civil Engineering, Chang'an University, Xi'an 710061, China
2 School of Civil Engineering, Southeast University, Nanjing 210096, China
3 School of Materials Science and Engineering, Chang'an University, Xi'an 710064, China
* Correspondence: zhangyongwang0825@163.com

Received: 7 July 2019; Accepted: 18 August 2019; Published: 20 August 2019

Abstract: To improve the energy dissipation capacity of rubber isolation bearings, it is important to find a new rubber material with good applicability and high damping properties. Two types of blends were prepared using nitrile rubber (NBR), brominated butyl rubber (BIIR) and ethylene-vinyl acetate copolymer (EVA): NBR/BIIR and NBR/BIIR/EVA. The vulcanization, mechanical and damping properties of the blends were analyzed. The results show that both blends exhibit excellent vulcanization plateaus and mechanical properties. For NBR/BIIR, as the BIIR content increases, the complementary effects of NBR and BIIR afforded by blending are enhanced. Two damping peaks appeared in the tanδ-T curve and shifted toward lower and higher temperatures, respectively, which clearly widened the effective damping temperature range. However, the damping value in the valley of the tanδ-T curve was as low as 0.39. For NBR/BIIR/EVA, the addition of EVA greatly increased damping in the valley of the tanδ-T curve to approximately 0.54. EVA was observed to be the optimal polymer for improving the compatibility of the NBR/BIIR blend. Moreover, hot air thermal aging tests showed that both blends demonstrated good stability.

Keywords: composite materials; mechanical properties; damping properties; stability

1. Introduction

As a highly unexpected and unpredictable natural disaster, earthquakes do serious damage to human lives and properties. Isolation bearings, an important device to reduce the seismic force transmitted to buildings with functions of energy dissipation, are set between the foundation and the building structure to increases the structure deformability and hysteretic damping ability of buildings, thereby protecting life and reducing economic loss [1–3]. In recent years, with the development and application of seismic isolation technology, the demand for high-damping seismic isolation bearings for buildings has become increasingly urgent [4]. Building seismic isolation bearings require not only high vertical bearing capacity but also good energy dissipation capacity and stability when large horizontal shear deformation occurs [5]. Excellent damping performance is a prerequisite for the isolation bearing materials. Compared with normal laminated rubber bearings, high-damping-isolation bearings usually exhibit better damping performance due to the use of composite rubber materials with high damping properties. Meanwhile, the bearings retain the horizontal and vertical mechanical properties of normal rubber bearings [6]. The composite materials used directly affect the mechanical properties, damping performance and durability of the bearings [7–9]. An ideal high-damping composite material should have excellent mechanical properties, a high dissipation factor, good anti-aging stability, and a wide effective damping temperature range [10–12]. Therefore, finding a suitable high-damping rubber material is the key to preparing high-damping-isolation bearings.

Generally, the blending of two or more types of rubbers is a useful and important way for the preparation and development of rubber blends with properties superior to those of individual constituents, which is also beneficial for improving damping and physical properties [13,14]. In recent years, much attention has been paid to prepare wide temperature range of damping materials by blending. Studies on the damping properties of chlorinated butyl rubber (CIIR) and nitrile rubber (NBR) blends by Tao et al. [15] indicate that the effective damping temperature range is approximately 130 °C. However, the compatibility of the two-phase materials is not particularly high, and the crosslink densities of the two phases are different. The damping value in the valley of the tanδ-T curve is low. The damping properties of natural rubber (NR)/butyl rubber (IIR) blends compatibilized by isobutylene-isoprene block copolymer (IIBC) were researched by Li et al. [16]. The results showed that the NR/IIR blends have typical incompatible sea island biphase structure and damping peak height is decreased while the damping temperature range is broadened. Furthermore, most of the modifications of blends by rubber blending focus on the improvement of damping properties, and only a few studies deal with the mechanical and aging properties of rubber blends, which is not enough research considering materials used for isolation bearings must have a good overall performance.

Among damping materials, polymer composites—especially the interpenetrating polymer networks (IPNs)—are extensively used as damping materials, for their high viscoelastic property around the glass transition temperature [17–20]. For example, Huang et al. [21] studied the damping properties of polydimethylsiloxane (PDMS) and polymethacrylate (PAC) sequential interpenetrating polymer networks. The results indicated that forced compatibility, resulting from the multilayer network from IPNs and the PDMS/PAC vulcanization system, is the key to obtaining a broad damping functional region. The physical and damping properties of pure polyurethane (PU)/epoxy resin (EP) graft interpenetrating polymer network (IPN) composites modified from introducing multi-walled carbon nanotubes (CNTs), hydroxy-terminated liquid nitrile rubber (HTLN) and nature clay montmorillonite (MMT) were further researched by Chen et al. [22–25]. Results reveal that the addition of CNTs, HTLN and MMT can improve the damping properties, tensile strength and impact strength of PU/EP IPN. Though there exist attractive properties of the IPN, the effective damping temperature range is still narrower for some extreme environments. In addition, complexity of the synthesis process and high cost of INP composites greatly limit the application of INP damping materials as structural damping materials in engineering applications (i.e., high-damping-isolation bearings).

The other way utilizes the introduction of other energy dissipation methods, such as the breaking and reorganizing of reversible process [26], the introduction of relaxation components [27] and so on. For example, Wu et al. [28] added hindered phenol into chlorinated polyethylene (CPE) and found a novel transition peak induced by the intermolecular hydrogen bonding. Zhao et al. [29] studied the damping and mechanical properties of hindered phenol AO-80/CIIR/NBR blend, and noted that the addition of AO-80 can effectively improve the damping in the valley of the blend materials. Qiao et al. [30] have systematically expounded the interaction mechanism of the significantly improved damping property of NBR contributed by the introduction of small molecules of AO-80, which provided some useful information to design the high-performance damping materials by adding small molecules. After that, abundant research based on hydrogen bonds has been implemented to prepare high damping material [31–33]. The introduction of hydrogen bonds can effectively improve the damping peak height, but the effective damping temperature range is slightly broadened. Besides, the damping stability is deteriorated due to the phase separation of matrix and additives.

Among damping rubbers, butyl rubber and NBR are most commonly used. BIIR is a type of modified halogenated butyl rubber that maintains the damping properties of butyl rubber while exhibiting better curing activity and polarity [34,35]. However, the rubber glass transition temperature (T_g) is approximately −40 °C, and the effective damping temperature range (corresponding to the temperature range in which tanδ ≥ 0.3 [36]) is mainly concentrated at low temperatures. NBR is known to exhibit numerous outstanding properties such as desirable physical and mechanical properties, and excellent aging resistance performance. The T_g of NBR is approximately −10 °C, and its damping performance is also excellent. As a damping material, NBR is suitable for high-temperature environments [37–39]. Considering the large T_g of NBR, the introduction of NBR into BIIR should be an effective way to improve the high-temperature damping property of BIIR (i.e., a wide effective damping temperature range). However, NBR/BIIR blends are slightly soluble and usually exhibit phase-separated morphology and poor interfacial adhesion between the phases. Therefore, it is necessary to introduce an interfacial compatibilizer to the binary system to improve the total performance. Ethylene-vinyl acetate copolymer (EVA) is a copolymer of non-polar ethylene (E) monomer and a polar vinyl acetate (VA) monomer, which contains both a non-polar ethylene segment and a polar vinyl acetate segment in its molecular chain [40–43]. It can be added as a compatibilizer to the NBR/BIIR blends to reduce the surface tension and improve the compatibility of the two phases. The EVA has the potential to improve the mechanical and damping properties simultaneously [44–46].

In the following section, the blending of strongly polar NBR and weakly polar BIIR to prepare NBR/BIIR blends is described in detail first. The vulcanization, mechanical and damping properties of the blends were analyzed using a moving die rheometer, a universal testing machine, and a dynamic mechanical analyzer. Based on the optimal blend ratio of NBR/BIIR, EVA copolymer, a material compatible with NBR/BIIR, was added to prepare NBR/BIIR/EVA. The same properties of NBR/BIIR/EVA were also systematically investigated. Furthermore, the aging properties of the two blends were studied using a hot air thermal aging test chamber. Through a comparison of the test results, the optimal mixture ratio of NBR/BIIR/EVA was determined, and all properties meet the requirements of the standard (Chinese Standard JG/T-118-2019). The results provide a theoretical basis for synthesizing composite rubber materials for high-damping-isolation bearings.

2. Experimental Research

2.1. Materials

The NBR used was JSRN230S (35% acrylonitrile content, Mooney viscosity of 42 ML (1 + 4) 100 °C), a Japanese JSR product (JSR Corporation, Tokyo, Japan), with the molecular formula illustrated in Figure 1a. The BIIR used was BB2255 (Mooney viscosity of 46 ML (1 + 4) 100 °C), an ExxonMobil Chemical Company product (Exxon Mobil Corporation, Ivan, Texas, USA), with the molecular formula illustrated in Figure 1b. The EVA copolymer used was 40 W, a US DuPont product (DuPont, Wilmington, Delaware, USA), with the molecular formula illustrated in Figure 1c. Carbon black (N330), sulfur (S), dicumyl peroxide (DCP), zinc oxide (ZnO), stearic acid, coumarone, dibutyl phthalate (DBP), N-cyclohexyl-2-benzothiazole sulfonamide (CZ), tetramethyl thiuram disulfide (TMTD), an antioxidant (4010NA), an anti-aging agent D (PBAN, $C_{16}H_{13}N$) and other reagents used were all commercially available products (Shanghai Deyin Chemical Company, Shanghai, China).

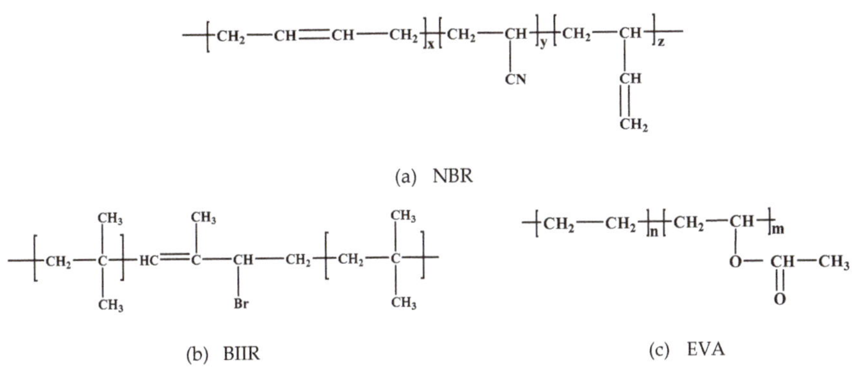

Figure 1. Molecular structural formulas: (a) nitrile rubber (NBR); (b) brominated butyl rubber (BIIR); (c) ethylene-vinyl acetate (EVA).

2.2. Basic Formulation

According to the design principle of rubber formula 3P (price + performance + process) +1C (circumstances), the basic formulation is designed by mathematical methods such as multi-factor optimization method (orthogonal experiment method) and variance analysis method. The basic formulation of NBR, BIIR and EVA masterbatch are shown in Table 1.

Table 1. Basic formulation of nitrile rubber (NBR), brominated butyl rubber (BIIR) and brominated butyl rubber (EVA) (100 phr).

Name	ZnO	Stearic Acid	S	DCP	2402 Resin	TMTD	CZ	DBP	N330	PBAN	4010NA
NBR	5–10	2–6	1–4	–	–	0.2–0.6	1–3	8–10	20–50	1.5–3	1–3
BIIR	5–10	2–6	–	–	5–10	0.2–0.4	1–1.5	8–10	20–50	1.5–3	1–3
EVA	5–10	2–6	–	1–2	–	0.2–0.4	1–1.5	8–10	20–50	1.5–3	1–3

2.3. Sample Preparation

1. NBR, BIIR and EVA masterbatch: The as-received NBR, BIIR and EVA were kneaded on a two-roll mill at room temperature for 3 min and then blended with compounding and cross-linking additives (the basic formulas of the recipes are shown in Table 1), respectively. Each compound was then kneaded on the two-roll mill at room temperature for 5 min to form the NBR, BIIR and EVA masterbatch.
2. NBR/BIIR and NBR/BIIR/EVA blends: The NBR, BIIR and EVA masterbatch was mixed in the corresponding proportions (the specific proportions of the blends are shown in Sections 4.1 and 4.2) in a certain order (i.e., for NBR/BIIR, first BIIR, then NBR; for NBR/BIIR/EVA, first BIIR, then NBR, and finally EVA) on the two-roll mill at room temperature for 5 min. After 16 h, a rheometer analyzer was used to determine the scorch time T_{10} and the vulcanization time T_{90} of the mixture. These mixtures were then hot pressed and vulcanized at 150 °C under a pressure of 15 MPa for T_{90} to obtain cross-linked NBR/BIIR and NBR/BIIR/EVA samples (Figure 2).

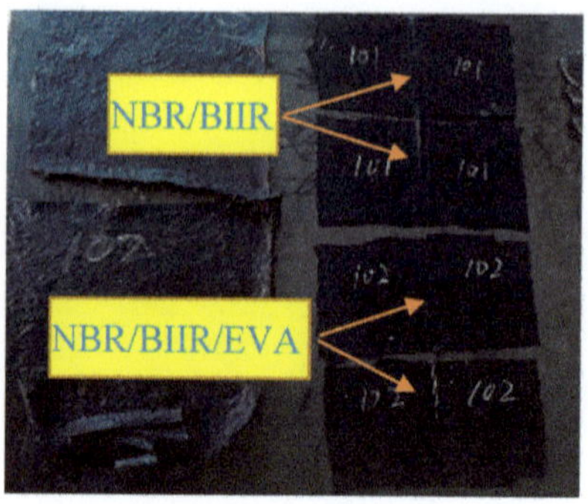

Figure 2. Test samples.

2.4. Performance Test

2.4.1. Vulcanization Characteristics

The samples were tested on an M-3000A moving die rheometer analyzer (Gotech Testing Machines Inc, Taiwan, China). The test temperature was 150 °C, and the test time was 30 min based on the ISO 6502:2016.

2.4.2. Mechanical Property Test

The samples were tested on a WDL-2500N universal tensile testing machine (Gotech Testing Machines Inc, Taiwan, China). The tensile test involved a standard sample, a tensile rate of 500 mm/min, and a test temperature of 23 °C based on the ISO 37:2017. A trouser tear test was carried out at a tensile rate of 100 mm/min according to ISO 34:2016.

2.4.3. Dynamic Mechanical Analysis (DMA)

A TA Instruments Q800 dynamic mechanical analyzer (TA Instruments, Wilmington, Delaware, USA) was adopted to measure the damping performance of the sample in accordance with the ISO 4664-1:2011. The sample dimensions were 35 mm × 12 mm × 2 mm (length × width × thickness), the test temperature range was −60–60 °C, the frequency was 10 Hz, and the heating rate was 3 °C/min.

2.4.4. Aging Characteristic Test

An R-PTH standard aging test box (Kerry Instruments Company, Shanghai, China) was used for accelerated aging of the pieces by heating in air according to ISO 188:2011. The aging condition was hot air at 100 °C × 72 h, as suggested by ISO 22762-3:2018. Sample performance was characterized based on the property variation percentage observed during aging (Figure 3).

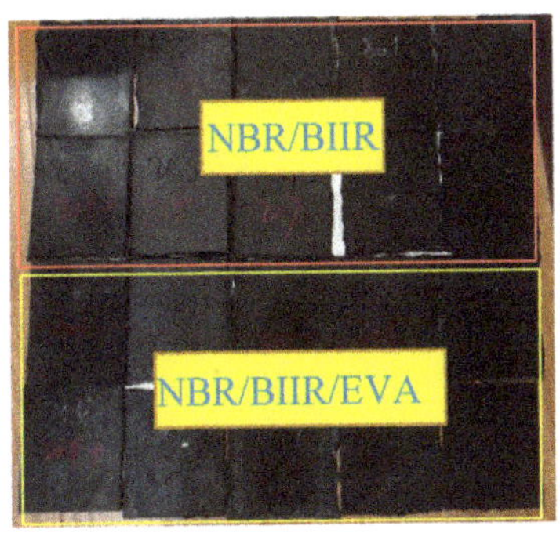

Figure 3. Samples after aging.

2.4.5. Shore Hardness Test

Sample preparation was made in accordance with ISO 23529: 2010. Hardness test was performed by using a Shore A durometer (LX-A; Jiangsu Mingzhu Testing Machinery, Yangzhou, Jiangsu, China) on the sample size of 15 mm × 15 mm × 6 mm according to ISO 48-4: 2018 at 25 ± 2 °C. Three measurements were involved at different positions on each side of a 6 mm thick specimen obtained by compression molding, and the median value was determined. The standard deviation for the Shore A hardness was 3%. The reading was taken 3 s after the pressure foot was in firm contact with the test piece.

3. Polymer Blend Theory

3.1. Phase Structure Mechanism of Polymer

The phase structure of a polymer is closely related to the compatibility of the basic materials. The better the compatibility between two basis materials is, the smaller the two-phases separation of the final material will be.

NBR/BIIR was prepared by blending strongly polar NBR with weakly polar BIIR base rubber. EVA copolymer was added to NBR/BIIR to promote the compatibility between NBR and BIIR phases and to further improve the overall performance of NBR/BIIR blends. The basis blending and cross-linking mechanism are shown in Figure 4.

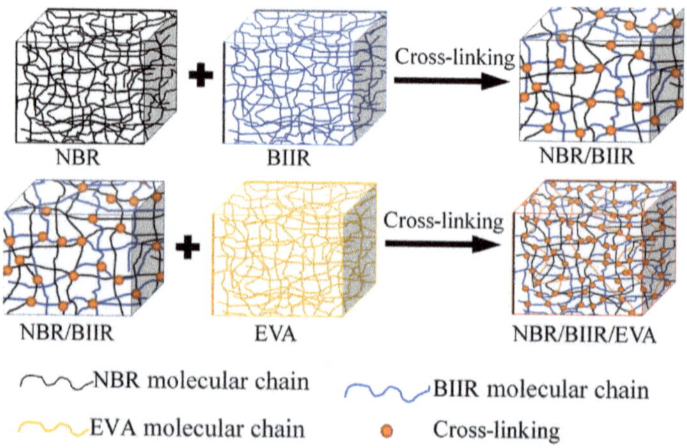

Figure 4. Basis blending and cross-linking mechanism.

3.2. Polymer Wide Temperature Range Mechanism

Two slightly soluble rubbers, NBR and BIIR, were blended, one component being BIIR with a glass transition temperature of T_{g1} (see Figure 5a) and the other component being NBR with a glass transition temperature of T_{g2} (see Figure 5b). The tanδ-T curve of the blends shows two external damping internal friction peaks, effectively expanding the scope of the damping temperature range. However, due to the low compatibility between NBR and BIIR, the damping value is low in the valley between the two internal friction peaks (see Figure 5c). By adding EVA copolymer, a blend with a high-damping value over a wide temperature range can be prepared (see Figure 5d). The dynamic mechanical analysis (DMA) temperature profile of the blend system varies with the miscibility of the constituent polymers, as shown in Figure 5.

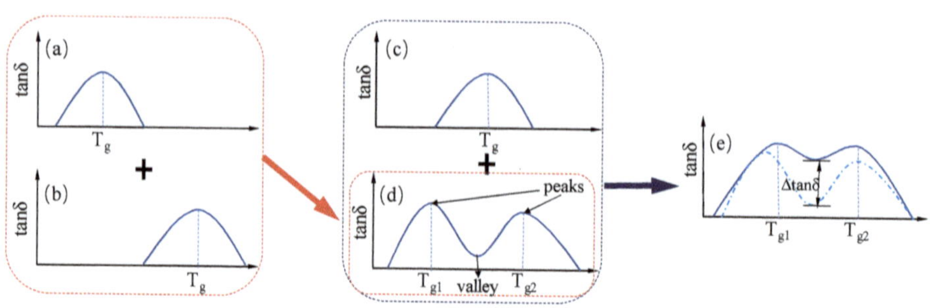

Figure 5. Variation of dynamic mechanical analysis (DMA) temperature spectrum of blends with miscibility of component polymers; **(a–c)** are the DMA temperature spectrum of the component high polymer, respectively; in **(d,e)**, the miscibility increases sequentially.

3.3. Polymer Damping Mechanism

NBR/BIIR and NBR/BIIR/EVA blends have different energy conversion modes (i.e., energy dissipation and energy storage) at different ambient temperatures during the deformation process. The entire polymer is in a glassy state when the temperature is below the lower limit of the glass transition range, and in a free state when above the upper limit of the glass transition range. In the two states, mechanical energy cannot be dissipated (i.e., energy storage). However, in the glass transition range, the chains of polymer molecules are in a critical state between "freezing" and "thawing," also called a viscoelastic state. When subjected to an external force, the polymer follows the energy dissipation mechanism of the viscoelastic material.

4. Results and Discussion

4.1. NBR/BIIR Results and Discussion

4.1.1. Fluidization Characteristics

The vulcanization curves and corresponding characteristics of NBR/BIIR with various blend ratios are shown in Figure 6 and Table 2.

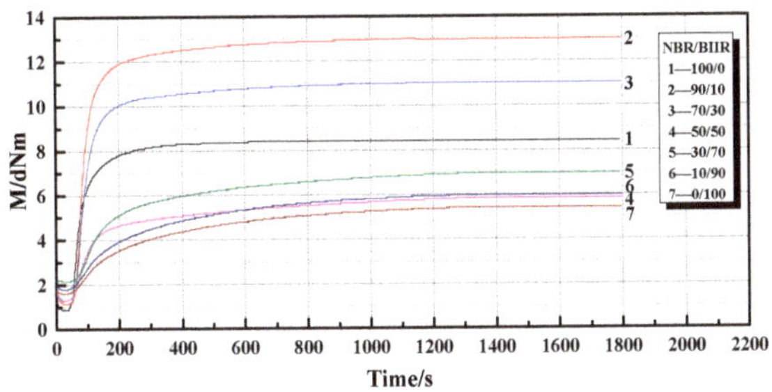

Figure 6. Curves of vulcanization characteristics for NBR/BIIR with various blend ratios.

Table 2. Vulcanization characteristics of NBR/BIIR with various blend ratios (150 °C).

Test Numbers	Rubber Name NBR/BIIR	MH dNm	ML dNm	T_{10} min	T_{90} min	tanPA (min)
1	100/0	8.5	0.9	0.6	3.2	0.08
2	90/10	13.0	1.2	1.1	3.1	0.03
3	70/30	11.1	1.4	1.1	3.4	0.04
4	50/50	6.0	1.3	1.0	13.2	0.07
5	30/70	7.1	2.0	1.1	12.5	0.08
6	10/90	6.1	1.8	1.2	13.2	0.05
7	0/100	5.4	1.6	1.2	13.5	0.05

Note MH: Maximum torque rating; ML: Minimum torque rating; T_{10}: Scorch time; T_{90}: Optimal vulcanization time; tanPA: Dynamic dissipation angle.

As shown in Figure 6, when a low-sulfur and high-accelerator vulcanization system was adopted, NBR/BIIR blends with various blend ratios exhibited good vulcanization plateau. Table 2 shows that when the amount of BIIR was increased from 0 to 90 phr, the vulcanization time and scorch time first increased and then gradually stabilized. The main reason is that the vulcanization and scorch times of BIIR are longer than those of NBR. When the BIIR content exceeded 50 phr, BIIR played a dominant role in determining the properties of NBR/BIIR, and the vulcanization time of the blend was greatly extended. Additionally, with the increase in BIIR content, the maximum torque of the blend increased first and then decreased. When the NBR/BIIR ratio of the blend was 50/50, the minimum torque was 6.0 dNm. The reason is that NBR has a stronger intermolecular interaction than BIIR.

4.1.2. Mechanical Properties

Mechanical properties, as indispensable performance indices of composite materials for high-damping-isolation bearings, must meet the requirements of ISO 22762-3:2010. The mechanical properties of NBR/BIIR with various blend ratios are shown in Figure 7.

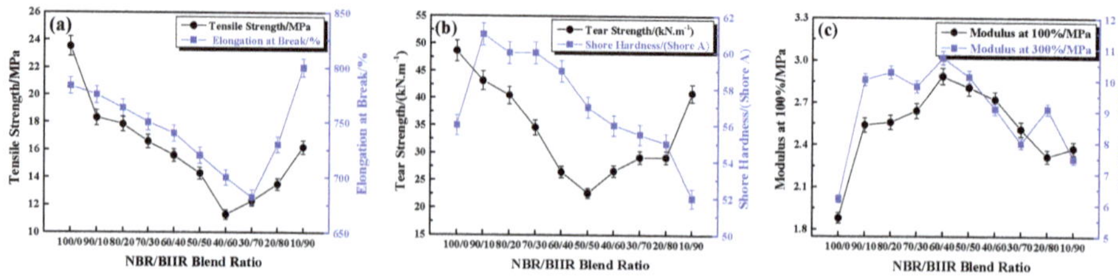

Figure 7. Mechanical properties of NBR/BIIR with various blend ratios: (**a**) Effect of tensile strength and elongation at break; (**b**) effect of tear strength and shore hardness and (**c**) effects of 100% and 300% module.

With the increase in BIIR content, the tensile strength and the elongation at break of the NBR/BIIR varied as a "V-shaped" pattern (Figure 7a). When the content of BIIR was 60 phr, the tensile strength and the elongation at the break were the lowest, but still reached 11.2 MPa and more than 675%, respectively. First, BIIR is more difficult to vulcanize than NBR, and the difference between the two-phase vulcanization speeds leads to different crosslink densities. In addition, because of the poor thermodynamic compatibility of the two rubbers, weak links were formed at the interface between the two phases, which reduced the mechanical properties of the blends. These two reasons resulted in a decrease in tensile strength and elongation at break.

With increasing BIIR, the tear strength decreased first and then increased as a result of the joint action of two factors (Figure 7b). The addition of BIIR increased the number of polysulfide bonds in the blend system, which provides high tear strength. However, increasing the cross-linking density decreased tear strength. When the blend ratio was 50/50, the tear strength reached a minimum of 22.6 MPa.

The modulus and hardness are both representatives of the rigidity of the blends. The modulus is related to large tensile deformation, while the hardness is related to smaller compressive deformation. As shown in Figure 7b,c, with an increase in the BIIR content, the change trends of the modulus and the hardness were roughly the same, both increasing first and then decreasing.

4.1.3. Damping Performance

The dissipation factor tanδ-T curves and damping parameter of NBR/BIIR with various blend ratios are shown in Figure 8 and Table 3.

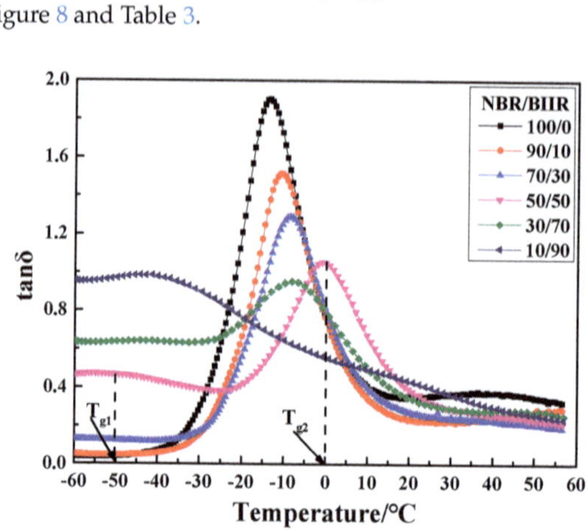

Figure 8. Damping properties versus temperature for NBR/BIIR with various blend ratios.

Table 3. Damping property parameters of NBR/BIIR with various blend ratios.

NBR/BIIR	Value	T_g (= T_{g2})/ °C	T_1/ °C	T_2/ °C	($T_2 - T_1$)/ °C
100/0	—	−14.3	−29.8	18.1	47.9
90/10	—	−8.3	−25.9	12.0	37.9
70/30	—	−5.1	−24.9	15.9	40.8
50/50	0.4	0.4	≤−60.0	27.0	≥87.0
30/70	0.6	−5.8	≤−60.0	25.6	≥85.6
10/90	1.0	−38.9	≤−60.0	38.5	≥98.5

Note: Values are those observed in the valley of the tan δ-T curve; T_1 and T_2 are the lowest and highest temperatures that satisfy tanδ ≥ 0.3, respectively; and T_2−T_1 is the effective temperature range.

Figure 8 shows that, with increasing BIIR, two damping peaks appeared in the tanδ-T curve. The low-temperature peak and the high-temperature peak correspond to T_{g1} and T_{g2} of BIIR and NBR, respectively. Moreover, the effective damping temperature range gradually expanded, but the damping peak decreased. This phenomenon was due to the large difference in polarity between NBR and BIIR, resulting in a certain phase separation tendency. When NBR was blended with BIIR, the moving ability of the NBR molecular segments was limited by surrounding and entangling actions from the bulky BIIR molecular segments. However, the slightly miscible BIIR reduced the intermolecular interactions of NBR, thus increasing the movement capacity of the NBR molecular chains. The two effects caused the T_g of NBR to move toward high temperatures. However, when the BIIR content increased to a certain limit, T_g moved in the opposite direction. At this point, BIIR played a dominant role in the blend system.

Table 3 shows that, as the BIIR content increased, the effective temperature range expanded and moved toward low temperatures. When the NBR/BIIR ratio was 50/50, the effective temperature range was at least 87.0 °C, but the damping at the bottom in the tanδ-T curve was low. Therefore, it was necessary to further improve the damping in the effective temperature range.

The storage modulus E′-T curves of NBR/BIIR are shown in Figure 9. The E′-T curves of both NBR/BIIR and pure NBR exhibit a transition peak. As the BIIR content increased, the storage modulus decreased significantly in the glassy region. When the BIIR content exceeded 50, the storage modulus tended to stabilize. This result shows that BIIR has a softening plasticizer effect on the NBR base rubber, making the NBR/BIIR blend softer and thus increasing the energy dissipation and improving the damping properties of the blend system.

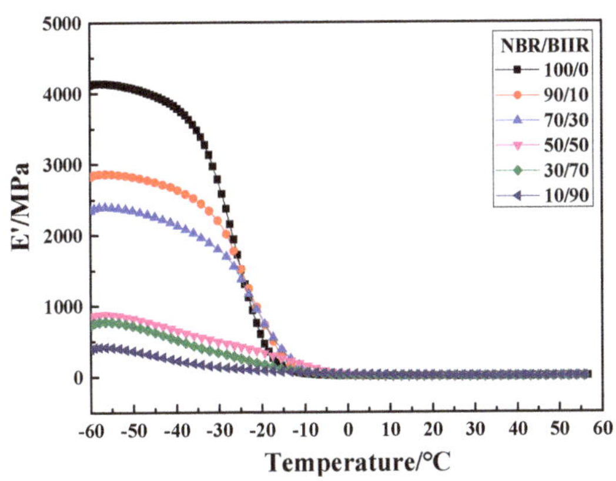

Figure 9. Storage modulus E′ versus temperature of NBR/BIIR with various blend ratios.

4.1.4. Aging Performance

The stability of the rubber material for rubber isolation bearings has a strong effect influence on the mechanical and damping performance of the bearings. Therefore, it was necessary to perform a further heat aging test based on the ISO 22762-3:2018.

1. Mechanical Properties

The mechanical properties and aging performance of NBR/BIIR with various blend ratios are shown in Figure 10 and Table 4.

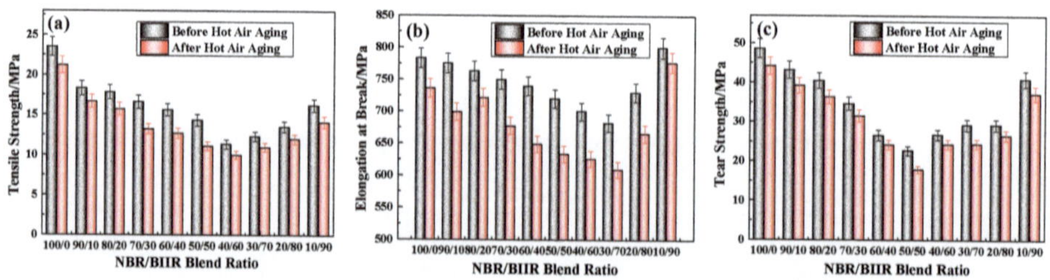

Figure 10. Comparison of mechanical properties before and after aging of NBR/BIIR with various blend ratios: (**a**) Tensile strength; (**b**) elongation at break, and (**c**) tear strength.

Table 4. Mechanical property parameters before and after aging of NBR/BIIR with various blend ratios.

NBR/BIIR	Tensile Strength/MPa	Elongation at Break/%	Tear Strength/kN/m
	Before and after Aging/%	Before and after Aging/%	Before and after Aging/%
100/0	9.9 ↓	6.0 ↓	9.1 ↓
90/10	9.1 ↓	9.9 ↓	9.1 ↓
80/20	12.2 ↓	5.5 ↓	10.2 ↓
70/30	20.4 ↓	9.8 ↓	9.1 ↓
60/40	18.8 ↓	12.4 ↓	9.2 ↓
50/50	22.7 ↓	12.1 ↓	16.3 ↓
40/60	11.8 ↓	10.7 ↓	9.6 ↓
30/70	11.4 ↓	10.7 ↓	15.4 ↓
20/80	11.5 ↓	8.9 ↓	10.7 ↓
10/90	13.6 ↓	2.9 ↓	9.0 ↓

Note: Property variation percentage during aging = (after aging − before aging)/before aging 100%; " ↓ " indicates decrease.

Figure 10 shows that the various mechanical properties of the NBR/BIIR blend decreased after aging and that the change trends of the mechanical properties with the change of BIIR content were roughly the same as those before. When the blend ratio was 50/50, the tensile strength and tear strength decreased by 22.7% and 16.3%, respectively, and the corresponding elongation at break decreased by 12.1% (Table 4). In general, the overall stability of the blends was more ideal.

2. Damping Performance

A DMA test was further performed after hot air thermal aging of the NBR/BIIR materials. The damping performance of the NBR/BIIR blends are shown in Figure 11.

The tanδ peak of NBR/BIIR with various blend ratios after aging was slightly lower than that before aging (Figure 8). Moreover, the overall change trend remained essentially consistent with that before aging. The rubber damping performance shows good stability for various ratios. When the blend ratio of NBR/BIIR was 50/50, the blend could meet the requirement of tanδ $\geq$ 0.3 and showed a wide temperature range of no less than 75.6 °C.

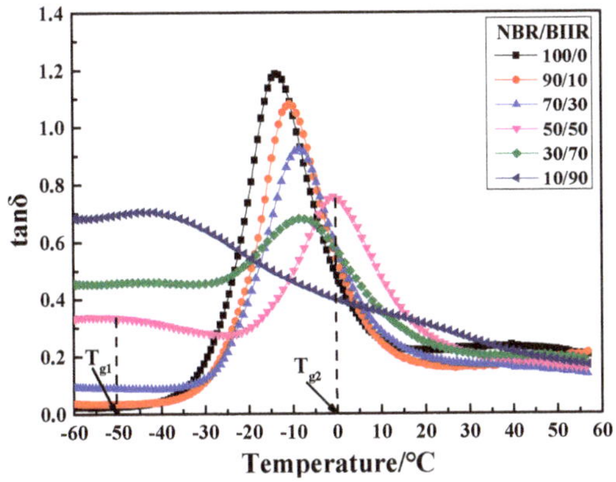

Figure 11. Damping properties versus temperature for NBR/BIIR with various blend ratios.

4.2. NBR/BIIR/EVA Results and Discussion

Based on the tests performed, we can conclude that when the NBR/BIIR blend ratio is 50/50, the blend achieves the best performance (mechanical and damping properties). However, the effective damping temperature region is concentrated in the low-temperature zone, and the damping in the valley of the tanδ-T curve is low.

To obtain a more ideal blend with a wider temperature range and higher damping, EVA copolymer was added to the blend with an NBR/BIIR blend ratio of 50/50. The effect of the EVA copolymer on the properties of NBR/BIIR is discussed in detail.

4.2.1. Fluidization Characteristics

The vulcanization curves of NBR/BIIR (50/50) after adding the EVA copolymer are shown in Figure 12 and Table 5.

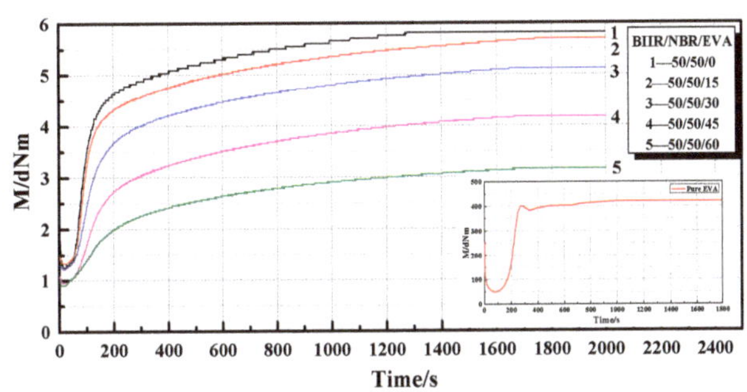

Figure 12. Curves of vulcanization characteristics for NBR/BIIR/EVA with various blend ratios.

As shown in Figure 12, NBR/BIIR/EVA exhibited good vulcanization plateau at various blend ratios. The addition of EVA decreased the maximum torque of the blends (Figure 6). Table 5 shows that with increasing EVA content, T_{90} increased. When the EVA content was 0 and 60 phr, the corresponding T_{90} values were 13.2 and 17.6 min, respectively.

Table 5. Vulcanization characteristic parameters of NBR/BIIR/EVA with various blend ratios (150 °C).

Test Number	Rubber Name NBR/BIIR/EVA	MH dNm	ML dNm	T_{10} min	T_{90} min	tanPA (min)
1	50/50/0	6.0	1.3	1.0	13.2	0.07
2	50/50/15	5.7	1.3	1.1	14.3	0.07
3	50/50/30	5.3	1.2	1.1	15.2	0.07
4	50/50/45	4.3	1.0	1.1	17.0	0.07
5	50/50/60	3.3	0.9	1.0	17.6	0.10
6	0/0/100	421	49.5	1.3	2.3	0.03

Note MH: Maximum torque rating; ML: Minimum torque rating; T_{10}: Scorch time; T_{90}: Optimal vulcanization time; tanPA: Dynamic dissipation angle.

This phenomenon is due to the different mutual cross-linking speeds caused by the different sensitivities of the EVA copolymer and the NBR/BIIR to the vulcanizing agent, thus resulting in different vulcanization times. As the EVA content increased, the rate of cross-linking within NBR/BIIR blends was further impeded, and the vulcanization rate of the blend was weakened.

4.2.2. Mechanical Properties

The effects of various blend ratios on the mechanical properties of NBR/BIIR/EVA are shown in Figure 13.

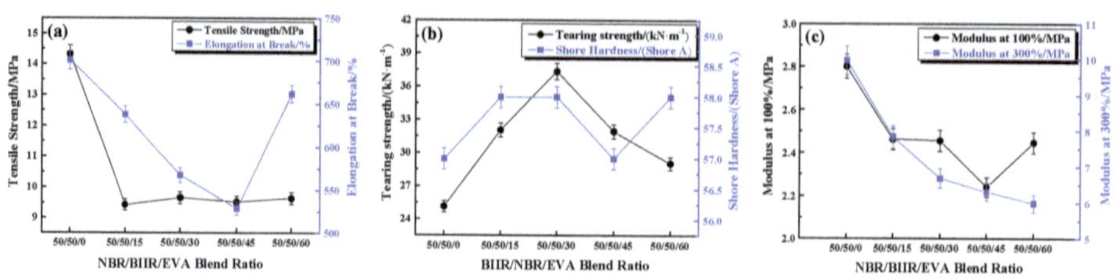

Figure 13. Mechanical properties of NBR/BIIR/EVA with various blend ratios: (**a**) Effect of tensile strength and elongation at break; (**b**) effect of tear strength and shore hardness, and (**c**) effects of 100% and 300% module.

Figure 13a shows that the tensile strength of the blend decreased when a small amount of EVA was added. As the EVA content was increased, the tensile strength remained nearly stable, and the minimum was 9.3 MPa. The elongation at break varied according to a "V-shaped" pattern. The minimum was 525% when the NBR/BIIR/EVA blend ratio was 50/50/45. The main reason is that the tensile strength and elongation at break of the EVA copolymer are lower than those of the rubber material, and the difference in the vulcanization rates results in an inconsistent crosslink density.

With increasing EVA, the tear strength increased first and then decreased as shown in Figure 13b. The addition of EVA increased the number of polysulfide bonds, the crosslink density and intermolecular force, which increased the tear strength of the blend. When the NBR/BIIR/EVA ratio was 50/50/30, the maximum tear strength reached 37.5 MPa.

As the EVA content increased, the modulus decreased and the hardness tended to stabilize, as shown in Figure 13c. Although the addition of EVA copolymer leads to a loss of mechanical properties, the resulting material could fulfill the requirements according to ISO 22762-3:2010.

4.2.3. Damping Performance

Further tests on the effect of the EVA copolymer on the dynamic mechanical properties of the NBR/BIIR blends were performed. The tanδ-T curve and damping parameters of the NBR/BIIR/EVA blend with various blend ratios are shown in Figure 14 and Table 6, respectively.

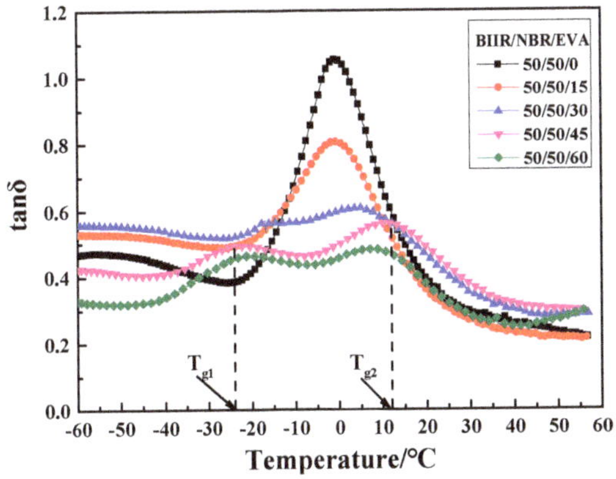

Figure 14. Damping properties versus temperature for NBR/BIIR/EVA with various blend ratios.

Table 6. Damping property parameters of NBR/BIIR/EVA with various blend ratios.

NBR/BIIR/EVA	Value	$T_g(=T_{g2})/\,^\circ$C	$T_1/\,^\circ$C	$T_2/\,^\circ$C	$(T_2-T_1)/\,^\circ$C
50/50/0	0.4	0.4	≤−60.0	27.0	≥87.0
50/50/15	0.5	3.2	≤−60.0	25.1	≥85.1
50/50/30	0.5	8.9	≤−60.0	27.9	≥87.9
50/50/45	0.4	12.5	≤−60.0	29.5	≥89.5
50/50/60	0.3	11.5	−48.6	21.7	69.4

Note: Values are those observed in the valley of the tan δ-T curve; T_1 and T_2 are the lowest and highest temperatures that satisfy tanδ ≥ 0.3, respectively; and $T_2 - T_1$ is the effective temperature range.

When the blend ratio of NBR/BIIR/EVA was 50/50/15, two peaks appeared in the tanδ-T curve. With increasing EVA, the damping peaks continuously decreased. The peak of T_{g2} moved toward a higher temperature, which effectively increased the damping in the high-temperature zone. The damping in the valley increased first and then decreased. When the EVA content was 30 phr, the damping in the valley reached up to 0.5, much higher than the former value of 0.4 observed for the NBR/BIIR (Figure 8).

Table 6 shows that when the ratio of NBR/BIIR/EVA was 50/50/45, the effective damping temperature range reached at least 89.5 °C (from −60 to 29.5 °C). The addition of EVA expanded the effective damping temperature range of the blend and effectively increased the damping at the valley of the tanδ-T curve (Table 3). One reason is that the EVA copolymer exhibits good compatibility with the basis rubbers. EVA enhances the solubility of NBR and BIIR, and strongly affects the interaction of the basic material. Thus, the phase structure of the blend changes from an "island" phase structure to a continuous phase, with good compatibility.

The storage modulus E′-T curves of NBR/BIIR/EVA are shown in Figure 15. The E′-T curves of NBR/BIIR/EVA all show transition peaks. As the ratio of EVA reached a certain value, the E′ peak changes remained relatively stable in the glassy region, indicating that the plasticizer of EVA was no longer distinct.

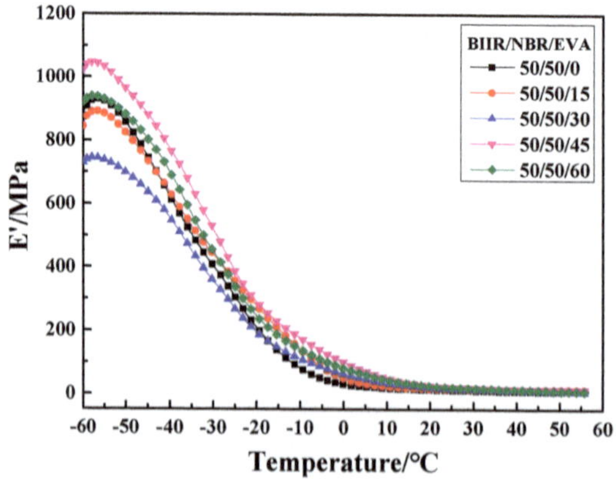

Figure 15. Storage modulus E′ versus temperature of NBR/BIIR/EVA with various blend ratios.

4.2.4. Aging Performance

Due to the mutual influence of the basis rubber, and the influence of the vulcanization system on aging, it was necessary to further study the anti-aging stability of the NBR/BIIR/EVA blends.

1. Mechanical properties

The mechanical properties of NBR/BIIR/EVA before and after hot air thermal aging tests are compared in Figure 16 and Table 7.

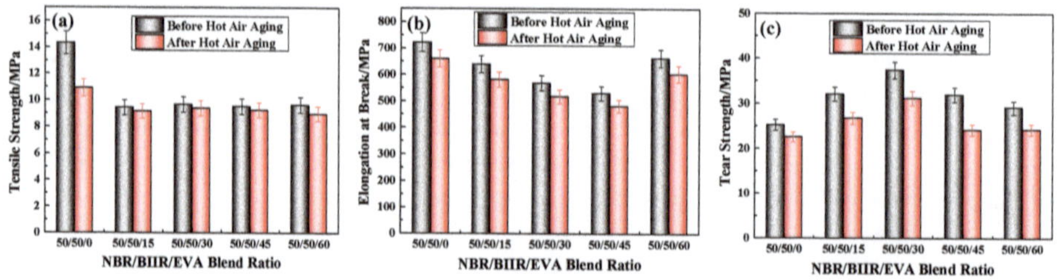

Figure 16. Comparison of mechanical properties before and after aging of NBR/BIIR/EVA with various blend ratios: (**a**) Tensile strength; (**b**) elongation at break, and (**c**) tear strength.

Table 7. Mechanical property parameters before and after aging of NBR/BIIR/EVA with various blend ratios.

NBR/BIIR/EVA	Tensile Strength/MPa	Elongation at Break/%	Tear Strength/ kN/m
	Before and after Aging/%	Before and after Aging/%	Before and after Aging/%
50/50/0	22.7↓	12.1↓	16.3↓
50/50/15	2.9↓	9.1↓	16.2↓
50/50/30	2.9↓	10.2↓	15.6↓
50/50/45	3.0↓	10.3↓	21.3↓
50/50/60	6.9↓	9.0↓	15.5↓

Note: Property variation percentage during aging = (before aging − after aging)/ before aging 100%, " ↓ " indicates decrease.

As shown in Figure 16, the mechanical properties of the NBR/BIIR/EVA blends deteriorated after the hot air thermal aging test, and the overall changing trends after aging remained essentially similar to those before. Table 7 shows that, with the addition of EVA, the aging resistance of tensile strength was significantly improved, but the effect on tear strength and elongation at the break was not distinct

(Table 4). The maximum deterioration in the tensile strength, elongation at break and tear strength were 6.9%, 10.2%, and 21.3%, respectively. Nevertheless, all of the results demonstrate that the blends exhibited strong anti-aging properties.

2. Damping Performance

Additionally, a DMA test was performed, and the damping performance is shown in Figure 17.

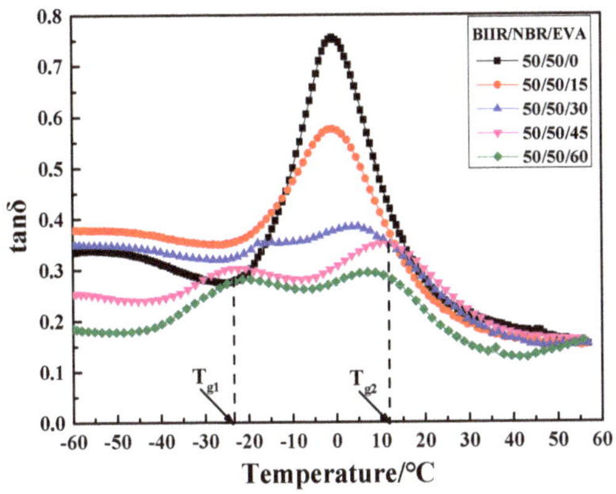

Figure 17. Damping properties versus temperature of NBR/BIIR/EVA with various blend ratios.

Due to the degradation of the material and the environmental conditions, the damping performance of the blends was reduced. The overall change was similar to that before aging, but the temperature range was not significantly reduced. As the EVA content was increased, the damping aging resistance of the blend decreased.

5. Conclusions

Based on tests of the blends NBR/BIIR and NBR/BIIR/EVA, the following conclusions were drawn:

- NBR/BIIR and NBR/BIIR/EVA possess good vulcanization and mechanical properties. The addition of EVA reduces the mechanical properties of NBR/BIIR. Nevertheless, the blends can still fulfill the standard requirements.
- The blending of NBR and BIIR is efficient in expanding the damping temperature range. When the NBR/BIIR ratio is 50/50, the effective temperature range is at least 87.0 °C. However, the damping in the valley is low.
- The addition of EVA to the NBR/BIIR effectively increases the damping in the valley of the tanδ-T curve (tanδ ≥ 0.3) and moves the effective temperature range toward high temperatures. Moreover, the NBR/BIIR/EVA with a 50/50/30 blend ratio is a comparatively ideal material.
- Hot air thermal aging tests of NBR/BIIR and NBR/BIIR/EVA show that the anti-aging stabilities of the two materials are not significantly altered.

Author Contributions: Conceptualization, T.L. and Y.-W.Z.; methodology, T.L.; software, Y.-W.Z.; validation, T.L., Y.-W.Z. and Y.-R.Y.; formal analysis, D.-L.K.; investigation, Y.-W.Z.; resources, T.L.; data curation, Y.-R.Y.; writing—original draft preparation, T.L.; writing—review and editing, D.-L.K.; visualization, T.L.; supervision, T.L.; project administration, Y.-W.Z.; funding acquisition, T.L.

Funding: This research was funded by National Natural Science Foundation of China, grant number (51578072, 51578077), and the Natural Science Foundation of Shaanxi Province, grant number (2016JM5037).

Acknowledgments: Y.-W.Z. feels grateful to Jian Li for their help in some of experiments and materials used for experiments.

Conflicts of Interest: The authors declare no conflict of interest.

References

1. Reggio, A.; De Angelis, M. Combined primary–secondary system approach to the design of an equipment isolation system with High-Damping Rubber Bearings. *J. Sound Vib.* **2014**, *333*, 2386–2403. [CrossRef]
2. Steelman, J.S.; Fahnestock, L.A.; Filipov, E.T.; LaFave, J.M.; Hajjar, J.F.; Foutch, D.A. Shear and Friction Response of Nonseismic Laminated Elastomeric Bridge Bearings Subject to Seismic Demands. *J. Bridg. Eng.* **2013**, *18*, 612–623. [CrossRef]
3. Zordan, T.; Liu, T.; Briseghella, B.; Zhang, Q. Improved equivalent viscous damping model for base-isolated structures with lead rubber bearings. *Eng. Struct.* **2014**, *75*, 340–352. [CrossRef]
4. Zhou, F.L. Seismic isolation, energy dissipation, and structural control system-Inevitable technological options for ending urban and rural earthquake disasters in China. *Cities Disaster Reduct.* **2016**, *5*, 1–10.
5. Japan Society of Architecture. *Recommendation for the Design of Base Isolated Buildings*, 3rd ed.; Japan Society of Architecture, Ed.; Earthquake Press: Beijing, China, 2006; pp. 35–64.
6. Kim, H.-S.; Oh, J. A Study on Isolation Performance of High Damping Rubber Bearing Through Shaking Table Test and Analysis. *J. Korea Acad. Coop. Soc.* **2016**, *17*, 601–611.
7. Shen, C.; Zhou, F.; Cui, J.; Huang, X.; Zhuang, X.; Ma, Y. Dependency test research of mechanical Performance of HDR and its parametric value analysis. *J. Earthq. Eng. Eng. Vib.* **2012**, *32*, 95–103. (In Chinese)
8. Bakhshi, A.; Jafari, M.H.; Tabrizi, V.V. Study on dynamic and mechanical characteristics of carbon fiber-and polyamide fiber-reinforced seismic isolators. *Mater. Struct.* **2014**, *47*, 447–457. [CrossRef]
9. Chen, Y.J.; Guo, K.M.; Li, Y.; Chen, F. Behavior of high damping seismic isolation rubber bearings for bridges. *J. Vib. Shock* **2015**, *34*, 136–141. (In Chinese)
10. Gu, H.S.; Itoh, Y. Ageing behavior of Natural Rubber and High Damping Rubber Materials Used in Bridge Rubber Bearings. *Adv. Struct. Eng.* **2010**, *13*, 1105–1113. [CrossRef]
11. Radhakrishnan, C.K. Dynamic mechanical properties of styrene butadiene rubber and poly (ethylene-co-vinyl acetate) blends. *J. Polym. Res.* **2008**, *15*, 161–171. [CrossRef]
12. He, W.; Xing, T.; Liao, G.X.; Lin, W.; Deng, F.; Jian, X.G. Dynamic Mechanical Properties of PPESK/Silica Hybrid Materials. *Polym. Technol. Eng.* **2009**, *48*, 164–169. [CrossRef]
13. Wang, J.; Zhao, X.; Wang, W.; Geng, X.; Zhang, L.; Guo, B.; Nishi, T.; Hu, G.-H. Significantly Improving Strength and Damping Performance of Nitrile Rubber via Incorporating Sliding Graft Copolymer. *Ind. Eng. Chem. Res.* **2018**, *57*, 16692–16700. [CrossRef]
14. Kaneko, H.; Inoue, K.; Tominaga, Y.; Asai, S.; Sumita, M. Damping performance of polymer blend/organic filler hybrid materials with selective compatibility. *Mater. Lett.* **2002**, *52*, 96–99. [CrossRef]
15. Tao, G.; Lu, X.; Guo, J.; Tian, M. Preparation and Properties of High Damping Rubber with Wide Temperature Range and Frequency Range. *Polym. Mater. Sci. Eng.* **2013**, *29*, 114–118. (In Chinese)
16. Li, J.-C.; Zhang, H.-S.; Zhao, X.-Y.; Jiang, J.-G.; Wu, Y.-X.; Lu, Y.-L.; Zhang, L.-Q.; Nishi, T. Development of high damping natural rubber/butyl rubber composites compatibilized by isobutylene-isoprene block copolymer for isolation bearing. *Express Polym. Lett.* **2019**, *13*, 686–696. [CrossRef]
17. Cristea, M.; Ibanescu, S.; Cascaval, C.N.; Rosu, D. Dynamic Mechanical Analysis of Polyurethane-Epoxy Interpenetrating Polymer Networks. *High Perform. Polym.* **2009**, *21*, 608–623. [CrossRef]
18. Qin, C.; Zhao, D.; Bai, X.; Zhang, X.; Zhang, B.; Jin, Z.; Niu, H.; Zhao, D. Vibration damping properties of gradient polyurethane/vinyl ester resin interpenetrating polymer network. *Mater. Chem. Phys.* **2006**, *97*, 517–524. [CrossRef]
19. Babkina, N.V.; Lipatov, Y.S.; Alekseeva, T.T. Damping properties of composites based on interpenetrating polymer networks formed in the presence of compatibilizing additives. *Mech. Compos. Mater.* **2006**, *42*, 385–392. [CrossRef]
20. Wang, Y.-B.; Huang, Z.-X.; Zhang, L.-M. Damping properties of silicone rubber/polyacrylate sequential interpenetrating networks. *Trans. Nonferr. Met. Soc. China* **2006**, *16*, s517–s520. [CrossRef]
21. Huang, G.-S.; Li, Q.; Jiang, L.-X.; Huang, G.; Jiang, L. Structure and damping properties of polydimethylsiloxane and polymethacrylate sequential interpenetrating polymer networks. *J. Appl. Polym. Sci.* **2002**, *85*, 545–551. [CrossRef]
22. Wang, T.M.; Chen, S.B.; Wang, Q.H.; Pei, X.Q. Damping analysis of polyurethane/epoxy graft interpenetrating polymer network composites filled with short carbon fiber and micro hollow glass bead. *Mater. Des.* **2010**, *31*, 3810–3815. [CrossRef]

23. Chen, S.; Wang, Q.; Wang, T. Damping, thermal, and mechanical properties of carbon nanotubes modified castor oil-based polyurethane/epoxy interpenetrating polymer network composites. *Mater. Des.* **2012**, *38*, 47–52. [CrossRef]
24. Chen, S.B.; Wang, Q.H.; Wang, T.M. Damping, thermal, and mechanical properties of montmorillonite modified castor oil-based polyurethane/epoxy graft IPN composites. *Mater. Chem. Phys.* **2011**, *130*, 680–684. [CrossRef]
25. Chen, S.B.; Wang, Q.H.; Wang, T.M. Hydroxy-terminated liquid nitrile rubber modified castor oil based polyurethane/epoxy IPN composites: Damping, thermal and mechanical properties. *Polym. Test.* **2011**, *30*, 726–731. [CrossRef]
26. Zhao, X.-Y.; Xiang, P.; Tian, M.; Fong, H.; Jin, R.; Zhang, L.-Q. Nitrile butadiene rubber/hindered phenol nanocomposites with improved strength and high damping performance. *Polymer* **2007**, *48*, 6056–6063. [CrossRef]
27. Xu, K.; Zhang, F.; Guo, J.; Wu, H. Molecular insights into the damping mechanism of poly(vinyl acetate)/hindered phenol hybrids by a combination of experiment and molecular dynamics simulation. *RSC Adv.* **2015**, *5*, 4200–4209. [CrossRef]
28. Wu, C.; Yamagishi, T.A.; Nakamoto, Y.; Ishida, S.; Nitta, K.H.; Kubota, S. Organic hybrid of chlorinated polyethylene and hindered phenol. I. Dynamic mechanical properties. *J. Polym. Sci. Part B Polym. Phys.* **2000**, *38*, 2285–2295. [CrossRef]
29. Zhao, X.Y.; Guo, C.M.; Zhang, D.Y.; Tian, M.; Zhang, L.Q. Preparation and damping properties of hindered phenol AO-80/chlorinated butyl rubber/nitrile rubber composites. *China. Synth. Rubber. Ind.* **2014**, *37*, 42–46. (In Chinese)
30. Qiao, B.; Zhao, X.; Yue, D.; Zhang, L.; Wu, S. A combined experiment and molecular dynamics simulation study of hydrogen bonds and free volume in nitrile-butadiene rubber/hindered phenol damping mixtures. *J. Mater. Chem.* **2012**, *22*, 12339. [CrossRef]
31. Zhao, X.; Zhang, G.; Lu, F.; Zhang, L.; Wu, S. Molecular-level insight of hindered phenol AO-70/nitrile-butadiene rubber damping composites through a combination of a molecular dynamics simulation and experimental method. *RSC Adv.* **2016**, *6*, 85994–86005. [CrossRef]
32. Zhang, J.H.; Wang, L.F.; Zhao, Y.F. Fabrication of novel hindered phenol/phenol resin/nitrile butadiene rubber hybrids and their long-period damping properties. *Polym. Compos.* **2012**, *33*, 2125–2133. [CrossRef]
33. Song, M.; Zhao, X.; Li, Y.; Hu, S.; Zhang, L.; Wu, S. Molecular dynamics simulations and microscopic analysis of the damping performance of hindered phenol AO-60/nitrile-butadiene rubber composites. *RSC Adv.* **2014**, *4*, 6719. [CrossRef]
34. Hu, W.; Ruxia; Zhang, Y.; Li, F.; He, X. Effects of sericite modified by macromolecular dispersant on the thermal, mechanical and electrical properties of the NR/SBR composites. *J. Macromol. Sci. Part A-Pure Appl. Chem.* **2011**, *48*, 962–968. [CrossRef]
35. Hintze, C.; Boldt, R.; Wiessner, S.; Heinrich, G. Influence of processing on morphology in short aramid fiber reinforced elastomer compounds. *J. Appl. Polym. Sci.* **2013**, *130*, 1682–1690. [CrossRef]
36. Zuo, K.C. Performance and Characterization of Nitrile Rubber Hybrid Damping. Ph.D. Thesis, Southwest Jiaotong University, Chengdu, China, 2016.
37. Lu, X.; Li, X.J.; Tian, M. Preparation of high damping elastomer with broad temperature and frequency ranges based on ternary rubber blends. *Polym. Adv. Technol.* **2014**, *25*, 21–28. (In Chinese) [CrossRef]
38. Jiang, Y.; Liu, Y.; Gong, J.; Li, C.; Xi, Z.; Cai, J.; Xie, H. Microstructures, thermal and mechanical properties of epoxy asphalt binder modified by SBS containing various styrene-butadiene structures. *Mater. Struct.* **2018**, *51*, 86. [CrossRef]
39. Zhang, J.; Zao, W.; Wang, L.; Zhao, Y. Preparation and characterization of low-temperature hydrogenated nitrile butadiene rubber hybrid with hydrogen bonds for the sealing applications. *Mater. Des.* **2013**, *52*, 896–904. [CrossRef]
40. Li, F.; Zhu, W.; Zhang, X.; Zhao, C.; Xu, M. Shape memory effect of ethylene–vinyl acetate copolymers. *J. Appl. Polym. Sci.* **1999**, *71*, 1063–1070. [CrossRef]
41. Jia, L.Y.; Fu, G.J.; Shi, X.Y. Foaming and Damping Properties of Ethylene Vinyl-Acetate Copolymer/Polylactic Acid Blends. *J. Macromol. Sci. Part B-Phys.* **2015**, *54*, 190–202. [CrossRef]

42. Zhang, X.; Yang, H.; Song, Y.; Zheng, Q. Influence of crosslinking on crystallization, rheological, and mechanical behaviors of high density polyethylene/ethylene-vinyl acetate copolymer blends. *Polym. Eng. Sci.* **2014**, *54*, 2848–2858. [CrossRef]
43. Lai, S.-M.; You, P.-Y. Preparation and Characterization of Ethylene Vinyl-Acetate Copolymer/Silicone Blends with Excellent Two-Way Shape Memory Properties. *Macromol. Res.* **2018**, *26*, 984–997. [CrossRef]
44. Hou, J.; Bai, H.; Li, D. Damping capacity measurement of elastic porous wire-mesh material in wide temperature range. *J. Mater. Process. Technol.* **2008**, *206*, 412–418. [CrossRef]
45. Wang, W.F.; Lin, X.Z.; Ma, Y.L. Viscoelastic Vibration Damping Materials for Application in a Temperature Range above 150 °C. *Key Eng. Mater.* **2017**, *730*, 569–573. [CrossRef]
46. Kan, L.; Ouyang, X.; Gao, S.; Li, R.; Ma, N.; Han, S.-H.; Wei, H.; Liu, L.-H. High damping and mechanical properties of hydrogen-bonded polyethylene materials with variable contents of hydroxyls: Effect of hydrogen bonding density. *Chin. J. Polym. Sci.* **2017**, *35*, 649–658. [CrossRef]

MDPI
St. Alban-Anlage 66
4052 Basel
Switzerland
Tel. +41 61 683 77 34
Fax +41 61 302 89 18
www.mdpi.com

Polymers Editorial Office
E-mail: polymers@mdpi.com
www.mdpi.com/journal/polymers

www.ingramcontent.com/pod-product-compliance
Lightning Source LLC
Chambersburg PA
CBHW061602220926
51850CB00005B/398
* 9 7 8 3 0 3 6 5 0 9 6 8 6 *